A Structural Introduction to

# Algebraic Topology

## A Reference Book

**Other World Scientific Titles by the Author**

---

*Homological Algebra: The Interplay of Homology with Distributive Lattices and Orthodox Semigroups*
ISBN: 978-981-4407-06-9

*Homological Algebra: In Strongly Non-Abelian Settings*
ISBN: 978-981-4425-91-9

*Category Theory and Applications: A Textbook for Beginners*
ISBN: 978-981-323-106-1

*Higher Dimensional Categories: From Double to Multiple Categories*
ISBN: 978-981-12-0510-1

*An Elementary Overview of Mathematical Structures: Algebra, Topology and Categories*
ISBN: 978-981-12-2031-9
ISBN: 978-981-98-2025-2 (pbk)

*Manifolds and Local Structures: A General Theory*
ISBN: 978-981-12-3399-9

*Category Theory and Applications: A Textbook for Beginners Second Edition*
ISBN: 978-981-12-3608-2
ISBN: 978-981-98-1407-7 (pbk)

*Algebraic Topology: A Structural Introduction*
ISBN: 978-981-12-4835-1
ISBN: 978-981-98-1405-3 (pbk)

A Structural Introduction to

# Algebraic Topology

## A Reference Book

Marco Grandis
Università di Genova, Italy

World Scientific

NEW JERSEY • LONDON • SINGAPORE • BEIJING • SHANGHAI • HONG KONG • TAIPEI • CHENNAI • TOKYO

*Published by*

World Scientific Publishing Co. Pte. Ltd.
5 Toh Tuck Link, Singapore 596224
*USA office:* 27 Warren Street, Suite 401-402, Hackensack, NJ 07601
*UK office:* 57 Shelton Street, Covent Garden, London WC2H 9HE

Library of Congress Control Number: 2026022881

**British Library Cataloguing-in-Publication Data**
A catalogue record for this book is available from the British Library.

**A STRUCTURAL INTRODUCTION TO ALGEBRAIC TOPOLOGY**
**A Reference Book**



ISBN 978-981-98-3044-2 (hardcover)
ISBN 978-981-98-3045-9 (ebook for institutions)
ISBN 978-981-98-3046-6 (ebook for individuals)

For any available supplementary material, please visit
https://www.worldscientific.com/worldscibooks/10.1142/14799#t=suppl

To my colleagues and friends

# Preface

Algebraic Topology is a system and strategy of partial translations, aiming to reduce difficult topological problems to algebraic facts that can be more easily solved. This interface role, between topology and algebra, is the source of its power and attraction.

Each of these translations views a topological space in a particular light, and codifies this picture by an algebraic structure – typically a group, or a graded abelian group, or a graded ring. In this view, a huge amount of information is dropped, and a manageable part is kept. Choosing an appropriate translation, we can hope to solve the topological problem we are analysing.

The main subject of this book is singular homology, the simplest of these translations. Studying this theory and its applications we also investigate parts of other disciplines, which form its underlying structural layout (and in part grew out of it):

- Homological Algebra, for tensor products, the Hom functor and their derived functors,
- Homotopy Theory, for the homotopy groups and fundamental groupoid,
- Category Theory, the grammar of translations between mathematical fields.

The cohomology theories of Alexander–Spanier, Čech and de Rham are also presented.

This book is an introduction to a complex domain, with references to its advanced parts and ramifications. It is written with minimal prerequisites – basic general topology and little else – and a moderate progression, starting with a very elementary beginning. A consistent part of the exposition is organised in the form of exercises, with suitable hints and solutions.

It can be used as a textbook for a first course on Algebraic Topology or self-study, and a reference book for further study.

It is based on a previous version [G6], which has been entirely reviewed, adding several presentations of advanced subjects, and exploring more deeply the interplay of homology and homotopy.

# Contents

# Introduction

## 0.1 Investigating spaces with algebraic structures

Algebraic Topology studies topological spaces and continuous mappings, transforming spaces into algebraic structures (like groups and rings) and maps into homomorphisms. The general aim is to reduce difficult topological problems to simpler algebraic facts, and investigate properties that are even hard to define by classical topological tools – like the topological dimension or the winding number of a map from the sphere to itself.

For instance, the following figure shows two surfaces, the sphere $\mathbb{S}^2$ and the 'torus' $\mathbb{T}$ (the surface of a lifebuoy)

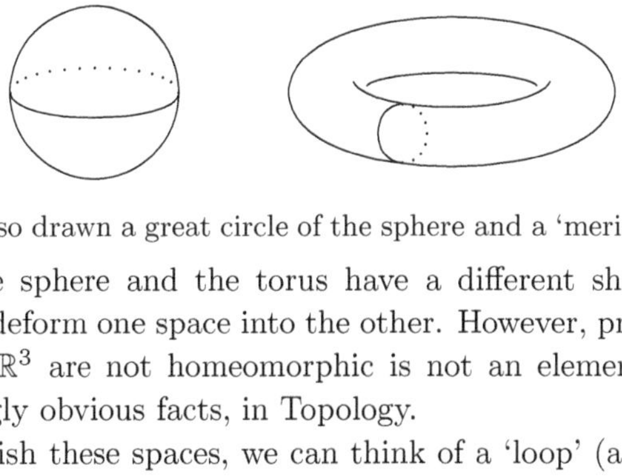

(0.1)

We have also drawn a great circle of the sphere and a 'meridian' of the torus.

Plainly, the sphere and the torus have a different shape: we cannot continuously deform one space into the other. However, proving that these subspaces of $\mathbb{R}^3$ are not homeomorphic is not an elementary thing, like many seemingly obvious facts, in Topology.

To distinguish these spaces, we can think of a 'loop' (a closed path) on these surfaces, as we have drawn above. On the sphere, every loop can be continuously shrunk to a point, but a tight string around the lifebuoy cannot be further contracted, in the surface we are considering.

All this can be made precise with elementary results of Algebraic Topology: the homology group $H_1(\mathbb{S}^2)$ is trivial, while $H_1(\mathbb{T})$ is not. This is sufficient to prove that the sphere and the torus are not homeomorphic: two homeomorphic spaces $X, Y$ have isomorphic homology groups $H_n(X) \cong H_n(Y)$, in every degree $n \geqslant 0$.

The homology groups of a space have a geometrical meaning, which is often fairly intuitive: for instance, $H_1(\mathbb{T})$ is a free abelian group with two generators, which can be realised as the homology classes $[a], [b]$ of two loops $a$ and $b$, around the 'equator' and a 'meridian' of the torus (with respect to the axis of rotation)

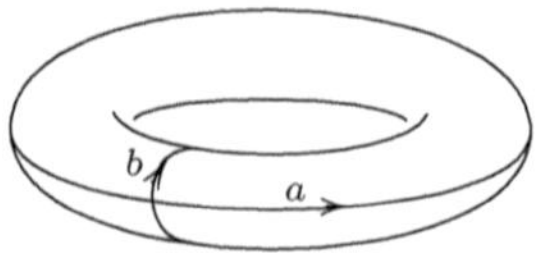


(0.2)

The homology class $3[a] + 2[b]$ can be realised as a loop which – globally – turns thrice around the equator and twice around a meridian. Negative coefficients turn up by reversing loops, and the trivial class corresponds to the loops which can be deformed into a trivial one. Finally, all the loops of the torus are classified by a pair of integers.

Loosely speaking, the homology group $H_1(X)$ explores 1-dimensional properties of a space, and can detect 1-dimensional holes, as we have seen. Similarly, the homology group $H_2(X)$ is related with 2-dimensional features. For the sphere and the torus, the groups $H_2(\mathbb{S}^2)$ and $H_2(\mathbb{T})$ are infinite cyclic, detecting a 2-dimensional 'cavity' in each space; moreover, the choice of a generator in $H_2(\mathbb{S}^2)$ or $H_2(\mathbb{T})$ amounts to choosing an orientation, in each of these manifolds.

The $n$-dimensional sphere $\mathbb{S}^n$ (for $n > 0$) has only two non trivial groups, namely $H_0(\mathbb{S}^n)$ and $H_n(\mathbb{S}^n)$, both isomorphic to the additive group of the integers. This shows that all the spheres are really distinct spaces: $\mathbb{S}^m$ and $\mathbb{S}^n$ are only homeomorphic if $m = n$, and justifies our idea of topological dimension. It is again an important fact, which looks obvious in low dimension, but is not easily proved with ordinary topological tools (after the easy cases, for $m \leqslant 1$ or $n \leqslant 1$).

There are diverse homology theories of topological spaces, but all of them coincide on 'good' spaces. More precisely, all homology theories satisfying Eilenberg–Steenrod axioms, with the same coefficient group, give the same result on every triangulable space, up to isomorphism (see Section 4.4).

## 0.2 Homology and cohomology theories

Essentially, a homology theory is a sequence $(H_n)_{n\geqslant 0}$ of transformations; each of them sends a topological space $X$ to an abelian group $H_n(X)$, and a map (or continuous mapping) $f\colon X \to Y$ to a homomorphism $H_n(f)\colon H_n(X) \to H_n(Y)$.

In this transformation, composition and identities are preserved:

- for consecutive maps $f\colon X \to Y$ and $g\colon Y \to Z$, we have: $H_n(gf) = H_n(g)H_n(f)$,
- for a space $X$ we have: $H_n(\mathrm{id}\, X) = \mathrm{id}\, H_n(X)$.

In the basic terminology of category theory, each $H_n$ is a 'covariant functor', from the category $\mathsf{Top}$ of topological spaces (and continuous mappings) to the category $\mathsf{Ab}$ of abelian groups (and homomorphisms). As an obvious consequence, each $H_n$ preserves invertible arrows: it sends a homeomorphism of spaces to an isomorphism of groups.

In fact, a homology theory turns up in a richer form: for every 'relative pair' $(X, A)$ of topological spaces (where $A$ is a subspace of $X$), and a fixed abelian group $G$, we have a sequence of groups $H_n(X, A; G)$ of relative homology, with coefficients in $G$. The basic case is absolute homology with integral coefficients: $H_n(X) = H_n(X, \emptyset; \mathbb{Z})$.

On the other hand, a cohomology theory $(H^n)$ sends a map $f\colon X \to Y$ to a sequence of homomorphisms $H^n(f)\colon H^n(Y) \to H^n(X)$, reversing arrows, reversing composition, and preserving identities.

$H^n$ is thus a 'contravariant functor', from $\mathsf{Top}$ to $\mathsf{Ab}$. Again, there is an enriched version, for relative pairs of topological spaces and a coefficient group.

An important fact appears: the cohomology groups with coefficients in a ring $R$ have a graded multiplication, and their family forms a graded $R$-algebra. We shall study singular cohomology (a by-product of singular homology), and explore or review other theories, namely Alexander–Spanier, Čech and de Rham cohomology, which naturally arise in contravariant form.

## 0.3 An outline

Chapter 1 is an introduction to the goals and methods of Algebraic Topology, with a brief analysis of the elementary issues of category theory involved in the interface between topology and algebra.

Chapters 2 and 3 deal with (absolute) singular homology, the simplest homology theory of general topological spaces, showing how to compute the homology groups, and get information on spaces and maps. Sections 2.4 and 2.5 are devoted to the homology groups of the spheres, and their consequences: for instance, the Theorem of Topological Dimension, the topological degree of a map $\mathbb{S}^n \to \mathbb{S}^n$, and the theorem about vector fields on even-dimensional spheres. Other computations of homology groups, including the compact surfaces and the projective spaces, can be found in Sections 3.1–3.4.

Singular homology is constructed in the cubical form, that – in the author's opinion – gives a simpler approach. The more usual simplicial form is briefly presented in Section 3.6; their equivalence is proved in Section 6.7.

In Chapter 4 we extend singular homology to relative pairs $(X, A)$. Section 4.3 shows how relative homology is able to investigate local features, for instance the local and global orientation of topological manifolds. The Eilenberg–Steenrod axioms for relative homology theories are examined in Section 4.4, and verified for relative singular homology. Alexander–Spanier cohomology, described in Section 4.5, satisfies the dual axioms, for a relative cohomology theory; its graded product is dealt with in Section 4.6. Similarly, de Rham cohomology of differentiable manifolds gives a graded algebra on the real field (see Section 4.8). Čech cohomology (with constant coefficients) and CW-complexes are briefly introduced in Sections 4.7 and 4.9.

Singular homology and cohomology with a coefficient group are studied in Chapters 5 and 6. Varying the coefficient group we may be able to get results that the ordinary theory (with integral coefficients) cannot obtain, as shown in Sections 5.4 and 6.3. All this requires some tools of Homological Algebra, namely the tensor and torsion products for homology, or the Hom and Ext functors for cohomology (in Sections 5.1 and 6.1–6.4).

Chapter 6 also explores the multiplicative structure of singular cohomology with coefficients in a ring (in Section 6.6), and the homology groups of a product of spaces (in Section 6.7). The latter are determined by Eilenberg–Mac Lane's Acyclic Model Theorem, which is also used to prove the equivalence of the cubical and simplicial constructions of singular homology.

Chapter 7 is an introduction to the fundamental groupoid, the fundamental group and the higher homotopy groups. The group-structure of the latter is based (again) on a cubical construction, related to singular homology by the Hurewicz theorems.

The links of homotopy and homology, for spaces and pointed spaces, are further analysed in Chapter 8, after a preliminary part on adjoint functors (Section 8.1). Reduced homology and cohomology are closely related to homotopy constructions like the suspension and the cofibre sequence of a map, built here in Section 8.2; in a dually similar way, the homotopy groups are related to loop spaces and the fibre sequence of a pointed map, as shown in Section 8.6. All this is based on homotopy constructions, like cones and suspension, or cocones and loop spaces.

Finally, Chapter 9 gathers most solutions to the exercises of the previous chapters.

Chapters 1, 2 and 7 can be used as a basic introduction to Algebraic

Topology, covering the beginning of singular homology and homotopy theory.

## 0.4 Translations and underlying affinities

As already said, Algebraic Topology is a 'discipline of translations', from topology to algebra. Each translation gives a partial view of spaces and maps, in a particular perspective. This simplified picture may be able to solve the problem we are studying; otherwise, another translation might be.

The grammar of this discipline is provided by elementary category theory, reviewed here in some sections of Chapter 1, as far as this makes clear how the translation works.

As an elementary instance, the fact that the functor $H_n$ transforms a disjoint union of spaces $\bigcup X_i$ into a direct sum $\bigoplus H_n(X_i)$ of abelian groups is the outer appearance of an underlying property, the preservation of categorical sums: a disjoint union of spaces and a direct sum of abelian groups satisfy the same categorical universal property (see 1.6.2).

The effectiveness of a translation likely depends on the affinity between the two languages involved: in our case the source category of topological spaces, on the one hand, and the target category of algebraic structures, on the other. In this regard, the category $\mathsf{Ab}$ of abelian groups is sufficiently flexible to offer 'covariant translations', by homology theories, and 'contravariant translations', by cohomology theories.

Less flexible categories, like those of rings and graded algebras, can only give a 'mirror-image', by a cohomology functor $H^*$; but this translation is far richer than viewing $H^*(X)$ as a mere graded abelian group. The roots of this phenomenon can be found in an affinity between $\mathsf{Top}$ and $\mathsf{Rng}^{\mathrm{op}}$, the dual of the category of rings – an affinity expressed in theorems, like Gelfand duality, and in formal categorical aspects sketched in 6.6.9.

We also note that covariant and contravariant theories are usually constructed in different ways. Typically, a covariant theory (like singular homology or the homotopy groups) explores a space $X$ by *probes* $M^n \to X$, that is, mappings defined on a family $M^n$ of standard models, one for each dimension $n \geqslant 0$ (for instance, all cubes, or tetrahedra, or spheres). On the other hand, a 'natural' contravariant theory (like the cohomology of Alexander–Spanier, or Čech, or de Rham) explores $X$ by *functions* defined on this space (or derived spaces), with values in a fixed algebraic structure. Obviously, a map $f\colon X \to Y$ takes a probe of $X$ to a probe of $Y$, but lifts a function of $Y$ to a function of $X$.

## 0.5 An inductive approach on structural foundations

This book follows three main guidelines.

(i) (*An inductive approach*) Notions will be presented in a concrete, 'inductive' way, starting from elementary examples. Whenever possible, the reader will be guided to build the theory through a series of exercises.

(ii) (*Structural foundations*) On the other hand, the roots of the interplay between Topology and Algebra, formalised in Homological Algebra and Category Theory, will be investigated more deeply than is usual in an introductory book. For instance, the method of 'acyclic models' and its main applications will be developed in detail; universal properties will be systematically highlighted.

(iii) (*A reference book*) Algebraic Topology is a complex domain, with ramifications in diverse fields. Many sections will end with (one or a few) star-marked subsections, which sketch further developments, giving extensive references for their study.

Some topics that take a consistent part in other textbooks, like CW-complexes and weak homotopy equivalences, are briefly elaborated here, at an informative level (in Sections 4.9 and 7.6). In fact, the important cohomology theories, like Alexander–Spanier's and Čech's (in Chapter 4), are not invariant up to weak homotopy, and there are interesting spaces that are not CW-complexes, even up to weak homotopy (see 4.9.6 and 7.6.2).

The exposition will be particularly elementary and detailed in Chapters 1 and 2: this is not a book on general topology or abelian groups, but the reader will be invited (and guided) to check the topological or algebraic ground on which we are building. This point made, we shall go on more quickly.

## 0.6 Prerequisites and literature

The only prerequisites are general topology and the very basic theory of groups, abelian groups and rings. Categories and functors are introduced when their need arises. The same holds for the part of Homological Algebra used here: exact sequences, chain complexes of abelian groups, tensor products and Hom-functors, their derived functors.

For the fascinating field of Algebraic Topology there are elementary textbooks, like Vick [Vi] and Massey [Mas3], and more advanced ones, like Eilenberg–Steenrod [EiS], Hilton–Wylie [HiW], Spanier [Sp2] and Hatcher [Ha] (the last is freely downloadable); many others will be cited for specific points. The history of this discipline is dealt with in Dieudonné [Dd].

For Homological Algebra we shall mostly refer to Cartan–Eilenberg [CE] and Mac Lane [M1]; for General Topology to Dugundji [Du], Kelley [Ke], Munkres [Mu] and Bourbaki [Bou2].

Category Theory is exposed in well known books, like Mac Lane [M2], Borceux [Bo1, Bo2, Bo3], Adámek, Herrlich and Strecker [AHS]. The author's [G5] is an elementary textbook, also devoted to applications in Algebra, Topology, Algebraic Topology and Homological Algebra.

## 0.7 Notation and conventions

Weak inclusion of sets is denoted by the symbol $\subset$, instead of the more frequent $\subseteq$. (Strict inclusion is a derived notion of little interest.)

As usual, the symbols $\mathbb{N}$, $\mathbb{Z}$, $\mathbb{Q}$, $\mathbb{R}$, $\mathbb{C}$ denote the sets of natural, integral, rational, real or complex numbers; $\mathbb{N}^*$ is the subset of the positive integers. $\mathbb{I}$ is the standard euclidean interval $[0,1]$; open and semiopen real intervals are also written with square brackets, like $]0,1[$ and $[0,1[$.

The identity mapping of a (possibly structured) set $X$ is written as $\mathrm{id}\,X$. A singleton is often written $\{*\}$. The equivalence class of an element $x$ with respect to a given equivalence relation is denoted as $[x]$, by default. A bullet in a diagram stands for an object. The two-valued index $\alpha$ (or $\beta$) takes values 0, 1, generally written $-, +$ in superscripts and subscripts.

A topological space is generally called a 'space', provided no ambiguity can arise. In this domain, a 'map' is a continuous mapping, and 'nbd' stays for neighbourhood. In a space $X$, the interior of a subset $A$ is written as $\mathrm{int}A$ or $\mathrm{int}_X A$, and the closure as $\mathrm{cl}A$, or $\mathrm{cl}_X A$, or $\overline{A}$; the subset $\partial A = \mathrm{cl}A \setminus \mathrm{int}A$ will be called the *border* of $A$ in $X$: the term 'boundary' is here reserved to chain complexes and manifolds with boundary.

A euclidean space is any subspace of some $\mathbb{R}^n$ with euclidean topology; the standard spheres and discs are written as $\mathbb{S}^n$ and $\mathbb{D}^n$ (see 1.4.1). A compact space is not assumed to be Hausdorff. The empty space is not excluded from the connected or path connected spaces. A ring is assumed to be unital, by default.

A part marked with * is out of the main line of exposition. It may refer to issues dealt with further on, or be addressed to readers with some knowledge of the subject which is being discussed, or give references for higher topics. Marginal remarks are written in small characters.

Each chapter and section has its own introduction. Many subsections end with a list of exercises; their solution is generally deferred to Chapter 9, but can follow the exercise, when it is important for the continuation. Very easy exercises are generally left to the reader.

## 0.8 Acknowledgements

This book is dedicated to the colleagues and dear friends with whom I have worked and am working.

In particular to the memory of Gabriele Darbo, my professor and friend. To Tim Porter and Ettore Carletti, with whom I discussed many topics which are developed here. To Bob Paré, George Janelidze and many others, recalling the pleasure of working together.

Diagrams and figures are composed with 'xy-pic', by K.H. Rose and R. Moore – a free package.

I would also like to thank Dr Lim Swee Cheng and Ms Angeline Husni, at World Scientific Publishing Company, for their kind help in the publication of this book. The cover, based on a photo of *Globularia cordifolia* taken in the Dolomite Alps, is due to Mr Loo Chuan Ming.

# 1
# Basic elements for Algebraic Topology

Algebraic Topology studies spaces and continuous mappings, transforming them into algebraic structures and homomorphisms, in order to reduce difficult topological problems to algebraic facts that can be more easily solved.

For instance, proving that the homology group $H_1$ of the circle is the additive group $\mathbb{Z}$ of the integers has various consequences, from straightforward ones (the circle is not a retract of the plane) to others, requiring additional work, as the Fundamental Theorem of Algebra. Computing the homology groups of the spheres has far-reaching results; several of them can be found in 1.1.5 and Sections 2.4, 2.5.

This introductory chapter is written in a very elementary way, and also reviews well known issues about topological spaces and abelian groups. The reader will know which parts to skip, or browse, or read.

## 1.1 Classifying spaces and maps

We begin by considering the shape and the 'homotopy shape' of a space, hinting at their investigation with homology groups. The homotopy relation will be further examined in Section 1.4.

### *1.1.1 The shape of a space*

Two topological spaces $X, Y$ are homeomorphic (written as $X \cong Y$) if there exists a homeomorphism $X \to Y$, i.e. a continuous mapping which has a continuous inverse. We also say that $X, Y$ *have the same (topological) shape*, or *the same homeomorphism-type*.

To prove that two given spaces have the same shape can require long computations, in order to build a homeomorphism between them, but is generally a 'confined' problem.

To prove that they are not homeomorphic can be quite difficult and enticing, even when we clearly 'see' that they have a different shape: we must prove that *there cannot exist* a homeomorphism between them.

The proof is generally based on a 'topological property' (invariant up to homeomorphism) which holds in one of them but not in the other, like connectedness, compactness, separation or countability axioms, higher forms of connectedness, etc. When these topological properties are explored throughout algebraic structures we enter in the domain of Algebraic Topology.

*Exercises and complements.* The following exercises show two families of spaces which is important to classify. The first classification, in (a) and (b), can be achieved with the usual means of General Topology; the second, in (c), will be completed later, using singular homology. The solutions can be found in Chapter 9.

(a) (*The shape of the intervals, I*) Any non-degenerate interval of the euclidean line is homeomorphic to $]0,1[$, or $[0,1[$, or $[0,1]$.

*Hints.* It is an easy exercise of Calculus, which can be solved using affine linear functions and some elementary transcendental functions. (One can also use rational functions and their pastings, but the argument would be longer.)

(b) (*The shape of the intervals, II*) These three homeomorphism-types are distinct.

Adding the degenerate cases, the intervals of $\mathbb{R}$ are thus classified in five types of homeomorphism:

(i) open non-degenerate intervals,

(ii) semiopen non-degenerate intervals,

(iii) compact non-degenerate intervals,

(iv) singletons, (v) the empty space.

*Hint:* consider the connected components of an interval deprived of a point.

(c) (*Topological Dimension in low degree*) Classifying the euclidean spaces $\mathbb{R}^n$, by proving that they represent distinct homeomorphism-types, is a crucial fact in Topology, without which there would be no notion of topological dimension.

This point will be examined in 1.1.5(b), but the reader can solve as of now the lowest degrees: the singleton $\mathbb{R}^0 = \{0\}$ and the euclidean line $\mathbb{R}$ are not homeomorphic to any higher euclidean space $\mathbb{R}^n$, for $n > 1$.

(d) An interval of the line is not homeomorphic to any sphere $\mathbb{S}^n$ ($n \geqslant 0$).

### 1.1.2 Classifying maps

Studying the maps $f\colon X \to Y$ between two topological spaces, we are also interested in classifying them, 'up to continuous deformation'.

Let us recall that two maps $f, g\colon X \to Y$ are said to be *homotopic* (written as $f \simeq g$) if there is a map $\varphi\colon X \times \mathbb{I} \to Y$ defined on the *cylinder* $I(X) = X \times \mathbb{I}$ (with the product topology), that coincides with $f$ on the lower basis of the cylinder and with $g$ on the upper one

$$\begin{aligned} &\varphi\colon X \times \mathbb{I} \to Y, \\ &\varphi(x,0) = f(x), \quad \varphi(x,1) = g(x) \qquad\qquad (\text{for } x \in X), \end{aligned} \tag{1.1}$$

forming a continuous deformation of $f$ into $g$. This is easily proved to be an equivalence relation (see 1.4.4). The homotopy will also be written as $\varphi\colon f \simeq g\colon X \to Y$.

A point $x \in X$ can be identified with the corresponding map $x\colon \{*\} \to X$, defined on the singleton space. A homotopy $a\colon x \simeq x'\colon \{*\} \to X$ is then the same as a *path* in $X$ from $x$ to $x'$, i.e. a map such that

$$a\colon \mathbb{I} \to X, \qquad\qquad a(0) = x, \quad a(1) = x'. \tag{1.2}$$

On the other hand, a homotopy $\varphi\colon f \simeq g\colon X \to Y$ gives a family of paths $\varphi(x,-)\colon \mathbb{I} \to Y$ from $f(x)$ to $g(x)$, indexed by $x \in X$ and varying continuously on $X$.

*Exercises and complements.* (a) The classification of the maps $\{*\} \to X$ up to homotopy amounts to the partition of $X$ in path components.

(b) We have two maps $f, g\colon X \to Y$ with values in a euclidean space $Y$ (i.e. a subspace of some $\mathbb{R}^n$).

If $Y$ is a convex subset of $\mathbb{R}^n$ (i.e. for all $y, y' \in Y$, the line segment from $y$ to $y'$ is contained in $Y$), the maps $f$ and $g$ are always homotopic. More generally, the same holds if, for every $x \in X$, the line segment from $f(x)$ to $g(x)$ is contained in $Y$.

*Hint.* We can use the linear structure of $\mathbb{R}^n$ to define the *affine homotopy*

$$\varphi\colon f \simeq g\colon X \to Y, \qquad \varphi(x,t) = (1-t)f(x) + tg(x), \tag{1.3}$$

which describes, at each $x \in X$, the line segment from $f(x)$ to $g(x)$, and is indeed an affine linear map in the variable $t \in \mathbb{I}$. Now we only have to verify the continuity of the mapping $\varphi$ on $X \times \mathbb{I}$.

### 1.1.3 Classifying the homotopy type

The homotopy of maps leads to a possibility of deforming spaces, called homotopy equivalence, which is far wider than homeomorphism.

Namely, a map $f\colon X \to Y$ is a *homotopy equivalence* if it admits a *weak inverse* $g\colon Y \to X$, up to homotopy equivalence: this means that $gf \simeq \mathrm{id}\, X$ and $fg \simeq \mathrm{id}\, Y$. Then we say that the spaces $X, Y$ are *homotopy equivalent*, or that they have *the same homotopy type*; it is again an equivalence relation between spaces (see 1.4.4), written as $X \simeq Y$.

A space is said to be *contractible* if it has the homotopy type of the singleton.

A subspace $A \subset X$ is a *deformation retract* of $X$ if there is a map $p$ satisfying the following conditions, where $m\colon A \to X$ is the inclusion

$$p\colon X \to A, \qquad pm = \mathrm{id}\, A, \qquad mp \simeq \mathrm{id}\, X. \tag{1.4}$$

Forgetting the homotopy $mp \simeq \mathrm{id}\, X$, we have a simple retract; the map $p$ is called a retraction of $m$, and $m$ is called a section of $p$.

We speak of a *strong deformation retract* if in (1.4) we can find a homotopy which does not move the points of $A$

$$\varphi\colon mp \simeq \mathrm{id}\, X, \qquad \varphi(x, t) = x \qquad (x \in A,\ t \in \mathbb{I}). \tag{1.5}$$

*Exercises and complements.* (a) Every euclidean space $\mathbb{R}^n$ is contractible, and admits $\{0\}$ as a strong deformation retract. (Any other point would do, but this is a general fact: see 1.4.3(b).)

(b) Every non-empty convex subspace $X \subset \mathbb{R}^n$ is contractible. More generally, the same is true of every subset $X \subset \mathbb{R}^n$ which is *starred* with respect to some $x_0 \in X$; this means that: if $x \in X$, the line segment from $x_0$ to $x$ is contained in $X$.

(c) Let $X$ be a non-empty space. The following conditions are equivalent:

(i) $X$ is contractible,
(ii) $\mathrm{id}\, X$ is homotopic to a constant endomap,
(iii) for any space $Y$, each map $Y \to X$ is homotopic to a constant map,
(iv) for any space $Y$, each map $X \to Y$ is homotopic to a constant map.

(d) The sphere $\mathbb{S}^n$ is a strong deformation retract of the pierced space $X = \mathbb{R}^{n+1} \setminus \{0\}$. *Hint:* use the normalisation map $N$

$$N\colon X \to \mathbb{S}^n, \qquad N(x) = x/||x|| \qquad (x \neq 0). \tag{1.6}$$

(e) The subspace $A \subset X$ is a deformation retract of $X$ if and only if there is a map $\varphi$ such that (for all $x \in X$, $a \in A$)

$$\varphi\colon X \times \mathbb{I} \to X, \qquad \varphi(x, 0) \in A, \quad \varphi(x, 1) = x, \quad \varphi(a, 0) = a. \tag{1.7}$$

For a strong deformation retract we replace the last condition with: $\varphi(a, t) = a$, for all $a \in A$, $t \in \mathbb{I}$.

### 1.1.4 Hints at singular homology

Singular homology will be defined in Chapter 2. The construction requires some work, and we try to explain why it is worthwhile.

Singular homology associates:

- to a topological space $X$ an abelian group $H_k(X)$, for each $k \geqslant 0$,
- to a map $f\colon X \to Y$ a homomorphism $H_k(f)\colon H_k(X) \to H_k(Y)$ of groups, for each $k \geqslant 0$,

so that composition and identities are preserved:

$$\text{(i)}\quad H_k(gf) = H_k(g)H_k(f), \quad \text{for all maps } f\colon X \to Y \text{ and } g\colon Y \to Z,$$

$$\text{(ii)}\quad H_k(\mathrm{id}\, X) = \mathrm{id}\; H_k(X), \quad \text{for any space } X.$$

Because of these properties the transformation $H_k$ is called a (covariant) 'functor', from spaces and maps to abelian groups and homomorphisms. $H_k(X)$ is called the *k-th group of singular homology* of the space $X$; $H_k(f)$ is called the *associated homomorphism*, and is usually written as $f_*\colon H_k(X) \to H_k(Y)$, or $f_{*k}$.

As a consequence of (i), (ii), if the spaces $X, Y$ are homeomorphic, the homology groups $H_k(X)$ and $H_k(Y)$ are isomorphic, for all $k \geqslant 0$.

In fact we have two maps, inverse to each other

$$f\colon X \rightleftarrows Y : g, \qquad gf = \mathrm{id}\, X, \quad fg = \mathrm{id}\, Y,$$

transformed by $H_k$ into homomorphisms $f_*\colon H_k(X) \rightleftarrows H_k(Y) : g_*$ which are again inverse to each other.

As a typical property of the transformations of Algebraic Topology, we shall see that the homology functors are *homotopy invariant*: this means that whenever two maps $f, f'\colon X \to Y$ are homotopic, then $H_k(f) = H_k(f')$, for every $k \geqslant 0$.

Therefore, a homotopy equivalence $f\colon X \rightleftarrows Y : g$, with $gf \simeq \mathrm{id}\, X$ and $fg \simeq \mathrm{id}\, Y$, is also transformed by $H_k$ into an isomorphism $H_k(X) \cong H_k(Y)$ of abelian groups.

The homology groups are thus *invariants of the homotopy type of a space*; directly, they cannot distinguish homotopy equivalent spaces. In particular, all the spaces $\mathbb{R}^n$ have the same homology groups as the singleton $\mathbb{R}^0 = \{*\}$, which will be (easily) computed as:

$$H_0(\mathbb{R}^n) \cong \mathbb{Z}, \qquad H_k(\mathbb{R}^n) = 0 \quad (k > 0), \tag{1.8}$$

the trivial group being (always) written as 0.

Yet, the distinction of the diverse $\mathbb{R}^n$ can be achieved, by suitable modifications of the spaces (which only preserve the homeomorphism type).

### 1.1.5 The homology of the spheres

Developing the theory, we shall learn to compute the homology groups and the associated homomorphisms. For instance, on the standard spheres $\mathbb{S}^n$ we get (in 2.4.1 and Theorem 2.4.2):

$$\begin{array}{lll} H_0(\mathbb{S}^0) \cong \mathbb{Z}^2, & H_k(\mathbb{S}^0) = 0 & (k > 0), \\ H_0(\mathbb{S}^n) \cong H_n(\mathbb{S}^n) \cong \mathbb{Z}, & H_k(\mathbb{S}^n) = 0 & (n > 0,\ 0 \neq k \neq n). \end{array} \tag{1.9}$$

Roughly speaking, $H_0(X)$ is related to path-connectedness, while a non-trivial group $H_k(X)$, for $k > 0$, detects 'holes' of dimension $k$ in the space.

All this has many important consequences, and we begin by presenting some straightforward ones.

(a) First, these results readily prove that no sphere is contractible. Moreover the sphere $\mathbb{S}^n$ is not homeomorphic, nor homotopy equivalent, to any other sphere $\mathbb{S}^m$ (for $m \neq n$).

In fact, if the space $X$ is homotopy equivalent to $\mathbb{S}^n$, then $H_k(X) \cong H_k(\mathbb{S}^n)$ (for all $k \geqslant 0$), and the integer $n$ is determined by the homology groups of $X$ as: $\max\{k \in \mathbb{N} \mid H_k(X) \neq 0\}$.

(b) (Theorem of Topological Dimension) *If the euclidean spaces $\mathbb{R}^m$ and $\mathbb{R}^n$ are homeomorphic, then $m = n$.*

In fact, the one-point compactifications $\mathbb{S}^m$ and $\mathbb{S}^n$ are homeomorphic.

Let us note the analogy with the theorem of *linear* dimension of vector spaces.

(c) We already remarked that we cannot get this result applying the homology functors, directly, to $\mathbb{R}^m$ and $\mathbb{R}^n$.

Loosely speaking, *the $n$-dimensional character of the topology of $\mathbb{R}^n$ appears more clearly in its Alexandrov compactification* – a process which preserves homeomorphism but does not preserve homotopy equivalence.

(d) The argument of (b) can also be achieved replacing Alexandrov compactification with a more elementary procedure, still invariant up to homeomorphism and not to homotopy equivalence: taking out a point.

Suppose for simplicity that $m, n > 0$. If the euclidean spaces $\mathbb{R}^m$ and $\mathbb{R}^n$ are homeomorphic, the pierced space $\mathbb{R}^m \setminus \{0\}$ is homeomorphic to a pierced space $\mathbb{R}^n \setminus \{x_0\}$, whence $\mathbb{S}^{m-1}$ and $\mathbb{S}^{n-1}$ are homotopy equivalent (by 1.1.3(d)), and again $m = n$.

For $m = 1$ we are reviewing the solution to Exercise 1.1.1(c): connectedness does not distinguish $\mathbb{R}$ from any other $\mathbb{R}^n$, but does distinguish the pierced line $\mathbb{R} \setminus \{0\}$ from any higher pierced space $\mathbb{R}^n \setminus \{x_0\}$ with $n > 1$.

(e) We already remarked, since the beginning of the general Introduction, how the homology groups of the sphere and the torus show that these spaces are not homeomorphic; we can now add that they are not even homotopy equivalent.

### *1.1.6 Studying retracts*

Showing that a subspace $A \subset X$ is a retract of $X$ is a 'confined' problem: we are to find a (continuous) retraction $p\colon X \to A$, such that $p(x) = x$ for all $x \in A$. Showing that it is *not*, i.e. that there is no (continuous) retraction, can be an important result.

For instance, saying that the subspace $\mathbb{S}^0 = \{-1, 1\}$ of the euclidean line $\mathbb{R}$ is not a retract is an obvious consequence of the Intermediate Value Theorem, and can be viewed as a way of stating the theorem. (The general statement is an elementary consequence of the previous property.)

More generally, saying that the sphere $\mathbb{S}^n$ is not a retract of $\mathbb{R}^{n+1}$ is intuitively clear (at least in low dimension), but not easy to prove with the usual tools of general topology. This fact has important consequences, like the Brouwer fixed-point theorem (see 2.4.4(a)).

*Exercises and complements.* (a) For $n > 0$, $\mathbb{S}^n$ is not a retract of $\mathbb{R}^{n+1}$, nor of any contractible space. *Hint:* use the functor $H_n$ and its values on our spaces, mentioned in (1.8) and (1.9).

(b) These facts can be easily extended: for a subspace $A \subset X$, the inclusion $m\colon A \to X$ gives a homomorphism $m_*\colon H_n(A) \to H_n(X)$ (for $n \geqslant 0$), and:

- this homomorphism need not be injective,
- if $A$ is a retract of $X$, then $m_*\colon H_n(A) \to H_n(X)$ has a left inverse, and is injective.

### *1.1.7 Topological embeddings and projections*

Even though general topology is a prerequisite of this book, we end this section by reviewing a few elementary issues of frequent use, to fix our terminology.

A map $f\colon X \to Y$ between topological spaces has a canonical factorisation

$$\begin{array}{ccc} X & \xrightarrow{\ f\ } & Y \\ {\scriptstyle p}\downarrow & & \uparrow{\scriptstyle m} \\ X/R_f & \xrightarrow[\ g\ ]{} & \operatorname{Im} f \end{array} \qquad f = mgp, \qquad (1.10)$$

where:

- $p$ is the canonical projection onto the topological quotient of $X$ modulo the equivalence relation associated to $f$ ($xR_f x'$ means that $f(x) = f(x')$),
- $m$ is the inclusion of the subspace $\operatorname{Im} f = f(X)$ into the codomain $Y$,
- $g\colon X/R_f \to \operatorname{Im} f$ is the bijective map defined by $g[x] = f(x)$ ($x \in X$).

If $f$ is open (or closed), the same holds for $g$, which is a homeomorphism.

We are also interested in the following cases, where $g$ is a homeomorphism.

(a) The map $f\colon X \to Y$ is said to be a *topological embedding* if it is an injective mapping and $X$ has the initial topology for $f$: the open sets of $X$ are (precisely) the preimages $f^{-1}V$ of the open sets of $Y$. Then $f$ restricts to a homeomorphism $g\colon X \to f(X)$ between $X$ and a subspace of $Y$

$$\begin{array}{ccc} X & \xrightarrow{\;f\;} & Y \\ & {}_{g}\searrow & \uparrow{\scriptstyle m} \\ & & \mathrm{Im}\, f \end{array} \qquad (1.11)$$

In this case we may want to 'identify' $X$ with its image in $Y$, so that it becomes a subspace. Plainly, topological embeddings are stable under composition. An injective map which is either open or closed is always a topological embedding. Moreover, an embedding is open or closed if and only if the same is true of its image.

(b) The map $f\colon X \to Y$ is said to be a *topological projection*, or a *quotient map*, if it is surjective and $Y$ has the final topology for $f$: the open sets of $Y$ are (precisely) the subsets $V$ whose preimage $f^{-1}V$ is open in $X$. Then $f$ induces a homeomorphism $g\colon X/R_f \to Y$ between a 'strict' quotient of $X$ and the space $Y$

$$\begin{array}{ccc} X & \xrightarrow{\;f\;} & Y \\ {\scriptstyle p}\downarrow & \nearrow_{g} & \\ X/R_f & & \end{array} \qquad (1.12)$$

Here we may want to identify $Y$ with this quotient of $X$. Topological projections are stable under composition, an advantage with respect to strict topological quotients. (Note that, in Set Theory, a quotient of a quotient of the set $X$ is not a quotient of $X$.)

A surjective map which is either open or closed is always a quotient map. In particular, a surjective map from a compact space to a Hausdorff space is closed, and a quotient map.

### *1.1.8 Cover Lemmas*

Let $(X_i)_{i\in I}$ be a *cover* of the space $X$, that is, $X = \bigcup X_i$. The continuity of a mapping $f\colon X \to Y$ with values in a topological space $Y$ can be reduced to the continuity of its restrictions $f_i\colon X_i \to Y$ in two main cases (a), (b), which have a common extension (c).

*Exercises and complements.* (a) (The Open Cover Lemma) *Let* $(X_i)_{i \in I}$ *be an* open cover *of the space* $X$ *(i.e. all* $X_i$ *are open subsets of* $X$*). Then* $f: X \to Y$ *is continuous if and only if all its restrictions* $f_i: X_i \to Y$ *are. The same holds if* $(X_i)_{i \in I}$ *is a* generalised open cover, *in the sense that* $X$ *is covered by the interior parts* $\mathrm{int}X_i$.

(b) (The Finite Closed Cover Lemma) *Let* $(X_i)_{i \in I}$ *be a* finite closed cover *of the space* $X$ *(i.e. all* $X_i$ *are closed subsets of* $X$*, and* $I$ *is finite). Then* $f: X \to Y$ *is continuous if and only if all its restrictions* $f_i: X_i \to Y$ *are.*

(c) The general situation, extending the previous ones, is a space $X$ and a cover $(X_i)_{i \in I}$ such that $X$ has the finest topology that makes all inclusions $u_i: X_i \to X$ continuous.

## 1.2 Transforming structures and problems

We want to transform topological spaces into algebraic structures (typically groups) and continuous mappings into homomorphisms.

Here we give a brief presentation of these topics, making informal use of issues that will be made precise later on.

The set of maps $X \to Y$ between two spaces is written $\mathsf{Top}(X, Y)$.

### *1.2.1 Hints at categories and functors*

We transform topological data into algebraic ones setting up a procedure $F$ that transforms

- a topological space $X$ into an abelian group $F(X)$ (or some other algebraic structure),

- a map $f: X \to Y$ into a homomorphism $F(f): F(X) \to F(Y)$,

so that composition and identities are preserved – as in 1.1.4, in the case of singular homology.

Such a procedure is denoted as $F: \mathsf{Top} \to \mathsf{Ab}$ and called a (covariant) *functor*, from the *category* $\mathsf{Top}$ *of topological spaces and continuous mappings* to the *category* $\mathsf{Ab}$ *of abelian groups and homomorphisms.* These issues will be defined in Section 1.3.

Here it will be sufficient to say that a category $\mathsf{C}$ consists of *objects* $X, Y, ...$ and *morphisms*, or *arrows*, $f: X \to Y, ...$ between objects; the latter have a (partial) *composition law*: for each pair of 'consecutive' morphisms $f: X \to Y$ and $g: Y \to Z$ there is a composed morphism $gf: X \to Z$.

This partial operation is associative (whenever composition is legitimate) and every object $X$ has an *identity*, written as $\mathrm{id}\,X: X \to X$, or $1_X$, which acts as a unit for legitimate compositions. The notation $\mathsf{C}(X, Y)$ denotes the set of morphisms $X \to Y$ in $\mathsf{C}$.

The prime example is the category $\mathsf{Set}$ *of sets* (and mappings), where:

- an object is a set,

- the morphisms $f\colon X \to Y$ between two sets are the (set-theoretical) mappings from $X$ to $Y$, and form the set $\mathsf{Set}(X, Y)$,

- the composition law is the usual composition of consecutive mappings: $(gf)(x) = g(f(x))$.

We also need the following categories of structured sets and structure-preserving mappings (with the usual composition):

- the category $\mathsf{Top}$ *of topological spaces* (and continuous mappings),

- the category $\mathsf{Ab}$ *of abelian groups* (and their homomorphisms),

- the category $\mathsf{Gp}$ *of groups* (and their homomorphisms),

- the category $R\,\mathsf{Mod}$ *of $R$-modules* on a commutative ring $R$ (and their homomorphisms),

- the category $\mathsf{Rng}$ *of rings* (and their homomorphisms),

- the category $\mathsf{Set}_\bullet$ *of pointed sets* (and pointed mappings),

- the category $\mathsf{Top}_\bullet$ *of pointed topological spaces* (and pointed maps).

Other categories of structured sets will be introduced when they are needed.

A functor $F\colon \mathsf{Top} \to \mathsf{Ab}$ is defined as above; we define in a similar way a functor $F\colon \mathsf{C} \to \mathsf{D}$, from a category $\mathsf{C}$ to a category $\mathsf{D}$. (Contravariant functors are introduced in 1.2.2(iii).)

*Remarks and complements.* (a) By default, a ring $R$ is assumed to be associative and unital; an $R$-module is assumed to be unitary.

(b) We recall that a *pointed set* is a pair $(X, x_0)$ consisting of a set $X$ and a *basepoint* $x_0 \in X$; a pointed mapping $f\colon (X, x_0) \to (Y, y_0)$ is a mapping $f\colon X \to Y$ such that $f(x_0) = y_0$; the composition is the usual one.

Similarly, a *pointed topological space* $(X, x_0)$ is a space with a basepoint, and a pointed map $f\colon (X, x_0) \to (Y, y_0)$ is a continuous mapping from $X$ to $Y$ such that $f(x_0) = y_0$. (We shall see that the fundamental group $\pi_1(X, x_0)$ is defined for a pointed topological space.)

It is often convenient to write a pointed set (or space) as $(X, 0)$, viewed as a set (or space) equipped with a zeroary operation $0\colon \{*\} \to X$ (a constant), under no axioms. Or just as $X$, leaving the zeroary operation understood, as is generally done in algebraic structures.

(c) When a category is named after its objects alone (e.g. the 'category of topological spaces'), this means that the morphisms are understood to be the obvious ones (in this case the continuous mappings), with the obvious composition law. Of course, different categories with the same objects are given different names.

### 1.2.2 *Exploring connectedness*

We begin by constructing some elementary functors $\mathsf{Top} \to \mathsf{D}$, with values in a category of algebraic structures. The important ones, the homology and homotopy functors, will require more work, but will appear to be 'higher dimensional versions' of some of the present ones, related to properties of 'higher connectedness'.

As in 1.1.4, a functor $F\colon \mathsf{Top} \to \mathsf{D}$ is *homotopy invariant* if, for every pair of homotopic maps $f \simeq g$, we have $F(f) = F(g)$.

(i) A set has a trivial algebraic structure, with no operations under no axioms, and all mappings preserve this structure: Set is itself a category of algebraic structures, the simplest.

There is an obvious *forgetful functor*

$$U\colon \mathsf{Top} \to \mathsf{Set}, \qquad U(X) = |X|, \qquad U(f) = |f|, \tag{1.13}$$

that sends each space $X$ to its underlying set $|X|$, and each map $f\colon X \to Y$ to the underlying mapping $|f|\colon |X| \to |Y|$.

Each category of structured sets has a similar *forgetful functor* with values in Set (see 1.3.7). Such functors are so obvious that we tend to forget *them*! The notation $|-|$ will be used when useful.

(ii) We also consider two less obvious functors with values in Set

$$\mathrm{Cc}\colon \mathsf{Top} \to \mathsf{Set}, \qquad \Pi_0\colon \mathsf{Top} \to \mathsf{Set}. \tag{1.14}$$

The former sends a space $X$ to the set $\mathrm{Cc}(X)$ of its connected components, and a map $f\colon X \to Y$ to the mapping $\mathrm{Cc}(f)\colon \mathrm{Cc}(X) \to \mathrm{Cc}(Y)$ which sends a component $A$ of $X$ to the component of $Y$ containing $f(A)$.

$\Pi_0$ is constructed in the same way, using the path components; it will be reviewed in (1.71).

(iii) (*A contravariant functor*) Here $\mathbb{Z}$ is the ring of integers equipped with the discrete topology. We form a *contravariant functor* (denoted by a dashed arrow)

$$\mathcal{A}\colon \mathsf{Top} \dashrightarrow \mathsf{Rng}, \qquad \mathcal{A}(X) = \mathsf{Top}(X, \mathbb{Z}). \tag{1.15}$$

It sends a space $X$ to the (commutative) ring of all maps $a\colon X \to \mathbb{Z}$, endowed with the pointwise operations:

$$(a+b)(x) = a(x) + b(x), \qquad (ab)(x) = a(x)\, b(x), \qquad 1(x) = 1. \tag{1.16}$$

A map $f\colon X \to Y$ is sent to a homomorphism of rings

$$\mathcal{A}(f)\colon \mathcal{A}(Y) \to \mathcal{A}(X), \qquad \mathcal{A}(f)(a\colon Y \to \mathbb{Z}) = (af\colon X \to \mathbb{Z}). \tag{1.17}$$

The reader will note that $\mathcal{A}$ *reverses the direction* of the map $f$; consistently, this transformation preserves the identities but reverses the composition: a map $g\colon Y \to T$ will give a homomorphism $\mathcal{A}(g)\colon \mathcal{A}(T) \to \mathcal{A}(Y)$, and

$$\begin{aligned} \mathcal{A}(gf) = \mathcal{A}(f)\mathcal{A}(g)\colon \mathsf{Top}(T, \mathbb{Z}) &\longrightarrow \mathsf{Top}(X, \mathbb{Z}), \\ (a\colon T \to \mathbb{Z}) &\mapsto (agf\colon X \to \mathbb{Z}). \end{aligned} \tag{1.18}$$

*Exercises and complements.* (a) The forgetful functor $U\colon \mathsf{Top} \to \mathsf{Set}$ is not homotopy invariant.

(b) The functors $\mathrm{Cc}\colon \mathsf{Top} \to \mathsf{Set}$, $\Pi_0\colon \mathsf{Top} \to \mathsf{Set}$ and $\mathcal{A}\colon \mathsf{Top} \dashrightarrow \mathsf{Rng}$ defined above are homotopy invariant.

(c) As already remarked in 1.1.2, the underlying set $U(X)$ can be identified with the set of maps $x\colon \{*\} \to X$, defined on the singleton space. In $\Pi_0(X)$ such maps are identified when they are homotopic, and in $\mathrm{Cc}(X)$ when they land in the same connected component.

Using the covariant forgetful functors $U, \mathrm{Cc}, \Pi_0\colon \mathsf{Top} \to \mathsf{Set}$ we are thus exploring spaces by *probes* defined on the 0-dimensional model $M^0 = \{*\}$. On the other hand, the contravariant functor $\mathcal{A}(X) = \mathsf{Top}(X, \mathbb{Z})$ explores spaces by *functions* defined on $X$, with values in a fixed ring.

All this agrees with our comments on covariant and contravariant theories, at the end of Section 0.4 of the Introduction, although in a basic 0-dimensional form.

(d) It will be interesting to evaluate $\mathcal{A}, U, \mathrm{Cc}$ and $\Pi_0$ on the one-point compactification of the discrete space $\mathbb{N}$

$$\mathbb{N}^\bullet = \mathbb{N} \cup \{\infty\}, \tag{1.19}$$

as we shall do in 1.2.3 and 1.2.4(c). This space will be called the (standard) *converging sequence*, because a map $x\colon \mathbb{N}^\bullet \to X$ with values in a topological space amounts to a sequence $(x_n)_{n\in\mathbb{N}}$ of $X$ converging to a point $x_\infty \in X$. We also note that $\mathbb{N}^\bullet$ is homeomorphic to the euclidean space

$$\mathbb{H}_0 = \{x \in \mathbb{R} \mid x = 0 \text{ or } x = 1/n, \text{ for } n > 0\}. \tag{1.20}$$

(e) More generally, one can consider a contravariant functor $\mathcal{A}(-; R)$, replacing $\mathbb{Z}$ by any topological ring $R$.

### 1.2.3 Other exercises

The reader may find it interesting to explore the contravariant functor $\mathcal{A}$ (defined in (1.15)), either independently or with the following exercises.

(a) A function $a\colon X \to \mathbb{Z}$ is continuous if and only if each preimage $X_k = a^{-1}\{k\}$ (for $k \in \mathbb{Z}$) is open in $X$, if and only the function $a$ is locally constant (i.e. constant on some neighbourhood of each point).

This implies that $a$ is constant on each connected component of $X$; the converse certainly holds if the connected components of $X$ are open, but plainly fails for the rational line $\mathbb{Q}$ whose connected components are singletons.

(b) A map $a\colon X \to \mathbb{Z}$ amounts to an indexed partition $(X_k)_{k\in\mathbb{Z}}$ of $X$ in disjoint open subsets (automatically closed in $X$). But this description is ineffective with respect to the operations of the ring $\mathcal{A}(X)$, defined in (1.16).

(c) The space $X$ is empty if and only if the ring $\mathcal{A}(X)$ is trivial.

(d) The following conditions on the space $X$ are equivalent:

(i) $X$ is non-empty and connected,

(ii) $\mathcal{A}(X) \cong \mathbb{Z}$ (as rings),

(iii) $\mathcal{A}(X)$ is an integral domain (i.e. a non-trivial commutative ring without proper zero divisors),

(iv) the underlying additive group of $\mathcal{A}(X)$ is infinite cyclic.

(e) Let $X = \bigcup_{i\in I} X_i$ be the partition of $X$ in connected components, and suppose that each of them is open in $X$ (which is certainly true if $I$ is finite). Then the ring $\mathcal{A}(X)$ is canonically isomorphic to the cartesian-power ring $\mathbb{Z}^I = \prod_{i\in I} \mathbb{Z}$. (Note that we can identify the index-set $I$ with the set $\mathrm{Cc}(X)$ of connected components of $X$.)

(f) Coming back to the space $\mathbb{N}^\bullet$ (see 1.2.2(d)), the ring $\mathcal{A}(\mathbb{N}^\bullet)$ can be embedded as the countable subring of $\mathbb{Z}^{\mathbb{N}}$ formed by the eventually constant sequences of $\mathbb{Z}$.

*(g) The ring $\mathcal{A}(\mathbb{Q})$ of the rational line (with euclidean topology) is uncountable. (In fact, it has the cardinal of the continuum, i.e. the cardinal of $\mathbb{R}$.)

### 1.2.4 Distinguishing shape and homotopy type

In a category $\mathsf{C}$, an *isomorphism* is a morphism $f\colon X \to Y$ which has an inverse, i.e. a morphism $g\colon Y \to X$ such that $gf = 1_X$ and $fg = 1_Y$. The inverse is determined by $f$ (by the usual verification) and written as $f^{-1}$.

We say that the objects $X, Y$ are isomorphic, written as $X \cong Y$, if there is an isomorphism $X \to Y$ (in the category that we are considering).

In the categories listed in 1.2.1 this definition gives the usual isomorphisms of each structure: bijective mappings in Set, homeomorphisms in Top, isomorphisms of groups in Gp and Ab, etc.

Plainly:

(i) a functor $F\colon \mathsf{C} \to \mathsf{D}$ takes any isomorphism of $\mathsf{C}$ to an isomorphism of $\mathsf{D}$,

(ii) a homotopy invariant functor $F\colon \mathsf{Top} \to \mathsf{D}$ transforms a homotopy equivalence $f\colon X \to Y$ into an isomorphism of the category $\mathsf{D}$; in other words, $X \simeq Y$ implies $F(X) \cong F(Y)$.

Therefore, a functor $F\colon \mathsf{Top} \to \mathsf{D}$ can be used *to distinguish spaces*, up to homeomorphism: two spaces $X, Y$ cannot be homeomorphic if $F(X)$ and $F(Y)$ are not isomorphic in $\mathsf{D}$. A homotopy invariant functor can be used *to distinguish homotopy types.*

Important results in this sense will come out of complex functors, like singular homology. The elementary examples of 1.2.2 give trivial or elementary results.

*Exercises and complements.* (a) Applying principle (i) to the functors $U$, Cc, $\Pi_0$ of 1.2.2 gives obvious facts, for which there would be no need of categories and functors: two homeomorphic spaces have equipotent underlying sets, the same number (or cardinal) of connected components, and the same number (or cardinal) of path components. We shall see that $\Pi_0$ is closely related to singular homology in degree zero.

(b) Applying principle (ii) to the homotopy invariant functors Cc and $\Pi_0$ gives elementary, less obvious facts: two homotopy equivalent spaces have the same number (or cardinal) of connected components, and the same number (or cardinal) of path components.

(c) The functor $\mathcal{A}$ is able to distinguish the 'converging sequence' $\mathbb{N}^\bullet$ (see 1.2.2(d)) from any infinite discrete space. On the other hand the functors $U, \mathrm{Cc}, \Pi_0$ give isomorphic results on $\mathbb{N}^\bullet$ and $\mathbb{N}$.

### *1.2.5 Free abelian groups*

We recall here some points about abelian groups, that will be frequently used throughout this book. (Everything can be extended to modules on a commutative ring $R$, that replaces the ring $\mathbb{Z}$ of integers: see 5.1.1.)

(a) In an abelian group $A$, a finite linear combination $\sum_{i \in I} \lambda_i x_i$ of elements $x_i \in A$ with integral coefficients $\lambda_i$ makes sense, without any ordering of the set $I$ of indices, because of the associative and commutative property of addition.

More generally, a linear combination $\sum_{i \in I} \lambda_i x_i$ makes sense for an arbitrary set of indices $I$, if the family of coefficients $(\lambda_i)_{i \in I} \in \mathbb{Z}^I$ is *essentially finite*: there are finitely many indices $i$ such that $\lambda_i \neq 0$; then the sum is computed on these indices.

(b) A *basis* of an abelian group $A$, is a family $(x_i)_{i \in I}$ of elements of $A$ such that every $x \in A$ can be uniquely written as a linear combination of $x_i$, with an essentially finite family of integral coefficients

$$x = \sum_{i \in I} \lambda_i x_i. \tag{1.21}$$

The abelian group $A$ is *free* (on the ring $\mathbb{Z}$) if it has a basis. The latter is an injective family, and we can embed $I$ in $A$, identifying $i = x_i$.

In fact, if $i \neq j$ in $I$, the difference $x_i - x_j = 1.x_i + (-1).x_j$ cannot be $0_A = 0.x_i + 0.x_j$.

The *universal property* of the basis says that every mapping $f \colon I \to B$ with values in an abelian group can be uniquely extended to a group-homomorphism $g \colon A \to B$

$$g(\textstyle\sum_{i \in I} \lambda_i x_i) = \sum_{i \in I} \lambda_i f(i), \tag{1.22}$$

computed by an (essentially finite) linear combination in $B$.

(c) Every set $I$ is the basis of a free abelian group. We construct the latter as the direct sum

$$\mathbb{Z}I = \textstyle\bigoplus_{i \in I} \mathbb{Z} \subset \mathbb{Z}^I, \tag{1.23}$$

namely the subgroup of $\mathbb{Z}^I$ formed by the essentially finite families $(\lambda_i)_{i \in I}$ of integers.

In fact $\mathbb{Z}I$ has a canonical basis $e_i$ $(i \in I)$, given by the Kronecker families

$$e_{ii} = 1, \qquad e_{ij} = 0 \qquad (j \neq i), \tag{1.24}$$

since the general element $(\lambda_i)_{i \in I} \in \mathbb{Z}I$ coincides with the essentially finite linear combination $\sum_{i \in I} \lambda_i e_i$.

Identifying any $i \in I$ with the element $e_i$ of the canonical basis, the elements of $\mathbb{Z}I$ can be written as *formal linear combinations* $\sum_{i \in I} \lambda_i i$ of the elements of $I$, with essentially finite integral coefficients.

(d) Finally we have the *free abelian group* functor

$$\begin{aligned} &F \colon \mathsf{Set} \to \mathsf{Ab}, &&F(I) = \mathbb{Z}I = \textstyle\bigoplus_{i \in I} \mathbb{Z}, \\ &F(f) \colon \mathbb{Z}I \to \mathbb{Z}J, &&F(f)(i) = f(i) \quad (\text{for } f \colon I \to J), \end{aligned} \tag{1.25}$$

where the homomorphism $F(f)$ is defined on the canonical basis $I$; its value on a general essentially finite family of $\mathbb{Z}I$ is:

$$F(f)(\lambda_i)_{i\in I} = \Sigma_i\, \lambda_i f(i) = (\Sigma_{f(i)=j}\, \lambda_i)_{j\in J}. \qquad (1.26)$$

The functorial properties of $F$ are readily verified on the bases: for instance, for a composed mapping $gf\colon I \to J \to K$

$$F(gf)(i) = (gf)(i) = g(f(i)) = F(g)F(f)(i).$$

(e) We have studied in 1.2.2 and 1.2.3 the contravariant functor $\mathcal{A} = \mathsf{Top}(-, \mathbb{Z})\colon \mathsf{Top} \dashrightarrow \mathsf{Rng}$.

Forgetting topology, we have a contravariant functor $P = \mathsf{Set}(-, \mathbb{Z})$ defined on sets:

$$\begin{aligned} &P = \mathsf{Set}(-, \mathbb{Z})\colon \mathsf{Set} \dashrightarrow \mathsf{Rng}, \qquad && P(I) = \mathbb{Z}^I = \Pi_{i\in I}\, \mathbb{Z}, \\ &P(f)\colon P(J) \to P(I), && P(f)(\lambda_j)_{j\in J} = (\lambda_{f(i)})_{i\in I}, \end{aligned} \qquad (1.27)$$

where $f\colon I \to J$ is a mapping. The functor $P$ is the same as the composite of the discrete-topology functor $D\colon \mathsf{Set} \to \mathsf{Top}$ with $\mathcal{A}\colon \mathsf{Top} \dashrightarrow \mathsf{Rng}$.

### *1.2.6 A structure theorem*

We shall also use some other classical results on abelian groups. This part can be skipped, until referred to. For integers $m, n$ the notation $m \mid n$ means that $m$ divides $n$, that is, $n = km$ for some $k \in \mathbb{Z}$.

(a) In an abelian group $A$, the elements of finite order form a subgroup

$$\mathrm{t}A = \{x \in A \mid \exists\, n \in \mathbb{N}^* \text{ such that } nx = 0\}, \qquad (1.28)$$

called the *torsion subgroup* of $A$. We say that $A$ is a *torsion group* if $A = \mathrm{t}A$; we say that it is *torsion free* if $\mathrm{t}A = 0$.

The quotient $A/\mathrm{t}A$ is obviously torsion free.

(b) An abelian group $A$ is *finitely generated* if there is a finite family $x_1, \ldots, x_n$ of elements that generates it: every $x \in A$ can be written as a linear combination $\Sigma_{i\in I}\, \lambda_i x_i$ of the generators, with integral coefficients.

We recall a classical theorem, that determines the structure of such groups. The proof can be found in any book on abelian groups (for instance [Fu]). Its extension to modules on principal ideal domains can be found in [Bou3], Section VII.4.4.

(c) (Structure Theorem of Finitely Generated Abelian Groups) *A finitely generated abelian group A is (non-canonically) isomorphic to a direct sum*

$$A \;\cong\; \mathbb{Z}^n \oplus \mathbb{Z}_{\tau_1} \oplus \ldots \oplus \mathbb{Z}_{\tau_k}, \qquad \tau_1 \mid \tau_2 \mid \ldots \mid \tau_{k-1} \mid \tau_k, \qquad (1.29)$$

*that determines the following natural numbers:*

- *the* rank $n = \mathrm{rk}A$ *of* $A$,
- *the* torsion coefficients $\tau_1, ..., \tau_k \in \mathbb{N}^*$, *ordered by divisibility* ($k \geqslant 0$).

*The sequence* $(n, \tau_1, ..., \tau_k)$ *classifies* $A$ *up to isomorphism; note that the torsion coefficients can be repeated.*

(d) A finitely generated abelian group $A$ can thus be decomposed (non-canonically) in the direct sum $A \cong \mathbb{Z}^n \oplus \mathrm{t}A$ of a free abelian group and the torsion subgroup

$$\mathrm{t}A \;\cong\; \mathbb{Z}_{\tau_1} \oplus ... \oplus \mathbb{Z}_{\tau_k}, \qquad \tau_1 \mid \tau_2 \mid ... \mid \tau_{k-1} \mid \tau_k. \tag{1.30}$$

Note that there is no canonical subgroup of $A$ isomorphic to $\mathbb{Z}^n$.

(e) A finitely generated abelian group $A$ which is torsion free is necessarily a free abelian group. All its bases have the same number of elements, namely $n = \mathrm{rk}A$.

(f) The additive group $\mathbb{Q}$ of rational numbers is not finitely generated. It is torsion free, and is not free: no element is a generator, but any two distinct elements are linearly dependent on $\mathbb{Z}$.

(g) The cyclic group $\mathbb{Z}_k$ has one torsion coefficient, namely $k$. The Klein four-element group $\mathbb{Z}_2 \oplus \mathbb{Z}_2$ has torsion coefficients $(2, 2)$, while $\mathbb{Z}_4$ has torsion coefficient 4. The group $\mathbb{Z}_2 \oplus \mathbb{Z}_3 \cong \mathbb{Z}_6$ has torsion coefficient 6.

## 1.3 The terminology of categories and functors

Categories and functors were introduced by Eilenberg and Mac Lane [EiM1] in 1945. We review here these issues, after the hints given in the previous section.

A difficulty of category theory, for a beginner, is the multi-layered structure of this discipline: three layers for the theory itself (categories, functors and their transformations), and two additional layers inside each category (objects and arrows). Starting from examples (as in the previous section) and working on exercises is the best way to master this feature.

Dealing with categories, one has to avoid the usual contradictions related to 'the set of all sets'. Here we make use of a particular set theory where there are *sets* and *classes*; every set is also a class, while a *proper* class is not a set: for instance the *class of all sets* and the *class of all groups* are proper (see [AHS] or the Appendix of [Ke]). This approach is followed in [Mit, Fr, AHS].

*When two levels, like sets and classes, are insufficient, one can introduce a third level of (say) 'hyperclasses'. Alternatively, a more flexible setting widely used for categories is ordinary Set Theory (Zermelo–Fraenkel with the Axiom of Choice) together with the assumption of the existence of a Grothendieck *universe*, or of a suitable hierarchy of universes (cf. [M2], Section I.6).*

### 1.3.1 Definition

A *category* $\mathsf{C}$ consists of the following data:

(a) a class $\mathrm{Ob}\,\mathsf{C}$, whose elements are called *objects* of $\mathsf{C}$,

(b) for every pair $X, Y$ of objects, a set $\mathsf{C}(X, Y)$ (called a *hom-set*) whose elements are called *morphisms* (or *arrows*) of $\mathsf{C}$ from $X$ to $Y$, and denoted as $f\colon X \to Y$,

(c) for every triple $X, Y, Z$ of objects of $\mathsf{C}$, a mapping of *composition*

$$\mathsf{C}(X,Y) \times \mathsf{C}(Y,Z) \to \mathsf{C}(X,Z), \qquad (f,g) \mapsto gf,$$

where $gf$ can also be written as $g.f$.

These data must satisfy the following axioms.

(i) (*Associativity*) Given three consecutive arrows $f\colon X \to Y$, $g\colon Y \to Z$ and $h\colon Z \to W$, one has: $h(gf) = (hg)f$.

(ii) (*Identities*) Every object $X$ has an endomorphism $e\colon X \to X$ which acts as an identity whenever composition makes sense: if $f\colon X' \to X$ and $g\colon X \to X''$, then $ef = f$ and $ge = g$.

One shows, in the usual way, that $e$ is determined by $X$; it is called the *identity* of $X$ and written as $\mathrm{id}\,X$, or $1_X$.

We generally assume that the following condition is also satisfied:

(iii) (*Separation*) For $X, X', Y, Y'$ objects of $\mathsf{C}$, the sets $\mathsf{C}(X, Y)$ and $\mathsf{C}(X', Y')$ are disjoint, unless $X = X'$ and $Y = Y'$.

Therefore a morphism $f\colon X \to Y$ has a well-determined *domain* and *codomain*

$$\mathrm{Dom}\,(f) = X, \qquad \mathrm{Cod}\,(f) = Y.$$

Concretely, when constructing a category, one can forget about condition (iii), since one can always satisfy it by redefining a morphism $\hat{f}\colon X \to Y$ as a triple $(X, Y, f)$ where $f$ is a morphism from $X$ to $Y$ in the original sense (possibly not satisfying the Separation axiom).

$\mathrm{Mor}\,\mathsf{C}$ denotes the class of all the morphisms of $\mathsf{C}$, i.e. the disjoint union of its hom-sets. Two morphisms $f, g$ of $\mathsf{C}$ are said to be *parallel* when they have the same domain and the same codomain. For every object $X$, the set $\mathsf{C}(X) = \mathsf{C}(X, X)$ of endomorphisms $X \to X$ forms a *monoid* (i.e. a unital semigroup).

If $\mathsf{C}$ is a category, the *opposite* (or *dual*) category, written as $\mathsf{C}^{\mathrm{op}}$, has the same objects as $\mathsf{C}$, reversed arrows and reversed composition

$$\mathsf{C}^{\mathrm{op}}(X,Y) = \mathsf{C}(Y,X), \qquad g * f = fg, \qquad \mathrm{id}^{\mathrm{op}}(X) = \mathrm{id}\, X. \qquad (1.31)$$

Every topic of category theory has a dual instance, which comes from the opposite category (or categories). A dual issue is generally distinguished by the prefix 'co-'.

The term 'map' is generally used as equivalent to 'morphism'; between topological spaces it will always mean 'continuous mapping'.

*Complements.* A category is *discrete* if the only arrows are its identities. A category is *indiscrete* if for any two objects $X, Y$ there is precisely one arrow $X \to Y$.

A category is *connected* if it is not empty and any two objects $X, Y$ are linked by some 'zigzag sequence' of arrows

$$X \to X_1 \leftarrow X_2 \to \ldots \leftarrow X_{n-1} \to Y. \qquad (1.32)$$

Excluding the empty structure from the connected categories is the usual option – which we are not following for connected and path connected spaces, as mentioned in Section 0.7.

### 1.3.2 Small and large categories

We have assumed that a category $\mathsf{C}$ has a *class* $\mathrm{Ob}\,\mathsf{C}$ of objects (e.g. the class of all sets, or the class of all topological spaces) and, for every pair $X, Y$ of objects, a *set* $\mathsf{C}(X,Y)$ of morphisms from $X$ to $Y$.

The categories of structured sets that we shall consider, like the examples of 1.2.1, are generally *large* categories, where the objects (or equivalently the morphisms) form a *proper class* (i.e. not a set). A category is said to be *small* if its class of objects is a set. A *finite* category is a small category whose set of morphisms is finite; then the same is true of its set of objects, since an object is determined by its identity.

*Examples and complements.* (a) A set $S$ will often be viewed as a discrete small category: its objects are the elements of $S$, and the only arrows are their (formal) identities; here $S^{\mathrm{op}} = S$.

(b) A preordered set $X$ is equipped with a relation $x \prec x'$ which is assumed to be reflexive and transitive. It will be viewed as a small category, where the objects are the elements of $X$ and the set $X(x,x')$ contains a unique arrow if $x \prec x'$ (which can be written as $(x,x')\colon x \to x'$), and no arrow otherwise.

Composition and identities are uniquely determined, as follows

$$(x', x'')\,(x, x') = (x, x''), \qquad \mathrm{id}\, x = (x, x). \tag{1.33}$$

In this sense, *categories generalise preordered sets.* Their dualities agree: the category associated to the opposite preordered set $X^{\mathrm{op}}$ (with reverse preordering) is dual to the category associated to $X$. Loosely speaking, the extension from preordered sets (or classes) to categories consists in allowing different arrows between two given objects.

In particular, each finite ordinal defines a category, which will be written as $\mathbf{0}, \mathbf{1}, \mathbf{2}$... Thus, $\mathbf{0}$ is the empty category; $\mathbf{1}$ is the *singleton category*, i.e. the discrete category on one object; $\mathbf{2}$ is the *arrow category*, with two objects (0 and 1), and one non-identity arrow, $0 \to 1$.

A (partial) *order* relation is a preordering which is anti-symmetric: $x \prec x'$ and $x' \prec x'$ implies $x = x'$.

(c) A monoid $M$ can be viewed as a small category with one object, say $*$, and one set of morphisms $M$, from $*$ to $*$; composition is the multiplication of $M$. In this sense, *categories generalise monoids.* Again, their dualities coincide: the opposite monoid $M^{\mathrm{op}}$ (with reverse multiplication) corresponds to the opposite category. Loosely speaking, a category can be viewed as a 'many-object' extension of a monoid.

(d) (*Commutative diagrams*) Objects and morphisms of a category are often represented by vertices and arrows in a *diagram*, as in the examples below

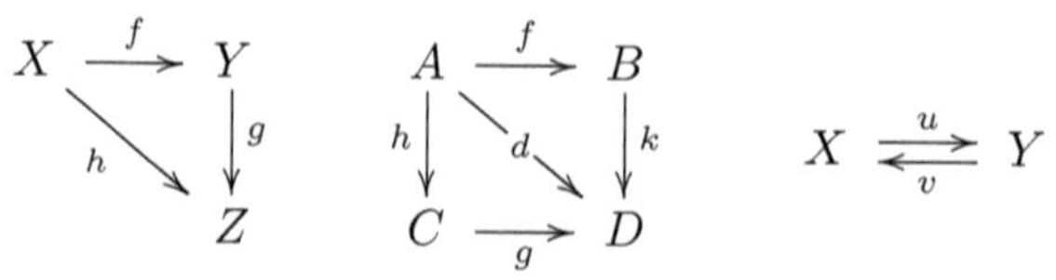


(1.34)

to make clear their relationship and which compositions are legitimate.

We say that such a diagram is *commutative* if:

(i) whenever we have two 'paths' of consecutive arrows, from a certain vertex to another, the two composed morphisms are the same,

(ii) whenever we have a 'loop' of consecutive arrows, from a vertex to itself, the composed morphism is the identity of the corresponding object.

Thus, the first diagram above is commutative if and only if $gf = h$. For the second, commutativity means that $kf = d = gh$. For the third, it means that $vu = \mathrm{id}\, X$ and $uv = \mathrm{id}\, Y$ (so that these morphisms are inverse to each other).

(e) A small category is (associated to) a preordered set (as in point (b)) if and only if all its diagrams commute.

### 1.3.3 Isomorphisms and retracts

As already said in 1.2.4, a morphism $f\colon X \to Y$ in a category $\mathsf{C}$ is said to be *invertible*, or an *isomorphism*, if it has an inverse, i.e. a morphism $g\colon Y \to X$ such that $gf = 1_X$ and $fg = 1_Y$. The latter is determined by $f$; it is called the *inverse* of $f$ and written as $f^{-1}$. (Isomorphisms of topological spaces are called homeomorphisms.)

The *isomorphism relation* $X \cong Y$ between objects of $\mathsf{C}$ (meaning that there is an isomorphism $X \to Y$) is an equivalence relation: an obvious consequence of identity arrows, inverting isomorphisms and composing them.

A *groupoid* is a category where every map is invertible. Every discrete or indiscrete category is a groupoid.

The fundamental groupoid of a space $X$ contains all the fundamental groups $\pi_1(X, x)$, for $x \in X$; it will be constructed in Section 7.1.

One-side inverses are also important. If, in a category $\mathsf{C}$, we have two morphisms $m$ and $p$ such that

$$m\colon A \rightleftarrows X : p, \qquad pm = \mathrm{id}\, A, \tag{1.35}$$

then $m$ is called a *section* (of $p$), $p$ is called a *retraction* (of $m$), and $A$ is called a *retract* of $X$.

In the usual categories of structured sets and structure-preserving mappings, $m$ is injective and $p$ is surjective: we can replace $A$ with its image $m(A) \subset X$, endowed with the restricted structure. (This is common practice, as in 1.1.3 and in the examples below.)

*Exercises and complements.* Retracts are easily characterised, below, in $\mathsf{Set}$ and $\mathsf{Ab}$. Examples in $\mathsf{Top}$ have been examined in 1.1.3 and 1.1.6; later, this issue will be studied with the help of homology functors.

(a) In $\mathsf{Set}$ a retract of a set $X \neq \emptyset$ is any non-empty subset. The empty set is only a retract of itself. In $\mathsf{Set}_\bullet$ a retract of an object $(X, x_0)$ is any pointed subset $(A, x_0)$.

(b) In $\mathsf{Ab}$, retracts coincide with 'direct summands'. More precisely, a subgroup $A$ of an abelian group $X$ is a retract if and only if there exists a subgroup $B \subset X$ such that $A \cap B = \{0\}$ and $A + B = X$. Then $X$ is said to be the *internal direct sum* of $A$ and $B$, and is canonically isomorphic to $A \oplus B$.

Note that $\mathbb{Z}$ is isomorphic to $2\mathbb{Z} \oplus 0$, but $2\mathbb{Z}$ is not a direct summand of $\mathbb{Z}$: the previous directed sum is not an internal one.

(c) A cyclic group $X$ of infinite or prime order has no other retracts than the trivial, unavoidable ones: $\{0\}$ and $X$.

(d) In $\mathsf{Top}$, suppose that $A$ is a retract of $X$, as defined in (1.35). Prove that the map $m$ is a topological embedding, i.e. an injective map $m\colon A \to X$ that makes $A$ homeomorphic to the subspace $m(A) \subset X$ (see 1.1.7(a)).

Identifying $A$ with this subspace we recover the usual definition of retracts in Topology: a subspace $A \subset X$ that admits a retraction, i.e. a map $p\colon X \to A$ such that $p(x) = x$ for all $x \in A$.

(e) In the same hypothesis, prove that $p$ is a quotient map (see 1.1.7(b)).

Identifying $A$ with this quotient, we can view a retract of $X$ as a quotient $p\colon X \to X/R$ that admits a (continuous) section, i.e. a map $m\colon X/R \to X$ such that $pm = \mathrm{id}\, X/R$.

### *1.3.4 Subcategories, quotients and products of categories*

(a) Let $\mathsf{C}$ be a category. A *subcategory* $\mathsf{D}$ is defined by assigning:

- a subclass $\mathrm{Ob}\,\mathsf{D} \subset \mathrm{Ob}\,\mathsf{C}$, whose elements are called *objects of* $\mathsf{D}$,

- for every pair of objects $X, Y$ of $\mathsf{D}$, a subset $\mathsf{D}(X,Y) \subset \mathsf{C}(X,Y)$, whose elements are called *morphisms of* $\mathsf{D}$, from $X$ to $Y$,

so that the following conditions hold:

(i) the composite in $\mathsf{C}$ of any pair of consecutive maps of $\mathsf{D}$ belongs to $\mathsf{D}$,

(ii) the identity in $\mathsf{C}$ of every object of $\mathsf{D}$ belongs to $\mathsf{D}$.

Then $\mathsf{D}$, equipped with the restricted composition law, is a category.

We say that $\mathsf{D}$ is a *full* subcategory of $\mathsf{C}$ if, for every pair of objects $X, Y$ of $\mathsf{D}$, we have $\mathsf{D}(X,Y) = \mathsf{C}(X,Y)$, so that $\mathsf{D}$ is determined by assigning a subclass of objects. We say that $\mathsf{D}$ is a *wide* subcategory of $\mathsf{C}$ if it has the same objects.

For instance, $\mathsf{Ab}$ is a full subcategory of $\mathsf{Gp}$. The open continuous mappings form a wide subcategory of $\mathsf{Top}$.

(b) A *congruence* $R = (R_{XY})$ in a category $\mathsf{C}$ consists of a family of equivalence relations $R_{XY}$ (also written as $R$) in each set of morphisms $\mathsf{C}(X,Y)$, that is consistent with composition:

(iii) if $f R_{XY} f'$ and $g\, R_{YZ}\, g'$, then $gf\, R_{XZ}\, g'f'$.

Equivalently (by transitivity) it is sufficient to say that:

(iii′) if $fRf'$ then $fh\, Rf'h$ and $kfRkf'$ (for all legitimate compositions).

Then one defines the *quotient category* $\mathsf{D} = \mathsf{C}/R$: the objects are those of $\mathsf{C}$, and $\mathsf{D}(X,Y) = \mathsf{C}(X,Y)/R_{XY}$; in other words, a morphism $[f]\colon X \to Y$ in $\mathsf{D}$ is an equivalence class of morphisms $X \to Y$ in $\mathsf{C}$. The composition is induced by that of $\mathsf{C}$, which is legitimate because of condition (iii):

$$[g][f] = [gf]. \tag{1.36}$$

For instance, in Top the homotopy relation $f \simeq f'$ is a congruence of categories: see 1.4.4.

The relation of isomorphism is wider in a quotient category, but it may coincide with the original one, also in a non-trivial quotient.

(c) If $\mathsf{C}$ and $\mathsf{D}$ are categories, one defines the *product category* $\mathsf{C} \times \mathsf{D}$. An object is a pair $(X, Y)$ where $X$ is in $\mathsf{C}$ and $Y$ in $\mathsf{D}$; a morphism is a pair of morphisms

$$(f, g)\colon (X, Y) \to (X', Y'), \qquad f \in \mathsf{C}(X, X'), \quad g \in \mathsf{D}(Y, Y'). \tag{1.37}$$

The composition of $(f, g)$ with a consecutive morphism $(f', g')$ is (obviously) defined componentwise: $(f', g')(f, g) = (f'f, g'g)$, and $\mathrm{id}\,(X, Y) = (\mathrm{id}\,X, \mathrm{id}\,Y)$.

More generally one defines the *cartesian product* $\mathsf{C} = \prod_{i \in I} \mathsf{C}_i$ of a family of categories $(\mathsf{C}_i)_{i \in I}$ indexed by a set $I$: an object of $\mathsf{C}$ is a family $(X_i)_{i \in I}$ where $X_i \in \mathrm{Ob}\,(\mathsf{C}_i)$ (for every $i$), and a morphism $f = (f_i)\colon (X_i) \to (Y_i)$ is a family of morphisms $f_i \in \mathsf{C}_i(X_i, Y_i)$; the composition is componentwise and $\mathrm{id}\,((X_i)_{i \in I}) = (\mathrm{id}\,X_i)_{i \in I}$. A product of groupoids is a groupoid.

Cartesian powers, like $\mathsf{C}^2 = \mathsf{C} \times \mathsf{C}$ and $\mathsf{C}^I = \prod_{i \in I} \mathsf{C}$, are a particular case.

### 1.3.5 Functors

A (covariant) *functor* $F\colon \mathsf{C} \to \mathsf{D}$ consists of the following data:

(a) a mapping $F_0\colon \mathrm{Ob}\,\mathsf{C} \to \mathrm{Ob}\,\mathsf{D}$, whose action is generally written as $X \mapsto F(X)$,

(b) for every pair of objects $X, X'$ in $\mathsf{C}$, a mapping

$$F_{XX'}\colon \mathsf{C}(X, X') \longrightarrow \mathsf{D}(F(X), F(X')),$$

whose action is generally written as $f \mapsto F(f)$.

Composition and identities must be preserved:

(i) if $f, g$ are consecutive maps in $\mathsf{C}$, then $F(gf) = F(g)\,F(f)$,

(ii) if $X$ is in $\mathsf{C}$, then $F(\mathrm{id}\,X) = \mathrm{id}\,(F(X))$.

For a second functor $G\colon \mathsf{D} \to \mathsf{E}$, one defines in the obvious way the *composed functor* $GF\colon \mathsf{C} \to \mathsf{E}$. The composition of consecutive functors is associative and has identities: the *identity functor* of each category

$$\mathrm{id}\,\mathsf{C}\colon \mathsf{C} \to \mathsf{C}, \qquad X \mapsto X, \qquad f \mapsto f.$$

A functor preserves commutative diagrams, isomorphisms, retracts, sections and retractions.

A *contravariant functor* $F\colon \mathsf{C} \dashrightarrow \mathsf{D}$ can be defined as a covariant functor $\mathsf{C}^{\mathrm{op}} \to \mathsf{D}$, or equivalently $\mathsf{C} \to \mathsf{D}^{\mathrm{op}}$. Composing two contravariant functors one gets a covariant one.

A *functor in two variables* is an ordinary functor $F\colon \mathsf{C} \times \mathsf{D} \to \mathsf{E}$ defined on the product of two categories. Fixing an object $X_0$ in $\mathsf{C}$, we have a functor $F(X_0, -)\colon \mathsf{D} \to \mathsf{E}$; and symmetrically.

Cat (resp. Gpd) will denote the category of small categories (resp. small groupoids) and their functors. (These categories also have a 2-dimensional structure, with higher morphisms $\varphi\colon F \to G$ between functors, dealt with in Section 1.5.)

The action of a functor $F$ on objects and morphisms will often be written without parentheses: $FX$ and $Ff$ for $F(X)$ and $F(f)$.

A functor (resp. a contravariant functor) between two preordered sets, viewed as categories, is a monotone (resp. antitone) mapping. A functor between two monoids, viewed as categories, is a homomorphism of monoids.

### 1.3.6 Isomorphic categories

An *isomorphism of categories* is a functor $F\colon \mathsf{C} \to \mathsf{D}$ which is invertible: there is a functor $G\colon \mathsf{D} \to \mathsf{C}$ such that $GF = \mathrm{id}\,\mathsf{C}$ and $FG = \mathrm{id}\,\mathsf{D}$. The functor $G$ is determined by $F$; it is called the *inverse* functor, and written $F^{-1}$.

An easy verification proves that the functor $F$ is an isomorphism if and only if all the mappings $F_0$ and $F_{XX'}$ considered above are bijective.

Being isomorphic categories is plainly an equivalence relation, written as $\mathsf{C} \cong \mathsf{D}$.

Categories linked by an obvious isomorphism are often viewed as 'the same thing'. For instance, everybody knows that an abelian group has a unique structure of module on the ring $\mathbb{Z}$ of integers (where $2x = x + x$, and so on), preserved by every homomorphism of abelian groups. This fact readily shows that the category Ab is canonically isomorphic to the category $\mathbb{Z}\,\mathsf{Mod}$ of modules on the ring of integers; one generally makes no distinction between these categories.

A topological space can be defined in many equivalent ways (based on open sets, closed sets, neighbourhoods, a closure operator, etc.); from a formal point of view, this yields isomorphic categories which is rarely convenient to distinguish.

### 1.3.7 Forgetful and structural functors

(a) Forgetting the structure, or part of it, yields various examples of functors between categories of structured sets, like the following obvious instances

$$\begin{array}{lll} \mathsf{Top} \to \mathsf{Set}, & \mathsf{Top}_\bullet \to \mathsf{Top}, & \mathsf{Top}_\bullet \to \mathsf{Set}_\bullet, \\ \mathsf{Set}_\bullet \to \mathsf{Set}, & \mathsf{Rng} \to \mathsf{Ab}, & \mathsf{Ab} \to \mathsf{Set}. \end{array} \qquad (1.38)$$

The first, for instance, forgets the topology; it was already considered in (1.13). The second forgets the basepoint; and so on.

These *forgetful functors* are often denoted by the letter $U$, which refers to the *underlying* set, or *underlying* space, or *underlying* abelian group, etc.

(b) A subcategory $\mathsf{D}$ of $\mathsf{C}$ has an *inclusion* functor $\mathsf{D} \to \mathsf{C}$, which we also write as $\mathsf{D} \subset \mathsf{C}$. For instance, $\mathsf{Ab} \subset \mathsf{Gp}$. These functors *forget properties*, of objects or morphisms, rather than structure.

(c) A congruence $R$ in a category $\mathsf{C}$ gives an obvious *projection functor* $P\colon \mathsf{C} \to \mathsf{C}/R$, which is the identity on objects and sends a morphism $f$ to its equivalence class $[f]$. For instance, we are interested in the projection functor $\mathsf{Top} \to \mathrm{ho}\mathsf{Top} = \mathsf{Top}/\simeq$ (see (1.63)).

(d) A product category $\mathsf{C} = \prod_{i\in I} \mathsf{C}_i$ has a family of projection functors $P_i\colon \mathsf{C} \to \mathsf{C}_i$.

### 1.3.8 Hom functors

(a) Every category $\mathsf{C}$ has a functor of morphisms, or *hom-functor*:

$$\mathrm{Mor}_{\mathsf{C}}\colon \mathsf{C}^{\mathrm{op}} \times \mathsf{C} \to \mathsf{Set}, \quad (X,Y) \mapsto \mathsf{C}(X,Y), \quad (h,k) \mapsto k(-)h, \qquad (1.39)$$

where, if $h\colon X' \to X$ and $k\colon Y \to Y'$ are in $\mathsf{C}$, the mapping $k(-)h$

$$k(-)h\colon \mathsf{C}(X,Y) \to \mathsf{C}(X',Y'), \qquad f \mapsto kfh, \qquad (1.40)$$

means pre-composing with $h$ and post-composing with $k$.

The hom-functor (1.39) is viewed as a functor on two variables of $\mathsf{C}$, 'contravariant on the first and covariant on the second'. Note that it takes values in $\mathsf{Set}$ because of our assumption, in 1.3.1(b), that all classes $\mathsf{C}(X,Y)$ be *sets*.

(b) Fixing the first variable we get a covariant functor on $\mathsf{C}$

$$\begin{array}{c} \mathsf{C}(X_0,-)\colon \mathsf{C} \to \mathsf{Set}, \\ k \mapsto k_\sharp = k(-)\colon \mathsf{C}(X_0,Y) \to \mathsf{C}(X_0,Y') \qquad (k\colon Y \to Y' \text{ in } \mathsf{C}), \end{array} \qquad (1.41)$$

that is said to be *representable*, and *represented by the object* $X_0$. This terminology will be extended to any functor $\mathsf{C} \to \mathsf{Set}$ isomorphic to $\mathsf{C}(X_0,-)$, as defined in 1.5.2. We already remarked, in 1.2.2(c), that the forgetful functor $U\colon \mathsf{Top} \to \mathsf{Set}$ is represented by the singleton space $\{*\}$.

Dually, fixing the second variable we get a functor

$$\begin{gathered} \mathsf{C}(-, Y_0)\colon \mathsf{C}^{\mathrm{op}} \to \mathsf{Set}, \\ h \mapsto h^\sharp = (-)h\colon \mathsf{C}(X, Y_0) \to \mathsf{C}(X', Y_0) \qquad (h\colon X' \to X \text{ in } \mathsf{C}), \end{gathered} \tag{1.42}$$

that is contravariant on $\mathsf{C}$, and *representable* as such. Again, we say the same of a functor $\mathsf{C}^{\mathrm{op}} \to \mathsf{Set}$ isomorphic to the former. Contravariant functors are often constructed in this way.

(c) In the category $\mathsf{Ab}$ of abelian groups, the set of homomorphisms $A \to B$ has a canonical structure of abelian group, by the pointwise addition $(f+g)(x) = f(x) + g(x)$. This abelian group is denoted as $\mathrm{Hom}(A, B)$.

The category $\mathsf{Ab}$ has thus an *enriched version* of the hom-functor, with values in the category itself, instead of $\mathsf{Set}$

$$\begin{gathered} \mathrm{Hom}\colon \mathsf{Ab}^{\mathrm{op}} \times \mathsf{Ab} \to \mathsf{Ab}, \\ (A, B) \mapsto \mathrm{Hom}(A, B), \qquad (h, k) \mapsto k(-)h. \end{gathered} \tag{1.43}$$

In fact one can easily verify that the mapping $k(-)h$ preserves the addition of homomorphisms: $k(f+g)h = kfh + kgh$.

(d) The representable contravariant functor $\mathcal{A} = \mathsf{Top}(-, \mathbb{Z})$ was also used in enriched form, from $\mathsf{Top}$ to $\mathsf{Rng}$, in Section 1.2.

## 1.4 Paths and homotopy

We review here the elementary issues of homotopy theory, also to fix the terminology we are using.

Typically, the 'invariants' of Algebraic Topology are invariant up to homotopy – a stronger property than invariance up to homeomorphism; a homology functor $H_k$ extracts from the complexity of a space, or a map, a limited information about $k$-dimensional connectedness, which we can hope to compute; choosing $k$, or collecting all the information provided by these functors (or others), we may be able to solve the problem we are studying.

The two-valued index $\alpha$ (or $\beta$) varies in the set $\{0, 1\}$, also written as $\{-, +\}$.

### *1.4.1 Euclidean spaces*

We use the following notation, for spaces which play an important role in Topology and Algebraic Topology.

(a) The space $\mathbb{R}^n$ of $n$-tuples of real numbers $x = (x_1, ..., x_n)$ is equipped with the euclidean topology, determined by the euclidean norm and the euclidean distance

$$||x|| = (x_1^2 + ... + x_n^2)^{1/2}, \qquad d(x, y) = ||x - y||, \tag{1.44}$$

or equivalently as a cartesian power of the euclidean line $\mathbb{R}$.

Identifying $\mathbb{R}^n$ with the hyperplane $\mathbb{R}^n \times \{0\} \subset \mathbb{R}^{n+1}$, these spaces form a nested family, and the singleton $\mathbb{R}^0 = \{0\}$ is included in all of them

$$\{0\} = \mathbb{R}^0 \subset \mathbb{R}^1 \subset \mathbb{R}^2 \subset ... \ \mathbb{R}^n \subset ... \tag{1.45}$$

(b) A *euclidean space* is any subspace of some $\mathbb{R}^n$, and an *open euclidean space* is any open subspace of some $\mathbb{R}^n$.

$\mathbb{I}$ denotes the standard compact interval $[0, 1]$ of $\mathbb{R}$, on which paths and homotopies are parametrised. Its cartesian power $\mathbb{I}^n$ is the *standard $n$-cube*, a subspace of $\mathbb{R}^n$, for $n \geqslant 0$. We are also interested in the *standard $n$-disc* (or closed ball) and the *standard $n$-sphere*

$$\mathbb{D}^n = \{x \in \mathbb{R}^n \mid ||x|| \leqslant 1\}, \qquad \mathbb{S}^n = \{x \in \mathbb{R}^{n+1} \mid ||x|| = 1\}. \tag{1.46}$$

When useful, $\mathbb{S}^n$ is given the basepoint $e_1 = (1, 0, ..., 0)$, corresponding to $1 \in \mathbb{R}$ in the previous nesting. In particular, $\mathbb{S}^0 = \{-1, 1\} \subset \mathbb{R}$ is a discrete space on two points, pointed at 1.

Adding a lower sphere $\mathbb{S}^{-1} = \emptyset$ as the sphere of $\mathbb{R}^0 = \{0\}$, the discs and the spheres form a nested family of subspaces of the family (1.45)

$$\begin{array}{ccccccccc} \{0\} = \mathbb{D}^0 & \subset & \mathbb{D}^1 & \subset & \mathbb{D}^2 & \subset ... & \mathbb{D}^n & \subset ... \\ \emptyset = \mathbb{S}^{-1} & \subset & \mathbb{S}^0 & \subset & \mathbb{S}^1 & \subset ... & \mathbb{S}^{n-1} & \subset ... \end{array} \tag{1.47}$$

where $\mathbb{S}^{n-1} = \underline{\partial}\mathbb{D}^n$ is the *border* of $\mathbb{D}^n$ in $\mathbb{R}^n$, for all $n \geqslant 0$.

We recall that the usual term 'boundary' is here reserved to other meanings, see 0.7. The convenience of adding the empty sphere $\mathbb{S}^{-1}$ will also appear elsewhere. Its topological dimension can be set at $-\infty$, to agree with the formula $\dim(X \times Y) = \dim X + \dim Y$.

(c) The sphere $\mathbb{S}^n$ is covered by the lower and upper hemispheres $\mathbb{D}^n_-$, $\mathbb{D}^n_+$

$$\begin{array}{ll} \mathbb{D}^n_- = \{x \in \mathbb{S}^n \mid x_{n+1} \leqslant 0\}, & \mathbb{D}^n_+ = \{x \in \mathbb{S}^n \mid x_{n+1} \geqslant 0\}, \\ \mathbb{D}^n_- \cup \mathbb{D}^n_+ = \mathbb{S}^n, & \mathbb{D}^n_- \cap \mathbb{D}^n_+ = \mathbb{S}^{n-1} \quad (n \geqslant 0). \end{array} \tag{1.48}$$

Each of them is homeomorphic to the $n$-disc, by orthogonal projection. We shall also use the inverse homeomorphism $u_n$ (an embedding into $\mathbb{S}^n$)

$$u_n \colon \mathbb{D}^n \to \mathbb{D}^n_+, \qquad u_n(x) = (x, \sqrt{1 - ||x||^2}). \tag{1.49}$$

(d) The circle $\mathbb{S}^1$ can also be viewed as the subspace $\{z \in \mathbb{C} \mid |z| = 1\}$ of the complex plane, which topologically is the same as $\mathbb{R}^2$.

The ordinary parametrisation of the circle on the standard interval

$$p\colon \mathbb{I} \to \mathbb{S}^1 \subset \mathbb{C}, \qquad p(t) = (\cos 2\pi t,\, \sin 2\pi t) = e^{2\pi i t}, \tag{1.50}$$

called the *exponential map*, covers the circle counterclockwise (in the oriented plane), identifying the two points of the border $\underline{\partial}\mathbb{I} = \{0, 1\}$ in $\mathbb{R}$.

Writing $\mathbb{I}/\underline{\partial}\mathbb{I}$ the quotient space, the map $p$ induces a bijection

$$p'\colon \mathbb{I}/\underline{\partial}\mathbb{I} \to \mathbb{S}^1, \qquad p'[t] = (\cos 2\pi t,\, \sin 2\pi t) = e^{2\pi i t}, \tag{1.51}$$

which is a homeomorphism (as any bijective map from a compact to a Hausdorff space).

(e) Generally, for a space $X$ and a non-empty subspace $A$, the notation $X/A$ denotes the quotient of $X$ modulo the equivalence relation that collapses $A$ to a single point. We have thus homeomorphisms

$$\mathbb{D}^n/\mathbb{S}^{n-1} \cong \mathbb{S}^n, \qquad \mathbb{I}^n/\underline{\partial}\mathbb{I}^n \cong \mathbb{S}^n \qquad (n > 0). \tag{1.52}$$

The case $n = 0$ can also be included: see 4.2.2.

### 1.4.2 Paths and loops

A *path* in the space $X$ is a map $a\colon \mathbb{I} \to X$ defined on the standard euclidean interval $\mathbb{I} = [0, 1]$.

The basic structure of the interval $\mathbb{I}$ consists of four maps, linking the interval to the singleton $\mathbb{I}^0 = \{*\}$, its 0-th cartesian power:

$$\begin{aligned}
&\partial^\alpha\colon \{*\} \rightleftarrows \mathbb{I}, && \partial^-(*) = 0,\ \ \partial^+(*) = 1 && (\textit{faces}),\\
&e\colon \mathbb{I} \to \{*\}, && e(t) = * && (\textit{degeneracy}),\\
&r\colon \mathbb{I} \to \mathbb{I}, && r(t) = 1 - t && (\textit{reversion}).
\end{aligned} \tag{1.53}$$

Identifying a point $x \in X$ with the corresponding map $x\colon \{*\} \to X$, this basic structure determines:

(i) the *endpoints* of a path $a\colon \mathbb{I} \to X$, namely the points $a\partial^- = a(0)$ and $a\partial^+ = a(1)$,

(ii) the *trivial path*, or constant path, at the point $x$, written as $e_x = xe$,

(iii) the *reversed* path of $a$, written as $a^\sharp = ar$.

We write as $P(X)$ the set of paths of the space $X$, and

$$P(X, x_0, x_1) = \{a \in P(X) \mid a(0) = x_0,\ a(1) = x_1\}, \tag{1.54}$$

will be the set of paths of $X$, *from* $x_0$ *to* $x_1$. (These sets are canonically equipped with the compact-open topology, reviewed in 8.3.2.)

Two paths $a, b\colon \mathbb{I} \to X$ are *consecutive* if $a(1) = b(0)$, that is, $a\partial^+ = b\partial^-$; then they have a *concatenated path* $a * b\colon \mathbb{I} \to X$ (see Exercise (a))

$$(a * b)(t) = \begin{cases} a(2t), & \text{for } 0 \leqslant t \leqslant 1/2, \\ b(2t-1), & \text{for } 1/2 \leqslant t \leqslant 1. \end{cases} \qquad (1.55)$$

This *partial* operation is not associative, generally; we shall see that it works well up to homotopy with fixed endpoints.

A *loop* at $x_0 \in X$ is a path from $x_0$ to $x_0$, called the *basepoint* of the loop. The set of loops of $X$ at $x_0$

$$\Omega(X, x_0) = P(X, x_0, x_0) = \{a \in \mathsf{Top}(\mathbb{I}, X) \mid a(0) = x_0 = a(1)\}, \qquad (1.56)$$

inherits a binary operation of concatenation, which again is not associative (except trivial cases). *It will be equipped with the compact-open topology, see 8.6.7.*

*Exercises and complements.* (a) Verify that the mapping $a*b\colon \mathbb{I} \to X$ is well defined in (1.55), and continuous. *Hint*: the Finite Closed Cover Lemma, in 1.1.8, is the standard tool for concatenating paths and homotopies.

(b) For two points $x, x' \in X$, we write $x \simeq x'$ if there is a path in $X$ from $x$ to $x'$. It is an equivalence relation, because of trivial paths, reversed paths and path-concatenation.

(c) The *path component* $[x]_P$ of the point $x \in X$ is its equivalence class for this relation. We recall that a space $X$ is *path connected* if any two points are linked by a path; this implies that $X$ is connected.

The empty space has no path component, and no connected component.

(d) A loop $a\colon \mathbb{I} \to X$ amounts to map $\hat{a}\colon \mathbb{S}^1 \to X$, where $\hat{a}(e^{2\pi i t}) = a(t)$.

We say that $a$ is a *simple loop* if $\hat{a}$ is injective; then, if $X$ is Hausdorff, the map $\hat{a}$ is a topological embedding (see 1.1.7) and gives a homeomorphism of the circle onto its image, a subspace of $X$.

A loop at $x_0$ amounts to a pointed map $\hat{a}\colon (\mathbb{S}^1, e_1) \to (X, x_0)$.

### *1.4.3 Cylinder and homotopies*

The *cylinder* $IX = X \times \mathbb{I}$ on the space $X$ has the product topology.

Given two maps $f, g\colon X \to Y$ (in $\mathsf{Top}$), a homotopy $\varphi\colon f \simeq g$ 'is' a map $\varphi\colon X \times \mathbb{I} \to Y$ defined on the cylinder $IX$, that coincides with $f$ on the lower basis and with $g$ on the upper one, forming a continuous deformation of $f$ into $g$

$$\begin{gathered} \varphi\colon X \times \mathbb{I} \to Y, \\ \varphi(x, 0) = f(x), \quad \varphi(x, 1) = g(x) \qquad (\text{for } x \in X). \end{gathered} \qquad (1.57)$$

The homotopy will also be written as $\varphi\colon f \simeq g\colon X \to Y$. The map (1.57) will be written as $\hat{\varphi}$ when we want to distinguish it from the homotopy which it represents.

All this relies on the basic structure of the cylinder $IX = X \times \mathbb{I}$, inherited from the structural maps (1.53) of the standard interval, and written as the latter

$$\begin{aligned}
&\partial^\alpha\colon X \to IX, \quad \partial^-(x) = (x,0),\ \partial^+(x) = (x,1) && (faces),\\
&e\colon IX \to X, \quad e(x,t) = x && (degeneracy), \qquad (1.58)\\
&r\colon IX \to IX, \quad r(x,t) = (x, 1-t) && (reversion).
\end{aligned}$$

This structure gives:

(i) the *faces* $\varphi\partial^\alpha\colon X \to Y$ of the homotopy, namely the maps $\varphi\partial^- = f$ and $\varphi\partial^+ = g$,

(ii) the *trivial homotopy* $e_f = fe$ of a map $f$,

(iii) the *reversed homotopy* $\varphi^\sharp = \varphi r$.

More precisely, the homotopy $\varphi^\sharp\colon g \simeq f$ is represented by the map $\hat{\varphi}r\colon IX \to Y$.

Identifying $I\{*\} = \mathbb{I}$, a path $a\colon \mathbb{I} \to X$ is the same as a homotopy $a\colon x \simeq x'\colon \{*\} \to X$ between its endpoints, consistently with the notation $x \simeq x'$ we are using. Moreover the structural maps of the standard interval coincide with the previous maps (1.58), when $X$ is the singleton.

Two consecutive homotopies $\varphi\colon f \simeq g$ and $\psi\colon g \simeq h$ of maps $X \to Y$ have a *concatenation* $\varphi * \psi\colon f \simeq h$

$$(\varphi * \psi)(x,t) = \begin{cases} \varphi(x, 2t), & \text{for } (x,t) \in X \times [0, 1/2], \\ \psi(x, 2t-1), & \text{for } (x,t) \in X \times [1/2, 1], \end{cases} \qquad (1.59)$$

which extends the concatenation of paths (see Exercise (a)).

Homotopy equivalences and deformation retracts have already been dealt with, at a basic level, in 1.1.3.

*Exercises and complements.* (a) Verify that the mapping $\varphi * \psi\colon X \times \mathbb{I} \to Y$ is well defined in (1.59), and continuous.

(b) Let $X$ be a contractible space, i.e. homotopy equivalent to the singleton (see 1.1.3). Then $X$ is path-connected and non-empty, and any singleton $\{x\}$ of $X$ is a deformation retract.

(c) The following conditions on a space $Y$ are equivalent:
- $Y$ is *totally path disconnected*, i.e. every path in $Y$ is trivial,
- for every space $X$, every homotopy $\varphi\colon f \simeq g\colon X \to Y$ is trivial.

*(d) The interval $\mathbb{I}$ (and therefore the cylinder $IX$) has a higher structure

$$\begin{aligned} &g^-: \mathbb{I}^2 \to \mathbb{I}, && g^-(t,t') = \max(t,t'), \\ &g^+: \mathbb{I}^2 \to \mathbb{I}, && g^+(t,t') = \min(t,t') \qquad (\textit{connections}), \\ &s: \mathbb{I}^2 \to \mathbb{I}^2, && s(t,t') = (t',t) \qquad (\textit{transposition}), \end{aligned} \tag{1.60}$$

which governs double homotopies (i.e. homotopies of homotopies) through the double cylinder $I^2(X) = X \times \mathbb{I}^2$: see [G5], Subsection 5.2.2.

*The whole structure satisfies the axioms of a 'diad' – a cubical version of a monad [G1]. Double homotopies will be frequently used in Chapter 7. The formal structure of the cylinder is also investigated in [KP].*

### 1.4.4 The homotopy category

Two parallel maps $f, g: X \to Y$ in $\mathsf{Top}$ are said to be *homotopic*, written as $f \simeq g$, if there exists a homotopy $\varphi: f \simeq g$. This is an equivalence relation in $\mathsf{Top}(X,Y)$, because of trivial homotopies, reversed homotopies and homotopy concatenation.

Furthermore, homotopies have a *whisker composition* with maps

$$X' \xrightarrow{h} X \underset{g}{\overset{f}{\underset{\downarrow\varphi}{\rightrightarrows}}} Y \xrightarrow{k} Y' \tag{1.61}$$

$$k\varphi h: kfh \to kgh: X' \to Y'.$$

where $k\varphi h$ is represented on the cylinder by the map:

$$k\,\hat{\varphi}\,I(h): IX' \to Y', \qquad (x',t) \mapsto k(\varphi(h(x'),t)), \tag{1.62}$$

for $(x',t) \in X' \times \mathbb{I}$. The homotopy relation is thus a congruence of categories in $\mathsf{Top}$.

The quotient category $\mathrm{ho}\mathsf{Top} = \mathsf{Top}/\!\simeq$ is called the *homotopy category of topological spaces*: its objects are the topological spaces, and a morphism $[f]: X \to Y$ is a homotopy class of continuous mappings $X \to Y$. We have a projection functor

$$P: \mathsf{Top} \to \mathrm{ho}\mathsf{Top}, \qquad P(X) = X, \quad P(f) = [f]. \tag{1.63}$$

The classical notation $[X,Y]$ stands for the set of these homotopy classes $[f]: X \to Y$

$$[X,Y] = \mathrm{ho}\mathsf{Top}(X,Y) = \mathsf{Top}(X,Y)/\!\simeq. \tag{1.64}$$

An isomorphism in $\mathrm{ho}\mathsf{Top}$ is the same as a homotopy equivalence in $\mathsf{Top}$, namely a pair of maps $f: X \rightleftarrows Y : g$ such that $gf \simeq \mathrm{id}\,X$ and $fg \simeq \mathrm{id}\,Y$.

A functor $F\colon \mathsf{Top} \to \mathsf{D}$ with values in an arbitrary category is said to be *homotopy invariant* if $F(f) = F(g)$ whenever $f \simeq g$ in $\mathsf{Top}$, so that $F$ induces a functor

$$\overline{F}\colon \mathrm{ho}\mathsf{Top} \to \mathsf{D}, \qquad \overline{F}(X) = F(X), \quad \overline{F}(f) = F[f], \tag{1.65}$$

determined by the relation $F = \overline{F}P$. Typically, this will be the case of all homology and homotopy functors. For simplicity, we may use the same symbol for $F$ and $\overline{F}$.

A homotopy invariant functor $\mathsf{Top} \to \mathsf{D}$ takes a homotopy equivalence $f\colon X \to Y$ in $\mathsf{Top}$ to an isomorphism in $\mathsf{D}$.

*Complements.* (a) The whisker composition of homotopies and maps has obvious associativity and unitarity properties:

$$k'(k\varphi h)h' = (k'k)\varphi(hh'), \quad (\mathrm{id}\, Y)\varphi(\mathrm{id}\, X) = \varphi, \quad ke_f h = e_{kfh}. \tag{1.66}$$

One can also define separately $k\varphi$ and $\varphi h$, and obtain $k\varphi h$ as $(k\varphi)h = k(\varphi h)$.

*(b) The *weak homotopy category* $\mathrm{Ho}\mathsf{Top}$ is defined by a more complex procedure (see [GaZ, Bo1]), formally inverting the *weak homotopy equivalences*, that is, the maps which induce isomorphisms between all homotopy sets and groups (see 7.6.1).

As we shall see in Section 7.6, singular homology is invariant up to weak homotopy. The homotopy groups are also, by definition, but Alexander–Spanier and Čech cohomology theories are not, and can distinguish spaces which are weakly homotopy equivalent.

### *1.4.5 Pointed homotopy*

The category $\mathsf{Top}_\bullet$ of pointed spaces has analogous issues. Given two parallel pointed maps $f, g\colon (X, x_0) \to (Y, y_0)$, a *pointed homotopy* $\varphi\colon f \simeq g$, or *homotopy of pointed maps*, is an ordinary homotopy that keeps the basepoint fixed

$$\begin{aligned} &\varphi\colon X \times \mathbb{I} \to Y, && \varphi(x, 0) = f(x), \quad \varphi(x, 1) = g(x), \\ &\varphi(x_0, t) = y_0 && (\text{for } x \in X,\ t \in \mathbb{I}). \end{aligned} \tag{1.67}$$

Also here, pointed homotopies present trivial instances, reversion, concatenation and whisker composition with pointed maps. The relation $f \simeq g$ of pointed homotopy, or *homotopy of pointed maps*, is a congruence of categories in $\mathsf{Top}_\bullet$, and the quotient category will be written as $\mathrm{ho}\mathsf{Top}_\bullet$.

A functor $F\colon \mathsf{Top}_\bullet \to \mathsf{D}$ which is *homotopy invariant* (up to pointed homotopy), induces a functor $\overline{F}$ (also written as $F$)

$$\overline{F}\colon \mathrm{ho}\mathsf{Top}_\bullet \to \mathsf{D}, \qquad \overline{F}(X) = F(X), \quad \overline{F}(f) = F[f]. \tag{1.68}$$

A pointed map $f\colon (X,x_0) \to (Y,y_0)$ is a (pointed) *homotopy equivalence* if it has a weak inverse $g$

$$f\colon (X,x_0) \rightleftarrows (Y,y_0) : g, \qquad gf \simeq \mathrm{id}\,(X,x_0), \quad fg \simeq \mathrm{id}\,(Y,y_0). \tag{1.69}$$

Then we say that $(X,x_0)$ and $(Y,y_0)$ are *homotopically equivalent* in $\mathsf{Top}_\bullet$, or as pointed spaces, and we write $(X,x_0) \simeq (Y,y_0)$. This amounts to isomorphism in $\mathrm{ho}\mathsf{Top}_\bullet$, and is an equivalence relation.

The symbol $[(X,x_0),(Y,y_0)]_\bullet$ will stand for the set of homotopy classes of pointed maps $[f]\colon (X,x_0) \to (Y,y_0)$

$$[(X,x_0),(Y,y_0)]_\bullet = \mathrm{ho}\mathsf{Top}_\bullet((X,x_0),(Y,y_0)), \tag{1.70}$$

written as $[X,Y]_\bullet$ when the basepoints are understood.

Working in $\mathsf{Top}_\bullet$ any map or homotopy is understood to be pointed, by default.

### *1.4.6 Basic homotopy functors*

We define various homotopy invariant functors, covariant or contravariant, on $\mathsf{Top}$ or $\mathsf{Top}_\bullet$. The variable space, or pointed space, is denoted as $X$; the variable map, or pointed map, as $f\colon X \to Y$.

(i) The homotopy invariant functor $\Pi_0\colon \mathsf{Top} \to \mathsf{Set}$ of path components (introduced in (1.14)) can be rewritten in the form

$$\begin{aligned} &\Pi_0\colon \mathsf{Top} \to \mathsf{Set}, \qquad \Pi_0(X) = |X|/{\simeq}, \\ &f_* = \Pi_0(f)\colon \Pi_0(X) \to \Pi_0(Y), \qquad f_*([x]_P) = [f(x)]_P, \end{aligned} \tag{1.71}$$

where $f\colon X \to Y$ is a map. In fact the quotient $|X|/{\simeq}$ is precisely the set $\Pi_0(X)$ of path components $[x]_P$ of $X$, and the map $\Pi_0(f)$ defined in 1.2.2 acts as above.

We say that the space $X$ is *0-connected* if $\Pi_0(X)$ is a singleton, which means that $X$ is path connected and non-empty. Note that $\Pi_0$ sharply distinguishes the empty space from all other path connected spaces – as is usually the case in Algebraic Topology.

(ii) We have a similar functor from pointed spaces to pointed sets:

$$\pi_0\colon \mathsf{Top}_\bullet \to \mathsf{Set}_\bullet, \qquad \pi_0(X) = (\Pi_0(X), [0_X]_P), \tag{1.72}$$

that acts as above on pointed maps. It will also be used in Chapter 7.

*Note.* We use the symbol $\Pi$ for action on spaces, and $\pi$ on pointed spaces.

(iii) Fixing a pointed space $S$, there is a *homotopy functor*

$$\begin{aligned} &\pi_S = [S,-]_\bullet\colon \mathsf{Top}_\bullet \to \mathsf{Set}_\bullet, \\ &f_*\colon [S,X]_\bullet \to [S,Y]_\bullet, \qquad f_*([a]) = [fa], \end{aligned} \tag{1.73}$$

which corresponds to a hom-functor $\mathrm{hoTop}_\bullet \to \mathsf{Set}_\bullet$ represented by the pointed space $S$, and is thus homotopy invariant. It explores pointed spaces by probes defined on $S$ (as considered in Section 0.4).

(iv) The pointed space $S$ also determines a *cohomotopy functor*

$$\begin{gathered}\pi^S = [-, S]_\bullet \colon \mathsf{Top}_\bullet{}^{\mathrm{op}} \to \mathsf{Set}_\bullet, \\ f^* \colon [S, Y]_\bullet \to [S, X]_\bullet, \qquad f^*([b]) = [bf],\end{gathered} \tag{1.74}$$

which 'is' a hom-functor $\mathrm{hoTop}_\bullet{}^{\mathrm{op}} \to \mathsf{Set}_\bullet$ represented by $S$, and is again homotopy invariant.

*Exercises and complements.* (a) The functor $\Pi_0 \colon \mathrm{hoTop} \to \mathsf{Set}$ is represented by the singleton $S = \{*\}$.

(b) The functor $\pi_0 \colon \mathrm{hoTop}_\bullet \to \mathsf{Set}_\bullet$ is represented by the sphere $\mathbb{S}^0 = \{-1, 1\}$, pointed at 1.

### *1.4.7 Homotopy of paths*

For paths $a, b$ in $X$, the general homotopy relation $a \simeq b$ of parallel maps is of little relevance (see Exercise (a)). We are interested in a stronger property, that will be useful to compute the homology group $H_1(X)$, in Chapter 2, and will be the starting point for the definition of the fundamental groupoid $\Pi_1 X$, in Chapter 7.

We say that the paths $a, b \colon \mathbb{I} \to X$ are *homotopic with fixed endpoints*, or *2-homotopic*, if there is a homotopy $h$ from $a$ to $b$ which does not move the endpoints, as shown in the figure below

$$\begin{aligned} &h \colon \mathbb{I} \times \mathbb{I} \to X, \qquad && h(s, 0) = a(s), \qquad && h(s, 1) = b(s), \\ &h(0, t) = x, && h(1, t) = y && (\text{for } s, t \in \mathbb{I}), \end{aligned} \tag{1.75}$$

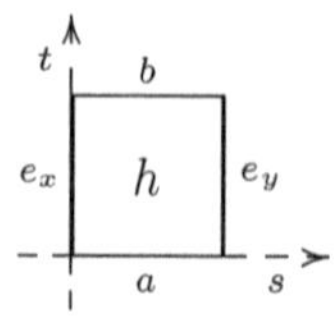


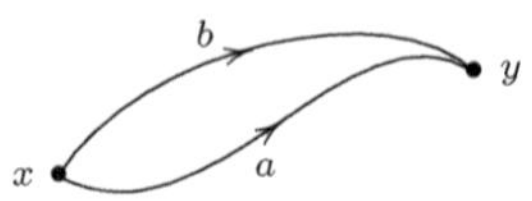


This requires that $a, b$ have the same endpoints, $x = a(0) = b(0)$ and $y = a(1) = b(1)$; $e_x$ and $e_y$ are the constant loops at them. The 2-homotopy $h$ will be denoted as $h \colon a \simeq_2 b \colon x \simeq y$.

In this context, we shall use $s$ as the variable of the given paths, and $t$ as the deformation variable. *The name of 2-homotopy and the notation $h \colon a \simeq_2 b \colon x \simeq y$ is suggested by the 2-cells in a 2-category, see 1.5.2(e), although here we have a lax version of such a structure.*

Loosely speaking, the property $a \simeq_2 b$ means that in $X$ *there is no 1-dimensional hole between the given paths.* The space $X$ is said to be *simply connected* if it is path connected and, for every pair $x, y \in X$, all paths in $X$ from $x$ to $y$ are 2-homotopic; it is said to be *1-connected* if it is simply connected and non-empty. We shall see that these properties are invariant up to homotopy equivalence, in 7.1.5(d); in particular, every contractible space is 1-connected.

A finite sequence $a_1, a_2, ..., a_n$ of consecutive paths in $X$ can be concatenated in various ways, by iterated binary concatenations, and it is important to show that all the paths we can obtain in this way are homotopic with fixed endpoints.

We write as $a_1 * a_2 * ... * a_n$ the *regular concatenation* of the sequence, based on the regular decomposition of the standard interval $\mathbb{I}$ in $n$ subintervals of length $1/n$, indexed by $k = 1, ..., n$

$$(a_1 * a_2 * ... * a_n)(t) = a_k(nt - k + 1), \qquad \text{for } (k-1)/n \leqslant t \leqslant k/n. \quad (1.76)$$

In general, any decomposition of $\mathbb{I}$ in $n$ closed subintervals meeting at the endpoints

$$\mathbb{I} = \cup_{k \geqslant 1} [t_{k-1}, t_k], \qquad 0 = t_0 < t_1 < ... < t_n = 1, \quad (1.77)$$

gives raise to *a* concatenation $c \colon \mathbb{I} \to X$ of the paths $a_k$ $(k = 1, ..., n)$

$$c(t) = a_k((t - t_{k-1})/(t_k - t_{k-1})), \qquad \text{for } t \in [t_{k-1}, t_k]. \quad (1.78)$$

Exercise (d) shows that all of them are homotopic with fixed endpoints.

All iterated binary concatenations of $a_1, a_2, ..., a_n$ can be obtained in this way: for instance, $a_1 * (a_2 * a_3)$ comes out of the decomposition $0 < 1/2 < 3/4 < 1$.

*Exercises and complements.* (a) Two paths $a, b \colon \mathbb{I} \to X$ are homotopic (in the general sense, as maps) if and only if their images belong to the same path component of $X$. This says nothing more on $X$ than the homotopy relation $x \simeq y$ of its points.

(b) Homotopy with fixed endpoints is an equivalence relation between paths (in a fixed space). Moreover, $a \simeq_2 b$ is equivalent to $a^\sharp \simeq_2 b^\sharp$.

(c) (*Reparametrised paths*) A (global) *reparametrisation* of the standard interval will be a continuous function

$$f \colon \mathbb{I} \to \mathbb{I}, \qquad f(0) = 0, \quad f(1) = 1, \quad (1.79)$$

which we are not assuming to be invertible.

For every path $a \colon \mathbb{I} \to X$, the *reparametrised path* $af \colon \mathbb{I} \to X$ is 2-homotopic to the former.

*Hint:* use a 2-homotopy $h \colon \mathrm{id}\,\mathbb{I} \simeq_2 f \colon \mathbb{I} \to \mathbb{I}$ 'of reparametrisation'.

(d) For a finite sequence $a_1, a_2, ..., a_n$ of consecutive paths in $X$, every concatenation $c$ (defined as in (1.78)) is 2-homotopic to the regular concatenation $a_1 * a_2 * ... * a_n$ defined in (1.76).

### 1.4.8 Homotopy of loops

For loops in $X$, we are interested in homotopy with fixed basepoint (still denoted as $a \simeq_2 b$), but also in a more general relation.

Two loops $a\colon x \simeq x$ and $b\colon y \simeq y$ in $X$ are *homotopic as (free) loops*, or *loop homotopic*, if there is a homotopy $h\colon a \simeq b$ such that each intermediate path $h_t = h(-,t)\colon \mathbb{I} \to X$ of the deformation is a loop

$$\begin{aligned} &h\colon \mathbb{I}\times\mathbb{I} \to X, \qquad h(s,0) = a(s), \qquad h(s,1) = b(s), \\ &h(0,t) = h(1,t) \qquad\qquad\qquad\qquad (\text{for } s,t \in \mathbb{I}), \end{aligned} \tag{1.80}$$

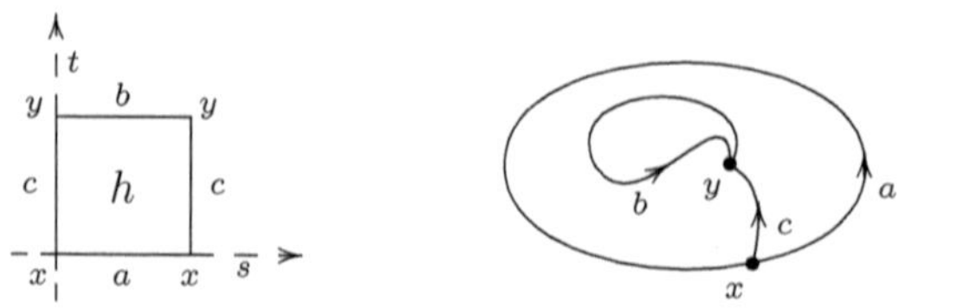


The basepoint $c(t) = h(0,t) = h(1,t)$ of the loop $h_t$ describes a path. The loop homotopy $h$ is a 2-homotopy $a \simeq_2 b$ (between two loops at the same basepoint) if and only if the path $c$ is constant.

*We shall see that loop homotopy is relevant in homology (in 2.2.8(c)), while the stronger condition of 2-homotopy of loops is relevant for the fundamental group (in Section 7.2). Two loops at the same basepoint can be homotopic as free loops, and not 2-homotopic: see Exercise 7.4.4(c).*

*Remarks.* (a) Loop homotopy in the space $X$ is an equivalence relation; the proof is the same as for 2-homotopy.

(b) The quotient of $\mathbb{I}\times\mathbb{I}$ modulo the equivalence relation that identifies $(0,t)$ and $(1,t)$ (for every $t \in \mathbb{I}$) is the cylinder $\mathbb{S}^1 \times \mathbb{I}$. The loop homotopy (1.80) amounts to an ordinary homotopy $\hat{a} \simeq \hat{b}\colon \mathbb{S}^1 \to X$, between the maps associated to $a$ and $b$, in 1.4.2(d).

### 1.4.9 The Lebesgue Number Lemma

(a) Studying maps $a\colon \mathbb{I}^n \to X$ defined on the standard cube (and paths, in particular), we shall have to decompose $a$ on small cubes $A \subset \mathbb{I}^n$, where $A$ is measured by its *diameter*, computed by the euclidean distance:

$$\mathrm{diam}(A) = \sup\{d(x,y) \mid x,y \in A\} \leqslant \sqrt{n}. \tag{1.81}$$

The decomposition will be based on the following classical result.

(b) (Lebesgue Number Lemma) *Let $X$ be a compact metric space and $\mathcal{U} = (U_i)_{i \in I}$ an open cover of $X$. There exists a number $\varepsilon > 0$ such that every subset $A \subset X$ with* $\mathrm{diam}(A) < \varepsilon$ *is contained in some element $U_i$ of the cover.* Such a number is called a *Lebesgue number of $\mathcal{U}$.*

(c) *A proof.* By compactness, we can suppose that the set $I$ is finite; we also assume that no $U_i$ is $X$ (or any positive number would do). For every index $i$, the real function $f_i(x) = d(x, X \setminus U_i)$ is continuous on $X$ and positive on $U_i$ (because $X \setminus U_i$ is closed). Therefore the continuous function $f = \max f_i$ is positive on $X$, and has a minimum $\varepsilon > 0$. A non-empty subset $A$ with $\mathrm{diam}(A) < \varepsilon$ is contained in the open disc $D_X(x, \varepsilon)$ centred at any $x \in A$; but there is an index $i$ such that $d(x, X \setminus U_i) \geqslant \varepsilon$ and therefore $A \subset D_X(x, \varepsilon) \subset U_i$.

## 1.5 Natural transformations and equivalence of categories

Between two functors $F, G \colon \mathsf{C} \to \mathsf{D}$ there can be 'higher arrows' $\varphi \colon F \to G$, called 'natural transformations' (and functorial isomorphisms, or natural isomorphisms, when they are invertible).

'Category', 'functor' and 'natural transformation' are the three basic terms of category theory, since the very beginning of the theory in [EiM1].

It is interesting to note that only the last term is taken from the common language: one can say that Eilenberg and Mac Lane introduced categories and functors because they wanted to formalise the natural transformations that they were encountering in Algebra and Algebraic Topology (as remarked in [M2], at the end of Section I.4). Much in the same way as a general theory of *continuity* (a familiar term for a familiar notion) requires the introduction of *topological spaces* (denoted by a theoretical term).

### *1.5.1 Natural transformations*

Given two *parallel* functors $F, G \colon \mathsf{C} \to \mathsf{D}$ (between the same categories), a *natural transformation* $\varphi \colon F \to G$, also written as $\varphi \colon F \to G \colon \mathsf{C} \to \mathsf{D}$, consists of the following data:

- for each object $X$ of $\mathsf{C}$, a morphism $\varphi X \colon FX \to GX$ in $\mathsf{D}$, called the *component* of $\varphi$ on $X$,

so that, for every arrow $f \colon X \to X'$ in $\mathsf{C}$, we have a commutative square in $\mathsf{D}$ (whose diagonal can be written as $\varphi(f)$)

$$\begin{array}{ccc} FX & \xrightarrow{\varphi X} & GX \\ {\scriptstyle Ff}\downarrow & & \downarrow{\scriptstyle Gf} \\ FX' & \xrightarrow[\varphi X']{} & GX' \end{array} \qquad (\varphi X')F(f) = G(f)(\varphi X) \ \ (= \varphi(f)). \qquad (1.82)$$

This is called the *naturality condition* of $\varphi$ on the morphism $f$. The component of $\varphi$ on the object $X$ is also written as $\varphi_X$, or $\varphi\colon FX \to GX$.

A natural transformation $\varphi\colon F \to G$ often arises from a 'canonical choice' of a family $(\varphi X)_X$ of morphisms, but these two aspects are not the same. There are canonical choices of such families that are not natural, and natural transformations generated by the axiom of choice (see Exercise 7.1.9(b)).

The *identity* of a functor $F\colon \mathsf{C} \to \mathsf{D}$ is the natural transformation $\mathrm{id}\,F\colon F \to F$, with components $(\mathrm{id}\,F)X = \mathrm{id}\,(FX)$.

*Exercises and complements.* (a) Let $F\colon R\,\mathsf{Mod} \to R\,\mathsf{Mod}$ be the identity functor of the category of modules on a commutative (unital) ring $R$. Every scalar $\lambda \in R$ gives a natural transformation $\lambda\colon F \to F$ whose component $\lambda_A\colon A \to A$ on a module $A$ is the multiplication $\lambda_A(x) = \lambda x$, by $\lambda$. Note that the commutativity of $R$ is required.

(b) Conversely, every natural transformation $\varphi\colon F \to F$ is of this form, determined by a unique $\lambda \in R$. *Hint:* use the component of $\varphi$ on $R$, as a module on itself.

### 1.5.2 Two operations

Two natural transformations $\varphi\colon F \to G$ and $\psi\colon G \to H$ between the same categories have a *vertical composition*, written as $\psi\varphi$ or $\psi.\varphi$

$$\mathsf{C} \;\underset{H}{\overset{F}{\underset{\downarrow\psi}{\overset{\downarrow\varphi}{\longrightarrow}}}}\; \mathsf{D} \qquad (\psi\varphi)(X) = \psi X.\varphi X\colon FX \to HX, \tag{1.83}$$

which is associative; the identities of functors act as units.

There is also a *whisker composition*, or *reduced horizontal composition*, of natural transformations with functors, written as $K\varphi H$, or $K\circ\varphi\circ H$

$$\mathsf{C}' \xrightarrow{H} \mathsf{C} \;\underset{G}{\overset{F}{\underset{\downarrow\varphi}{\longrightarrow}}}\; \mathsf{D} \xrightarrow{K} \mathsf{D}' \tag{1.84}$$

$$K\varphi H\colon KFH \to KGH, \qquad (K\varphi H)(X') = K(\varphi(HX')).$$

An *isomorphism of functors*, or *functorial isomorphism*, or *natural isomorphism*, is a natural transformation $\varphi\colon F \to G$ which is invertible for the vertical composition. Then we write $F \cong G$; it is obviously an equivalence relation between parallel functors. Two examples can be found in the solution of Exercises 1.4.6(a), (b).

It is easy to verify that the natural transformation $\varphi\colon F \to G\colon \mathsf{C} \to \mathsf{D}$ is invertible if and only if all its components $\varphi X\colon FX \to GX$ are invertible in

D: if this is the case, the inverse components $\psi X = (\varphi X)^{-1}\colon GX \to FX$ form a natural transformation $\psi\colon G \to F$ which is inverse to $\varphi$.

As a consequence, every natural transformation $\varphi\colon F \to G\colon \mathsf{C} \to \mathsf{D}$ with values in a groupoid D is invertible.

*Exercises and complements* (The calculus of natural transformations). The reader should prove the following properties, which will often be used in computations. The (easy) solution of point (d) can be found in Chapter 9; the others are straightforward.

(a)
$$\chi(\psi\varphi) = (\chi\psi)\varphi,$$
$$K'(K\varphi H)H' = (K'K)\varphi(HH') \qquad (\textit{associativity}),$$

(b)
$$\varphi.\mathrm{id}\, F = \varphi = \mathrm{id}\, G.\varphi,$$
$$1_Y\circ\varphi\circ 1_X = \varphi, \qquad K\circ\mathrm{id}\, F\circ H = \mathrm{id}\,(KFH) \qquad (\textit{identities}),$$

(c)
$$K(\psi\varphi)H = (K\psi H).(K\varphi H) \qquad (\textit{distributivity}),$$

(d)
$$\psi G.H\varphi = K\varphi.\psi F \qquad (\textit{interchange}).$$

The last formula is concerned with two 'horizontally consecutive' natural transformations, as in the following diagram

$$\mathsf{C} \underset{G}{\overset{F}{\rightrightarrows}}\!\!{\scriptstyle\downarrow\varphi}\; \mathsf{D} \underset{K}{\overset{H}{\rightrightarrows}}\!\!{\scriptstyle\downarrow\psi}\; \mathsf{E}. \qquad (1.85)$$

*(e) The category Cat of small categories and functors, enriched with the natural transformations as second order morphisms $\varphi\colon F \to_2 G$ between ordinary arrows, and the previous operations, becomes a two-dimensional structure called a 2-*category*. This is the beginning of higher dimensional Category Theory; one can find an introduction to this domain in [M3], Chapter XII, or [G5], Chapter 7.

### 1.5.3 The cylinder functor

As we have seen in 1.4.3, homotopies of maps of topological spaces are based on the cylinder $X \times \mathbb{I}$. This transformation $I = - \times \mathbb{I}$ acts on a general map $f\colon X \to Y$ in the obvious way (already used above), forming the *cylinder functor* of topological spaces

$$\begin{gathered} I\colon \mathsf{Top} \to \mathsf{Top}, \qquad IX = X \times \mathbb{I}, \\ If = f \times \mathrm{id}\,\mathbb{I}\colon X \times \mathbb{I} \to Y \times \mathbb{I}, \qquad (If)(x,t) = (f(x),t), \end{gathered} \qquad (1.86)$$

where $x \in X$ and $t \in \mathbb{I}$. The map $f \times \mathrm{id}\,\mathbb{I}$ is also written as $f \times \mathbb{I}$.

Now, the faces ($\partial^\alpha\colon X \to IX$), degeneracy ($e\colon IX \to X$) and reversion ($r\colon IX \to IX$) described in (1.58) are the components on the space $X$ of four natural transformations of endofunctors

$$\begin{aligned} &\partial^\alpha\colon \mathrm{id} \to I\colon \mathsf{Top} \to \mathsf{Top} && (\textit{faces}),\\ &e\colon I \to \mathrm{id}: \mathsf{Top} \to \mathsf{Top} && (\textit{degeneracy}),\\ &r\colon I \to I\colon \mathsf{Top} \to \mathsf{Top} && (\textit{reversion}). \end{aligned} \tag{1.87}$$

As in 1.4.3, the object $X$ is often left understood in the notation of a component, writing $\partial^\alpha\colon X \to IX$ for $\partial^\alpha X\colon X \to IX$, and so on.

We already mentioned in 1.4.3(d) the higher structure of the cylinder functor, related to double homotopies.

*Exercises and complements.* (a) Prove the naturality of the transformations in (1.87).

(b) If $F, G\colon \mathsf{Top} \to \mathsf{D}$ are isomorphic functors and $F$ is homotopy invariant, so is $G$.

*(c) Homotopies can be equivalently based on the path endofunctor $P$: $\mathsf{Top} \to \mathsf{Top}$: this will be examined in 8.3.4.

### *1.5.4 Categories of functors*

Let $\mathsf{S}$ be a small category. A *diagram* in the category $\mathsf{C}$, *of type* $\mathsf{S}$ (or *based on* $\mathsf{S}$) will be any functor $X: \mathsf{S} \to \mathsf{C}$.

It can be written in *index notation*, with indices $i \in \mathrm{Ob}\,\mathsf{S}$ and $a\colon i \to j$ in $\mathrm{Mor}\,\mathsf{S}$:

$$X: \mathsf{S} \to \mathsf{C}, \qquad i \mapsto X_i, \quad a \mapsto (X_a\colon X_i \to X_j), \tag{1.88}$$

or as a system $((X_i), (X_a))$.

We write as $\mathsf{C}^\mathsf{S}$ the category whose objects are the functors $X: \mathsf{S} \to \mathsf{C}$, and whose morphisms are the natural transformations $X \to Y\colon\ \mathsf{S} \to \mathsf{C}$, with vertical composition. It is also written as $\mathsf{Cat}(\mathsf{S}, \mathsf{C})$ (even if $\mathsf{C}$ is not assumed to be small).

In particular the arrow category $\mathbf{2}$, with two objects (0 and 1) and one non-identity arrow, $0 \to 1$ (see 1.3.2(b)) gives the category $\mathsf{C}^\mathbf{2}$ *of morphisms* of $\mathsf{C}$, where an object is a morphism $x\colon X_0 \to X_1$ of $\mathsf{C}$, and a map $f = (f_0, f_1)\colon x \to y$ is a commutative square of $\mathsf{C}$, as in the left diagram below

$$\begin{array}{ccc} X_0 & \xrightarrow{f_0} & Y_0 \\ {\scriptstyle x}\downarrow & & \downarrow{\scriptstyle y} \\ X_1 & \xrightarrow[f_1]{} & Y_1 \end{array} \qquad\qquad \begin{array}{ccccc} X_0 & \xrightarrow{f_0} & Y_0 & \xrightarrow{g_0} & Z_0 \\ {\scriptstyle x}\downarrow & & \downarrow{\scriptstyle y} & & \downarrow{\scriptstyle z} \\ X_1 & \xrightarrow[f_1]{} & Y_1 & \xrightarrow[g_1]{} & Z_1 \end{array} \tag{1.89}$$

These maps are composed *by pasting* commutative squares, as in the right diagram above. The identity of $x$ is: $\mathrm{id}\,(x) = (\mathrm{id}\,X_0, \mathrm{id}\,X_1)\colon x \to x$.

Writing an object of $\mathsf{C}^\mathsf{S}$ in index notation, as above, the *diagonal functor*

$$D\colon \mathsf{C} \to \mathsf{C}^\mathsf{S},$$
$$(DA)_i = A, \qquad (DA)_a = \mathrm{id}\,A \qquad (i \in \mathrm{Ob}\,\mathsf{S},\ a \in \mathrm{Mor}\,\mathsf{S}), \tag{1.90}$$

sends an object $A$ to the constant functor at $A$, and a morphism $f\colon A \to B$ to the natural transformation $Df\colon DA \to DB\colon \mathsf{S} \to \mathsf{C}$ with constant components $(Df)_i = f$.

*Exercises and complements.* (a) $\mathsf{C}^\mathbf{0}$ is the singleton category $\mathbf{1}$ (also when $\mathsf{C} = \mathbf{0}$). $\mathsf{C}^\mathbf{1}$ is isomorphic to $\mathsf{C}$. $\mathsf{C}^\mathbf{2}$ is described above. $\mathsf{C}^\mathbf{3}$ is the *category of pairs of consecutive arrows* of $\mathsf{C}$. $\mathsf{C}^{\mathbf{2}\times\mathbf{2}}$ is the *category of commutative squares* of $\mathsf{C}$. For a set $I$ (a discrete category), $\mathsf{C}^I$ is the power category $\prod_{i\in I} \mathsf{C}$.

Several interesting categories of diagrams in $\mathsf{Top}$ are obtained in this way.

(b) The diagonal functor $D\colon \mathsf{C} \to \mathsf{C}^\mathsf{S}$ is an embedding, provided that $\mathsf{S}$ is not empty.

(c) A natural transformation $\varphi$ between functors $\mathsf{C} \to \mathsf{D}$ can be viewed as a functor $\mathsf{C} \times \mathbf{2} \to \mathsf{D}$, or equivalently as a functor $\mathsf{C} \to \mathsf{D}^\mathbf{2}$.

*The cylinder functor $I = - \times \mathbf{2}\colon \mathsf{Cat} \to \mathsf{Cat}$ represents natural transformations as directed homotopies between functors: cf. [G1], Section 1.1.6.*

### 1.5.5 *Equivalence of categories*

Categories have a notion of equivalence, more general than isomorphism, that plays an important role in the theory of categories, but will play a restricted role in this book. (It will be used for fundamental groupoids, in 7.1.4).

An *equivalence of categories* is a functor $F\colon \mathsf{C} \to \mathsf{D}$ which is invertible up to functorial isomorphism: there is a functor $G\colon \mathsf{D} \to \mathsf{C}$ such that $GF \cong \mathrm{id}\,\mathsf{C}$ and $FG \cong \mathrm{id}\,\mathsf{D}$. The functor $G$ can be called a *weak inverse* of $F$.

One says that the categories $\mathsf{C}$, $\mathsf{D}$ are *equivalent*, written as $\mathsf{C} \simeq \mathsf{D}$, if there exists an equivalence of categories between them. This is indeed an equivalence relation, as can be easily proved.

The reader will note a formal analogy, where

(i) topological spaces, maps, homotopies $\varphi\colon f \simeq g\colon X \to Y$, their operations, and homotopy equivalence,

correspond to:

(ii) categories, functors, natural isomorphisms $\varphi\colon F \to G\colon \mathsf{C} \to \mathsf{D}$, their operations, and categorical equivalence.

*This analogy becomes sharper in Directed Algebraic Topology, where *directed homotopies* need not be reversible and correspond to general *natural transformations*: see [G1].*

*Exercises and complements.* (a) A category $\mathsf{C}$ is equivalent to the singleton category $\mathbf{1}$ if and only if it is non-empty and indiscrete, i.e. each hom-set $\mathsf{C}(X, Y)$ has precisely one element (see 1.3.1).

*(b) The calculus of homotopies is more complex than the calculus of natural transformations: in (i), part of the properties 1.5.2(a)–(d) are only satisfied up to higher homotopies.

### 1.5.6 *Presheaves and sheaves

(a) A functor $\mathsf{S}^{\mathrm{op}} \to \mathsf{C}$, defined on the opposite category $\mathsf{S}^{\mathrm{op}}$, is also called a *presheaf* of $\mathsf{C}$ on the (small) category $\mathsf{S}$. They form the *presheaf category* $\mathrm{Psh}(\mathsf{S}, \mathsf{C}) = \mathsf{C}^{\mathsf{S}^{\mathrm{op}}}$.

In particular, we have the category $\mathrm{Psh}(\mathsf{S}, \mathsf{Set}) = \mathsf{Set}^{\mathsf{S}^{\mathrm{op}}}$ *of presheaves of sets*, on $\mathsf{S}$. The category $\mathsf{S}$ is canonically embedded in the latter, by the *Yoneda embedding*

$$y\colon \mathsf{S} \to \mathsf{Set}^{\mathsf{S}^{\mathrm{op}}}, \qquad y(i) = \mathsf{S}(-, i)\colon \mathsf{S}^{\mathrm{op}} \to \mathsf{Set}, \tag{1.91}$$

which sends every object $i \in \mathrm{Ob}\,\mathsf{S}$ to the presheaf $y(i)$ *represented by* $i$.

Categories of presheaves are often used in homology and homotopy; cubical sets will be introduced in 2.2.2, and simplicial sets in 3.6.6.

(b) For a topological space $X$, *a presheaf of abelian groups on* $X$ is a functor $\Gamma\colon \mathcal{O}(X)^{\mathrm{op}} \to \mathsf{Ab}$, where $\mathcal{O}(X)$ is the set of open sets of $X$, ordered by inclusion. It is generally assumed that $\Gamma(\emptyset) = 0$.

Explicitly, we are assigning an abelian group $\Gamma(U)$ to every open set $U$ of $X$, and a homomorphism $\rho^U_V\colon \Gamma(U) \to \Gamma(V)$, called a *restriction*, when $U \supset V$, under the obvious axioms:

$$\rho^V_W\, \rho^U_V = \rho^U_W, \qquad \rho^U_U = \mathrm{id}\,\Gamma(U) \qquad (\text{for } U \supset V \supset W), \tag{1.92}$$

to which one generally adds the condition $\Gamma(\emptyset) = 0$. One defines similarly a presheaf of $R$-modules, or rings, or $R$-algebras (cf. 4.6.1).

(c) *An example.* For $X = \mathbb{R}^n$, let $C^k(U, \mathbb{R})$ be the $\mathbb{R}$-algebra *of real functions of class* $C^k$ on the open set $U$ (for $0 \leqslant k \leqslant \infty$). This gives a presheaf $C^k(X, \mathbb{R})$ of $\mathbb{R}$-algebras, with obvious restrictions of functions $\rho^U_V(f) = f|V$, from any open set to a smaller one.

(d) $C^k(X, \mathbb{R})$ is actually a *sheaf*: for any family of open sets $(U_i)_{i \in I}$ and every 'consistent' family of functions $f_i \in C^k(U_i, \mathbb{R})$ (where $f_i$ and $f_j$ have the same restriction to $U_i \cap U_j$), there is precisely one function $f \in C^k(\bigcup_i U_i, \mathbb{R})$ which restricts to $f_i$ on each $U_i$.

For a sheaf $\Gamma$ of abelian groups (or algebras), the condition $\Gamma(\emptyset) = 0$ is a consequence of the previous property, since $\emptyset$ is the union of the empty family.

The theory of sheaves is studied in [MaM, Bo3, Go].

## 1.6 Products, sums and universal properties

Cartesian products are a basic, elementary issue of topological spaces and algebraic structures. Yet, it will be useful to unify these issues in a single categorical concept, based on a universal property.

This is even more evident for the dual notion, a sum of objects, which has dissimilar realisations in the usual categories of structured sets: disjoint unions of sets and spaces, direct sums of abelian groups, free products of groups, tensor products of commutative rings, etc.

Moving from one category to another, for instance from topological spaces to abelian groups, every homology functor $H_n \colon \mathsf{Top} \to \mathsf{Ab}$ 'preserves sums', transforming disjoint unions of spaces into direct sums of abelian groups; this statement makes a precise sense with respect to the universal properties of these structures.

Universal properties have a general formalisation in category theory, that unifies diverse topics, from free algebraic structures to cartesian products and sums, from universal compactification to metric completion. This is briefly examined in 1.6.6 and 1.6.7; the global aspect of this issue – adjoint functors – will be outlined in Section 8.1.

### *1.6.1 Products*

In the categories we are considering (and many other categories of structured sets) we have cartesian products, constructed with the cartesian product of the underlying sets, equipped with a 'natural' structure.

All these procedures can be unified: we have a family $(X_i)_{i \in I}$ of objects of the category $\mathsf{C}$, indexed by a set $I$, and we want to find an object $X$ equipped with a family of morphisms $p_i \colon X \to X_i$ $(i \in I)$, called *cartesian projections*, which satisfies the following *universal property*:

$$\begin{array}{ccc} X & \xrightarrow{p_i} & X_i \\ {\scriptstyle f}\uparrow & \nearrow_{f_i} & \\ Y & & \end{array} \qquad (1.93)$$

(i) for every object $Y$ and every family of morphisms $f_i \colon Y \to X_i$ there is a unique morphism $f \colon Y \to X$ such that, for all $i \in I$, $p_i f = f$.

The morphisms $f_i$ are called the *components* of $f$; the latter is often written as $(f_i)$ – although it is not the same as the family of its components, of course.

The product of a family of objects need not exist. If it does, it is determined up to a unique *coherent* isomorphism, in the sense that if also $Y$ is a product of the family $(X_i)_{i\in I}$ with projections $q_i\colon Y \to X_i$, then the unique morphism $f\colon X \to Y$ which commutes with all projections (i.e. $q_i f = p_i$, for all $i$) is an isomorphism.

In fact there is also a unique morphism $g\colon Y \to X$ such that $p_i g = q_i$; moreover $gf = \mathrm{id}\, X$ (because $p_i(gf) = p_i(\mathrm{id}\, X)$, for all $i$) and $fg = \mathrm{id}\, Y$.

We speak thus of *the* product of the family $(X_i)$, denoted as $\Pi_i X_i$ or $X_1 \times ... \times X_n$ in a finite case. We say that a category $\mathsf{C}$ *has products* (resp. *finite products*) if every family (resp. finite family) of objects has a product in $\mathsf{C}$.

In particular the product of the empty family of objects $\emptyset \to \mathrm{Ob}\,\mathsf{C}$ means an object $X$ (equipped with no projections) such that for every object $Y$ (equipped with no arrows) there is a unique morphism $f\colon Y \to X$ (satisfying no condition). The solution is called the *terminal* object of $\mathsf{C}$; again, it need not exist, but is determined up to a unique isomorphism. It can be written as $\top$.

*Examples and complements.* (a) In $\mathsf{Set}$, $\mathsf{Top}$, $\mathsf{Ab}$, $\mathsf{Gp}$, $\mathsf{Rng}$ all products exist and are the usual cartesian ones (built on the product of the underlying sets). The terminal object is the singleton, with the appropriate structure.

We recall that a topological product $X$ with projections $p_i\colon X \to X_i$ has the coarsest topology making all projections continuous. This structure is less obvious than the other product structures we have mentioned; the universal property confirms that it is indeed the 'right' choice, in $\mathsf{Top}$.

The reader will also note that the projections $p_i\colon X \to X_i$ of a product of sets, or topological spaces, need not be surjective: this fails if (and only if) some factors are empty and others are not.

(b) Products in $\mathsf{Set}_\bullet$ and $\mathsf{Top}_\bullet$ are readily constructed with the unpointed analogues

$$\Pi_i\,(X_i, \overline{x}_i) = (\Pi_i\, X_i, (\overline{x}_i)). \qquad (1.94)$$

(c) In the category $X$ associated to a preordered set, the categorical product of a family of points $x_i \in X$ amounts to $\inf x_i$ (the greatest lower bound), and the terminal object amounts to the greatest element of $X$. These elements need not exist; they are determined up to the equivalence relation associated to our preorder, and uniquely determined in an ordered set.

(d) Products in $\mathsf{Cat}$ have been considered in 1.3.4; the terminal object is the singleton category $\mathbf{1}$. Category theory can 'operate on itself', so to say.

(e) If the category C has (all) binary products, there is a functor

$$-\times -\colon \mathsf{C}\times\mathsf{C}\to\mathsf{C}, \qquad (X,Y)\mapsto X\times Y,$$
$$(f,g)\ \mapsto\ f\times g\colon X\times Y\to X'\times Y', \tag{1.95}$$

where $f\colon X\to X'$, $g\colon Y\to Y'$ and the morphism $f\times g$ has the following components, with respect to the projections $p,q$ and $p',q'$ of our products

$$p'(f\times g)=fp\colon X\times Y\to X', \qquad q'(f\times g)=gq\colon X\times Y\to Y'. \tag{1.96}$$

(f) Similarly, if the category C has products, for every set $I$ there is a functor

$$\Pi_{i\in I}\colon \mathsf{C}^I\to\mathsf{C}, \qquad (X_i)\mapsto \Pi_i X_i, \qquad (f_i)\mapsto \Pi_i f_i, \tag{1.97}$$

where $f=\Pi_i f_i\colon \Pi_i X_i\to \Pi_i Y_i$ is characterised by: $p_i f = f_i p_i$, for $i\in I$ (writing the projections of both products by the same symbols).

(g) If the terminal object exists, it acts as a unit for the cartesian product, in the sense that $X\times\top$ and $\top\times X$ are canonically isomorphic to $X$. In fact, one easily verifies that $X$ satisfies the universal property of the product $X\times\top$, with projections $p=\mathrm{id}\,X$ and $q\colon X\to\top$.

In the previous concrete categories this is obvious: $\top$ is the appropriate singleton.

### 1.6.2 Sums

The dual notion, the sum of a family $(X_i)_{i\in I}$ of objects, has dissimilar realisations in the usual categories of structured sets, but can be recognised as a single issue in category theory.

The *sum*, or *coproduct*, of a family $(X_i)_{i\in I}$ of objects of C is an object $X$ equipped with a family of morphisms $u_i\colon X_i\to X$ $(i\in I)$, called *injections*, which satisfy the following universal property:

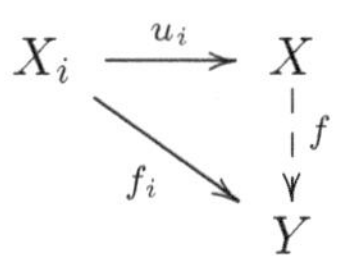


(1.98)

(i*) for every object $Y$ and every family of morphisms $f_i\colon X_i\to Y$, there is a unique morphism $f\colon X\to Y$ such that, for all $i\in I$, $fu_i=f_i$.

The map $f$ will be written as $[f_i]$, by its *co-components*.

Again, if the sum of the family $(X_i)$ exists, it is determined up to a unique coherent isomorphism, and denoted as $\Sigma_i X_i$ (or $\amalg X_i$), and $X_1+\ldots+X_n$ in a finite case.

The sum of the empty family is the *initial* object $X$: this means that every object $Y$ has precisely one morphism $X \to Y$. It can be written $\bot$.

*Examples and complements.* (a) Sums in $\mathsf{Set}$ are realised as 'disjoint unions'. There is a standard way of doing this, replacing each summand $X_i$ by an isomorphic copy $X_i \times \{i\}$

$$\Sigma_i X_i = \cup_i X_i \times \{i\}, \qquad u_i(x) = (x, i) \qquad (\text{for } x \in X_i). \qquad (1.99)$$

The initial object is the empty set.

(b) In $\mathsf{Top}$ we do the same, equipping the disjoint union $X$ with the finest topology that makes all the injections continuous; in other words, the open subsets of $X$ are arbitrary unions of open subsets of the summands. The initial object is the empty space.

(c) In $\mathsf{Ab}$ and $R\,\mathsf{Mod}$ categorical sums are realised as direct sums

$$\bigoplus A_i = \{(a_i) \in \prod A_i \mid a_i = 0 \text{ except for a finite set of indices}\}, \quad (1.100)$$

where the family $(a_i)$ is essentially finite. Now, for every family of homomorphisms $f_i\colon A_i \to B$ we let $f((a_i)_{i\in I}) = \Sigma f_i(a_i)$, an essentially finite sum.

The initial object is the singleton. Direct summands and internal direct sums of modules are dealt with as in $\mathsf{Ab}$: see 1.3.3(b).

(d) In $\mathsf{Set}_\bullet$ and $\mathsf{Top}_\bullet$ the initial object is the singleton. A binary sum is usually written as $(X, x_0) \vee (Y, y_0)$, it is called a *wedge* and realised in $(X, x_0) \times (Y, y_0)$ as the subset or subspace $X \times y_0 \cup x_0 \times Y$, pointed at $(x_0, y_0)$; the finite case is similar. Arbitrary sums can be similarly constructed in $\mathsf{Set}_\bullet$, but in $\mathsf{Top}_\bullet$ we would not get the correct topology: they will be constructed in 3.1.4.

(e) In the category $X$ associated to a preordered set, the categorical sum of a family of points $x_i \in X$ amounts to $\sup x_i$ (the least upper bound), while the initial object amounts to the least element of $X$.

(f) In $\mathsf{Cat}$ a sum $\Sigma\,\mathsf{C}_i$ of categories is their obvious disjoint union. The initial object is the empty category $\mathbf{0}$.

(g) If the category $\mathsf{C}$ has binary sums, or arbitrary sums, we can build functors corresponding to those of 1.6.1(e) and 1.6.1(f).

(h) In $\mathsf{Gp}$ categorical sums are classically known as 'free products' (see 7.4.1). *In the category of commutative rings, categorical sums are realised as tensor products on $\mathbb{Z}$.*

### 1.6.3 Comments

(a) The categorical definitions of product and coproduct are dual to each other: each of them is obtained from the other by reversing all arrows and compositions, and each theorem about them has a dual version

This only makes sense within category theory, because the dual of a given category, formed by reversing its arrows and partial composition law, is a formal construction: the dual of a category of structured sets and mappings is *not* a category of the same kind – although, *in certain cases*, it may essentially be, as a result of some duality theorem. For instance, $\mathsf{Ab}^{\mathrm{op}}$ is equivalent to the category of compact Hausdorff abelian groups, by Pontryagin duality [Mo]; another example will be cited in 6.6.9(a)).

Similarly, the main topics of Homological Algebra that we shall build in categories of modules have 'dual versions', reversing arrows. This vague duality becomes a theoretical issue in the selfdual framework of abelian categories (see [M2], Chapter VIII, or [G5], Section 6.4), where proving a statement automatically proves its dual.

(b) We shall see that the singular homology functor $H_n\colon \mathsf{Top} \to \mathsf{Ab}$ *preserves sums*, in the sense that it transforms a sum $X = \sum X_i$ of topological spaces into the direct sum of their homology groups.

This means that the group $H_n(X)$ is canonically isomorphic to the direct sum $\bigoplus H_n(X_i)$: we should not expect these groups to coincide, in a literal sense.

The precise meaning of the property we are considering is that the family of structural injections $u_i\colon X_i \to X$ is transformed into a family

$$(H_n(X_i) \to H_n(X))_{i\in I}$$

*that makes* $H_n(X)$ into the sum of the groups $H_n(X_i)$. The concrete realisation of sums is of a marginal interest.

(c) It is easy to see that the contravariant functor $\mathcal{A}(X) = \mathsf{Top}(X, \mathbb{Z})$ introduced in (1.15) takes a sum of spaces (a product in $\mathsf{Top}^{\mathrm{op}}$) to a product of rings. This is a typical property of cohomology theories.

(d) A 'categorical definition' *is based on morphisms and their composition*, while the objects only step in as domains and codomains of arrows.

If we want to understand what unifies the product of a family $(X_i)$ of sets, or topological spaces, or groups, we should forget the nature of the objects, and think of the family of cartesian projections $p_i : X \to X_i$, together with the previous property. Then – in each category we are interested in – we come back to the objects in order to prove that a solution exists. Fixing a particular one (typically, with the help of set theory) can be useful; but it can also be a hindrance when functors are involved, as in point (b).

From a structural point of view, a category only 'knows' its objects by their morphisms and composition.

### *1.6.4 Finite direct sums*

The reader may have remarked that, in $R\,\mathsf{Mod}$, a *finite* product $\Pi A_i$ and the corresponding finite sum $\Sigma A_i$ are realised as the same object, namely $A = \bigoplus A_i$.

Essentially, we are saying two things. First, this category has a *zero object*, which is both terminal (the empty product) and initial (the empty sum): this is the trivial module 0.

Secondly, for a pair of modules $A, B$, the direct sum $C = A \oplus B$ is determined by a *butterfly diagram* $(C; u, v, p, q)$ satisfying the following equivalent properties – where we use the notation by components and co-components of 1.6.1–1.6.2, and the pointwise addition of parallel homomorphisms

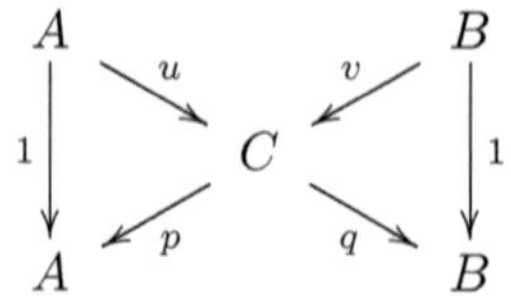


(1.101)

(i) $(C; p, q)$ is the product of $A, B$ and $u = (\mathrm{id}\,A, 0)$, $v = (0, \mathrm{id}\,B)$,

(i*) $(C; u, v)$ is the sum of $A, B$ and $p = [\mathrm{id}\,A, 0]$, $q = [0, \mathrm{id}\,B]$,

(ii) $pu = \mathrm{id}\,A, \quad qv = \mathrm{id}\,B, \qquad up + vq = \mathrm{id}\,C.$

The equivalence is (easily) proved in Exercise (a). The object $C$ is thus a product and a coproduct, by distinct structural arrows (of course): the projections $p, q$ and the injections $u, v$. It is also called the *biproduct* of $A$ and $B$.

A finite biproduct $A = \bigoplus A_i$ of modules is similarly analysed: it has projections $p_i \colon A \to A_i$ and injections $u_i \colon A_i \to A$ satisfying respectively the universal properties of product and sum, and the equations:

$$p_i u_i = \mathrm{id}\,A_i, \qquad \Sigma_i\, u_i p_i = \mathrm{id}\,A. \tag{1.102}$$

For the empty biproduct, condition (1.102) reduces to $\mathrm{id}\,A = 0$ (the zero endomorphism), which indeed characterises the trivial module.

*Exercises and complements.* (a) Prove the equivalence of properties (i), (i*) and (ii).

(b) The binary product and sum also coincide on morphisms.

We have thus the *biproduct functor*

$$\begin{aligned} &-\oplus-\colon R\,\mathsf{Mod}^2 \to R\,\mathsf{Mod}, \qquad (A,B) \mapsto A\oplus B, \\ &(f,g) \mapsto f\oplus g, \qquad (f\oplus g)(a,b) = (f(a),g(b)). \end{aligned} \tag{1.103}$$

### 1.6.5 Biproducts and additive categories

Biproducts and the addition $f+g$ of parallel morphisms are analysed in a general way, in category theory. In order to keep the present exposition elementary and concrete, we only hint here at this topic; an interested reader can see [M2], Section VIII.2, or [G5], Section 6.4.

First, a *zero object* in a category $\mathsf{C}$, often written as 0, is both initial and terminal. This exists in $\mathsf{Ab}$, $R\,\mathsf{Mod}$, $\mathsf{Gp}$, $\mathsf{Set}_\bullet$ and $\mathsf{Top}_\bullet$, where the zero object is the appropriate singleton; it does not exist in $\mathsf{Set}$, $\mathsf{Top}$ and $\mathsf{Cat}$ where the initial and terminal object are distinct.

A category with zero object is said to be *pointed*; then each pair of objects $A, B$ has a *zero-morphism* $0_{AB}\colon A \to B$ (also written as 0), namely the composite $A \to 0 \to B$.

If $\mathsf{C}$ is a pointed category, the *biproduct* of a family of objects $(A_i)$ is an object $A$ satisfying:

- the universal property of the product, by a family of projections $p_i\colon A \to A_i$,
- the universal property of the sum, by a family of injections $u_i\colon A_i \to A$,
- the equations $p_iu_i = \mathrm{id}\,A_i$ and $p_ju_i = 0\colon A_i \to A_j$ (for $i \neq j$).

The solution can be written as $\bigoplus A_i$. The empty biproduct is the zero object.

As a related notion, a *preadditive*, or *$\mathbb{Z}$-linear*, category is a category $\mathsf{C}$ where every hom-set $\mathsf{C}(A,B)$ is equipped with a structure of abelian group, generally written as $f+g$, so that composition is additive in each variable (or bilinear over $\mathbb{Z}$)

$$(f+g)h = fh+gh, \qquad k(f+g) = kf+kg \qquad (\textit{distributivity}), \tag{1.104}$$

where $h\colon A' \to A$ and $k\colon B \to B'$.

Finally, an *additive category* $\mathsf{C}$ is a preadditive category with finite products, or equivalently with finite sums, which are then biproducts. The equivalence is proved in the references cited above, where we also see that, *in this case*, the addition of parallel maps is determined by the categorical structure.

Typical examples are again $\mathsf{Ab}$ and $R\,\mathsf{Mod}$, where $f+g$ is the pointwise addition of homomorphisms (see 1.3.8(c)). The categories of chain complexes introduced in the following chapter are also additive categories.

*Exercises and complements.* (a) In a preadditive category $\mathsf{C}$, an object $Z$ is initial if and only if $\mathsf{C}(Z,Z)$ is the trivial group, if and only if $Z$ is terminal (and a zero object). If the zero object exists, the zero-morphism $0_{AB}\colon A \to B$ is the same as the additive identity of the abelian group $\mathsf{C}(A,B)$.

(b) A preadditive category need not have a zero object. *Hint*: use a subcategory of $\mathsf{Ab}$.

(c) The category $\mathsf{Rng}$ of (unital) rings does not have a zero object. The larger category $\mathsf{Rng}'$ *of possibly non-unital rings* has a zero object.

(d) Let $\mathsf{C}$ be pointed category with binary sums and products; these can be distinct, as in $\mathsf{Set}_\bullet$ and $\mathsf{Top}_\bullet$, but there always is a canonical morphism $f\colon X+Y \to X \times Y$, which is invertible when $X$ or $Y$ is 0.

(e) A preadditive category on a single object is the same as a (unital) ring.

*(f) Arbitrary biproducts exist in rather peculiar cases: for instance in the (selfdual) category of relations of sets, where they are realised as disjoint unions.

### *1.6.6 Universal arrows*

We have encountered diverse universal properties, and many more will be seen later. There is a general way of formalising them, based on a functor $U\colon \mathsf{A} \to \mathsf{X}$ and an object $X$ of the codomain category $\mathsf{X}$. It presents two forms, dual to each other.

(a) A *universal arrow from the object $X$ to the functor $U$* is a pair $(A, \eta\colon X \to UA)$ consisting of an object $A$ of $\mathsf{A}$ and an arrow $\eta$ of $\mathsf{X}$ which is universal, in the sense that every similar pair $(B, f\colon X \to UB)$ factorises uniquely through $(A,\eta)$. In other words, there exists a unique morphism $g\colon A \to B$ in $\mathsf{A}$ such that the following triangle commutes in $\mathsf{X}$

$$\begin{array}{ccc} X & \xrightarrow{\ \eta\ } & UA \\ & \searrow_{f} & \downarrow{\scriptstyle Ug} \\ & & UB \end{array} \qquad\qquad Ug.\eta = f. \qquad (1.105)$$

The pair $(A,\eta)$ may exist or not, but is determined up to a unique isomorphism: if $(A', \eta'\colon X \to UA')$ is also a solution, the unique map $g\colon A \to A'$ such that $Ug.\eta = \eta'$ is an isomorphism of $\mathsf{A}$.

In fact, by the usual argument, there is also a unique map $h\colon A' \to A$ in $\mathsf{A}$ such that $(Uh)\eta' = \eta$. The composite $hg\colon A \to A$ coincides with $\mathrm{id}\, A$, because $U(hg)\eta = \eta = U(\mathrm{id}\, A)\eta$. In the same way $gh = \mathrm{id}\, A'$, so that $g$ and $h$ are inverse to each other, in $\mathsf{A}$.

(b) Dually, a *universal arrow from the functor $U$ to the object $X$* is a pair $(A, \varepsilon\colon UA \to X)$ consisting of an object $A$ of $\mathsf{A}$ and an arrow $\varepsilon$ of $\mathsf{X}$ such that every similar pair $(B, f\colon UB \to X)$ factorises uniquely through $(A, \varepsilon)$: there exists a unique $g\colon B \to A$ in $\mathsf{A}$ such that the following triangle commutes in $\mathsf{X}$

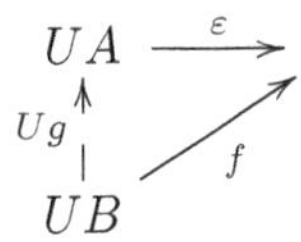


$$\varepsilon.Ug = f. \tag{1.106}$$

(c) Let us point out that, from a formal point of view, *existence* and *uniqueness* of solutions are independent facts: existence means that some solution exists, while uniqueness means that, *if* there are two solutions, *then* they coincide (maybe up to some equivalence relation, as here).

Nevertheless, we often express uniqueness in a 'psychologically prudent' way, saying that the solution is unique *provided it exists* – a formally useless condition.

*(d) Universal arrows compose: for a composed functor $UV\colon \mathsf{B} \to \mathsf{A} \to \mathsf{X}$, given

- a universal arrow $(A, \eta\colon X \to UA)$ from an object $X$ to $U$,
- a universal arrow $(B, \zeta\colon A \to VB)$ from the previous object $A$ to $V$,

it is easy to verify that we have a universal arrow from $X$ to $UV$, constructed as follows

$$(B, U\zeta.\eta\colon X \to UA \to UVB). \tag{1.107}$$

Dually, universal arrows from functors to objects can also be composed.

### *1.6.7 Exercises and complements*

(a) (*Free abelian groups*) The free abelian group $\mathbb{Z}X$ on the set $X$ comes equipped with a mapping $\eta\colon X \to |\mathbb{Z}X|$ (the insertion of the basis) satisfying the universal property stated in (1.22). This means that the pair $(\mathbb{Z}X, \eta)$ is a universal arrow from the object $X$ (a set) to the forgetful functor $U\colon \mathsf{Ab} \to \mathsf{Set}$.

(b) (*Free objects*) In general, for any category $\mathsf{A}$ equipped with a functor $U\colon \mathsf{A} \to \mathsf{Set}$, the *free* $\mathsf{A}$*-object over the set $X$* (with respect to the functor $U$) is defined as a universal arrow $(A, \eta\colon X \to UA)$, from the set $X$ to the forgetful functor $U$.

This terminology is typically used for a category $\mathsf{A}$ of algebraic structures, with respect to its forgetful functor $UA = |A|$ (see 1.3.7). We shall also see the free monoid $M(X)$, below, the free $R$-module $RX$, in (5.5), and the free group $G(X)$, in 7.4.1(c).

*The solution always exists if $\mathsf{A}$ is a category of 'equational algebras': see [G5], Chapter 4. Free commutative rings are constructed with polynomial rings; free rings with 'non-commutative polynomials'.*

(c) (*Free monoids*) Prove that the free monoid $M(X)$ on the set $X$ can be constructed as the disjoint union $\sum_{n\in\mathbb{N}} X^n$ of the cartesian powers of the set $X$. This will be useful further on.

(d) (*Products and sums*) Products and sums in a category $\mathsf{C}$ can be viewed as universal arrows. For a set $I$, a family $(X_i)_{i\in I}$ of objects of $\mathsf{C}$ is an object of the cartesian-power category $\mathsf{C}^I = \prod_{i\in I}\mathsf{C}$ (see 1.3.4). The diagonal functor (defined in 1.90)

$$D\colon \mathsf{C}\to\mathsf{C}^I, \qquad D(X) = (X)_{i\in I}, \quad D(f) = (f)_{i\in I}, \tag{1.108}$$

sends objects and arrows of $\mathsf{C}$ to constant families.

Now, the product of the family $(X_i)_{i\in I}$, defined by the universal property 1.6.1(i), is a universal arrow $(X, \varepsilon\colon DX \to (X_i)_{i\in I})$ from the functor $D$ to the object $(X_i)_{i\in I}$. The morphism $\varepsilon = (p_i\colon X\to X_i)_{i\in I}$ gives the family of projections.

Dually, the sum of the family $(X_i)_{i\in I}$, defined by the universal property 1.6.2(i*), is a universal arrow $(X, \eta\colon (X_i)_{i\in I} \to DX)$ from the object $(X_i)_{i\in I}$ to the functor $D$. The morphism $\eta = (u_i\colon X_i \to X)_{i\in I}$ gives the family of injections.

## 1.7 Pullbacks and pushouts, limits and colimits

Pullbacks and pushouts intervene in diverse fields of mathematics; they are closely related to binary products $X_1\times X_2$ and binary sums $X_1 + X_2$, respectively. In particular, pushouts will be frequently used here to 'paste' spaces, groups and groupoids.

After these issues, we introduce the general notion of limit and colimit in the sense of Category Theory (due to D.M. Kan [Ka]). Limits extend products, pullbacks, and the classical 'projective limits' of Algebra and Topology, casting all of them in the same mould. Dually, colimits extend and unify sums, pushouts, and the classical 'inductive limits'. The latter are dealt with in 1.7.6–1.7.8. *The reader can prefer to skip this part and come back when these notions are required.*

*Literature.* Categorical limits and colimits are dealt with in all textbooks on categories, often including their applications in Algebra, Topology and Algebraic Topology.

### 1.7.1 Pullbacks and pushouts

(a) (*Pullbacks*) In the category C, the *pullback* of a pair of morphisms $f_i\colon X_i \to X_0$ $(i = 1, 2)$, with the same codomain, is an object $X$ equipped with two morphisms $u_i\colon X \to X_i$ which form a commutative square $f_1u_1 = f_2u_2$, in a universal way:

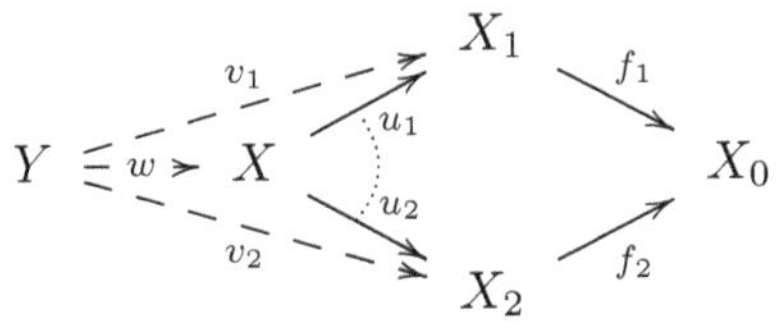

(1.109)

- for every triple $(Y, v_1, v_2)$ such that $f_1v_1 = f_2v_2$, there is a unique morphism $w\colon Y \to X$ such that $u_1w = v_1$ and $u_2w = v_2$.

As with products, the solution is determined up to a unique isomorphism coherent with the structural arrows $u_i$: if $(Y, v_1, v_2)$ is also a pullback of $(f_1, f_2)$, the previous morphism $w\colon Y \to X$ is an isomorphism.

The square diagram $f_1u_1 = f_2u_2$ is called a *pullback square*; it can be marked as above, in the corner of the pullback-object $X$; the latter is also called a *fibred product* over $X_0$, and written as $X_1 \times_{X_0} X_2$ (leaving the maps $f_i$ understood). If C has a terminal object $\top$, a fibred product $X_1 \times_\top X_2$ is the same as a cartesian product $X_1 \times X_2$.

In Set (resp. Top, Ab, Gp) the pullback-object can be realised as a subset (resp. subspace, subgroup, subgroup) of the product $X_1 \times X_2$:

$$X = \{(x_1, x_2) \in X_1 \times X_2 \mid f_1(x_1) = f_2(x_2)\}. \tag{1.110}$$

*A pullback is a particular categorical limit, as we shall see in 1.7.4(b): its structural arrows $(u_i)$ go *from* the solution *to* the given data, as in a product.*

(b) (*Pushouts*) The dual construction will be more frequently used here. In the category C, the *pushout* of a pair of morphisms $f_i\colon X_0 \to X_i$ $(i = 1, 2)$, with the same domain, is an object $X$ equipped with two morphisms $u_i\colon X_i \to X$ $(i = 1, 2)$ which form a commutative square $u_1f_1 = u_2f_2$, in a universal way:

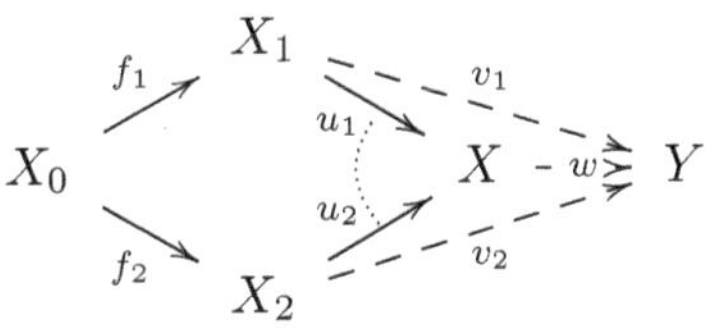

(1.111)

- for every triple $(Y, v_1, v_2)$ such that $v_1f_1 = v_2f_2$ there exists a unique morphism $w\colon X \to Y$ such that $wu_1 = v_1$ and $wu_2 = v_2$.

The pushout corner can be marked as above. The pushout-object $X$ is also called a *pasting* over $X_0$ and written as $X_1 +_{X_0} X_2$. A pushout over the initial object is the same as a categorical sum.

In Set, a pasting $X_1 +_{X_0} X_2$ is a quotient of the sum $X_1 + X_2$ modulo the equivalence relation generated by all the pairs $(f_1(x), f_2(x))$ (viewed in the sum), for $x \in X_0$.

In Top the solution is constructed as in Set and equipped with the quotient topology of the sum, that is, the final topology for the structural maps $u_i$. It is a useful way of constructing new spaces, that will be used in particular for CW-complexes, in Section 4.9, and mapping cylinders, in Chapter 9.

Pushouts of groups (or groupoids) will be constructed in Section 7.4, and used for the van Kampen Theorem on the fundamental group (or groupoid).

*A pushout is a particular colimit, as we shall see in 1.7.4(b): its structural arrows $(u_i)$ go *from* the given data *to* the solution, as in a sum.*

### 1.7.2 Exercises and complements

(a) Pullbacks and pushouts can be viewed as universal arrows.

A pair of morphisms $f_i \colon X_i \to X_0$ is a functor $(f_1, f_2) \colon \wedge \to \mathsf{C}$ defined on the category $1 \to 0 \leftarrow 2$, namely an object of the category $\mathsf{C}^{\wedge}$. Its pullback is a universal arrow from the diagonal functor $D \colon \mathsf{C} \to \mathsf{C}^{\wedge}$ (see (1.90)) to the object $(f_1, f_2)$.

Dually, a pair of morphisms $f_i \colon X_0 \to X_i$ is a functor $(f_1, f_2) \colon \vee \to \mathsf{C}$ defined on the opposite category $1 \leftarrow 0 \to 2$. Its pushout is a universal arrow from the object $(f_1, f_2)$ to the diagonal functor $D \colon \mathsf{C} \to \mathsf{C}^{\vee}$.

(b) (*Pasting diagrams*) Pullbacks and pushouts enjoy various diagrammatic properties: one can see [G5], Section 2.3. We shall use the following 'pasting property': if the two squares below are pullbacks (resp. pushouts), so is their 'pasting', i.e. the outer rectangle

$$\begin{array}{ccccc} X & \xrightarrow{\;g\;} & Y & \xrightarrow{\;h\;} & Z \\ {\scriptstyle f}\downarrow & & \downarrow{\scriptstyle u} & & \downarrow{\scriptstyle v} \\ X' & \xrightarrow[\;u'\;]{} & Y' & \xrightarrow[\;v'\;]{} & Z' \end{array} \qquad (1.112)$$

*(c) There is also a 'depasting property', equally easy to prove and left to an interested reader: if, in this diagram, the outer rectangle and the right (resp. left) square are pullbacks (resp. pushouts), so is the other square.

(d) (*Pullbacks and preimages*) In Set, the pullback of a mapping $f \colon X \to Y$

with the inclusion $B \to Y$ of a subset is the preimage $f^{-1}(B) \subset X$. There are similar results in Top, Ab and many concrete categories.

### 1.7.3 Limits and colimits

We now want to unify the universal constructions we have been considering in this section, in a more general issue with two dual forms.

We are working in a general category C, where a diagram 'of data' is given, as a functor $X\colon \mathsf{S} \to \mathsf{C}$ defined on an auxiliary small category S – the shape of the diagram. The functor $X$ is written in index notation (as in (1.88)), for $i \in I = \mathrm{Ob}\,\mathsf{S}$ and $a\colon i \to j$ in S

$$X\colon \mathsf{S} \to \mathsf{C}, \qquad i \mapsto X_i, \qquad a \mapsto (X_a\colon X_i \to X_j). \tag{1.113}$$

If the category S is discrete, this amounts to a family $(X_i)_{i\in I}$ of objects.

(a) *(Limits)* A *cone* of the diagram $X$ is an object $A$ of C (the *vertex* of the cone) equipped with a family of maps $(f_i\colon A \to X_i)_{i\in I}$ in C, such that the following triangles commute, for $a\colon i \to j$ in S

$$\begin{array}{ccc} A & \xrightarrow{f_i} & X_i \\ & \searrow^{f_j} & \downarrow X_a \\ & & X_j \end{array} \qquad X_a\, f_i = f_j. \tag{1.114}$$

The *limit* of $X\colon \mathsf{S} \to \mathsf{C}$ is a universal cone $(L, (u_i\colon L \to X_i)_{i\in I})$. This means a cone of $X$ such that every cone $(A, (f_i\colon A \to X_i)_{i\in I})$ factorises uniquely through the former; in other words, there is a unique arrow $f\colon A \to L$ such that, for all $i \in I$, $u_i f = f_i$.

The solution is determined up to isomorphism. If it exists, the limit object $L$ is denoted as $\mathrm{Lim}\, X$.

The uniqueness is proved in the usual way; it also follows from the presentation of limits as universal arrows, in (c).

(b) *(Colimits)* Dually, a *cocone* $(A, (f_i\colon X_i \to A)_{i\in I})$ of the diagram $X$ satisfies the following condition, for $a\colon i \to j$ in S

$$\begin{array}{ccc} X_i & \xrightarrow{f_i} & A \\ X_a \downarrow & \nearrow_{f_j} & \\ X_j & & \end{array} \qquad f_j\, X_a = f_i. \tag{1.115}$$

The *colimit* of the functor $X$ is a universal cocone $(L', (u_i\colon X_i \to L')_{i\in I})$: the universal property says now that for every cocone $(A, (f_i))$ of $X$ there is a unique map $f\colon L' \to A$ such that $fu_i = f_i$ (for all $i$). The colimit object, if it exists, is denoted as $\mathrm{Colim}\, X$.

In [M2] a cocone is called a 'cone'; the context makes the difference.

(c) Limits and colimits can be defined by universal arrows (see 1.6.6), with respect to the diagonal functor $D\colon \mathsf{C} \to \mathsf{C}^{\mathsf{S}}$, introduced in (1.90)

$$(DA)_i = A, \qquad (DA)_a = \mathrm{id}\, A \qquad (i \in \mathsf{S},\ a \in \mathrm{Mor}\, \mathsf{S}). \tag{1.116}$$

In fact, a cone of $X$ of vertex $A$ is the same as a natural transformation $DA \to X\colon \mathsf{S} \to \mathsf{C}$. Therefore, the limit of $X$ in $\mathsf{C}$ is the same as a universal arrow $(L, \varepsilon\colon DL \to X)$ from the diagonal functor $D$ to the object $X$ of $\mathsf{C}^{\mathsf{S}}$ (as defined in 1.6.6).

Dually, a cocone of $X$ of vertex $A$ is a natural transformation $X \to DA$, and the colimit of $X$ in $\mathsf{C}$ is the same as a universal arrow $(L', \eta\colon X \to DL')$ from the object $X$ of $\mathsf{C}^{\mathsf{S}}$ to the functor $D$.

*(d) A *weak limit* (resp. a *weak colimit*) of $X$ is a cone (resp. a cocone) that is only assumed to satisfy the existence part of the universal property.

These issues are important in Homotopy Theory. For instance, a homotopy pushout of topological spaces (in Section 8.2) gives a weak pushout in the homotopy category hoTop.

### *1.7.4 Examples and complements*

(a) The product $\prod X_i$ of a family $(X_i)_{i\in I}$ of objects of $\mathsf{C}$ is the limit of the corresponding functor $X\colon \mathsf{S} \to \mathsf{C}$, defined on the discrete category whose objects are the elements of $I$. The sum $\sum X_i$ is the colimit of the same functor.

(b) A pullback in $\mathsf{C}$ is the limit of a functor $\wedge \to \mathsf{C}$ defined on the category $1 \to 0 \leftarrow 2$, already used in 1.7.2(a). Dually, a pushout is the colimit of a functor $\vee \to \mathsf{C}$ defined on the opposite category $1 \leftarrow 0 \to 2$.

(c) The *equaliser* in $\mathsf{C}$ of a pair of parallel morphisms $f, g\colon X \to Y$ is the limit of the obvious functor defined on the category $0 \rightrightarrows 1$ by these morphisms.

In Set, the equaliser of a pair of mappings $f, g\colon X \to Y$ is the subset $E$ of $X$ on which they coincide

$$E = \{x \in X \mid f(x) = g(x)\}, \tag{1.117}$$

or, more precisely, the inclusion $E \to X$.

In Top the solution is the same, with the subspace topology on $E$. We have similar constructions in the usual categories of structured sets, putting on $E$ the structure induced by $X$.

(d) The *coequaliser* in $\mathsf{C}$ of a pair of parallel morphisms $f, g\colon X \to Y$ is the colimit of the same functor. (Note that the category $0 \rightrightarrows 1$ is selfdual.)

In Set, the coequaliser of $f, g$ is the quotient $Y/\mathcal{R}$ modulo the least equivalence relation containing all pairs $(f(x), g(x))$, for $x \in X$; more precisely, the colimit is the projection $Y \to Y/\mathcal{R}$.

In Top we have the same solution, with the quotient topology on $Y/\mathcal{R}$. There are similar constructions in Gp, Ab, $R$ Mod ..., where $\mathcal{R}$ is the congruence of the current algebraic structure, generated by all pairs $(f(x), g(x))$. For instance, in Gp, the coequaliser can be obtained as $Y/K$ where $K$ is the normal subgroup generated by the elements $f(x)g(x^{-1})$, for $x \in X$.

(e) Conversely, in all these categories, every quotient (in the usual sense) is a coequaliser.

In Set the projection $p\colon X \to X/\mathcal{R}$ modulo an equivalence relation $\mathcal{R} \subset X \times X$ is the coequaliser of the restrictions $f_1, f_2\colon \mathcal{R} \rightrightarrows X$ of the cartesian projections $p_i\colon X \times X \to X$.

This also works in Top, Gp, Ab and $R$ Mod, viewing $\mathcal{R}$ as a subspace, or a subgroup, or a submodule of the cartesian product.

### *1.7.5 Completeness*

A category C is said to be *complete* if it has all limits (based on small categories). Dually, C is said to be *cocomplete* if it has all colimits.

Many usual categories of structured sets are complete and cocomplete, as one can prove using point (c), below.

*Exercises and complements.* (a) A category that has binary products and equalisers has all pullbacks. Formula (1.110) can suggest the construction.

Dually, pushouts can be constructed with binary sums and coequalisers (when they exist), extending what we have seen in 1.7.1 for sets and topological spaces (or we shall see in 7.4.2 for groups).

(b) We have seen that binary products can be constructed with pullbacks and the terminal object. However, a category having all pullbacks (resp. pushouts) need not have binary products (resp. sums), nor the terminal (resp. initial) object. For instance, in a preordered set $X$, the existence of pullbacks only requires the existence of the product $x_1 \times x_2 = \inf\{x_1, x_2\}$ for each pair of elements *having an upper bound* $x_0$.

*(c) Point (a) has an important, easy extension that will not be used here, but can be found in any book on categories: for instance, [M2], Theorem V.2.1, or [G5], Theorem 2.2.4.

All limits in C can be constructed with arbitrary products and equalisers, provided all these exist. Moreover, if C is complete, a functor $F\colon \mathsf{C} \to \mathsf{D}$ with values in an arbitrary category preserves all limits (i.e. takes each universal cone of C to a universal cone of D) if and only if it preserves products and equalisers. Dually, cocompleteness is equivalent to the existence of arbitrary sums and coequalisers, again in a constructive way.

### 1.7.6 Inductive limits and filtered colimits

This is again a classical notion, used in many concrete categories of algebraic or topological kind, to construct new structures.

(a) A *filtered (preordered) set* $I$ is a preordered set where every finite subset has an upper bound. Or, equivalently, a non-empty preordered set such that each pair $i, j \in I$ has an upper bound $k \geqslant i, j$. It is also called a *directed* set.

Definitely the case $I = \emptyset$ should not be included here, as it will be evident in 1.7.7(c).

An *inductive system* $X$ indexed by $I$, in the category $\mathsf{C}$, is a functor $X\colon I \to \mathsf{C}$ defined on the filtered set $I$. It consists of a family of objects $(X_i)$ and a family of morphisms $(u_{ij})$ such that:

$$\begin{aligned} &X = ((X_i), (u_{ij})), \qquad u_{ij}\colon X_i \to X_j, \qquad \text{for } i \leqslant j \text{ in } I,\\ &u_{ii} = \mathrm{id}\, X_i, \qquad u_{jk}u_{ij} = u_{ik}, \qquad \text{for } i \leqslant j \leqslant k \text{ in } I. \end{aligned} \tag{1.118}$$

If $i \leqslant j \leqslant i$, we have thus a pair of isomorphisms $u_{ij}\colon X_i \rightleftarrows X_j : u_{ji}$. It is often sufficient to consider filtered *ordered* sets, but the preordered case will be needed for Čech cohomology, in Section 4.7.

(b) The *inductive limit* of the system, denoted as $\varinjlim_i X_i$, or $\varinjlim_I X_i$, is the categorical *colimit* in $\mathsf{C}$: an object $L$ equipped with a universal cocone $(u_i)$

$$u_i\colon X_i \to L, \qquad u_j u_{ij} = u_i, \qquad \text{for } i \leqslant j \text{ in } I, \tag{1.119}$$

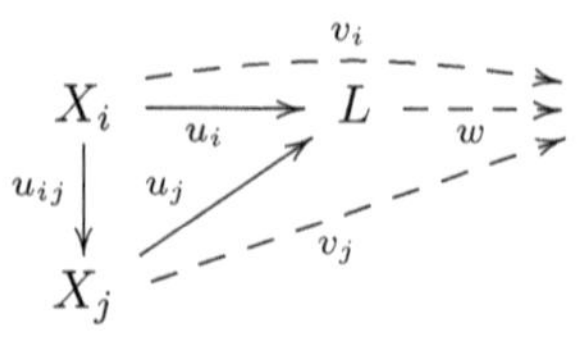


- for every system $(Y, (v_i))$ satisfying the same coherence condition (that is, $v_j u_{ij} = v_i$, for $i \leqslant j$) there is a unique arrow $w\colon L \to Y$ such that $wu_i = v_i$, for all indices $i$.

(c) *Functoriality.* There are given two inductive limits in the category $\mathsf{C}$

$$(L, (u_i)) = \varinjlim_I ((X_i), (u_{ii'})). \qquad (M, (v_j)) = \varinjlim_J ((Y_j), (v_{jj'})).$$

We also have an increasing function $j\colon I \to J$, and a coherent family of morphisms $f_i\colon X_i \to Y_{j(i)}$ $(i \in I)$, in the obvious sense: $f_{i'}u_{ii'} = v_{jj'}f_i$, for $i \leqslant i'$ in $I$ and $j = j(i)$, $j' = j(i')$ in $J$.

There is then a unique morphism $f\colon L \to M$ (called the *inductive limit* of the morphisms $f_i$) such that $fu_i = v_{j(i)}f_i\colon X_i \to M$, for all $i \in I$.

In fact, the family $v_{j(i)}f_i\colon X_i \to M$ $(i \in I)$ is a cocone of the system $X$, and factorises uniquely through $L$.

*Comments and literature.* The term 'inductive limit' is the classical one, predating categories. In category theory an inductive limit is a kind of colimit, written $\mathrm{Colim}\, X$, and has a natural generalisation to a *filtered colimit.* One can find an exposition in [M2], Section IX.1, or [G5], Section 2.6.

### 1.7.7 Exercises and complements

Generally speaking, inductive limits in a concrete category $\mathsf{C}$ are easy to construct, and an important tool well related to the forgetful functor $U\colon \mathsf{C} \to \mathsf{Set}$. The reader should find it interesting to complete the details of the following exercises, possibly without using the solutions given below.

$X = ((X_i), (u_{ij}))$ is always an inductive system in a category, indexed by a filtered set $I$.

(a) In $\mathsf{Set}$ the inductive limit $\varinjlim X$ always exists and has a simple construction: loosely speaking, we put together all the elements of the system $(X_i)$, identifying them when they eventually coincide along $I$. Therefore, if all the morphisms $u_{ij}$ of the system are injective, so are the morphisms $u_i\colon X_i \to \varinjlim X$.

(b) $\mathsf{Ab}$ has a similar construction, preserved by the forgetful functor $U\colon \mathsf{Ab} \to \mathsf{Set}$ – an important property of inductive limits. (We recall that $U$ does not preserve sums nor coequalisers, which are also categorical colimits.)

(c) The previous construction in $\mathsf{Ab}$ fails for $I = \emptyset$, which is not a filtered set.

(d) The construction in (b), based on sets, works quite similarly in many categories of 'algebras', like $\mathsf{Gp}$, $\mathsf{Rng}$, $R\,\mathsf{Mod}$, $\mathsf{Set}_\bullet$, etc.

*It works in every category of equational algebras: see [G5], Theorem 4.4.1.*

(e) In $\mathsf{Top}$ one takes again the inductive limit $L$ of the underlying system $UX\colon I \to \mathsf{Set}$, and endows $L$ with the finest topology making all mappings $u_i\colon X_i \to L$ continuous: a subset $A \subset L$ is open (resp. closed) if and only if every preimage $u_i^{-1}(A)$ is open (resp. closed) in $X_i$.

An inductive limit of discrete (resp. indiscrete) spaces is discrete (resp. indiscrete).

*(f) The dual notion, classically called a 'projective limit', is the categorical limit of a functor $X\colon I \to \mathsf{C}$ defined on a cofiltered set $I$ (which means that the preordered set $I^{\mathrm{op}}$ is filtered).

This is less important in the usual categories of structured sets, where all limits are already 'well behaved'; for instance they are preserved by the forgetful functor $\mathsf{C} \to \mathsf{Set}$.

*Solutions.* (a) We begin by forming the sum $S = \Sigma X_i$, whose elements are written as pairs $(x, i)$, with $i \in I$ and $x \in X_i$, as in 1.6.2(a). One can now (easily) prove that the relation $(x, i) \sim (y, j)$ in $S$ defined by

$$\text{there exists } k \geqslant i, j \text{ in } I \text{ such that } u_{ik}(x) = u_{jk}(y) \text{ in } X_k, \qquad (1.120)$$

is an equivalence relation; $\varinjlim X$ can be obtained as the quotient set:

$$L = S/\!\sim, \qquad u_i \colon X_i \to L, \quad u_i(x) = [x, i]. \qquad (1.121)$$

(b) For a functor $X \colon I \to \mathsf{Ab}$, one takes the colimit $L$ of the underlying system $UX \colon I \to \mathsf{Set}$. The reader will check that $L$ has a structure of abelian group, defined as follows

$$[x, i] + [y, j] = [u_{ik}(x) + u_{jk}(y), k], \qquad \text{for } i, j \leqslant k \text{ in } I. \qquad (1.122)$$

Let us note that this is the unique structure of abelian group of $L$ that makes all mappings $u_i \colon X_i \to L$ into homomorphisms. With this structure $(L, (u_i))$ is the colimit of $X$ in $\mathsf{Ab}$.

This construction can be easily adapted for exercises (d) and (e).

(c) For $I = \emptyset$ the universal property we are considering gives the initial object, that is, $\emptyset$ in $\mathsf{Set}$ and $\{0\}$ in $\mathsf{Ab}$: the solution in $\mathsf{Ab}$ is not constructed on the former.

(e) The first statement and the discrete case are obvious. If all $X_i$ are indiscrete, and $U \neq \emptyset$ is open in the inductive limit $L$, then $U$ has a non-empty open preimage in some $X_i$, and $u_i^{-1}(U) = X_i$. Then $u_j^{-1}(U) = X_j$, for all $j \geqslant i$; but these sets cover the space $L$, and $U = L$.

**1.7.8** ***Exercises and complements*** (Inductive limits of 'parts')

(a) In $\mathsf{Set}$ every object $X$ is the inductive limit of the filtered set $\mathcal{P}_{\mathrm{f}}(X)$ of its finite subsets, ordered by inclusion. The reader is invited to make precise and prove this statement; one can check the solution in Chapter 9.

(b) In $\mathsf{Ab}$ every object $A$ is the inductive limit of the filtered set $\mathrm{Sub}_{\mathrm{fg}}(A)$ of its *finitely generated* subgroups, ordered by inclusion.

(c) *Finite* subgroups would not give the same, generally. Characterise the inductive limit of the finite subgroups of an abelian group $A$.

(d) The reader can easily extend point (b) to various categories of 'algebras', like $\mathsf{Gp}$, $\mathsf{Rng}$, $R\,\mathsf{Mod}$, $\mathsf{Set}_\bullet$, etc.

(e) (*Infinite sums as inductive limits*) Let $\mathsf{C}$ be a category with finite sums and inductive limits. The sum of an arbitrary family $(X_i)_{i \in I}$ of objects exists, and can be constructed as the inductive limit of the finite sums $X_J = \Sigma_{i \in J} X_i$, over the filtered set $\mathcal{P}_\mathrm{f}(I)$ of the finite subsets of $I$.

*(f) The infinite dimensional spaces $\mathbb{R}^\infty$, $\mathbb{S}^\infty$ and $\mathbb{P}^\infty$ will be introduced in Section 3.4, as inductive limits of finite dimensional spaces.

# 2

# Absolute singular homology

Singular homology is the simplest homology theory of general topological spaces. In this theory, a space $X$ is explored by studying, for every $n \geqslant 0$, all the maps (or *probes*) $a\colon M^n \to X$ defined on a standard model of dimension $n$; these maps are not supposed to be topological embeddings (or nearly so), as in the initial approaches; allowing 'singularities' was the origin of the name.

We adopt the *cubical* construction (where $M^n$ is the standard euclidean cube $\mathbb{I}^n$), rather than the *simplicial* one (where $M^n$ is an $n$-dimensional tetrahedron): in various cases, this allows simpler geometrical constructions, essentially because cubes are closed under cartesian products, while the products of tetrahedra must be covered with tetrahedra. (This aspect will be evident in the proof of the homotopy invariance of singular homology, in Theorem 2.2.6.)

This subject is developed here and in the next chapter. Relative singular homology is deferred to Chapter 4, singular homology and cohomology with a coefficient group to Chapter 5.

The two-valued index $\alpha$ (or $\beta$) varies in the cardinal set $2 = \{0, 1\} = \{-, +\}$. The trivial group is always written 0.

*Literature.* Two elementary textbooks, by Vick [Vi] and Massey [Mas3], follow respectively the simplicial and the cubical construction. More advanced texts on Algebraic Topology, cited in the general Introduction, generally use the simplicial form. Hilton–Wylie [HiW] uses both, proving their equivalence.

## 2.1 Chain complexes and their homology

Before dealing with topological spaces, we have to fix some preliminary algebraic tools: exact sequences and chain complexes of abelian groups.

We are here in the domain of Homological Algebra, at its beginning. This subject arose in Algebraic Topology, when the importance of 'exact sequences' of abelian groups was remarked (by Hurewicz [Hur], in 1941) and soon acquired a central role, registered in Eilenberg–Steenrod's axioms of homology theory ([EiS], in 1952).

An exact sequence is a quite concrete notion, defined by a simple condition: $\mathrm{Im}\, f = \mathrm{Ker}\, g$, for a pair $(f, g)$ of consecutive homomorphisms of abelian groups $A \to B \to C$. But it will pay to examine this issue in a slightly more formal way, where subgroups are replaced by monomorphisms, quotient groups are replaced by epimorphisms, and the image $\mathrm{Im}\, f \subset B$ of $f \colon A \to B$ is often replaced by the 'cokernel' $\mathrm{Cok}\, f = B/f(A)$, a quotient of $B$ that plays a role symmetrical to $\mathrm{Ker}\, f$.

The reader is invited to follow this formalisation, without loosing contact with its concrete interpretation, although it may initially seem that we are expressing simple things in a complicated way.

Exact sequences and chain complexes of $R$-modules are dealt with in the same way, but will not be used before Section 4.6.

### *2.1.1 Kernels and cokernels*

Let $f \colon A \to B$ be a homomorphism of abelian groups, also called a morphism in $\mathsf{Ab}$. We recall that $f$ is said to be a *monomorphism* (or *mono*) if it is injective, and an *epimorphism* (or *epi*) if it is surjective.

An arrow $\rightarrowtail$ will always denote a monomorphism, while $\twoheadrightarrow$ stands for an epimorphism.

A symmetric role will be played, on the one hand by the kernel of $f$ (a subgroup of the domain $A$), and on the other hand by the cokernel of $f$ (a quotient of the codomain $B$, less used in Algebra)

$$\mathrm{Ker}\, f = \{x \in A \mid f(x) = 0\}, \qquad \mathrm{Cok}\, f = B/f(A). \tag{2.1}$$

More precisely, we shall use the embedding of $\mathrm{Ker}\, f$ in $A$, and the projection of $B$ onto $\mathrm{Cok}\, f$, written as

$$\ker f \colon \mathrm{Ker}\, f \rightarrowtail A, \qquad \mathrm{cok}\, f \colon B \twoheadrightarrow \mathrm{Cok}\, f, \tag{2.2}$$

and characterised in Exercises 2.1.2 by universal properties that will be frequently used.

Concretely, the image $\mathrm{Im}\, f = f(A) \subset B$ comes first, and the cokernel of $f$ is a derived notion. Yet, in Homological Algebra, the cokernel has a primitive role, symmetric to the role of the kernel, and $\mathrm{Im}\, f$ tends to come out as $\mathrm{Ker}\,(\mathrm{cok}\, f)$, the kernel object of the cokernel morphism.

This is already evident in the *canonical factorisation* of the homomorphism $f = mgp$ (or first Noether isomorphism theorem), analysed in the following commutative diagram, where:

$$\begin{array}{ccccccc} \mathrm{Ker}\, f & \xrightarrow{\ker f} & A & \xrightarrow{\quad f \quad} & B & \xrightarrow{\mathrm{cok}\, f} & \mathrm{Cok}\, f \\ & & {\scriptstyle p}\downarrow & & \uparrow{\scriptstyle m} & & \\ & & \mathrm{Coim}\, f & \xrightarrow[\cong]{g} & \mathrm{Im}\, f & & \end{array} \tag{2.3}$$

- $\mathrm{Coim}\, f = A/\mathrm{Ker}\, f$ is the *coimage* of $f$ and $p = \mathrm{cok}\,(\ker f)\colon A \twoheadrightarrow \mathrm{Coim}\, f$ is the projection,

- $\mathrm{Im}\, f = \mathrm{Ker}\,(\mathrm{cok}\, f)$ is the image of $f$ and $m = \ker\,(\mathrm{cok}\, f)\colon \mathrm{Im}\, f \rightarrowtail B$ is the inclusion,

- $g\colon \mathrm{Coim}\, f \to \mathrm{Im}\, f$ is the induced Noether isomorphism.

### *2.1.2 Exercises and complements*

The following elementary results on kernels and cokernels in Ab are important, and of frequent use.

(a) The kernel morphism $\ker f\colon \mathrm{Ker}\, f \rightarrowtail A$ of $f\colon A \to B$ is characterised up to a coherent isomorphism by the following *universal property*:

(i) $f(\ker f) = 0$, and every morphism $h$ such that $fh = 0$ factorises uniquely through $\ker f$

$$\begin{array}{ccccc} \mathrm{Ker}\, f & \xrightarrow{\ker f} & A & \xrightarrow{f} & B \\ & {\scriptstyle u}\nwarrow & \uparrow{\scriptstyle h} & & \\ & & \bullet & & \end{array} \tag{2.4}$$

which means that there is a unique morphism $u$ such that $h = (\ker f)u$.

Loosely speaking, the universal property says that $\ker f$ is the 'best morphism' which annihilates $f$ on its domain; the property does not assume that $\ker f$ be injective: this is a consequence.

A morphism $k\colon K \to A$ that satisfies this property will be said to be *a kernel* of $f$; then there is a unique morphism $u\colon K \to \mathrm{Ker}\, f$ such that $(\ker f)u = k$, and it is an isomorphism.

An 'exact functor' preserves kernels in this generalised sense, and need not preserve subobjects as subsets: see 5.2.2(a).

(b) The cokernel morphism $\mathrm{cok}\, f\colon B \twoheadrightarrow \mathrm{Cok}\, f$ is characterised up to a coherent isomorphism by the following universal property:

(i*) $(\operatorname{cok} f)f = 0$, and every morphism $h$ such that $hf = 0$ factorises uniquely through $\operatorname{cok} f$

$$\begin{array}{ccccc} A & \xrightarrow{f} & B & \xrightarrow{\operatorname{cok} f} & \operatorname{Cok} f \\ & & {\scriptstyle h}\downarrow & \swarrow {\scriptstyle u} & \\ & & \bullet & & \end{array} \tag{2.5}$$

i.e. there is a unique morphism $u$ such that $h = u(\operatorname{cok} f)$.

A morphism $c\colon B \to C$ that satisfies this property will be said to be *a cokernel* of $f$; then there is a unique morphism $u\colon \operatorname{Cok} f \to C$ such that $u(\operatorname{cok} f) = c$, and it is an isomorphism.

The universal property says that $\operatorname{cok} f$ is the 'best morphism' which annihilates $f$ on its codomain. Again, an 'exact functor' will preserve cokernels in this wider sense.

(c) A commutative square $vf = gu$ has a unique extension to the following commutative diagram

$$\begin{array}{ccccccc} \operatorname{Ker} f & \xrightarrow{\ker f} & A & \xrightarrow{f} & B & \xrightarrow{\operatorname{cok} f} & \operatorname{Cok} f \\ {\scriptstyle u'}\downarrow & & {\scriptstyle u}\downarrow & & \downarrow{\scriptstyle v} & & \downarrow{\scriptstyle v'} \\ \operatorname{Ker} g & \xrightarrow[\ker g]{} & C & \xrightarrow[g]{} & D & \xrightarrow[\operatorname{cok} g]{} & \operatorname{Cok} g \end{array} \tag{2.6}$$

We say that $u'$ is *induced by* $u$ (on the monomorphisms $\ker f$ and $\ker g$), and that $v'$ is *induced by* $v$ (on the epimorphisms $\operatorname{cok} f$ and $\operatorname{cok} g$). Both kinds of induction are consistent with composition. In the first case we also speak of a *restricted* morphism.

(d) Starting again from the commutative square $vf = gu$, the canonical factorisations of $f$ and $g$ gives a commutative diagram – the *canonical (horizontal) factorisation* of the given square

$$\begin{array}{ccccccccccc} \operatorname{Ker} f & \rightarrowtail & A & \twoheadrightarrow & \operatorname{Coim} f & \xrightarrow{f'} & \operatorname{Im} f & \rightarrowtail & B & \twoheadrightarrow & \operatorname{Cok} f \\ {\scriptstyle u'}\downarrow & & {\scriptstyle u}\downarrow & & {\scriptstyle u''}\downarrow & & \downarrow{\scriptstyle v''} & & \downarrow{\scriptstyle v} & & \downarrow{\scriptstyle v'} \\ \operatorname{Ker} g & \rightarrowtail & C & \twoheadrightarrow & \operatorname{Coim} g & \xrightarrow[g']{} & \operatorname{Im} g & \rightarrowtail & D & \twoheadrightarrow & \operatorname{Cok} g \end{array} \tag{2.7}$$

where $f'$ and $g'$ are the isomorphisms associated to $f$ and $g$, $u''$ is induced by $u$ and $v''$ by $v$.

(e) Every monomorphism $f\colon A' \to A$ is *normal* in $\mathsf{Ab}$, which means that it is a kernel of some morphism, and in particular of the projection $\operatorname{cok} f\colon A \twoheadrightarrow A/(\operatorname{Im} f)$.

This is also shown in the left diagram below, a particular case of the canonical factorisation of $f$ (since $\operatorname{Ker} f = 0$)

$$\begin{array}{ccccc} A' & \overset{f}{\rightarrowtail} & A & \twoheadrightarrow & \operatorname{Cok} f \\ & \underset{\cong}{\searrow} & \uparrow{\scriptstyle m} & & \\ & & \operatorname{Im} f & & \end{array} \qquad \begin{array}{ccccc} \operatorname{Ker} f & \rightarrowtail & A & \overset{f}{\twoheadrightarrow} & A'' \\ & & {\scriptstyle p}\downarrow & \underset{\cong}{\nearrow} & \\ & & \operatorname{Coim} f & & \end{array} \tag{2.8}$$

Symmetrically, every epimorphism $f\colon A \twoheadrightarrow A''$ is *normal* in Ab, which means that it is a cokernel of some morphism, and in particular of the embedding $\ker f\colon \operatorname{Ker} f \rightarrowtail A$, as shown in the right diagram above.

*There is a sharp distinction between the (abelian) category Ab and the (non-abelian) category Gp of all groups, where monomorphisms need not be normal (but all epimorphisms still are).*

(f) Kernels and cokernels can be defined in any category with zero object, by the universal properties (i) and (i*). $\mathsf{Set}_\bullet$ and $\mathsf{Top}_\bullet$ have kernels and cokernels. Exact sequences $(f, g)$ of pointed sets are defined by the usual condition $\operatorname{Im} f = \operatorname{Ker} g$, and are important in homotopy theory; we remark that the forgetful functor $\mathsf{Gp} \to \mathsf{Set}_\bullet$ preserves and reflects exact sequences.

### 2.1.3 Exact sequences of abelian groups

Exact sequences are a main algebraic tool for homology and homotopy theory. A pair $(f, g)$ of consecutive homomorphisms, in Ab

$$A \xrightarrow{f} B \xrightarrow{g} C \tag{2.9}$$

is said to be *exact* (at $B$) if $\operatorname{Im} f = \operatorname{Ker} g$ (or equivalently if $\operatorname{Cok} f = \operatorname{Coim} g$). A sequence of consecutive homomorphisms (bounded or not) is said to be *exact* if it is in all locations where the condition makes sense, i.e. excluding the first and the last position (if any).

For instance, the upper row of the canonical factorisation diagram (2.3) is exact, at $A$ and $B$.

*Exercises and complements.* (a) The sequence (2.9) is said to be *of order two* if $gf = 0$, or equivalently $\operatorname{Im} f \subset \operatorname{Ker} g$. The sequence is exact if and only if the other inclusion holds too.

(b) (*Detecting trivial objects*) The sequence $0 \to A \to 0$ is exact if and only if $A = 0$.

(c) (*Detecting monos, epis and isos*) The sequence $0 \to A \xrightarrow{f} B \to 0$

- is exact at $A$ if and only if $f$ is mono,
- is exact at $B$ if and only if $f$ is epi,
- is exact (at $A$ and $B$) if and only if $f$ is an isomorphism.

The reader will note that this terminology can be ambiguous: the objects $A, B$ could be the same! In such a case one refers to locations in the sequence.

(d) (*Detecting kernels and cokernels*) The sequence

$$0 \longrightarrow A \xrightarrow{f} B \xrightarrow{g} C \longrightarrow 0 \tag{2.10}$$

- is exact at $A, B$ if and only if $f$ is a kernel of $g$,
- is exact at $B, C$ if and only if $g$ is a cokernel of $f$.

(e) *Short exact sequences* are of particular interest. By definition, this is a sequence of form (2.10), that is exact (at $A, B, C$).

This means that the sequence satisfies the following equivalent conditions:

(i) $f$ is a monomorphism, $g$ is an epimorphism and $\operatorname{Im} f = \operatorname{Ker} g$,

(ii) $f$ is a kernel of $g$, and $g$ is a cokernel of $f$.

A short exact sequence will often be written in the form $A \rightarrowtail B \twoheadrightarrow C$.

(f) In the short exact sequence (2.10) we can replace $A$ with the subobject $B' = m(A)$ of $B$, and replace $C$ with the quotient $B'' = B/B'$. In this way any short exact sequence can be replaced, up to isomorphism, by a 'standard' one

$$B' \overset{m}{\rightarrowtail} B \overset{p}{\twoheadrightarrow} B'' \tag{2.11}$$

where $m$ is the inclusion of a subobject and $p$ is the projection on a quotient.

(g) (*Short exact sequences detect exactness*) A sequence $(f, g)$ of consecutive homomorphisms is exact if and only if it can be inserted in the commutative diagram below, with slanting short exact sequences

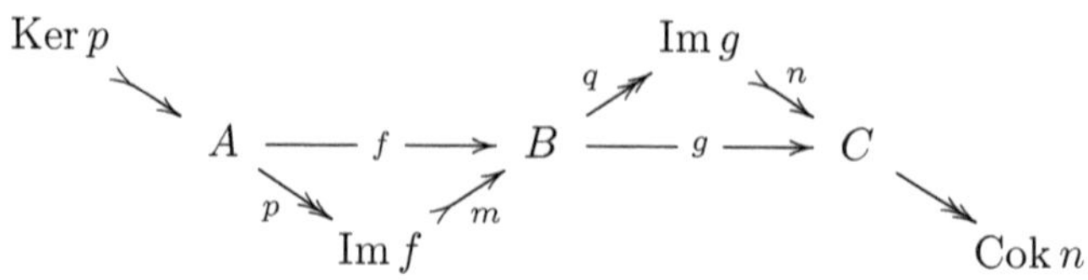


(2.12)

(h) In the following exact sequence of abelian groups:

$$A \xrightarrow{f} B \xrightarrow{g} C \xrightarrow{h} D \tag{2.13}$$

- $f$ is epi $\Leftrightarrow$ $g = 0$ $\Leftrightarrow$ $h$ is mono,
- $g$ is iso $\Leftrightarrow$ $f$ and $h$ are zero morphisms.

### 2.1.4 Induction Lemma

*We have a solid diagram in* Ab, *with short exact rows*

$$\begin{array}{ccccc} A' & \overset{m}{\rightarrowtail} & A & \overset{p}{\twoheadrightarrow} & A'' \\ {\scriptstyle f'}\downarrow & & \downarrow{\scriptstyle f} & & \downarrow{\scriptstyle f''} \\ B' & \underset{n}{\rightarrowtail} & B & \underset{q}{\twoheadrightarrow} & B'' \end{array} \qquad (2.14)$$

*The following conditions are equivalent:*

*(i) there is a (unique) morphism $f' \colon A' \to B'$ making the left square commutative,*

*(ii) there is a (unique) morphism $f'' \colon A'' \to B''$ making the right square commutative.*

*When this holds true, the morphisms $f'$ and $f''$ are induced by $f$ (on the monomorphisms $m, n$, or respectively on the epimorphisms $p, q$).*

*If the rows are short exact sequences in standard form, condition (i) can be rewritten as*

*(i′) $f(A') \subset B'$.*

*Moreover, $f'$ and $f''$ are computed as: $f'(x) = f(x)$, $f''[x] = [f(x)]$.*

*Proof* It is a straightforward application of Exercise 2.1.2(c), where the terminology of induced morphisms has been introduced. □

### 2.1.5 The category of chain complexes

A (positive) *chain complex* of abelian groups is a diagram in Ab

$$A = ((A_n)_{n \geqslant 0}, (\partial_n)_{n \geqslant 1})$$

consisting of a sequence of consecutive homomorphisms

$$\dots\ A_{n+1} \xrightarrow{\partial_{n+1}} A_n \xrightarrow{\partial_n} A_{n-1} \quad \dots \quad A_1 \xrightarrow{\partial_1} A_0 \dashrightarrow 0\ \dots \qquad (2.15)$$

with $\partial_n \partial_{n+1} = 0$ for $n \geqslant 1$.

The homomorphisms $\partial_n$ are called *boundaries* or *differentials*. When useful, all terms are extended at the right with trivial groups $A_n$ (for $n < 0$) and zero homomorphisms $\partial_n$ (for $n \leqslant 0$).

The name of 'boundary' is related to the singular chain complex of a space (see (2.49)); the name of 'differential' is related to the cochain complex of differential forms (see Section 4.8).

A *morphism of chain complexes* $f \colon A \to B$, or *chain morphism*, is a family $(f_n\colon A_n \to B_n)_{n\geqslant 0}$ of homomorphisms in Ab (its *components*), that commute with the boundaries:

$$\begin{array}{ccc} A_n & \xrightarrow{\partial_n} & A_{n-1} \\ {\scriptstyle f_n}\downarrow & & \downarrow{\scriptstyle f_{n-1}} \\ B_n & \xrightarrow[\partial_n]{} & B_{n-1} \end{array} \qquad \partial_n f_n = f_{n-1}\partial_n \qquad (n > 0). \tag{2.16}$$

Again, these components are extended with zero homomorphisms in negative degree, when useful; note also that we are using the same notation for the differentials of $A$ and $B$.

With the componentwise composition we get the category $\mathrm{Ch}_+\mathsf{Ab}$ of *chain complexes of abelian groups*. For every $n \geqslant 0$ we have a forgetful functor

$$U_n\colon \mathrm{Ch}_+\mathsf{Ab} \to \mathsf{Ab}, \qquad A \mapsto A_n, \quad f \mapsto f_n, \tag{2.17}$$

associating to any complex (or morphism) its *component* in degree $n$. On the other hand, there is a canonical embedding

$$J\colon \mathsf{Ab} \to \mathrm{Ch}_+\mathsf{Ab}, \qquad (JA)_0 = A, \qquad (JA)_n = 0, \text{ for } n > 0. \tag{2.18}$$

The category $\mathrm{Ch}_+\mathsf{Ab}$ has a zero object, the trivial chain complex 0 whose components are trivial groups. It also has a componentwise addition of parallel morphisms $f, g\colon A \to B$

$$f + g\colon A \to B, \qquad (f + g)_n = f_n + g_n, \tag{2.19}$$

making $\mathrm{Ch}_+\mathsf{Ab}$ a preadditive category (see 1.6.5). Actually, it is an additive category, as (nearly obvious and) written down in 2.1.8.

*Remarks.* (a) $\mathrm{Ch}_+\mathsf{Ab}$ is a full subcategory of the category of diagrams $\mathsf{Ab}^{\boldsymbol{\omega}^{\mathrm{op}}}$, where $\boldsymbol{\omega}$ is the first infinite ordinal (i.e. the ordered set of natural numbers), and an object $A = ((A_n)_{n\geqslant 0}, (u_n)_{n\geqslant 1})$ is a sequence of consecutive homomorphisms of Ab (under no condition). Cochain complexes, in 4.5.1, will form a full subcategory of $\mathsf{Ab}^{\boldsymbol{\omega}}$. For unbounded chain complexes, see 3.6.1.

(b) The family of functors $(U_n)_{n\geqslant 0}$ is *jointly faithful*: if $f, g$ are parallel chain morphism and $U_n(f) = U_n(g)$, for every $n \geqslant 0$, then $f = g$. This family can also be viewed as a faithful functor $U\colon \mathrm{Ch}_+\mathsf{Ab} \to \mathsf{Ab}^{\mathbb{N}}$ with values in a cartesian power of Ab, or a full embedding $\mathrm{Ch}_+\mathsf{Ab} \to \mathsf{Ab}^{\boldsymbol{\omega}^{\mathrm{op}}}$. (Here we are considering $\mathbb{N}$ as a mere set, and a discrete category, while $\boldsymbol{\omega}$ and $\boldsymbol{\omega}^{\mathrm{op}}$ are ordered sets.)

(c) The differential of chain complexes gives a natural transformation $\partial_n\colon U_n \to U_{n-1}$.

### 2.1.6 Chain homology

For a chain complex $A$, an element $a \in A_n$ is called a *chain of degree* $n$ (or dimension $n$, when appropriate). We are interested in two subgroups of $A_n$

$$\begin{aligned} Z_n(A) &= \mathrm{Ker}\,(\partial_n \colon A_n \to A_{n-1}) \;\subset\; A_n, \\ B_n(A) &= \mathrm{Im}\,(\partial_{n+1} \colon A_{n+1} \to A_n) \;\subset\; A_n, \end{aligned} \tag{2.20}$$

whose elements are called *n-cycles* and *n-boundaries*, respectively. Again, these names will acquire a concrete sense in Section 2.2.

This includes $Z_0(A) = \mathrm{Ker}\,(\partial_0 \colon A_0 \to 0) = A_0$. The condition $\partial_n\partial_{n+1} = 0$ is equivalent to saying that $B_n(A) \subset Z_n(A)$ (for $n \geqslant 0$): *every boundary is a cycle.*

We have thus a quotient, the *n-th homology group* of the complex $A$

$$H_n(A) = Z_n(A)/B_n(A) = \mathrm{Ker}\,(\partial_n)/\mathrm{Im}\,(\partial_{n+1}) \quad (n \geqslant 0). \tag{2.21}$$

Every cycle $z \in Z_n(A)$ has a *homology class* $[z] \in H_n(A)$, written as $[z]_H$ when useful. Two cycles $z, z' \in A_n$ are *homologous* if $z - z'$ is a boundary, i.e. $[z] = [z']$.

Saying that $H_n(A) = 0$ amounts to the condition $\mathrm{Im}\,\partial_{n+1} = \mathrm{Ker}\,\partial_n$, also expressed saying that the complex $A$, as a sequence in $\mathsf{Ab}$ (see (2.15)), is exact at $A_n$, or – more precisely – in degree $n$. Chain homology measures *the failure of exactness* of a chain complex.

It is easy to see that we have functors

$$Z_n \colon \mathrm{Ch}_+\mathsf{Ab} \to \mathsf{Ab}, \qquad B_n \colon \mathrm{Ch}_+\mathsf{Ab} \to \mathsf{Ab} \qquad (n \geqslant 0), \tag{2.22}$$

where $Z_n(f)$ and $B_n(f)$ are restrictions of the component $f_n \colon A_n \to B_n$ (see Exercise (a)).

The morphism $f = (f_n) \colon A \to B$ gives thus a commutative diagram (for each $n \geqslant 0$)

$$\begin{array}{ccccc} B_n(A) & \rightarrowtail & Z_n(A) & \twoheadrightarrow & H_n(A) \\ {\scriptstyle B_n(f)}\downarrow & & {\scriptstyle Z_n(f)}\downarrow & & \downarrow{\scriptstyle H_n(f)} \\ B_n(B) & \rightarrowtail & Z_n(B) & \twoheadrightarrow & H_n(B) \end{array} \tag{2.23}$$

where the (short exact) rows are formed of inclusions and projections, while the left and right columns are induced homomorphisms (Exercise (b)).

Finally we have defined the *chain homology functor* of degree $n$

$$H_n \colon \mathrm{Ch}_+\mathsf{Ab} \to \mathsf{Ab}, \qquad H_n(A) = (\mathrm{Ker}\,\partial_n)/(\mathrm{Im}\,\partial_{n+1}). \tag{2.24}$$

The induced homomorphism $H_n(f) \colon H_n(A) \to H_n(B)$ is traditionally written $f_{*n}$.

*The whole construction can be extended from Ab to every abelian category A, giving an abelian category $\mathrm{Ch}_+(\mathsf{A})$, with homology functors $H_n\colon \mathrm{Ch}_+(\mathsf{A}) \to \mathsf{A}$.*

*Exercises and complements.* (a) The homomorphism $f_n\colon A_n \to B_n$ does have restrictions $Z_n(f)$ and $B_n(f)$, as defined in 2.1.2.

(b) The homomorphism $H_n(f)$ is well defined, in diagram (2.23).

(c) A (positive) *graded abelian group* $A = (A_n)_{n\geqslant 0}$ is a sequence of abelian groups, and can be identified with a chain complex with trivial differentials. These objects form a full subcategory $\mathrm{Gr}_+\mathsf{Ab} \subset \mathrm{Ch}_+\mathsf{Ab}$.

Chain homology can be viewed as a single functor $H_+\colon \mathrm{Ch}_+\mathsf{Ab} \to \mathrm{Gr}_+\mathsf{Ab}$.

### *2.1.7 Chain homotopy*

The category $\mathrm{Ch}_+\mathsf{Ab}$ has a notion of homotopy, well related to topological homotopy (as we shall see). For two morphisms $f, g\colon A \to B$ in $\mathrm{Ch}_+\mathsf{Ab}$, a *homotopy* $\varphi\colon f \simeq g$ is a sequence of homomorphisms $\varphi_n\colon A_n \to B_{n+1}$ $(n \geqslant 0)$ such that

$$\partial_{n+1}\varphi_n + \varphi_{n-1}\partial_n = g_n - f_n, \tag{2.25}$$

where, as usual, we are letting $\varphi_{-1} = 0$.

Then the morphisms $f, g\colon A \to B$ are said to be *homotopic* in $\mathrm{Ch}_+\mathsf{Ab}$, written as $f \simeq g$.

More precisely, we should define the homotopy as the triple $(f, \varphi, g)$, because the family $(\varphi_n)$ does not determine its faces $f, g$; this will become clearer in 2.1.9.

*Exercises and complements.* (a) The homotopy relation $f \simeq g$ is a congruence of categories, in $\mathrm{Ch}_+\mathsf{Ab}$.

(b) (*Homotopy invariance of chain homology*) If two morphisms $f, g\colon A \to B$ in $\mathrm{Ch}_+\mathsf{Ab}$ are homotopic, then $H_n(f) = H_n(g)$, for all $n \geqslant 0$.

### *2.1.8 Direct sum of chain complexes*

A family of chain complexes

$$A^i = ((A^i_n)_{n\geqslant 0}, (\partial^i_n)_{n\geqslant 1}) \qquad (i \in I), \tag{2.26}$$

has an obvious cartesian product and direct sum, defined componentwise in Ab

$$B = \textstyle\prod_{i\in I} A^i, \qquad B_n = \prod_i A^i_n, \qquad \partial_n = \prod_i \partial^i_n, \tag{2.27}$$

$$A = \textstyle\bigoplus_{i\in I} A^i, \qquad A_n = \bigoplus_i A^i_n, \qquad \partial_n = \bigoplus_i \partial^i_n. \tag{2.28}$$

These are obviously the categorical product and sum in $\mathrm{Ch}_+\mathsf{Ab}$; they coincide on finite families. Taking into account the addition $f+g$ of parallel morphisms (in (2.19)), $\mathrm{Ch}_+\mathsf{Ab}$ is an additive category (as defined in 1.6.5).

Here we are mainly interested in arbitrary direct sums.

*An exercise.* (a) Chain homology preserves direct sums: in the previous situation, there is a canonical isomorphism, for each $n \geqslant 0$

$$H_n(A) \;\cong\; \bigoplus\nolimits_{i\in I} H_n(A^i), \qquad [(z^i)_{i\in I}] \mapsto ([z^i])_{i\in I}. \tag{2.29}$$

The reader is invited to carefully write down the proof, which is perhaps too easy, and may lead to overlooking a point. This is why we write a solution here.

*Solution.* Let $n \geqslant 0$. Plainly the functors $B_n$ and $Z_n$ preserve direct sums

$$\begin{aligned} Z_n(A) &= \{(z^i) \in A \mid \partial^i_n(z^i) = 0,\ \forall\, i \in I\} = \bigoplus\nolimits_{i\in I} Z_n(A^i), \\ B_n(A) &= \{(b^i) \in A \mid \exists\, (c^i) \in A\colon \partial^i_n(c^i) = b^i,\ \forall\, i \in I\} \\ &= \bigoplus\nolimits_{i\in I} B_n(A^i). \end{aligned} \tag{2.30}$$

The obvious epimorphism

$$p\colon Z_n(A) \to \bigoplus\nolimits_{i\in I} H_n(A^i), \qquad p((z^i)_{i\in I}) = ([z^i])_{i\in I}, \tag{2.31}$$

gives a sequence

$$0 \longrightarrow B_n(A) \xrightarrow{\;u\;} Z_n(A) \xrightarrow{\;p\;} \bigoplus\nolimits_{i\in I} H_n(A^i) \longrightarrow 0 \tag{2.32}$$

where $u$ is the inclusion. The sequence is short exact in each component of index $i$, and short exact as a whole. Therefore $u$ is the kernel of $p$, which induces the isomorphism (2.29).

Reconsidering the proof, after (2.30) one is tempted to jump to the conclusion, writing

$$Z_n(A)/B_n(A) \;\cong\; \bigoplus\nolimits_{i\in I} (Z_n(A^i)/B_n(A^i)),$$

which is correct *if* we note that the formulas (2.30) describe $Z_n(A)$ and $B_n(A)$ as *subgroups* of $A_n$, not as 'independent' groups.

Indeed, $2\mathbb{Z}$ and $3\mathbb{Z}$ are isomorphic groups, but $\mathbb{Z}/2\mathbb{Z}$ and $\mathbb{Z}/3\mathbb{Z}$ are not!

Alternatively, and more formally, one can prove that the canonical injections $A^i \to A$ induce a family of homomorphisms $(H_n(A^i) \to H_n(A))_{i\in I}$ that satisfies the universal property of the categorical sum, in $\mathsf{Ab}$.

### 2.1.9 *The cylinder functor

Homotopies of chain complexes can be defined by a cylinder endofunctor

$$
\begin{aligned}
&I\colon \mathrm{Ch}_+\mathsf{Ab} \to \mathrm{Ch}_+\mathsf{Ab}, \qquad (IA)_n = A_n \oplus A_{n-1} \oplus A_n,\\
&\partial_n(a,x,b) = (\partial_n a - x,\ -\partial_{n-1}x,\ \partial_n b + x), \qquad\qquad (2.33)\\
&(If)_n = f_n \oplus f_{n-1} \oplus f_n \qquad (\text{for } f\colon A \to B \text{ in } \mathrm{Ch}_+\mathsf{Ab}),
\end{aligned}
$$

as verified in the exercises below.

This functor has a structure analogous to that of the cylinder functor of $\mathsf{Top}$, in 1.4.3, with the following *faces*, *degeneracy* and *reversion*:

$$
\begin{aligned}
&\partial^\alpha\colon A \to IA, \qquad \partial^-(a) = (a,0,0), \quad \partial^+(a) = (0,0,a),\\
&e\colon IA \to A, \qquad e(a,x,b) = a+b, \qquad\qquad (2.34)\\
&r\colon IA \to IA, \qquad r(a,x,b) = (b,-x,a).
\end{aligned}
$$

*This structure is studied in [G1], Section 4.4, in its natural setting: chain complexes over an additive category.*

*Exercises and complements.* (a) The differential defined in (2.33) satisfies the condition $\partial_n\partial_{n+1} = 0$.

(b) For a chain morphism $f\colon A \to B$, the components $(If)_n$ commute with the differential.

(c) The definitions (2.34) give indeed chain morphisms.

(d) A chain morphism $\Phi\colon IA \to B$ amounts to a chain homotopy from $f = \Phi\partial^-$ to $g = \Phi\partial^+$.

## 2.2 The singular homology groups

We now define the singular homology groups of a space $X$, exploring it by *cubical probes* $\mathbb{I}^n \to X$. This requires three steps:

(i) a functor Cub: $\mathsf{Top} \to$ CubSet, which transforms spaces into cubical sets,

(ii) a functor $\mathrm{Ch}_+$: CubSet $\to \mathrm{Ch}_+\mathsf{Ab}$, which transforms the latter into chain complexes,

(iii) the chain homology functors $H_n\colon \mathrm{Ch}_+\mathsf{Ab} \to \mathsf{Ab}$, already defined in 2.1.6.

After introducing the first two steps, we define singular homology (in 2.2.4) and begin studying its basic properties.

The Betti numbers and torsion coefficients of a space (in 2.2.7) are numerical invariants, derived from the homology groups (under convenient hypotheses). We already remarked that the historical development followed the reversed path: these invariants came before, and were later reorganised in algebraic structures. Emmy Noether played an important role, in the first half of the 20th century, promoting the importance of these structures.

In Section 3.6 we shall see that singular homology can also be constructed using the tetrahedral models $\Delta^n$, instead of the cubical ones.

### 2.2.1 The geometry of the cubes

We come back to the basic structure of the standard interval $\mathbb{I} = [0,1]$, formed of two faces and a degeneracy, and already considered in 1.4.2 (with a different notation):

$$\delta^\alpha \colon \{*\} \rightleftarrows \mathbb{I} : \varepsilon, \qquad \delta^\alpha(*) = \alpha, \qquad \varepsilon\delta^\alpha = \mathrm{id} \qquad (\alpha = 0,1). \tag{2.35}$$

Higher faces and degeneracies of the standard cubes $\mathbb{I}^n$ are built with the previous ones, by means of the cartesian product, for $1 \leqslant i \leqslant n$ and $\alpha = 0,1$

$$\begin{aligned} \delta^\alpha_{ni} &= \mathbb{I}^{i-1} \times \delta^\alpha \times \mathbb{I}^{n-i} \colon \mathbb{I}^{n-1} \longrightarrow \mathbb{I}^n, \\ \delta^\alpha_{ni}(t_1, ..., t_{n-1}) &= (t_1, ..., t_{i-1}, \alpha, t_i, ..., t_{n-1}), \\ \varepsilon_{ni} &= \mathbb{I}^{i-1} \times \varepsilon \times \mathbb{I}^{n-i} \colon \mathbb{I}^n \longrightarrow \mathbb{I}^{n-1}, \\ \varepsilon_{ni}(t_1, ..., t_n) &= (t_1, ..., t_{i-1}, t_{i+1}, ..., t_n). \end{aligned} \tag{2.36}$$

We shall generally omit the degree $n$ in these indices. The standard square $\mathbb{I}^2$ has thus four faces $\delta^\alpha_i \colon \mathbb{I} \to \mathbb{I}^2$, which are parametrisations of its edges on the standard interval

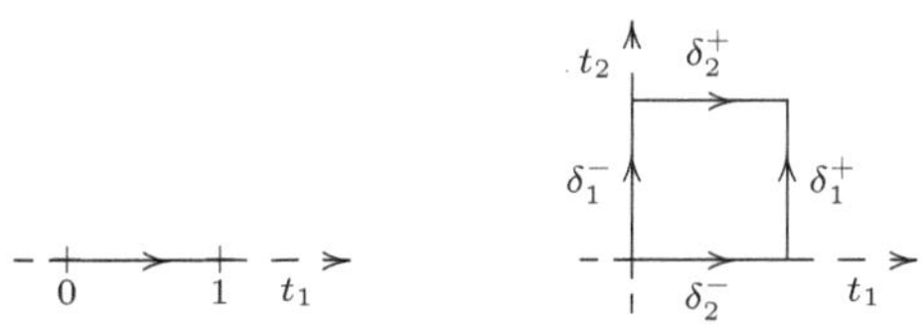


(2.37)

Faces and degeneracies satisfy the *cocubical relations* (proved in an exercise below):

$$\begin{aligned} \delta^\beta_j \delta^\alpha_i &= \delta^\alpha_{i+1}\delta^\beta_j, && \text{for } j \leqslant i, \\ \varepsilon_i\varepsilon_j &= \varepsilon_j\varepsilon_{i+1}, && \text{for } j \leqslant i, \\ \varepsilon_j\delta^\alpha_i &= \begin{cases} \delta^\alpha_{i-1}\varepsilon_j, & \text{for } j < i, \\ \mathrm{id}, & \text{for } j = i, \\ \delta^\alpha_i\varepsilon_{j-1}, & \text{for } j > i. \end{cases} \end{aligned} \tag{2.38}$$

The system $((\mathbb{I}^n), (\delta_i^\alpha), (\varepsilon_i))$ is called a *cocubical object* in $\mathsf{Top}$.

*Exercises and comments.* (a) Prove the cocubical relations.

(b) Geometrically, the cocubical relation of the faces express the fact that each vertex of a square belongs to two edges, each edge of a 3-cube belongs to two faces, and so on. The other relations have also a clear geometrical interpretation.

### 2.2.2 The singular cubical set of a space

For a space $X$, we apply the contravariant functor

$$\mathsf{Top}(-, X)\colon \mathsf{Top}^{\mathrm{op}} \to \mathsf{Set}$$

to the system $((\mathbb{I}^n), (\delta_i^\alpha), (\varepsilon_i))$, obtaining the *singular cubical set* $\mathrm{Cub}(X)$ of the space $X$.

This is a system of sets and mappings, still called *faces* and *degeneracies*

$$\begin{aligned} &\mathrm{Cub}(X) = ((\mathrm{Cub}_n X), (\partial_i^\alpha), (e_i)), && \mathrm{Cub}_n X = \mathsf{Top}(\mathbb{I}^n, X), \\ &\partial_i^\alpha\colon \mathrm{Cub}_n X \to \mathrm{Cub}_{n-1} X, && \partial_i^\alpha(a\colon \mathbb{I}^n \to X) = a\delta_i^\alpha, \\ &e_i\colon \mathrm{Cub}_{n-1} X \to \mathrm{Cub}_n X, && e_i(a\colon \mathbb{I}^{n-1} \to X) = a\varepsilon_i. \end{aligned} \tag{2.39}$$

The arrows satisfy now the *cubical relations*, dual to the cocubical ones (as we are applying a contravariant functor)

$$\begin{aligned} &\partial_i^\alpha \partial_j^\beta = \partial_j^\beta \partial_{i+1}^\alpha, && \text{for } j \leqslant i, \\ &e_j e_i = e_{i+1} e_j, && \text{for } j \leqslant i, \\ &\partial_i^\alpha e_j = \begin{cases} e_j \partial_{i-1}^\alpha, & \text{for } j < i, \\ \mathrm{id}, & \text{for } j = i, \\ e_{j-1} \partial_i^\alpha, & \text{for } j > i. \end{cases} \end{aligned} \tag{2.40}$$

For instance: $\partial_i^\alpha \partial_j^\beta(a) = a\delta_j^\beta \delta_i^\alpha = a\delta_{i+1}^\alpha \delta_j^\beta = \partial_j^\beta \partial_{i+1}^\alpha(a)$, for $j \leqslant i$.

In general a *cubical set* $K = ((K_n), (\partial_i^\alpha), (e_i))$ is a sequence of sets $(K_n)_{n \geqslant 0}$ equipped with faces and degeneracies

$$\partial_i^\alpha\colon K_n \to K_{n-1}, \quad e_i\colon K_{n-1} \to K_n \qquad (i = 1, ..., n,\ \alpha = 0, 1), \tag{2.41}$$

$$K_0 \;\substack{\longleftarrow \\ \longleftarrow \\ \longrightarrow}\; K_1 \;\substack{\longleftarrow \\ \longleftarrow \\ \longleftarrow \\ \longleftarrow \\ \longrightarrow \\ \longrightarrow}\; K_2 \quad \ldots$$

that satisfy the cubical relations (2.40).

When useful, we use a more explicit notation: $\partial_{ni}^\alpha$ and $e_{ni}$, without omitting the degree $n$.

The elements of $K_n$ are called *n-cubes*, and *vertices* or *edges* when $n$

is 0 or 1, respectively. Every $n$-cube $a$ has $2^n$ vertices: for instance a 3-cube $a\colon \mathbb{I}^3 \to X$ has eight of them, namely $\partial_1^\alpha\partial_2^\beta\partial_3^\gamma(a) = a(\alpha,\beta,\gamma)$, for $\alpha,\beta,\gamma = 0,1$.

A *morphism* of cubical sets $f = (f_n)\colon K \to L$ is a sequence of mappings $f_n\colon K_n \to L_n$ that commute with faces and degeneracies, giving commutative squares (for $n \geqslant 1$)

$$\begin{array}{ccc} K_n & \xrightarrow{f_n} & L_n \\ {\scriptstyle\partial_i^\alpha}\downarrow & & \downarrow{\scriptstyle\partial_i^\alpha} \\ K_{n-1} & \xrightarrow[f_{n-1}]{} & L_{n-1} \end{array} \qquad \begin{array}{ccc} K_{n-1} & \xrightarrow{f_{n-1}} & L_{n-1} \\ {\scriptstyle e_i}\downarrow & & \downarrow{\scriptstyle e_i} \\ K_n & \xrightarrow[f_n]{} & L_n \end{array} \tag{2.42}$$

Cubical sets and their morphisms, with the obvious componentwise composition, form the category $\mathsf{CubSet}$.

Finally, we have the *singular cubical functor*, which acts on a space $X$ as we have seen, and on a map $f\colon X \to Y$ by covariant composition

$$\begin{gathered} \mathrm{Cub}\colon \mathsf{Top} \to \mathsf{CubSet}, \qquad X \mapsto \mathrm{Cub}(X), \\ \mathrm{Cub}(f) = f_\sharp\colon \mathrm{Cub}(X) \to \mathrm{Cub}(Y), \\ f_{\sharp n}(a\colon \mathbb{I}^n \to X) = fa\colon \mathbb{I}^n \to Y. \end{gathered} \tag{2.43}$$

This achieves the first step to define singular homology.

**Complements.* Cubical sets can be defined as presheaves $K\colon \underline{\mathbb{I}}^{\mathrm{op}} \to \mathsf{Set}$ on the *basic cubical site* $\underline{\mathbb{I}}$. The latter is the small subcategory of $\mathsf{Set}$ consisting of the *elementary cubes* $2^n = \{0,1\}^n$ (for $n \geqslant 0$), together with the mappings $2^m \to 2^n$ which delete some coordinates and insert some 0's and 1's (without modifying the order of the remaining coordinates). This can be seen in [G5], Section 5.4, or in the original paper [GM], in a more extended form.

### 2.2.3 The chain complex of a cubical set

For the second step, we construct the chain complex $\mathrm{Ch}_+(K)$ associated to a cubical set $K$.

First, an $n$-cube $a \in K_n$ is said to be *degenerate* if $a = e_i(b)$, for some cube $b \in K_{n-1}$ and some index $i$

$$\mathrm{Deg}_n K = \cup_i \,\mathrm{Im}\,(e_i\colon K_{n-1} \to K_n), \qquad \mathrm{Deg}_0 K = \emptyset \qquad (n > 0), \tag{2.44}$$

which means that $a$ does not depend on the $i$-th coordinate. Then, because of the cubical relations, $\partial_i^\alpha e_j(b)$ is degenerate for $i \neq j$, while $\partial_i^\alpha e_i(b) = b$. We only register that:

$$a \in \mathrm{Deg}_n K \;\Rightarrow\; (\partial_i^\alpha a \in \mathrm{Deg}_{n-1} K \;\text{ or }\; \partial_i^- a = \partial_i^+ a). \tag{2.45}$$

Now the *(normalised) chain complex* $\mathrm{Ch}_+(K) = ((C_n(K)), (\partial_n))$ is defined as follows

$$
\begin{gathered}
C_n(K) \xrightarrow{\partial_n} C_{n-1}(K) \ \ldots\ C_1(K) \xrightarrow{\partial_1} C_0(K) \\
C_n(K) = (\mathbb{Z}K_n)/(\mathbb{Z}\mathrm{Deg}_n K), \qquad \partial_n \colon C_n(K) \to C_{n-1}(K), \\
\partial_n(\hat{a}) \ = \ \Sigma_{i,\alpha}\, (-1)^{i+\alpha}\, (\partial_i^\alpha a)\hat{} \qquad (a \in K_n).
\end{gathered} \tag{2.46}
$$

Thus $C_n(K)$ is the quotient of the free abelian group $\mathbb{Z}K_n$ generated by the set $K_n$ (see 1.2.5(c)), modulo the subgroup generated by the degenerate $n$-cubes. For an $n$-cube $a \in K_n$ we write as $\hat{a}$ its *normalised class* in $C_n(K)$, up to degenerate cubes, called a *normalised cube.*

As specified above, the boundary $\partial_n$ takes the class $\hat{a}$ to a linear combination of the classes of its faces $\partial_i^\alpha a$, which are normalised cubes of dimension $n-1$. The exercises below prove that this is correct, and that $\partial_n \partial_{n+1} = 0$.

In computations, we shall usually write the class $\hat{a}$ as $a$, in order to simplify notation; of course we have to check that a definition involving this class is invariant up to adding degenerate cubes.

The second step is now complete: we have a functor

$$
\begin{gathered}
\mathrm{Ch}_+ \colon \mathsf{CubSet} \to \mathrm{Ch}_+\mathsf{Ab}, \\
K \mapsto \mathrm{Ch}_+(K) = ((C_n(K))_{n\geqslant 0}, (\partial_n)_{n \geqslant 1}), \\
f_{\sharp n} = C_n(f) \colon C_n(K) \to C_n(L), \quad \Sigma\, \lambda_i a_i \mapsto \Sigma\, \lambda_i f_n(a_i).
\end{gathered} \tag{2.47}
$$

On a morphism $f \colon K \to L$ of cubical sets, the $n$-component of the chain morphism $\mathrm{Ch}_+(f)$ is induced by the linear extension of $f_n$; this is legitimate, because a degenerate $n$-cube $a = e_i b$ is sent to the degenerate cube $f_n(a) = e_i f_n(b)$. Plainly, this procedure preserves composition and identities.

Composing with the functor $\mathrm{Cub} \colon \mathsf{Top} \to \mathsf{CubSet}$ of (2.43), we get the functor *of the singular (cubical) chain complex*:

$$
\begin{gathered}
C_+ \colon \mathsf{Top} \to \mathrm{Ch}_+\mathsf{Ab}, \qquad C_+(X) = \mathrm{Ch}_+(\mathrm{Cub}(X)), \\
f_{\sharp n}(a) = fa \qquad (\text{for } f \colon X \to Y \text{ and } a \colon \mathbb{I}^n \to X \text{ in } \mathsf{Top}).
\end{gathered} \tag{2.48}
$$

The name of 'boundary' has here a geometrical sense: for a map $a \colon \mathbb{I}^n \to X$, the boundary $\partial_n a$ is a (formal) linear combination of the faces of $a$, as

in the following figure, for $n = 2$

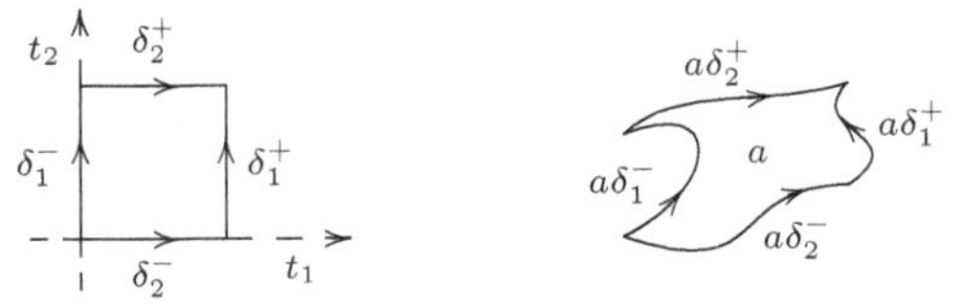


(2.49)

$$\partial_2 a = a\delta_2^- + a\delta_1^+ - a\delta_2^+ - a\delta_1^- .$$

Taking account of signs, we can view the chain $\partial_2 a$ as representing a path around $a$.

*Exercises and complements.* (a) The boundary homomorphism $\partial_n$ is well defined in (2.46). *Hint:* use the Induction Lemma 2.1.4.

(b) Prove that $\partial_n \partial_{n+1} = 0$, in (2.46). *Hint:* use the cubical relations of the faces.

(c) $C_n(K)$ is isomorphic to the subgroup of $\mathbb{Z}K_n$ generated by the non-degenerate cubes $a \in K_n \setminus \mathrm{Deg}_n K$, and *is a free abelian group.*

This fact will be important later, but is not relevant here. In fact, the faces $\partial_i^\alpha a$ of a non-degenerated cube $a$ can be degenerate (for instance, a non-degenerate 2-cube can have constant edges), and the boundary $\partial_n \colon \mathbb{Z}K_n \to \mathbb{Z}K_{n-1}$ *cannot be restricted* to the subgroups spanned by the non-degenerated cubes; we really have to take an induced morphism modulo degenerated cubes.

### 2.2.4 Singular homology

We are now ready to define the singular homology functor $H_n \colon \mathsf{Top} \to \mathsf{Ab}$ as the composite of three functors, defined in (2.43), in (2.47) and in (2.24)

$$\begin{aligned}
H_n &= (\mathsf{Top} \xrightarrow{\mathrm{Cub}} \mathsf{CubSet} \xrightarrow{\mathrm{Ch}_+} \mathrm{Ch}_+\mathsf{Ab} \xrightarrow{H_n} \mathsf{Ab}), \\
H_n(X) &= H_n(\mathrm{Ch}_+(\mathrm{Cub}(X))) = H_n(C_+(X)), \\
f_{*n} = H_n(f) &\colon H_n(X) \to H_n(Y), \qquad f_{*n}([z]) = [f_{\sharp n}(z)],
\end{aligned} \tag{2.50}$$

for a map $f \colon X \to Y$ and a cycle $z \in C_n(X)$.

The $n$-th *singular homology group* $H_n(X)$ of the space $X$ is thus a homology group of the (normalised) chain complex associated to the cubical set $\mathrm{Cub}(X)$.

We write:

$$\begin{aligned}
Z_n(X) &= Z_n(C_+(X)) = \mathrm{Ker}\,(\partial_n\colon C_n(X) \to C_{n-1}(X)) \\
&\qquad\qquad\qquad\qquad\qquad\qquad \text{(the group of } n\text{-}cycles\text{)}, \\
B_n(X) &= B_n(C_+(X)) = \mathrm{Im}\,(\partial_{n+1}\colon C_{n+1}(X) \to C_n(X)) \qquad (2.51)\\
&\qquad\qquad\qquad\qquad\qquad \text{(the group of } n\text{-}boundaries\text{)}, \\
H_n(X) &= Z_n(X)/B_n(X).
\end{aligned}$$

A *singular $n$-chain* $c \in C_n(X)$ is a linear combination of $n$-cubes $\mathbb{I}^n \to X$ (up to degenerate cubes); it is an *$n$-cycle* if $\partial_n(c) = 0$ in $C_{n-1}(X)$, and an *$n$-boundary* if $c \in \mathrm{Im}\,\partial_{n+1}$. Every cycle $z$ determines a homology class $[z] \in H_n(X)$, also written as $[z]_H$, which vanishes if and only if $z$ is a boundary. Two $n$-cycles $z, z'$ are homologous if $z - z'$ is a boundary.

As already stated in (2.50), a map $f\colon X \to Y$ in Top has an associated homomorphism $H_n(f)\colon H_n(X) \to H_n(Y)$, usually written as $f_{*n}$.

*Remarks and complements.* (a) A 0-chain is a formal linear combination $\Sigma\,\lambda_i x_i$ of points $x_i\colon \{*\} \to X$; it is always a 0-cycle.

A 1-chain is a formal linear combination of paths $\mathbb{I} \to X$. The boundary $\partial_1(a) = a(1) - a(0)$ of a path is the formal difference of its endpoints; the path $a$ is a 1-cycle if and only if it is a loop.

(b) A chain $c = \Sigma\,\lambda_i a_i \in C_n(X)$ of the space $X$ has a *support*, the (compact) subspace

$$\mathrm{supp}(c) = \bigcup \mathrm{Im}\, a_i \subset X, \qquad (2.52)$$

where we are supposing that all $a_i$ are distinct non-degenerate cubes, and all $\lambda_i$ are non-zero. Plainly

$$\mathrm{supp}(\partial_n c) \subset \mathrm{supp}(c), \qquad (2.53)$$

because the image of each face $\partial_i^\alpha a = a\delta_i^\alpha\colon \mathbb{I}^{n-1} \to X$ of a cube $a$ is contained in $\mathrm{Im}\, a$.

*(c) The set $\mathcal{K}$ of compact subspaces of $X$, ordered by inclusion, is a filtered set. The previous point shows that $H_n(X)$ is the inductive limit (see 1.7.6) of the system of abelian groups $H_n(K)$ ($K \in \mathcal{K}$), with respect to the system of homomorphisms $H_n(K) \to H_n(K')$ induced by the inclusions. This property is expressed saying that *singular homology has compact supports.*

### 2.2.5 *Exercises and complements* (Basic results)

A few very elementary computations of homology groups can be done directly. We shall use the partition of a space $X$ in path components

$$X = \bigcup_{i\in I} X_i, \tag{2.54}$$

and we refer to $I$ as the *set of path components* of $X$ (written as $\Pi_0(X)$ in (1.71)).

(a) The homology groups of the empty space and the singleton are:

$$H_k(\emptyset) = 0 \qquad (k \geqslant 0), \tag{2.55}$$

$$H_0(\{*\}) \cong \mathbb{Z}, \qquad H_k(\{*\}) = 0 \quad (k > 0). \tag{2.56}$$

(b) (*Path-connected spaces*) For a path-connected space $X \neq \emptyset$ we have

$$H_0(X) \cong \mathbb{Z}. \tag{2.57}$$

More precisely, $H_0(X)$ is the free abelian group generated by the homology class $[x]$ of any point $x \in X$: all these classes are the same.

(c) Using the partition (2.54) in path components, there are canonical isomorphisms

$$\begin{aligned} &\textstyle\bigoplus_{i\in I} C_+(X_i) \to C_+(X), \qquad (c_i)_{i\in I} \mapsto \sum_{i\in I} c_i,\\ &\textstyle\bigoplus_{i\in I} H_n(X_i) \to H_n(X), \qquad ([z_i])_{i\in I} \mapsto \sum_{i\in I} [z_i], \end{aligned} \tag{2.58}$$

where a cube $a\colon \mathbb{I}^n \to X_i$ is identified with its 'extension' to $X$, and a chain in $X_i$ with the corresponding chain in $X$. (This issue will be better examined in 2.3.4.)

More formally, the inclusions $u_i\colon X_i \to X$ give homomorphisms

$$u_{i\sharp}\colon C_+(X_i) \to C_+(X), \qquad u_{i*}\colon H_n(X_i) \to H_n(X),$$

which are the injections of a categorical sum, for every $n \geqslant 0$.

(d) Loosely speaking, the functor $H_0$ 'counts' the path components of the space: the group $H_0(X)$ is isomorphic to the free abelian group $\mathbb{Z}I = \bigoplus_{i\in I} \mathbb{Z}$ generated by the set $I$ of path components of $X = \bigcup_{i\in I} X_i$. More precisely, $H_0(X)$ has a canonical basis $[x_i]_H$ $(i \in I)$, where $x_i$ is any point of the path component $X_i$.

There is thus a natural isomorphism

$$\begin{aligned} &\varphi\colon H_0 \to F\Pi_0\colon \mathsf{Top} \to \mathsf{Ab},\\ &\varphi X\colon H_0(X) \to F\Pi_0(X), \qquad [x]_H \mapsto [x]_P, \end{aligned} \tag{2.59}$$

with values in the composite of $\Pi_0\colon \mathsf{Top} \to \mathsf{Set}$ (defined in (1.14)) with the free-abelian-group functor $F\colon \mathsf{Set} \to \mathsf{Ab}$, $F(I) = \mathbb{Z}I$ (in (1.25)).

(e) A space $X$ whose path components (if any) are singletons has $H_n(X) = 0$ for all $n > 0$, while $H_0(X)$ is determined in (d).

This includes all discrete spaces, the rational line $\mathbb{Q}$ and the converging sequence $\mathbb{N}^\bullet \cong \mathbb{H}_0$. Note that singular homology does not distinguish the euclidean spaces $\mathbb{Q}$, $\mathbb{H}_0$ and $\mathbb{Z}$.

### 2.2.6 Homotopy Invariance Theorem

*(a) The functors $H_n\colon \mathsf{Top} \to \mathsf{Ab}$ are homotopy invariant: if $f \simeq g\colon X \to Y$ in $\mathsf{Top}$, then $H_n(f) = H_n(g)\colon H_n(X) \to H_n(Y)$ (for all $n \geqslant 0$).*

*(b) If the spaces $X, Y$ are homotopy equivalent, then $H_n(X) \cong H_n(Y)$, for all $n \geqslant 0$.*

*(c) Contractible spaces have the same homology groups as the singleton, see (2.56). In particular, this applies to all euclidean spaces $\mathbb{R}^m$.*

*Proof* It is sufficient to prove (a). We want to prove that a homotopy $\varphi\colon f \to g$ between two maps $f, g\colon X \to Y$ has an associated chain homotopy $\Phi$ between the corresponding morphisms $f_\sharp$ and $g_\sharp$ of chain complexes

$$\begin{aligned} &\Phi\colon f_\sharp \simeq g_\sharp\colon \mathrm{Ch}_+(\mathrm{Cub}(X)) \longrightarrow \mathrm{Ch}_+(\mathrm{Cub}(Y)), \\ &(\partial_{n+1}\Phi_n + \Phi_{n-1}\partial_n)(a) = ga - fa \qquad (\text{for } a\colon \mathbb{I}^n \to X). \end{aligned} \tag{2.60}$$

This will imply that $H_n(f) = H_n(g)$ (for $n \geqslant 0$), by the homotopy invariance of chain homology (in 2.1.7(b)).

We write the homotopy $\varphi\colon f \to g\colon X \to Y$ as a continuous mapping $\varphi\colon \mathbb{I} \times X \to Y$. For every $n \geqslant 0$ we define the homomorphism

$$\Phi_n\colon C_n(X) \to C_{n+1}(Y), \qquad \Phi_n(a) = \varphi(\mathbb{I} \times a), \tag{2.61}$$

sending an $n$-cube $a\colon \mathbb{I}^n \to X$ to the composite $\varphi(\mathbb{I} \times a)\colon \mathbb{I}^{n+1} \to Y$, an $(n+1)$-cube of $Y$. This is legitimate, because a degenerate cube $a = b\varepsilon_i\colon \mathbb{I}^n \to \mathbb{I}^{n-1} \to X$ is sent to the cube

$$\varphi(\mathbb{I} \times b\varepsilon_i) = \varphi(\mathbb{I} \times b)\varepsilon_{i+1}\colon \mathbb{I}^{n+1} \to Y,$$

which is also degenerate. Finally (with $\alpha = 0, 1$ in each indexed sum)

$$\begin{aligned} \partial_{n+1}\Phi_n(a) &= \partial_{n+1}(\varphi(\mathbb{I} \times a)) = \textstyle\sum_{\alpha, 1 \leqslant i \leqslant n+1} (-1)^{i+\alpha}\, \varphi(\mathbb{I} \times a)\delta_i^\alpha \\ &= \textstyle\sum_\alpha (-1)^{1+\alpha}\, \varphi(\delta_1^\alpha \times a) + \sum_{\alpha, 2 \leqslant i \leqslant n+1} (-1)^{i+\alpha}\, \varphi(\mathbb{I} \times a\delta_{i-1}^\alpha) \\ &= ga - fa - \textstyle\sum_{\alpha, 1 \leqslant i \leqslant n} (-1)^{i+\alpha}\, \varphi(\mathbb{I} \times a\delta_i^\alpha) \\ &= ga - fa - \Phi_{n-1}\partial_n(a). \end{aligned}$$

*Remarks.* (i) If we use the cylinder $X \times \mathbb{I}$ (instead of $\mathbb{I} \times X$) the computation is slightly more complex: we should single out the last faces instead of the first ones, and modify $\Phi$ by a sign change depending on the degree $n$.

(ii) Let us note that the construction of $\Phi$ is easy because the cylinder functor $\mathbb{I} \times -$: Top $\to$ Top takes cubes to cubes (from dimension $n$ to dimension $n+1$). *The corresponding simplicial construction will be sketched in 3.6.9(a).* □

### *2.2.7 Numerical invariants*

Let $X$ be a space whose singular homology groups are finitely generated, as is often the case when homology theory is used.

Applying the Structure Theorem recalled in 1.2.6(c), the $n$-th homology group $H_n(X)$ gives a numerical invariant (up to homotopy equivalence, by Theorem 2.2.6)

$$\beta_n(X) = \mathrm{rk}(H_n(X)) \in \mathbb{N} \qquad (n \geqslant 0). \tag{2.62}$$

called the *$n$-th Betti number* of $X$, or the *Betti number of degree $n$*. In particular, $\beta_0(X)$ is the number of path components of $X$.

One can organise the Betti numbers in a polynomial: see 6.7.7(c).

If we also assume that these numbers are eventually zero (as will often be the case, here), we can define another important numerical invariant, the *Euler–Poincaré characteristic* of $X$

$$\chi(X) \;=\; \Sigma_{n\geqslant 0}\, (-1)^n\, \beta_n(X) \in \mathbb{Z}. \tag{2.63}$$

In fact, the Structure Theorem of finitely generated abelian groups gives, in each dimension $n$, a finite sequence of numerical invariants, that characterise the group $H_n(X)$ up to isomorphism: the Betti number $\beta_n(X)$, together with the *torsion coefficients of $X$ of degree $n$* (possibly none)

$$\tau_{n1}(X) = \tau_1(H_n(X)), \quad \tau_{n2}(X) = \tau_2(H_n(X)), \; ... \tag{2.64}$$

### *2.2.8 Paths and homology in degree 1*

Let $X$ be a space. We are not yet able to compute $H_1(X)$, except very simple cases, but we can note that this group is formed of equivalence classes $[\Sigma\, \lambda_i a_i]$ of 1-cycles of $X$

$$0 = \partial_1(\Sigma\, \lambda_i a_i) = \Sigma\, \lambda_i a_i(1) - \Sigma\, \lambda_i a_i(0). \tag{2.65}$$

Here each $a_i\colon \mathbb{I} \to X$ is a path; by abuse of notation, it represents the associated normalised 1-cube $\hat{a}_i \in C_1(X) = \mathbb{Z}\mathrm{Cub}_1(X)/\mathbb{Z}\mathrm{Deg}_1(X)$, up to constant paths.

*Exercises and complements.* All the following exercises are about paths in $X$, written as $a\colon x \simeq y$ to denote the endpoints.

(a) We already remarked that a single path $a\colon x \simeq y$ is a cycle if and only if

$$\partial(a) = \partial_1^+ a - \partial_1^- a = y - x = 0,$$

if and only if $a$ is a loop. In this case there is a homology class $[a]$, which can be of interest or trivial – the zero element of $H_1(X)$. Let us note that the class (2.65) can only be written as $\Sigma\lambda_i[a_i]$ when all the paths $a_i$ are loops.

(b) If the paths $a, b\colon x \simeq y$ are homotopic with fixed endpoints (as defined in 1.4.7), the chain $a - b \in C_1(X)$ is a boundary: it belongs to $B_1(X)$. Therefore, in every cycle $z = \sum \lambda_i a_i$ where $a$ is present, we can replace $a$ with $b$, obtaining a homologous cycle $z'$.

(c) If two loops $a, b$ are homotopic as loops (as defined in 1.4.8), then $[a] = [b]$ in $H_1(X)$.

(d) If the loop $a$ is loop homotopic to a constant loop, then $[a] = 0$ in $H_1(X)$. On the other hand, if $[a] \neq 0$, the loop $a$ cannot be contracted to a point. A fortiori, it is not 2-homotopic to a constant loop and $X$ is not simply connected.

(e) If $c = a * b$ is the concatenation of two consecutive paths in $X$ (see (1.55)), then $a + b - c$ is a boundary: in every cycle in which $a + b$ occurs, we can replace it with $c$, without changing the homology class; or vice versa.

(f) The reversed path $a^\sharp$ of $a$ (see 1.4.2(iii)) gives a chain $a + a^\sharp$ which is a boundary: $[a + a^\sharp] = 0$. In every cycle where $a^\sharp$ occurs, we can replace it with the chain $-a$, obtaining a cycle with the same homology class; or vice versa.

## 2.3 Mayer–Vietoris and subdivision

The exact sequence of Mayer–Vietoris is the main tool to compute the homology of a space, knowing the homology of suitable subspaces.

We begin by introducing some further elements of basic Homological Algebra, about exact sequences of chain complexes, culminating in Theorem 2.3.3 on the exact sequence of chain homology.

The Subdivision Theorem, in 2.3.5, says that homology can be computed by 'small chains' – an essential property. It opens the way to the Mayer–Vietoris sequence, in 2.3.7, and to basic properties of relative homology, in Chapter 4.

The trivial group and the trivial chain complex are always written 0.

### 2.3.1 Exact sequences of chain complexes

All issues about exactness can be lifted from $\mathsf{Ab}$ to $\mathrm{Ch}_+\mathsf{Ab}$, using the family of forgetful functors $U_n \colon \mathrm{Ch}_+\mathsf{Ab} \to \mathsf{Ab}$, $A \mapsto A_n$ of (2.17).

For a chain morphism $f \colon A \to B$, the kernel and cokernel are defined componentwise, on a chain morphism $f \colon A \to B$

$$\mathrm{Ker}\, f = ((\mathrm{Ker}\, f_n), (\partial_n)), \qquad \mathrm{Cok}\, f = ((\mathrm{Cok}\, f_n), (\partial_n)), \tag{2.66}$$

where the differentials are induced by the differentials of $A$ and $B$, applying 2.1.2(c)

$$\begin{array}{ccccccc} \mathrm{Ker}\, f_n & \rightarrowtail & A_n & \xrightarrow{f_n} & B_n & \twoheadrightarrow & \mathrm{Cok}\, f_n \\ {\scriptstyle\partial}\downarrow & & {\scriptstyle\partial}\downarrow & & \downarrow{\scriptstyle\partial} & & \downarrow{\scriptstyle\partial} \\ \mathrm{Ker}\, f_{n-1} & \rightarrowtail & A_{n-1} & \xrightarrow[f_{n-1}]{} & B_{n-1} & \twoheadrightarrow & \mathrm{Cok}\, f_{n-1} \end{array} \tag{2.67}$$

The chain complexes $\mathrm{Coim}\, f$ and $\mathrm{Im}\, f$ are similarly defined, applying 2.1.2(d)

$$\mathrm{Coim}\, f = ((\mathrm{Coim}\, f_n), (\partial_n)), \qquad \mathrm{Im}\, f = ((\mathrm{Im}\, f_n), (\partial_n)). \tag{2.68}$$

The chain morphism $f \colon A \to B$ is said to be:

- a *monomorphism* (or *mono*) if each component $f_n$ is mono, or equivalently $\mathrm{Ker}\, f = 0$,

- an *epimorphism* (or *epi*) if each component $f_n$ is epi, or equivalently $\mathrm{Cok}\, f = 0$ (i.e. $\mathrm{Im}\, f = B$).

The arrows $\rightarrowtail$ and $\twoheadrightarrow$ will also denote monomorphisms and epimorphisms of chain complexes, respectively.

The *exactness* of a sequence $A \to B \to C$ in $\mathrm{Ch}_+\mathsf{Ab}$ is defined as in $\mathsf{Ab}$, and amounts to the exactness of all its components $A_n \to B_n \to C_n$.

A short exact sequence $A \rightarrowtail B \twoheadrightarrow C$ of chain complexes is again detected by its components. Up to isomorphism, it can always be realised in standard form, taking:

- a *chain subcomplex* $A \subset B$, which means a family of subgroups $A_n \subset B_n$ such that $\partial_n(A_n) \subset A_{n-1}$, so that we can define the differentials of $A$ as restrictions of the differentials of $B$,

- the *quotient chain complex* $C = B/A$, with components $C_n = B_n/A_n$

and induced differentials

$$\begin{array}{ccccc} A_n & \rightarrowtail & B_n & \twoheadrightarrow & C_n \\ \downarrow \partial & & \downarrow \partial & & \downarrow \partial \\ A_{n-1} & \rightarrowtail & B_{n-1} & \twoheadrightarrow & C_{n-1} \end{array} \tag{2.69}$$

*Exercises and complements.* Kernels and cokernels of chain complexes agree with the abstract definition of 2.1.2, as shown by the following easy points, left to the reader.

(a) The embedding $\ker f\colon \mathrm{Ker}\, f \rightarrowtail A$ satisfies the universal property 2.1.2(i), in $\mathrm{Ch}_+\mathsf{Ab}$.

(b) The projection $\mathrm{cok}\, f\colon B \twoheadrightarrow \mathrm{Cok}\, f$ satisfies the universal property 2.1.2(i*), in $\mathrm{Ch}_+\mathsf{Ab}$.

(c) The Induction Lemma 2.1.4, written in $\mathsf{Ab}$, can be readily extended to $\mathrm{Ch}_+\mathsf{Ab}$.

### 2.3.2 A category of short exact sequences

We form the category $\mathrm{ShCh}_+\mathsf{Ab}$ *of short exact sequences of chain complexes*, as a full subcategory of the category $(\mathrm{Ch}_+\mathsf{Ab})^{\mathbf{3}}$ of pairs of consecutive chain morphisms.

An object is a pair $(f, g)$ forming a short exact sequence in $\mathrm{Ch}_+\mathsf{Ab}$, as in the upper line of the following diagram. A morphism $(u, v, w)\colon (f, g) \to (f', g')$ is a commutative diagram with short exact rows

$$\begin{array}{ccccc} A & \overset{f}{\rightarrowtail} & B & \overset{g}{\twoheadrightarrow} & C \\ \downarrow u & & \downarrow v & & \downarrow w \\ A' & \underset{f'}{\rightarrowtail} & B' & \underset{g'}{\twoheadrightarrow} & C' \end{array} \tag{2.70}$$

We are also interested in the functors

$$\begin{array}{ll} P'\colon \mathrm{ShCh}_+\mathsf{Ab} \to \mathrm{Ch}_+\mathsf{Ab}, & P'''\colon \mathrm{ShCh}_+\mathsf{Ab} \to \mathrm{Ch}_+\mathsf{Ab}, \\ P'(A \rightarrowtail B \twoheadrightarrow C) = A, & P'''(A \rightarrowtail B \twoheadrightarrow C) = C, \\ P'(u, v, w) = u, & P'''(u, v, w) = w, \end{array} \tag{2.71}$$

which project a short exact sequence to its first or third issue.

**2.3.3 Theorem** (The exact sequence of chain homology)
*A short exact sequence of chain complexes, in* $\mathrm{Ch}_+\mathsf{Ab}$

$$A \overset{f}{\rightarrowtail} B \overset{g}{\twoheadrightarrow} C \tag{2.72}$$

*has an associated* long exact homology sequence *in* Ab

$$
\begin{array}{ccccc}
 & \cdots & \longrightarrow & H_{n+1}(C) & \\
 & \swarrow D_{n+1} & & & \\
H_n(A) & \xrightarrow{f_{*n}} & H_n(B) & \xrightarrow{g_{*n}} & H_n(C) \\
 & \swarrow D_n & & & \\
H_{n-1}(A) & \longrightarrow & \cdots & \longrightarrow & H_0(C) \longrightarrow 0
\end{array}
\qquad (2.73)
$$

*natural for morphisms of short exact sequences of chain complexes.*

*The homomorphisms* $D_n \colon H_n(C) \to H_{n-1}(A)$ *are called* connecting morphisms. *For* $[c] \in H_n(C)$, *the value* $D_n[c] = [a] \in H_{n-1}(A)$ *is obtained by the following diagram (made clear in the proof), however we choose the representative* $n$*-cycle* $c$ *and the elements* $b \in B_n$, $a \in A_{n-1}$, *under the following constraints*

$$
\begin{array}{ccccc}
 & & b & \xmapsto{g_n} & c \\
 & & \downarrow \partial & & \downarrow \partial \\
a & \xmapsto{f_{n-1}} & \partial b & \xmapsto{g_{n-1}} & 0 \\
\downarrow \partial & & \downarrow \partial & & \\
\partial a & \xmapsto{f_{n-2}} & 0 & &
\end{array}
\qquad
\begin{array}{l}
g_n(b) = c, \\[2em]
f_{n-1}(a) = \partial_n(b).
\end{array}
\qquad (2.74)
$$

*Note.* The naturality of the sequence amounts to saying that the connecting morphisms $D_n \colon H_n(C) \to H_{n-1}(A)$ are the components of a natural transformation:

$$D_n \colon H_n P''' \to H_{n-1} P' \colon \mathrm{ShCh}_+\mathsf{Ab} \longrightarrow \mathsf{Ab}, \qquad (2.75)$$

where $P'''$ and $P'$ are the projections defined in (2.71).

*Proof* The argument uses a procedure called *diagram chasing*, exemplified in diagram (2.74) and in many similar ones, below.

We note that, after the last term $H_0(C)$ of sequence (2.73), there is nothing special to prove: by the usual extension of chain complexes on negative degrees, all homology groups $H_k(-)$ are zero for $k < 0$, and the surjectivity of $g_{*0}$ will be a consequence of exactness at $H_0(C)$.

(a) We begin by proving that the connecting homomorphism $D_n$ is well defined, constructing diagram (2.74) from the upper right corner to the lower left one.

We start from the cycle $c \in Z_n(C)$, with $\partial_n(c) = 0$ in $C_{n-1}$. The homomorphism $g_n$ is surjective, and there is some $b \in B_n$ such that $g_n(b) = c$. Now $g_{n-1}\partial_n(b) = \partial_n g_n(b) = \partial_n(c) = 0$, which means that $\partial_n(b) \in \operatorname{Ker} g_{n-1} = \operatorname{Im} f_{n-1}$: there is some $a \in A_{n-1}$ such that $f_{n-1}(a) = \partial_n(b)$ (and in fact only one, because $f_{n-1}$ is mono). This element $a$ is a cycle, because $f_{n-2}\partial_{n-1}(a) = \partial_{n-1}f_{n-1}(a) = \partial_{n-1}\partial_n(b) = 0$, and $f_{n-2}$ is mono. We have obtained a homology class $[a] \in H_{n-1}(A)$.

We prove now that this class does not depend on the choice of the representative $c$ (in its homology class), and of the element $b$.

Suppose we have a homologous cycle $c' \in Z_n(C)$, with $c - c' = \partial_{n+1}(\hat{c})$. We choose a new $b' \in B_n$ and form a second commutative diagram

$$\begin{array}{ccccccc} & & b' & \mapsto & c' & & (n) \\ & & \downarrow & & \downarrow & & \\ a' & \mapsto & \partial b' & \mapsto & 0 & & (n-1) \end{array} \tag{2.76}$$

omitting the names of arrows, but writing the degree in the right-hand column.

We have to prove that $a - a'$ is a boundary. In the right part below, $g_n(b-b') = c-c'$. Moreover there is some $\hat{b} \in B_{n+1}$ such that $g_{n+1}(\hat{b}) = \hat{c}$. Its boundary need not be $b - b'$, but $g_n\partial_{n+1}(\hat{b}) = \partial_{n+1}(\hat{c}) = g_n(b - b')$. Thus $g_n(b - b' - \partial_{n+1}\hat{b}) = 0$

$$\begin{array}{ccccccccc} & & & & \hat{b} & \longmapsto & \hat{c} & & (n+1) \\ & & & & & & \downarrow & & \\ \hat{a} & \mapsto & b - b' - \partial\hat{b} & & b - b' & \mapsto & c - c' & & (n) \\ \downarrow & & \downarrow & & & & & & \\ \partial\hat{a} & \longmapsto & \partial(b - b') & & & & & & (n-1) \end{array} \tag{2.77}$$

Moving to the left part of the previous diagram, exactness in degree $n$ gives some $\hat{a} \in A_n$ such that $f_n(\hat{a}) = b - b' - \partial_{n+1}(\hat{b})$. Since $f$ commutes with differentials, $f_{n-1}\partial_n(\hat{a}) = \partial_n(b-b') = f_{n-1}(a-a')$, and $\partial_n(\hat{a}) = a-a'$, so that $[a] = [a']$.

Finally, the mapping $D_n$ is a homomorphism of abelian groups: for two cycles $c, c' \in Z_n(C)$, we construct $D_n[c] = [a]$ and $D_n[c'] = [a']$ as in (2.74) and (2.76). All the arrows of these diagrams are homomorphisms, and the following diagram says that $D_n[c + c'] = [a + a'] = [a] + [a']$

$$\begin{array}{ccccccc} & & b + b' & \longmapsto & c + c' & & (n) \\ & & \downarrow & & \downarrow & & \\ a + a' & \mapsto & \partial(b + b') & \longmapsto & 0 & & (n-1) \end{array} \tag{2.78}$$

(b) The exactness of the homology sequence is to be verified at $H_n(A)$, $H_n(B)$ and $H_n(C)$, for each $n \geqslant 0$. We prove the first claim; the rest will be the subject of two exercises, in 2.3.8.

To show that $\operatorname{Im} D_{n+1} \subset \operatorname{Ker} f_{*n}$ it is sufficient to note that the composite $f_{*n}D_{n+1}\colon H_{n+1}(C) \to H_n(B)$ is zero. In fact diagram (2.74), lifted one degree up, shows that

$$f_{*n}D_{n+1}[c] = f_{*n}[a] = [\partial_{n+1}(b)] = 0.$$

Suppose now that $a \in Z_n(A)$ and $f_{*n}[a] = [f_n(a)] = 0$. Thus $f_n(a) = \partial_{n+1}(\hat{b})$, for some $\hat{b} \in B_{n+1}$, and we form the diagram

$$\begin{array}{ccccc} & & \hat{b} & \mapsto & \hat{c} & \qquad (n+1) \\ & & \downarrow & & \downarrow & \\ a & \mapsto & b & \mapsto & 0 & \qquad (n) \end{array} \tag{2.79}$$

where $\hat{c}$ is a cycle (because $g_n f_n(a) = 0$), and $[a] = D_{n+1}[\hat{c}]$ belongs to $\operatorname{Im} D_{n+1}$.

(c) We end by checking the naturality of the homology sequence, for a morphism in $\mathrm{ShCh}_+\mathsf{Ab}$

$$(u, v, w) \colon (A \rightarrowtail B \twoheadrightarrow C) \to (A' \rightarrowtail B' \twoheadrightarrow C'),$$

as in diagram (2.70). We are to prove the commutativity of three squares (for $n \geqslant 0$)

$$\begin{array}{ccccccc} H_n(A) & \xrightarrow{f_{*n}} & H_n(B) & \xrightarrow{g_{*n}} & H_n(C) & \xrightarrow{D_n} & H_{n-1}(A) \\ \downarrow{\scriptstyle u_{*n}} & & \downarrow{\scriptstyle v_{*n}} & & \downarrow{\scriptstyle w_{*n}} & & \downarrow{\scriptstyle u_{*n-1}} \\ H_n(A') & \xrightarrow[f'_{*n}]{} & H_n(B') & \xrightarrow[g'_{*n}]{} & H_n(C') & \xrightarrow[D_n]{} & H_{n-1}(A') \end{array} \tag{2.80}$$

This is automatically true for the left and central square, applying the functor $H_n$ to the commutative diagram (2.70). For the right square, we take a class $[c] \in H_n(C)$ and construct $D_n[c] = [a]$ as in diagram (2.74)

$$\begin{array}{ccccc} & & b & \mapsto & c \\ & & \downarrow & & \downarrow \\ a & \mapsto & \partial b & \mapsto & 0 \end{array} \qquad\qquad \begin{array}{ccccc} & & v_n b & \mapsto & w_n c & \qquad (n) \\ & & \downarrow & & \downarrow & \\ u_{n-1}a & \mapsto & \partial v_n b & \mapsto & 0 & \qquad (n-1) \end{array}$$

Applying the three homomorphisms $u_{n-1}$, $v_n$ and $w_n$, the right diagram above shows that: $D_n w_{*n}[c] = D_n[w_n c] = [u_{n-1}a] = u_{*n-1}D_n[c]$. □

### 2.3.4 Small chains

Let $X$ be a topological space. It will be useful to fix notation for chains of $X$ 'living' in a subspace: in fact, as established in the next theorem, 'small chains are sufficient to compute homology'.

(i) Let $U$ be a subspace of $X$ (not necessarily open), with inclusion $u \colon U \to X$.

It can be convenient to view a cube $a\colon \mathbb{I}^n \to U$ of $U$ as a cube of $X$ with image contained in $U$ (formally, the map $ua$), and a chain $c = \sum \lambda_i a_i$ of $U$ as a $U$-*small* chain of $X$, with support contained in $U$ (see (2.52)).

This is correct, because a cube of $U$ is degenerate if and only if it is degenerate as a cube of $X$.

In this way, $C_+(U)$ is viewed as a subcomplex of $C_+(X)$, with inclusion $u_\sharp\colon C_+(U) \to C_+(X)$.

On the other hand, the homomorphism $u_{*n}\colon H_n(U) \to H_n(X)$ induced in homology need not be injective (as already remarked in 1.1.6(b)): our abuse of notation on chains requires a more specific notation for homology. Thus, for a cycle $z \in C_n(U)$, we shall distinguish its homology class $[z]_U \in H_n(U)$ from the homology class $[z]_X = u_{*n}[z]_U \in H_n(X)$.

Actually, for an $n$-chain $c$ of $U$:

- $c$ is a cycle in $U$ if and only if it is in $X$,

- if $c$ is a boundary $\partial\hat{c}$ in $U$, then it is also a boundary in $X$, but the converse fails: the support of any such $\hat{c} \in C_{n+1}(X)$ need not be contained in $U$.

(ii) Extending this issue, the next theorem is concerned with a generalised open cover $\mathcal{U} = (U_i)$ of the space $X$, which means that $X = \bigcup_i \mathrm{int}(U_i)$ (see 1.1.8(a)).

Again, we are not assuming that the subspaces $U_i$ are open in $X$; this will generally be the case in the applications of this chapter and the next, but will not be in the applications of Chapter 4.

We denote as $C_+(X;\mathcal{U})$ the subcomplex of $C_+(X)$ of $\mathcal{U}$-*small chains* (or *small chains*, for short): $C_n(X;\mathcal{U})$ is generated by the cubes $a\colon \mathbb{I}^n \to X$ whose image is contained in some $U_i$ (up to degenerate small cubes).

*An exercise.* (a) $C_+(X;\mathcal{U})$ is indeed a chain subcomplex of $C_+(X)$. The obvious verification is left to the reader.

### *2.3.5 Subdivision Theorem*

*For a space $X$ and a generalised open cover $\mathcal{U} = (U_i)$ of $X$, the inclusion $j\colon C_+(X;\mathcal{U}) \to C_+(X)$ induces isomorphism in homology, in every degree:*

$$j_{*n}\colon H_n(X;\mathcal{U}) \to H_n(X).$$

*Proof* The general idea is to subdivide cubes, replacing them by $\mathcal{U}$-small chains. The proof, which is long and interesting, is deferred to 2.3.9.

This result will be applied, in Theorems 2.3.7 and 4.2.1, to a generalised open cover $(U, V)$ formed of two subspaces, but this particular case would reduce the interest of the statement, without simplifying the proof. □

### 2.3.6 Pasting two spaces

The next theorem links the homology of a space $X$ with that of three subspaces $U, V, A$, where

$$X = \operatorname{int} U \cup \operatorname{int} V, \qquad A = U \cap V. \tag{2.81}$$

Again, the subspaces $U, V$ need not be open in $X$. The inclusions between these spaces form a commutative square $ui = vj$ in Top

$$\begin{array}{ccc} A & \xrightarrow{i} & U \\ {\scriptstyle j}\downarrow & & \downarrow{\scriptstyle u} \\ V & \xrightarrow[v]{} & X \end{array} \qquad\qquad \begin{array}{ccc} C_+(A) & \xrightarrow{i_{\sharp n}} & C_+(U) \\ {\scriptstyle j_{\sharp n}}\downarrow & & \downarrow{\scriptstyle u_{\sharp n}} \\ C_+(V) & \xrightarrow[v_{\sharp n}]{} & C_+(X) \end{array} \tag{2.82}$$

and we agreed in 2.3.4 to consider the associated monomorphisms of $\mathrm{Ch}_+\mathsf{Ab}$, in the right square above, as inclusions.

We already noted that a cycle $z \in C_n(A)$ has also a homology class in the other spaces

$$[z]_U = i_{*n}[z]_A, \qquad [z]_V = j_{*n}[z]_A, \qquad [z]_X = u_{*n}[z]_U = v_{*n}[z]_V,$$

while a cycle $z$ of $U$ or $V$ also has a homology class $[z]_X \in H_n(X)$.

*An exercise.* (a) The square $ui = vj$ in (2.82) is a pullback and a pushout.

### 2.3.7 Theorem (The Mayer–Vietoris sequence)

*Let $X$ be a topological space, and $U$, $V$ subspaces of $X$ such that $X = \operatorname{int} U \cup \operatorname{int} V$, and $A = U \cap V$.*

*There is an exact sequence of singular homology groups*

$$\ldots \to H_n(A) \xrightarrow{h_n} H_n(U) \oplus H_n(V) \xrightarrow{k_n} H_n(X) \xrightarrow{D_n} H_{n-1}(A) \to \ldots \tag{2.83}$$

*called the* exact sequence of Mayer–Vietoris, *or* MV-sequence, *of the space $X$ with respect to the generalised open cover $(U, V)$.*

*With the notation of 2.3.6, these homomorphisms are computed as:*

$$\begin{aligned} h_n[z]_A &= ([z]_U, [z]_V), \\ k_n([z]_U, [z']_V) &= [z]_X - [z']_X = [z - z']_X, \\ D_n[z]_X &= [\partial_n c]_A = [\partial_n c']_A, \end{aligned} \tag{2.84}$$

*where $z \in Z_n(X)$ is expressed as $z = c - c'$, with $c \in C_n(U)$ and $c' \in C_n(V)$.*

*The sequence is natural for continuous mappings* $f\colon X \to X'$, *where*

$$X' = \operatorname{int} U' \cup \operatorname{int} V', \qquad f(U) \subset U', \qquad f(V) \subset V'.$$

*Notes.* The definition of the connecting morphism $D_n$ will be shown to make sense in the proof; but we can already note that, in the given conditions, $\partial_n c = \partial_n c'$ is a cycle in $A$ and need not be a boundary there: $c$ and $c'$ are chains in $U$ and $V$, respectively, and need not be chains of $A$.

Typically, this sequence will be used to compute the homology of $X$, knowing the homology of the other spaces. The reader can keep in mind a simple case, where $X$ is the circle covered by two open (non-total) arcs $U, V$ which meet at $A$, the union of two disjoint arcs. This will be our first application, in 2.4.1.

*Proof* (a) We use the notation of 2.3.6 for the inclusion maps $i, j, u, v$

$$\begin{array}{ccc} A & \xrightarrow{i} & U \\ {\scriptstyle j}\downarrow & & \downarrow{\scriptstyle u} \\ V & \xrightarrow[v]{} & X \end{array} \qquad\qquad \begin{array}{ccc} C_+(A) & \xrightarrow{i_{\sharp n}} & C_+(U) \\ {\scriptstyle j_{\sharp n}}\downarrow & & \downarrow{\scriptstyle u_{\sharp n}} \\ C_+(V) & \xrightarrow[v_{\sharp n}]{} & C_+(X) \end{array} \qquad (2.85)$$

and view the associated monomorphisms of chain complexes as inclusions: a chain of $A$ is identified with its image in $C_+(U)$, and so on (as in 2.3.4(i)).

(b) We form a short exact sequence of chain complexes, where $\mathcal{U}$ is the generalised open cover $(U, V)$ of $X$, and the direct sum of chain complexes is defined componentwise (see 2.1.8)

$$0 \longrightarrow C_+(A) \xrightarrow{h'} C_+(U) \oplus C_+(V) \xrightarrow{k'} C_+(X;\mathcal{U}) \longrightarrow 0 \qquad (2.86)$$

$$h'_n(c) = (i_{\sharp n}(c), j_{\sharp n}(c)) = (c, c), \quad k'_n(c, c') = u_{\sharp n}(c) - v_{\sharp n}(c) = c - c'.$$

*The exactness of this sequence amounts to saying that the right square (2.85), modified replacing $C_+(X)$ by $C_+(X;\mathcal{U})$, is both a pullback and a push-out, in the abelian category $\mathsf{Ch}_+\mathsf{Ab}$: cf. [G5], Exercises 6.4.8(b), (b*).*

In fact:

- every component $h'_n$ is obviously mono,

- every component $k'_n$ is epi: a small cube $a\colon \mathbb{I}^n \to X$ of $C_n(X;\mathcal{U})$ has image contained in $U$ or $V$, and can be obtained as $k'_n(a, 0)$ or $k'_n(0, -a)$ (in both ways if $\operatorname{Im} a \subset A$),

- plainly, $k'_n h'_n = 0$ (for all $n$).

The last point, $\operatorname{Ker} k' \subset \operatorname{Im} h'$, is slightly more complex. Take a pair $(c, c') \in \operatorname{Ker} k'_n$: then $c$ and $c'$ are the same chain of $X$, and their support is contained in $U \cap V = A$; in other words, $c = c' \in C_n(A)$ and $(c, c') = h'_n(c)$.

(c) Applying the Subdivision Theorem 2.3.5, the inclusion of $C_+(X;\mathcal{U})$ into $C_+(X)$ induces isomorphism in homology, in every degree. Taking a class

$[z] \in H_n(X)$, we can always replace the cycle $z$ with a homologous cycle $c - c' \in C_n(X;\mathcal{U})$, where $c \in C_n(U)$ and $c' \in C_n(V)$; neither of them has to be a cycle.

From now on, we shall identify the homology groups of $C_+(X;\mathcal{U})$ and $C_+(X)$.

(d) We also identify the group $H_n(C_+(U) \oplus C_+(V))$ with $H_n(U) \oplus H_n(V)$, using the canonical isomorphism (2.29).

Applying Theorem 2.3.3 to the sequence (2.86), we get the exact sequence of chain homology

$$... \to H_n(A) \xrightarrow{h_n} H_n(U) \oplus H_n(V) \xrightarrow{k_n} H_n(X) \xrightarrow{D_n} H_{n-1}(A) \to ... \quad (2.87)$$

where the homomorphisms $h_n$ and $k_n$, induced in chain homology by $h'$ and $k'$, act as claimed in the statement.

(e) We still have to compute $D_n[z]_X$ for a cycle $z \in Z_n(X)$; as remarked above, we can assume that the representative $z$ can be rewritten as $c - c'$, with $c \in C_n(U)$ and $c' \in C_n(V)$. Following the definition in diagram (2.74)

$$\begin{array}{ccccccc} & & (c,c') & \longmapsto & z & & (n) \\ & & \downarrow & & \downarrow & & \\ \partial c & \mapsto & (\partial c, \partial c') & \mapsto & 0 & & (n-1) \end{array} \quad (2.88)$$

we lift $z$ to the pair $(c,c') \in C_n(U) \oplus C_n(V)$. The relation

$$\partial_n c - \partial_n c' = k'_{n-1}\partial_n(c,c') = 0 \in C_{n-1}(X)$$

means that $\partial_n c = \partial_n c' \in C_{n-1}(A)$, and

$$h'_{n-1}(\partial_n c) = (\partial_n c, \partial_n c) = (\partial_n c, \partial_n c'),$$

proving that $D_n[z]_X = [\partial_n c] = [\partial_n c']$.

(f) Finally, given a generalised open cover $\mathcal{U}' = (U', V')$ of $X'$, and a map respecting the given covers:

$$f\colon X \to X', \qquad f(U) \subset U', \qquad f(V) \subset V',$$

we also have $f(A) \subset A' = U' \cap V'$. The map $f$ (with its restrictions, still written as $f$) gives a morphism of short exact sequences

$$\begin{array}{ccccc} C_+(A) & \rightarrowtail & C_+(U) \oplus C_+(V) & \twoheadrightarrow & C_+(X;\mathcal{U}) \\ \downarrow f_\sharp & & \downarrow f_\sharp & & \downarrow f_\sharp \\ C_+(A') & \rightarrowtail & C_+(U') \oplus C_+(V') & \twoheadrightarrow & C_+(X';\mathcal{U}') \end{array} \quad (2.89)$$

and this gives a natural transformation of the corresponding homology sequences. □

***2.3.8 Exercises*** (The exact sequence of chain homology, continued) For a short exact sequence of chain complexes $A \rightarrowtail B \twoheadrightarrow C$, we have formed the homology sequence (2.73). There are still two points to prove.

(a) The homology sequence is exact at $H_n(B)$, for each $n \geqslant 0$.

(b) The homology sequence is exact at $H_n(C)$, for each $n \geqslant 0$.

### 2.3.9 *Proof of the Subdivision Theorem

The general idea is to subdivide cubes, as many times as needed, to replace an $n$-chain of the space $X$ by a $\mathcal{U}$-small chain; this is based on the Lebesgue Number Lemma 1.4.9, applied to the compact metric space $\mathbb{I}^n$.

We give a nearly complete account of the proof, which has many interesting aspects.

(a) First we build the *subdivision operator*, a natural morphism of chain complexes

$$\begin{aligned} &\mathrm{Sd}\colon C_+(X) \to C_+(X), \quad \mathrm{Sd}_n(a) = \Sigma_v\, au_v \quad (v \in \{0,1\}^n), \\ &u_v\colon \mathbb{I}^n \to \mathbb{I}^n, \qquad u_v(t) = (t+v)/2 \qquad (t \in \mathbb{I}^n), \end{aligned} \tag{2.90}$$

which subdivides any $n$-cube into a chain of $2^n$ cubes, indexed by the vertices $v \in \{0,1\}^n$ of $\mathbb{I}^n$, as in the following examples in dimension 2

(2.91)

$$u_{00}(t_1,t_2) = (t_1/2, t_2/2), \qquad u_{01}(t_1,t_2) = (t_1/2, (t_2+1)/2).$$

The fact that Sd commutes with the differential is geometrically evident, looking at the previous picture: in $\partial_n\, \mathrm{Sd}_n(a)$ the 'internal' faces cancel while the 'external' ones give $\mathrm{Sd}_{n-1}\, \partial_n(a)$. (One can find the details in [Mas3], Section II.7.)

Repeating subdivision, $\mathrm{Sd}^k(a)$ is a chain of cubes of $X$, each of them parametrised on a 'subcube' of $\mathbb{I}^n$, of edge $2^{-k}$.

(b) This morphism Sd is homotopic to $\mathrm{id}\, C_+(X)$, by a chain homotopy $\varphi = (\varphi_n)$

$$\varphi_n\colon C_n(X) \to C_{n+1}(X), \qquad \mathrm{Sd}_n - \mathrm{id}_n = \partial_{n+1}\varphi_n + \varphi_{n-1}\partial_n, \tag{2.92}$$

that preserves small chains:

$$\varphi_n(C_n(X;\mathcal{U})) \subset C_{n+1}(X;\mathcal{U}). \tag{2.93}$$

We shall say something of this construction in the last point (e).

(c) We now prove that the homomorphism $j_{*n}\colon H_n(X;\mathcal{U}) \to H_n(X)$ is surjective.

For every cube $a\colon \mathbb{I}^n \to X$, we consider the following open cover of $\mathbb{I}^n$

$$V_i = a^{-1}(\operatorname{int} U_i) \qquad (i \in I). \tag{2.94}$$

Applying the Lebesgue Number Lemma on open covers of compact metric spaces, there is some $k \in \mathbb{N}$ such that any 'subcube' $K$ of $\mathbb{I}^n$ with edge $2^{-k}$ is contained in some $V_{i_K}$. It follows that

$$a(K) \subset a(V_{i_K}) \subset \operatorname{int} U_{i_K} \subset U_{i_K}, \quad \mathrm{Sd}^k(a) \in C_n(X;\mathcal{U}). \tag{2.95}$$

We take a cycle $z \in C_n(X)$. For every cube $a\colon \mathbb{I}^n \to X$ which appears in $z$, we can proceed as above. There is thus some $k \in \mathbb{N}$ such that $z' = \mathrm{Sd}^k(z) \in C_n(X;\mathcal{U})$. The composed chain homotopy $\psi\colon \mathrm{Sd}^k \simeq \mathrm{id} : C_+(X) \to C_+(X)$ gives in $C_n(X)$

$$z - z' = \partial_{n+1}\psi_n(z) + \psi_{n-1}\partial_n(z) = \partial_{n+1}\psi_n(z), \tag{2.96}$$

so that $[z] = j_{*n}[z']$.

(d) We prove that the induced homomorphism $j_{*n}\colon H_n(X;\mathcal{U}) \to H_n(X)$ is injective.

We take a cycle $z \in C_n(X;\mathcal{U})$ which vanishes in $H_n(X)$:

$$z = \partial_{n+1}c, \quad \text{for some chain } c \in C_{n+1}(X), \tag{2.97}$$

and we prove that $z$ is also a boundary of small chains.

As above, there is some $k \in \mathbb{N}$ such that $c' = \mathrm{Sd}^k(c) \in C_{n+1}(X;\mathcal{U})$.

The composed chain homotopy $\psi\colon \mathrm{Sd}^k \simeq \mathrm{id} : C_+(X) \to C_+(X)$ gives

$$c - c' = \partial_{n+2}\psi_{n+1}(c) + \psi_n\partial_{n+1}(c) = \partial_{n+2}\psi_{n+1}(c) + \psi_n(z). \tag{2.98}$$

We have proved that $z = \partial_{n+1}c = \partial_{n+1}c' + \partial_{n+1}\psi_n(z)$. This is a boundary in $C_n(X;\mathcal{U})$, because $\varphi$ sends $C_n(X;\mathcal{U})$ to $C_{n+1}(X;\mathcal{U})$, by (2.93), and also the composed homotopy $\psi$ does.

(e) Finally, we describe the chain homotopy $\varphi\colon \mathrm{id} \simeq \mathrm{Sd}\colon C_+(X) \to C_+(X)$, as constructed in [Mas3], Section II.7.

The homomorphism $\varphi_n$ is obtained by a family of maps $\eta_v\colon \mathbb{I}^{n+1} \to \mathbb{I}^n$ indexed by the vertices $v = (v_1, ..., v_n) \in \{0,1\}^n$ of $\mathbb{I}^n$

$$\varphi_n\colon C_n(X) \to C_{n+1}(X), \qquad \varphi_n(a) = (-1)^{n+1} \textstyle\sum_v a\eta_v. \tag{2.99}$$

The mappings $\eta_v$ are defined as follows.

- For $n = 0$, $\eta\colon \mathbb{I} \to \{*\}$ is the degeneracy, and $\varphi_0 = 0\colon C_0(X) \to C_1(X)$.

- For $n = 1$: $\eta_0(t_1, t_2) = t_1/(2 - t_2)$,

$$\eta_1(t_1, t_2) = \begin{cases} (t_1 + 1)/(2 - t_2), & \text{if } t_1 + t_2 \leqslant 1, \\ 1, & \text{if } t_1 + t_2 \geqslant 1. \end{cases}$$

- For $n \geqslant 2$:

$$\eta_v(t_1, t_2, ..., t_{n+1}) = (\eta_{v_1}(t_1, t_{n+1}), \eta_{v_2}(t_2, t_{n+1}), ..., \eta_{v_n}(t_n, t_{n+1})).$$

Condition (2.93) on the preservation of small chains is plainly satisfied, because the support of the chain $\varphi_n(a)$ in (2.99) is contained in $\mathrm{Im}\, a$. We still have to verify the identity (2.92), for $n \geqslant 0$.

For $n = 1$, we check that $\partial_2\varphi_1(a) = \mathrm{Sd}_1(a) - a$, on every 1-cube $a\colon \mathbb{I} \to X$

$$\begin{aligned} \partial_2\varphi_1(a) &= \partial_2(a\eta_0 + a\eta_1) \\ &= a\eta_0(-\delta_1^- + \delta_1^+ + \delta_2^- - \delta_2^+) + a\eta_1(-\delta_1^- + \delta_1^+ + \delta_2^- - \delta_2^+) \\ &= a\eta_0\delta_2^- - a\eta_0\delta_2^+ + a\eta_1\delta_2^- = au_0 - a + au_1 = \mathrm{Sd}_1(a) - a, \end{aligned}$$

taking out three degenerate cubes and cancelling two terms. This follows from the following computations of the faces $\eta_j\delta_i^\alpha\colon \mathbb{I} \to \mathbb{I}$

- $\eta_0\delta_1^-(t) = \eta_0(0, t) = 0$ is constant, and $a\eta_0\delta_1^-(t)$ is a degenerate 1-cube,
- $\eta_0\delta_1^+(t) = 1/(2 - t) = \eta_1\delta_1^-(t)$, so that $a\eta_0\delta_1^+ - a\eta_1\delta_1^- = 0$,
- $\eta_0\delta_2^-(t) = \eta_0(t, 0) = t/2 = u_0(t)$,
- $\eta_0\delta_2^+(t) = t$,
- $\eta_1\delta_1^+(t) = \eta_1(1, t) = 1$ and $\eta_1\delta_2^+(t) = 1$ are constant,
- $\eta_1\delta_2^-(t) = \eta_1(t, 0) = (t + 1)/2 = u_1(t)$.

The general case is similarly proved. An interested reader can fill the missing points, with the help of [Mas3], p. 35. □

## 2.4 The homology of the spheres

We compute the homology groups of the spheres $\mathbb{S}^n$, anticipated in (1.9), and explore several consequences.

The reader should find it useful to work on the exercises and applications. Easy arguments about homotopy equivalence and deformation retracts will now be left as understood.

### 2.4.1 The homology of the circle

We already know, from the basic results of 2.2.5, that the 0-dimensional sphere $\mathbb{S}^0 = \{-1, 1\}$ has homology

$$H_0(\mathbb{S}^0) \cong \mathbb{Z} \oplus \mathbb{Z}, \qquad H_k(\mathbb{S}^0) = 0 \quad (k > 0). \tag{2.100}$$

More precisely, $H_0(\mathbb{S}^0)$ is the free abelian group generated by the two 0-cubes, identified with their homology classes

$$p, q \colon \mathbb{I}^0 \to \mathbb{S}^0, \qquad p(*) = 1, \quad q(*) = -1. \tag{2.101}$$

A general element of $H_0(\mathbb{S}^0)$ is a linear combination $\lambda p + \mu q$, with $\lambda, \mu \in \mathbb{Z}$. (Of course, it is not convenient to write the points $p, q$ as $\pm 1$.)

Now we want to compute the homology of the circle $\mathbb{S}^1$. We choose an open cover formed by two open non-total arcs $U, V$, e.g. the complement of the South pole $\underline{s} = (0, -1)$ and the complement of the North pole $\underline{n} = (0, 1)$

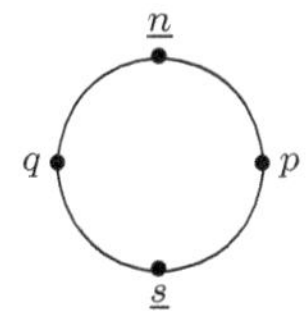


$$U = \mathbb{S}^1 \setminus \{\underline{s}\} \simeq \{*\}, \qquad V = \mathbb{S}^1 \setminus \{\underline{n}\} \simeq \{*\}. \tag{2.102}$$

$U$ and $V$ are contractible spaces; their intersection $A = \mathbb{S}^1 \setminus \{\underline{s}, \underline{n}\}$ admits $\mathbb{S}^0 = \{p, q\}$ as a deformation retract.

*Exercises and complements.* The following exercises compute the groups $H_k(\mathbb{S}^1)$, using the MV-sequence of the space $X = \mathbb{S}^1$, with respect to the open cover $(U, V)$. A solution can be found below.

(a) $H_0(\mathbb{S}^1) \cong \mathbb{Z}, \qquad H_k(\mathbb{S}^1) = 0,$ for $k > 1$.

(b) $H_1(\mathbb{S}^1) \cong \mathbb{Z}$.

These results say that the circle is path connected, has a hole of dimension 1 and no holes of higher dimension.

(c) As an infinite cyclic group, $H_1(\mathbb{S}^1)$ has a generator $\zeta = [z]$ of infinite order. A cycle $z \in Z_1(\mathbb{S}^1)$ satisfying this condition can be obtained from the definition of the connecting morphism $D_1$, in (2.84).

(d) One can find a simpler expression of the same generator $\zeta = [a]$, using the standard counterclockwise loop of the circle: $a(t) = e^{2\pi i t}$, for $t \in \mathbb{I}$.

(e) We have now a clear geometrical meaning for the elements of $H_1(\mathbb{S}^1)$. A homology class is expressed as $k\zeta = [ka]$, for a unique $k \in \mathbb{Z}$. But it can also be expressed as $k\zeta = [a_k]$, using the loop

$$a_k \colon \mathbb{I} \to \mathbb{S}^1, \qquad a_k(t) = e^{2k\pi i t}, \tag{2.103}$$

that turns $k$ times around the origin.

It is understood that $-n$ times means $n$ times clockwise, for $n > 0$. Adding homology classes amounts to adding the number of turns (in $\mathbb{Z}$).

(f) It is also evident that choosing a generator of $H_1(\mathbb{S}^1)$, either $\zeta$ or $-\zeta$, amounts to choosing an orientation of the circle. We shall see how the orientation of manifolds can be based on homology, in Section 4.3.

*Solutions.* (a) For $H_0(\mathbb{S}^1)$ it is sufficient to apply Exercise 2.2.5(b), on the 0-th homology group of a path-connected space. (The MV-sequence would also give this result for $H_0(\mathbb{S}^1)$, as the cokernel of the homomorphism $h_0$ computed below, in (2.106).)

For $H_k(\mathbb{S}^1)$, with $k > 1$, we use a portion of the sequence of Mayer–Vietoris

$$H_k(U) \oplus H_k(V) \to H_k(\mathbb{S}^1) \to H_{k-1}(A). \tag{2.104}$$

As the left and the right term are 0, the central term is also (by 2.1.3(b)).

(b) For $H_1(\mathbb{S}^1)$ we use another portion of the sequence

$$H_1(U) \oplus H_1(V) \xrightarrow{k_1} H_1(\mathbb{S}^1) \xrightarrow{D_1} H_0(A) \xrightarrow{h_0} H_0(U) \oplus H_0(V). \tag{2.105}$$

Here $H_1(U) \oplus H_1(V) = 0$, whence $D_1$ is mono, and induces an isomorphism

$$D_1'\colon H_1(\mathbb{S}^1) \to \operatorname{Im} D_1 = \operatorname{Ker} h_0.$$

The homomorphism $h_0$ is computed as follows, on each point $x \in A$:

$$h_0\colon H_0(A) \to H_0(U) \oplus H_0(V), \qquad h_0[x]_A = ([x]_U, [x]_V).$$

Now $H_0(A) \cong H_0(\mathbb{S}^0)$ is the free abelian group on two generators, $[p]_A$ and $[q]_A$, while $H_0(U)$ is a free abelian group on a single generator, $[p]_U = [q]_U$; similarly for $V$. Thus:

$$h_0(\lambda[p]_A + \mu[q]_A) = ((\lambda + \mu)[p]_U, (\lambda + \mu)[p]_V). \tag{2.106}$$

The outcome is 0 if and only if $\lambda + \mu = 0$. Finally the group

$$H_1(\mathbb{S}^1) \cong \operatorname{Ker} h_0 = \{\lambda([p]_A - [q]_A) \in H_0(A) \mid \lambda \in \mathbb{Z}\},$$

is isomorphic to the cyclic subgroup of $H_0(A)$ generated by the chain $[p]_A - [q]_A$. This is non-zero (because the 0-cubes $p, q$ belong to different components of $A$), and infinite (because $H_0(A)$ is torsion-free).

(c) By the previous point, $H_1(\mathbb{S}^1)$ is generated by the homology class $[z]$ such that $D_1[z]$ is the generator $[q]_A - [p]_A$ of $\operatorname{Ker} h_0$. To compute $D_1[z]$ we write the 1-cycle $z$ as

$$z = c + c', \quad c \in C_1(U), \quad c' \in C_1(V), \quad \partial_1 c + \partial_1 c' = 0.$$

Here this form is more convenient than the form $c - c'$ used in (2.84); of course, it gives $D_1[c + c'] = [\partial_1 c]_A = [-\partial_1 c']_A$.

Taking two semicircles $c, c' \colon \mathbb{I} \to \mathbb{S}^1$, with image in $U, V$, respectively

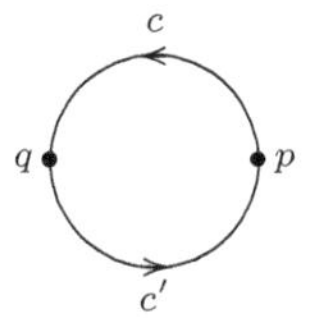


(2.107)

$$c(t) = (\cos \pi t, \sin \pi t), \qquad \partial_1 c = q - p,$$
$$c'(t) = (\cos(\pi(1+t)), \sin(\pi(1+t)), \qquad \partial_1 c' = p - q,$$

we do get a cycle $c + c'$ with

$$D_1[c + c'] = [\partial_1 c]_A = [q]_A - [p]_A,$$

proving that $\zeta = [c + c']$ is a generator of the infinite cyclic group $H_1(\mathbb{S}^1)$.

(d) By Exercise 2.2.8(e) we can replace $c + c'$ with the concatenation $a = c * c'$, which is the standard counterclockwise loop of the circle $a(t) = e^{2\pi i t}$.

(e) We want to show that the loop $a_k(t) = e^{2k\pi i t}$ represents the general homology class $[a_k] = k[a]$, for $k \in \mathbb{Z}$.

The case $k = 0$ is obvious, because $a_0$ is a degenerate loop.

If $k > 0$, $a_k$ is the regular concatenation $a * a * ... * a$ (see (1.76)) of $k$ loops equal to $a$. This loop is 2-homotopic to the iterated concatenation $(...((a * a) * a)...) * a$ (by Exercise 1.4.7(d)), which is homologous to the chain $a + a + ... + a = ka$ (by Exercise 2.2.8(e)). Therefore

$$[a_k] = [a * a * ... * a] = [(...((a * a) * a)...) * a] = [ka] = k[a].$$

The case $h = -1$ gives the reversed loop $a_{-1} = a^\sharp$, whose homology class is indeed $-[a]$, by Exercise 2.2.8(f). The case $k < 0$ is a consequence of the previous ones.

***2.4.2 Theorem*** (The homology of the spheres)

*For $n \geqslant 1$, the singular homology of $\mathbb{S}^n$ is:*

$$H_0(\mathbb{S}^n) \cong H_n(\mathbb{S}^n) \cong \mathbb{Z}, \qquad H_k(\mathbb{S}^n) = 0 \qquad (0 \neq k \neq n). \tag{2.108}$$

*Moreover, for $n \geqslant 2$ there is an isomorphism*

$$D'_n \colon H_n(\mathbb{S}^n) \to H_{n-1}(\mathbb{S}^{n-1}), \tag{2.109}$$

*natural for maps $f \colon \mathbb{S}^n \to \mathbb{S}^n$ satisfying the conditions*

$$f(\mathbb{S}^{n-1}) \subset \mathbb{S}^{n-1}, \qquad f^{-1}\{\underline{s}\} = \{\underline{s}\}, \quad f^{-1}\{\underline{n}\} = \{\underline{n}\}. \tag{2.110}$$

*Notes.* $\mathbb{S}^{n-1}$ is embedded as the equator of $\mathbb{S}^n$, in the hyperplane $t_{n+1} = 0$, as in (1.47). The points $\underline{s} = (0, ..., 0, -1)$ and $\underline{n} = (0, ..., 0, 1)$ are the poles of $\mathbb{S}^n$. The 'naturality' of $D'_n$ is made clear in diagram (2.113) of the proof. A generator of $H_n(\mathbb{S}^n)$ will be constructed in 2.5.5.

*Proof* We know that $H_0(\mathbb{S}^n) \cong \mathbb{Z}$, and prove the other results of (2.108) by induction. The initial step, for $n = 1$, has been proved in 2.4.1. We suppose that (2.108) is true for $n-1$, and prove that it also holds for $n \geqslant 2$.

We choose an open cover of $\mathbb{S}^n$ formed by two open subsets $U, V$, the complements of the two poles; the figure represents the sphere $\mathbb{S}^2$, with the equator $\mathbb{S}^1$ in the plane $t_3 = 0$

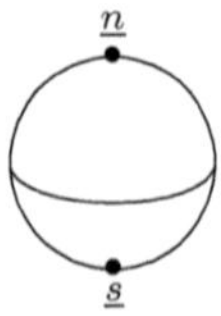


$$U = \mathbb{S}^n \setminus \{\underline{s}\} \simeq \{*\},$$
$$V = \mathbb{S}^n \setminus \{\underline{n}\} \simeq \{*\}, \qquad (2.111)$$
$$A = U \cap V = \mathbb{S}^n \setminus \{\underline{s}, \underline{n}\} \simeq \mathbb{S}^{n-1}.$$

$U$ and $V$ are contractible spaces (they are homeomorphic to $\mathbb{R}^n$, by stereographic projection). The inclusion $u \colon \mathbb{S}^{n-1} \to A$ is a deformation retract, with retraction

$$p \colon A \to \mathbb{S}^{n-1}, \qquad p(t_1, ..., t_{n+1}) = N(t_1, ..., t_n),$$

where $N \colon \mathbb{R}^n \setminus \{0\} \to \mathbb{S}^{n-1}$ is the normalisation map: indeed $t_{n+1} \neq \pm 1$ in $A$, and $(t_1, ..., t_n) \neq 0$.

We use the sequence of Mayer–Vietoris for these data.

(a) For $k > 2$, $k \neq n$, the portion of the sequence around $H_k(\mathbb{S}^n)$

$$H_k(U) \oplus H_k(V) \;\to\; H_k(\mathbb{S}^n) \;\to\; H_{k-1}(A)$$

has trivial groups in the left and the right term, by the inductive assumption and $A \simeq \mathbb{S}^{n-1}$. Thus the central term is also the trivial group.

(b) For $H_1(\mathbb{S}^n)$ we use the portion

$$0 \;\to\; H_1(\mathbb{S}^n) \xrightarrow{D_1} H_0(A) \xrightarrow{h_0} H_0(U) \oplus H_0(V)$$

where $h_0$ is mono (by 3.1.2(c)) and $D_1 = 0$, so that $H_1(\mathbb{S}^n) = 0$.

(c) For $H_n(\mathbb{S}^n)$ we use the portion

$$H_nU \oplus H_nV \xrightarrow{k_n} H_n\mathbb{S}^n \xrightarrow{D_n} H_{n-1}A \xrightarrow{h_{n-1}} H_{n-1}U \oplus H_{n-1}V.$$

The first and the last term are again trivial groups, and $D_n$ is an isomorphism. Composing the latter with the isomorphism $p_{*n-1} = (u_{*n-1})^{-1}$: $H_{n-1}A \to H_{n-1}(\mathbb{S}^{n-1})$, we have the isomorphism required in (2.109)

$$D'_n = p_{*n-1}D_n = (u_{*n-1})^{-1}D_n \colon H_n(\mathbb{S}^n) \to H_{n-1}(\mathbb{S}^{n-1}). \qquad (2.112)$$

By the inductive assumption, $H_n(\mathbb{S}^n) \cong \mathbb{Z}$.

(d) Finally we prove the naturality of $D'_n$ with respect to a map $f\colon \mathbb{S}^n \to \mathbb{S}^n$ satisfying the conditions (2.110), which imply that $f(U) \subset U$, $f(V) \subset V$ and $f(A) \subset A$.

More precisely, letting $f'\colon A \to A$ and $f''\colon \mathbb{S}^{n-1} \to \mathbb{S}^{n-1}$ be the restrictions of $f$, we prove the commutativity of the square

$$\begin{array}{ccc} H_n(\mathbb{S}^n) & \xrightarrow{D'_n} & H_{n-1}(\mathbb{S}^{n-1}) \\ {\scriptstyle f_{*n}}\downarrow & & \downarrow{\scriptstyle f''_{*n-1}} \\ H_n(\mathbb{S}^n) & \xrightarrow[D'_n]{} & H_{n-1}(\mathbb{S}^{n-1}) \end{array} \qquad (2.113)$$

working on the following diagram

$$\begin{array}{ccccc} H_n(\mathbb{S}^n) & \xrightarrow{D_n} & H_{n-1}(A) & \xleftarrow{u_{*n-1}} & H_{n-1}(\mathbb{S}^{n-1}) \\ {\scriptstyle f_{*n}}\downarrow & & \downarrow{\scriptstyle f'_{*n-1}} & & \downarrow{\scriptstyle f''_{*n-1}} \\ H_n(\mathbb{S}^n) & \xrightarrow[D_n]{} & H_{n-1}(A) & \xleftarrow{u_{*n-1}} & H_{n-1}(\mathbb{S}^{n-1}) \end{array} \qquad (2.114)$$

The left square commutes, by the naturality of the MV-sequence (in Theorem 2.3.7). The right square commutes, because we are applying the functor $H_{n-1}$ to the commutative square $f'u = uf''$, in Top. Reversing the isomorphism $u_{*n-1}$ in the two horizontal arrows, the new square still commutes, and the commutativity of (2.113) is proved.

Let us note that the commutativity of this new square comes from the commutativity of the old one, in (2.114): the maps $f'$ and $f''$ need not agree with the retraction $p\colon A \to \mathbb{S}^{n-1}$, or any retraction of $u$. □

### 2.4.3 Applications, I

The rest of this chapter investigates applications of the homology groups of the spheres, computed above. We begin by reviewing the main consequences of these results, already surveyed in 1.1.5 and 1.1.6, or obvious.

(a) If the spheres $\mathbb{S}^m$ and $\mathbb{S}^n$ are homotopy equivalent, then $m = n$.

(b) (Theorem of Topological Dimension) *If the spaces $\mathbb{R}^m$ and $\mathbb{R}^n$ are homeomorphic, then $m = n$.*

(c) The sphere $\mathbb{S}^n$ is not a retract of any contractible space.

(d) The Euler–Poincaré characteristic of $\mathbb{S}^n$ (defined in (2.63)) is

$$\chi(\mathbb{S}^n) = 1 + (-1)^n \in \mathbb{Z} \qquad (n \geqslant 0), \qquad (2.115)$$

that is, either 0 (for $n$ odd) or 2 (for $n$ even).

The parity of the dimension of the spheres will also turn up in Theorems 2.5.3 and 2.5.4.

### *2.4.4 Exercises and complements* (Applications, II)

The following results are important. The solutions are not straightforward, except (b) and (d), and can be found below.

(a) (The Brouwer Fixed-point Theorem) *Every endomap* $f\colon \mathbb{D}^n \to \mathbb{D}^n$ *of the compact disc has at least a fixed point.*

(b) (Surjectivity Lemma) *Every map* $f\colon \mathbb{S}^n \to \mathbb{S}^n$ *homotopic to the identity is surjective.* (This point will be better analysed in 2.5.2.)

(c) (Intermediate Value Theorem on the Cube) *Let* $f\colon \mathbb{I}^n \to \mathbb{I}^n$ *be a continuous mapping which sends each face to itself*

$$f(\delta_i^\alpha \mathbb{I}^{n-1}) \subset \delta_i^\alpha \mathbb{I}^{n-1} \qquad (\alpha = \pm,\ i = 1, ..., n).$$

*Then* $f$ *is surjective and sends each face* onto *itself.*

(d) For $n = 2$ this statement can be concretely visualised. We have a square cloth, and a square frame of lesser or equal dimension; their edges are numbered (say) clockwise. We arrange the cloth on the frame (this represents the map $f\colon \mathbb{I}^2 \to \mathbb{I}^2$), so that each edge is arranged on the corresponding edge of the frame; we can make any folds we want in the cloth ($f$ is not assumed to be injective).

However we proceed, each vertex of the cloth (belonging to two edges) goes to the corresponding vertex of the frame; therefore each edge of the cloth covers the corresponding edge of the frame, and the cloth itself covers the whole frame.

(e) (Intermediate Value Theorem on the Disc) *Let* $f\colon \mathbb{D}^n \to \mathbb{D}^n$ *be a continuous mapping which sends the border* $\mathbb{S}^{n-1}$ *to itself, so that the restriction* $g\colon \mathbb{S}^{n-1} \to \mathbb{S}^{n-1}$ *is homotopic to the identity. Then* $f$ *is surjective.*

*Solutions.* (a) We suppose, for a contradiction, that $f$ has no fixed point, and construct a retraction $p\colon \mathbb{D}^n \to \mathbb{S}^{n-1}$ of $\mathbb{S}^{n-1}$ which cannot exist.

For every $x \in \mathbb{D}^n$ we are assuming that $f(x) \neq x$, and define $p(x)$ as the intersection of $\mathbb{S}^{n-1}$ with the half-line from $f(x)$ to $x$

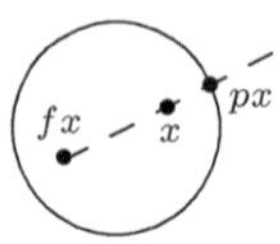


(2.116)

so that $p(x) = x$ on the sphere. The continuity of $p$ is geometrically obvious. It can be verified analytically, determining the coordinates of $p(x)$ in $\mathbb{R}^n$ – an exercise of Analytic Geometry.

(b) If $n = 0$, $\mathbb{S}^0$ is discrete and two homotopic maps $\mathbb{S}^0 \to \mathbb{S}^0$ coincide.

For $n > 0$, suppose that $f(\mathbb{S}^n) \subset \mathbb{S}^n \setminus \{x_0\}$, a contractible space (by stereographic projection); then $f$ is homotopic to a constant map $g$ (by 1.1.3(c)), and $f_{*n} = g_{*n} = 0\colon H_n(\mathbb{S}^n) \to H_n(\mathbb{S}^n)$, as the map $g$ takes any $n$-cube to a degenerate one. But $H_n(\mathbb{S}^n) \neq 0$, and $f$ cannot be homotopic to $\mathrm{id}\,\mathbb{S}^n$.

(c) This statement is trivial for $n = 0$, and amounts to the classical Intermediate Value Theorem for $n = 1$.

We suppose that the statement holds for $n-1$, and prove it for $n \geqslant 1$. The inductive hypothesis can be applied to each face (because each face of $\mathbb{I}^{n-1}$ is the intersection of two faces of $\mathbb{I}^n$); it follows that $f(\delta_i^\alpha \mathbb{I}^{n-1}) = \delta_i^\alpha \mathbb{I}^{n-1}$ and $f(\underline{\partial}\mathbb{I}^n) = \underline{\partial}\mathbb{I}^n$: we register that $f$ covers the border of $\mathbb{I}^n$, and we have to prove that it also covers the interior points.

Now, $f\colon \mathbb{I}^n \to \mathbb{I}^n$ is homotopic to the identity by an affine homotopy (see (1.3)) $h\colon \mathbb{I}^n \times \mathbb{I} \to \mathbb{I}$; moreover $h$ moves inside each face of $\mathbb{I}^n$ (because $f$ and $\mathrm{id}\,\mathbb{I}^n$ both do, and each face is convex). Globally, $h$ moves inside the border of $\mathbb{I}^n$, in the sense that $h(\underline{\partial}\mathbb{I}^n \times \mathbb{I}) \subset \underline{\partial}\mathbb{I}^n$.

Collapsing $\underline{\partial}\mathbb{I}^n$ to a point, $f$ induces a map $g\colon \mathbb{S}^n \to \mathbb{S}^n$, and $h$ induces a homotopy $h'\colon g \simeq \mathrm{id}\,\mathbb{S}^n$

$$\begin{array}{ccc} \mathbb{I}^n & \xrightarrow{f} & \mathbb{I}^n \\ {\scriptstyle p}\downarrow & & \downarrow{\scriptstyle p} \\ \mathbb{S}^n & \xrightarrow[g]{} & \mathbb{S}^n \end{array} \qquad\qquad \begin{array}{ccc} \mathbb{I}^n \times \mathbb{I} & \xrightarrow{h} & \mathbb{I}^n \times \mathbb{I} \\ {\scriptstyle p\times\mathbb{I}}\downarrow & & \downarrow{\scriptstyle p\times\mathbb{I}} \\ \mathbb{S}^n \times \mathbb{I} & \xrightarrow[h']{} & \mathbb{S}^n \times \mathbb{I} \end{array} \qquad (2.117)$$

In fact, the cartesian product $- \times \mathbb{I}$ (by a locally compact space) preserves topological quotients, as proved in 8.3.3. (Alternatively, we can remark that $\mathbb{I}^n \times \mathbb{I}$ is compact and $\mathbb{S}^n \times \mathbb{I}$ is Hausdorff, whence the surjective map $p \times \mathbb{I}$ is necessarily a quotient map.)

Therefore $g$ is surjective (by Exercise (b)), which implies that $f$ covers the interior part $\mathbb{I}^n \setminus \underline{\partial}\mathbb{I}^n$ (on which $p$ is injective). Finally, $f$ is surjective.

(d) Again, the statement is trivial for $n = 0$, and amounts to the Intermediate Value Theorem for $n = 1$. We suppose that $n \geqslant 2$.

In our hypotheses, the restriction $g\colon \mathbb{S}^{n-1} \to \mathbb{S}^{n-1}$ is surjective (by Exercise (b)). If $f$ is not surjective, then $\mathrm{Im}\, f \subset \mathbb{D}^n \setminus \{x_0\}$, for an *internal* point $x_0$ of $\mathbb{D}^n$. Then $\mathbb{S}^{n-1}$ is a deformation retract of $\mathbb{D}^n \setminus \{x_0\}$ (which is homeomorphic to $\mathbb{D}^n \setminus \{0\}$), and we have a commutative diagram

$$\begin{array}{ccccc} \mathbb{S}^{n-1} & \xrightarrow{g} & \mathbb{S}^{n-1} & & \\ {\scriptstyle m}\downarrow & & {\scriptstyle m'}\downarrow & \searrow^{\mathrm{id}} & \\ \mathbb{D}^n & \xrightarrow[f]{} & \mathbb{D}^n \setminus \{x_0\} & \xrightarrow[p]{} & \mathbb{S}^{n-1} \end{array} \qquad (2.118)$$

Applying $H_{n-1}$ we get a contradiction: the homomorphism $g_{*n-1}\colon H_{n-1}(\mathbb{S}^{n-1}) \to H_{n-1}(\mathbb{S}^{n-1})$ is the identity of a non-trivial group, but factorises through $H_{n-1}(\mathbb{D}^n) = 0$.

### *2.4.5 The suspension*

For a non-empty space $X$, the *suspension* $\Sigma X$ is defined as the topological quotient

$$\Sigma X = (X \times \mathbb{I})/R, \tag{2.119}$$

which collapses, independently, the bases of the cylinder $IX$ to a lower and an upper *vertex* (or *pole*), $v^-$ and $v^+$.

Plainly, the suspension of $\mathbb{S}^n$ is canonically homeomorphic to $\mathbb{S}^{n+1}$, and we shall identify often them

$$\Sigma\mathbb{S}^n = \mathbb{S}^{n+1}, \qquad \mathbb{S}^n = \Sigma^n\,\mathbb{S}^0 \qquad (n \geqslant 0), \tag{2.120}$$

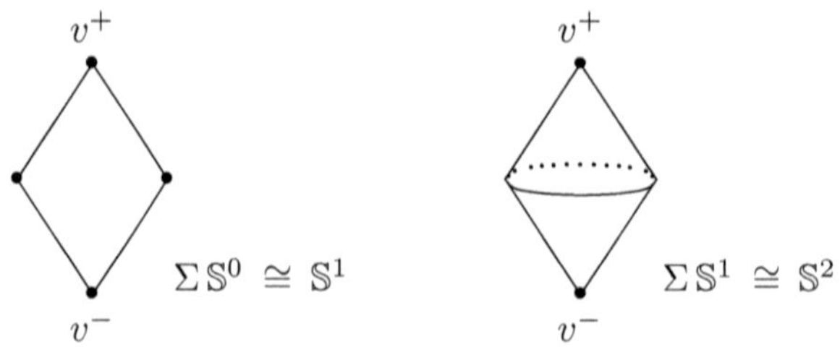


*Exercises and complements.* (a) Coming back to the general case, there is a natural isomorphism

$$\delta_n X\colon H_{n+1}(\Sigma X) \to H_n(X) \qquad n \geqslant 1, \tag{2.121}$$

given by the connecting morphism of the MV-sequence of $\Sigma X$, for the open cover

$$U = \Sigma X \setminus \{v^-\} \simeq \{*\}, \qquad V = \Sigma X \setminus \{v^+\} \simeq \{*\}, \tag{2.122}$$

with $A = U \cap V \simeq X$.

(b) If $X$ is path connected, non-empty, the group

$$H_1(\Sigma X) = 0 \tag{2.123}$$

is not isomorphic to $H_0(X)$. Reduced homology will give a better result, in (3.6.3).

(c) The suspension plays an important role in homotopy theory, that will be examined in Section 8.2.

Its structural definition will give an endofunctor $\Sigma\colon \mathsf{Top} \to \mathsf{Top}$ (in 8.2.5) defined everywhere, with $\Sigma\,\emptyset = \{v^-, v^+\}$. This completes the previous relationship with the spheres:

$$\Sigma\,\mathbb{S}^n = \mathbb{S}^{n+1} \qquad (n \geqslant -1), \tag{2.124}$$

where $\mathbb{S}^{-1} = \emptyset$, as in 1.4.1.

## 2.5 Complements

This last section is about the classification of the endomaps of the spheres, and a detailed study of the generator of $H_n(\mathbb{S}^n)$.

### *2.5.1 The winding number*

Let $f\colon \mathbb{S}^n \to \mathbb{S}^n$ be an endomap of the $n$-sphere, for $n > 0$. The associated homology homomorphism

$$f_{*n}\colon H_n(\mathbb{S}^n) \to H_n(\mathbb{S}^n) \tag{2.125}$$

is an endomorphism of an infinite cyclic group. It is thus the multiplication by an integral number $k \in \mathbb{Z}$, called the (topological) *degree*, or *winding number* of $f$, and written as $\deg f$.

Its properties are investigated in the (easy) exercises below.

*Notes.* Loosely speaking, the winding number represents how many times the sphere is wound-up on itself by $f$, preserving or reversing an arbitrary orientation. This is easily visualised on $\mathbb{S}^1$, as shown in Exercise (e).

We are exploiting the fact that the domain and codomain of $f$ are the *same* sphere, and the domain and codomain of $f_{*n}$ are the same group. If we are only given a map $f\colon X \to Y$ between two spaces homeomorphic to $\mathbb{S}^n$, we can only define the degree of $f$ 'up to sign change' (or as a natural number). One cannot prove Theorem 2.5.4 in this way.

### *2.5.2 Exercises and complements*

Let $f, g\colon \mathbb{S}^n \to \mathbb{S}^n$.

(a) The degree is invariant up to homotopy: $f \simeq g$ implies $\deg f = \deg g$.

*The converse is also true: *the degree classifies the homotopy classes of maps $f\colon \mathbb{S}^n \to \mathbb{S}^n$*. This important result is harder to prove; here, it will only be proved for $n = 1$, in Theorem 6.4.7.*

(b) We have:

$$\deg(fg) = (\deg f)(\deg g), \qquad \deg(\mathrm{id}\,\mathbb{S}^n) = 1. \tag{2.126}$$

The degree is a homomorphism of monoids, from $\mathsf{Top}(\mathbb{S}^n, \mathbb{S}^n)$ to the multiplicative monoid $\mathbb{Z}$.

(c) A homotopy equivalence $f: \mathbb{S}^n \to \mathbb{S}^n$ has degree $\pm 1$, and any weak inverse of $f$ has the same degree.

(d) Extending 2.4.4(b), every map $f: \mathbb{S}^n \to \mathbb{S}^n$ with $\deg f \neq 0$ is surjective. (One can similarly extend 2.4.4(e), assuming that $\deg g \neq 0$.)

(e) For $n = 1$, the standard map that winds-up $k$ times the circle on itself

$$f_k : \mathbb{S}^1 \to \mathbb{S}^1, \qquad f_k(z) = z^k, \tag{2.127}$$

(in complex variable) has topological degree $k$; all the integers are thus covered by the degree.

(f) (*Degree of a suspension*) For every map $f: \mathbb{S}^n \to \mathbb{S}^n$, the suspension $\Sigma f : \mathbb{S}^{n+1} \to \mathbb{S}^{n+1}$ has the same degree

$$\deg(\Sigma f) = \deg f. \tag{2.128}$$

Applying (e), there is thus a map $\Sigma^{n-1} f_k : \mathbb{S}^n \to \mathbb{S}^n$ of arbitrary degree $k$, for every $n \geqslant 1$.

Formula (2.128) is easily visualised for $n = 1$: a map $f: \mathbb{S}^1 \to \mathbb{S}^1$ is extended to a map $\Sigma f : \mathbb{S}^2 \to \mathbb{S}^2$, which coincides with $f$ on the equator and moves each meridian according to $f$; the poles $v^-, v^+$ stay fixed.

*(g) (The Fundamental Theorem of Algebra) *Every non-constant polynomial with coefficients in the complex field $\mathbb{C}$ has a root in $\mathbb{C}$.*

An elegant proof based on $H_2(\mathbb{S}^2)$ is given in [EiS], Section XI.5. Essentially, a polynomial function $f: \mathbb{C} \to \mathbb{C}$ of (algebraic) degree $k$ can be extended to a map $f^\bullet : \mathbb{S}^2 \to \mathbb{S}^2$ on the one-point compactification of $\mathbb{C}$; this map is proved to have topological degree $k$. If $k > 0$, $f^\bullet$ is surjective (by (d)) and $f$ is too.

Taking out a point, instead of adding it, one can give a proof based on $H_1(\mathbb{S}^1)$, or equivalently on the fundamental group of the circle: see [Ha], Theorem 1.8 or [Mu], Chapter 9, Theorem 56.1.

### *2.5.3* ***Theorem*** (The antipodal map)

*(a) The* antipodal map *of the n-sphere (for $n > 0$)*

$$T: \mathbb{S}^n \to \mathbb{S}^n, \qquad T(x) = -x, \tag{2.129}$$

*has winding number* $\deg T = (-1)^{n+1}$.

*(b) Suppose we have two maps $f, g: \mathbb{S}^n \to \mathbb{S}^n$ such that $f(x) \neq Tg(x)$, for all $x \in \mathbb{S}^n$. Then $f \simeq g$ and* $\deg f = \deg g$.

*Proof* (a) A reader will find it useful, and hopefully interesting, to write the proof following this outline (completed below).

First, the map $T$ is written as a composite of sign changes, in each variable. We prove that these $n+1$ maps have the same degree. On $\mathbb{S}^1$ this degree is proved to be $-1$; by induction, this result is extended to arbitrary dimension, making use of the natural isomorphism $D'_n \colon H_n(\mathbb{S}^n) \to H_{n-1}(\mathbb{S}^{n-1})$ of (2.109). Finally $\deg T = (-1)^{n+1}$.

Carrying out this schedule, $T$ is the composite of the following $n+1$ maps (in any order)

$$i_j \colon \mathbb{S}^n \to \mathbb{S}^n, \quad i_j(t_1, ..., t_{n+1}) = (t_1, ..., -t_j, ..., t_{n+1}), \tag{2.130}$$

where $j = 1, ..., n+1$. To prove that all $i_j$ have the same degree, we remark that the involution $h$ that interchanges the coordinates $t_1$ and $t_j$ gives a commutative diagram

$$\begin{array}{ccc} \mathbb{S}^n & \xrightarrow{h} & \mathbb{S}^n \\ {\scriptstyle i_1}\downarrow & & \downarrow{\scriptstyle i_j} \\ \mathbb{S}^n & \xrightarrow[h]{} & \mathbb{S}^n \end{array}$$

and the relation $(\deg i_1)(\deg h) = (\deg h)(\deg i_j)$. But $\deg h = \pm 1$ (by 2.5.2(c)) can be cancelled, and $\deg i_1 = \deg i_j$.

For $n = 1$, the degree of $i_1 \colon \mathbb{S}^1 \to \mathbb{S}^1$ is easily computed using the generator $\zeta = [c + c'] \in H_1(\mathbb{S}^1)$ of (2.107). In fact, $i_1$ reverses the paths $c, c'$, so that $i_{1*}[c + c'] = [i_1 c + i_1 c'] = -\zeta$ (by Exercise 2.2.8(f)).

Finally, we suppose by induction that $f' = i_1 \colon \mathbb{S}^{n-1} \to \mathbb{S}^{n-1}$ has degree $-1$, and prove that the same holds for $f = i_1 \colon \mathbb{S}^n \to \mathbb{S}^n$.

Noting that $f$ only acts on the first coordinate, the conditions (2.110) are satisfied

$$f(\mathbb{S}^{n-1}) \subset \mathbb{S}^{n-1}, \qquad f^{-1}\{\underline{s}\} = \{\underline{s}\}, \quad f^{-1}\{\underline{n}\} = \{\underline{n}\},$$

and $D'_n$ is natural on $f$, giving a commutative square

$$\begin{array}{ccc} H_n(\mathbb{S}^n) & \xrightarrow{D'_n} & H_{n-1}(\mathbb{S}^{n-1}) \\ {\scriptstyle f_{*n}}\downarrow & & \downarrow{\scriptstyle f'_{*n}} \\ H_n(\mathbb{S}^n) & \xrightarrow[D'_n]{} & H_{n-1}(\mathbb{S}^{n-1}) \end{array}$$

On any homology class $\zeta \in H_n(\mathbb{S}^n)$ the inductive hypothesis gives:

$$D'_n(f_{*n}(\zeta)) = f'_{*n} D'_n(\zeta) = -D'_n(\zeta) = D'_n(-\zeta).$$

Cancelling the isomorphism $D'_n$, we get $f_{*n}(\zeta) = -\zeta$ and $\deg f = -1$. Finally, $T = i_1 i_2 ... i_{n+1}$ has degree $(-1)^{n+1}$.

(b) Suppose we have two maps $f, g\colon \mathbb{S}^n \to \mathbb{S}^n$ such that $f(x) \neq Tg(x)$, for all $x \in \mathbb{S}^n$.

The affine homotopy $\varphi\colon f \simeq g\colon \mathbb{S}^n \to \mathbb{R}^{n+1}$ (see (1.3)) describes, for every $x \in \mathbb{S}^n$, the line segment from $f(x)$ to $g(x)$; the latter does not pass through the origin, because $f(x)$ and $g(x)$ cannot be antipodal

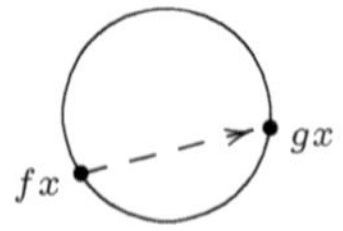


(2.131)

We have thus a homotopy $\varphi\colon f \simeq g\colon \mathbb{S}^n \to \mathbb{R}^{n+1} \setminus \{0\}$, to which we can apply the normalisation map $N\colon \mathbb{R}^{n+1} \setminus \{0\} \to \mathbb{S}^n$, getting a homotopy $N\varphi$ between $Nf = f$ and $Ng = g$. □

**2.5.4 Theorem** (Vector fields on spheres)

*(a) If $n > 0$ is even, every tangent vector field on $\mathbb{S}^n$ vanishes at least in a point. Equivalently, there is no field of tangent versors on $\mathbb{S}^n$.*

*(b) On the other hand, every odd dimensional sphere has fields of tangent versors.*

Informally, one says that the even dimensional spheres 'cannot be combed'. In dimension 2, for instance, one cannot coherently comb the whole surface of a hairy ball.

*Proof* The geometry of the sphere is so simple that Differential Geometry is not needed here: giving a field $v(x)$ of tangent versors on $\mathbb{S}^n$ is equivalent to giving a map $f\colon \mathbb{S}^n \to \mathbb{S}^n$ such that $f(x) \perp x$ (the orthogonality relation), for all $x \in \mathbb{S}^n$, as shown in the following figure (for $n = 1$)

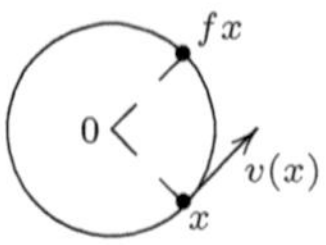


(2.132)

In fact, the tangent versor $v(x) \in T_x\mathbb{S}^n$ (the $\mathbb{R}$-linear space of all vectors at $x$, tangent to the $n$-sphere) corresponds to a versor $f(x) \perp x$, in the vector space $T_0\mathbb{R}^{n+1}$, which is viewed as a point of $\mathbb{S}^n$.

(a) The main part of the theorem is a corollary of Theorem 2.5.3. We suppose that there is a map $f\colon \mathbb{S}^n \to \mathbb{S}^n$ with $f(x) \perp x$, everywhere. It follows that $f(x) \neq Tx$ and $f(x) \neq x = TTx$ (where $Tx = -x$ is the antipodal map of $\mathbb{S}^n$).

Applying 2.5.3(b), we deduce that $f \simeq \mathrm{id}\,(\mathbb{S}^n)$ and $f \simeq T$. Thus $T \simeq \mathrm{id}\,\mathbb{S}^n$ and $\deg T = 1$. But we know that $\deg T = (-1)^{n+1}$ (by 2.5.3), and $n$ must be odd.

(b) On $\mathbb{S}^1$ the following map, that corresponds to figure (2.132)

$$f\colon \mathbb{S}^1 \to \mathbb{S}^1, \qquad f(t_1, t_2) = (-t_2, t_1),$$

gives a scalar product $(x \mid f(x))$ everywhere zero: $-t_1t_2 + t_2t_1 = 0$.

In general, provided that $n+1$ is even, the following map gives the same result

$$f\colon \mathbb{S}^n \to \mathbb{S}^n, \qquad f(t_1, ..., t_{n+1}) = (-t_2, t_1, -t_4, t_3, ..., -t_{n+1}, t_n).$$

□

### 2.5.5 Homology generators of the spheres

We want explicit descriptions of the generator of the group $H_n(\mathbb{S}^n) \cong \mathbb{Z}$, for $n \geqslant 1$.

*The case $n = 0$ can also be included, using the reduced homology group $\tilde{H}_0(\mathbb{S}^0) \cong \mathbb{Z}$: see Section 3.6.*

To simplify the notation, we begin by replacing (up to homeomorphism) the standard disc $\mathbb{D}^{n+1}$ with the standard cube $\mathbb{I}^{n+1}$, and the sphere $\mathbb{S}^n = \partial\mathbb{D}^{n+1}$ with the border $S^n = \partial\mathbb{I}^{n+1}$ of the cube.

The identity map $f_{n+1}\colon \mathbb{I}^{n+1} \to \mathbb{I}^{n+1}$ has a boundary in $C_n(\mathbb{I}^{n+1})$ whose $n$-cubes are (parametrised) faces of $\mathbb{I}^{n+1}$, with image in $S^n$. We have thus an $n$-cycle of $S^n$ that involves $2n+2$ cubes (represented in the following picture for $n = 1$)

$$z_n = \partial_{n+1} f_{n+1} \in C_n(S^n), \qquad \partial_n z_n = 0, \tag{2.133}$$

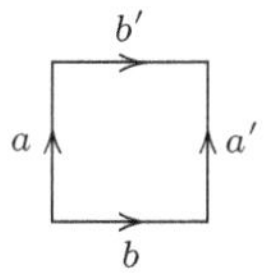


$$z_1 = \partial_2 f_1 = -a + a' + b - b'.$$

The following exercises prove that its homology class $[z_n]$ is a generator of $H_n(S^n)$, and transfer this result to the standard sphere. We also deduce other expressions of the generator, involving two cubes or only one.

*Exercises and complements.* We recall that $n \geqslant 1$.

(a) Prove that $[z_1]$ is a generator of $H_1(S^1)$, as is obvious from the previous figure.

(b) Prove that $[z_n]$ is a generator of $H_n(S^n)$, by induction on $n \geqslant 1$.

*Hint:* use the MV-sequence of $S^n$ with respect to the open subsets

$$\begin{aligned} U &= S^n \setminus \{\underline{s}\}, & \underline{s} &= (1/2, ..., 1/2, -1), \\ V &= S^n \setminus \{\underline{n}\}, & \underline{n} &= (1/2, ..., 1/2, 1). \end{aligned} \tag{2.134}$$

(c) Define a homeomorphism:

$$h_n \colon \mathbb{I}^n \to \mathbb{D}^n, \qquad h(\partial \mathbb{I}^n) = \partial \mathbb{D}^n = \mathbb{S}^{n-1}. \tag{2.135}$$

The previous points show that $[\partial_{n+1} h_{n+1}]$ is a generator of $H_n(\mathbb{S}^n)$, for the standard sphere. It is a chain formed of $2n + 2$ cubes of dimension $n$ that cover the $n$-sphere, enfolding its cavity.

(d) Form a generator $[z'_n]$ involving only two $n$-cubes

$$\begin{aligned} z'_n = a_n - i_{n+1} a_n \in C_n(\mathbb{S}^n), \quad a_n &= u_n h_n \colon \mathbb{I}^n \to \mathbb{D}^n \to \mathbb{S}^n, \\ u_n(x) &= (x, \sqrt{1 - ||x||^2}), \\ i_{n+1}(t_1, ..., t_{n+1}) &= (t_1, ..., t_n, -t_{n+1}), \end{aligned} \tag{2.136}$$

where $u_n \colon \mathbb{D}^n \to \mathbb{S}^n$ embeds the disc as the upper hemisphere, as in (1.49), and restricts to the canonical embedding $\mathbb{S}^{n-1} \to \mathbb{S}^n$ on its border. The sign change $i_{n+1}$, already used in (2.130), turns the latter into the lower hemisphere, without modifying the boundary of $a_n$, which shows that $z'_n$ is indeed a cycle.

(e) The same generator $[z'_n]$ can be induced by a single $n$-cube with constant faces

$$b_n = \sigma_n a_n \colon \mathbb{I}^n \to \mathbb{S}^n \to \mathbb{S}^n, \tag{2.137}$$

using an endomap $\sigma_n \colon \mathbb{S}^n \to \mathbb{S}^n$ that collapses the lower hemisphere to a point.

*This last point cannot be extended to $\tilde{H}_0(\mathbb{S}^0)$, which has no generator induced by a singleton.*

# 3
# Computing singular homology

We go on computing singular homology for several spaces, including the compact surfaces, the projective spaces and the standard infinite dimensional spaces. The simplicial form is presented in Section 3.6, and will be proved to be equivalent to the cubical one in Section 6.7.

Homology always means (absolute) singular homology. For a non-empty path connected space $X$ we know, and shall not repeat, that $H_0(X)$ is an infinite cyclic group, generated by the homology class of any point.

## 3.1 Combining circles and spheres

This section is, essentially, a list of exercises with the goal of:

- developing intuition on the topological meaning of the homology groups and their generators,
- learning how to extract information from an exact sequence.
- showing that all sequences of finitely generated abelian groups can be realised in homology (see 3.1.7 and 3.1.9).

Sums of pointed spaces are widely used.

### *3.1.1 The figure-eight space*

We want to compute the homology of the *figure-eight space*, the subspace $X$ of the plane formed of two circles tangent at a point

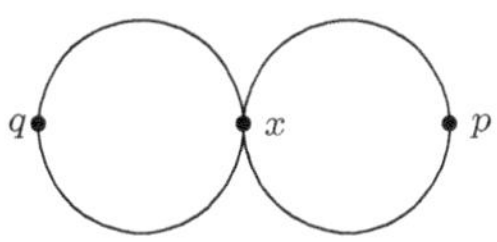


$$
\begin{aligned}
U &= X \setminus \{p\} \simeq \mathbb{S}^1, \\
V &= X \setminus \{q\} \simeq \mathbb{S}^1, \\
A &= U \cap V = X \setminus \{p, q\} \simeq \{*\}.
\end{aligned} \tag{3.1}
$$

*An exercise.* (a) We may guess that $H_1(X)$ is the free abelian group generated by two elements, the homology classes of two simple loops around each circle; and that the higher homology groups of $X$ are zero.

The reader is invited to write down the proof, which is not difficult and written below. One can use the open cover $(U, V)$ described above; the homology groups of $U, V$ and $A$ are already known.

*Solution.* All homology groups $H_k(X)$ are zero for $k > 1$, because they occur between two trivial groups, in an exact sequence

$$H_k(U) \oplus H_k(V) \to H_k(X) \to H_{k-1}(A). \tag{3.2}$$

For $H_1(X)$ we use a segment of the MV-sequence around this group

$$H_1A \xrightarrow{h_1} H_1U \oplus H_1V \xrightarrow{k_1} H_1X \xrightarrow{D_1} H_0A \xrightarrow{h_0} H_0U \oplus H_0V. \tag{3.3}$$

Here $H_1(A) = 0$, and $k_1$ is a monomorphism.

$H_0(A)$, $H_0(U)$ and $H_0(V)$ are free abelian groups on a single generator, the homology class of a point of $A$ – for instance the meeting point $x$ of the circles. The homomorphism $h_0\colon H_0A \to H_0U \oplus H_0V$ is computed as $h_0[x]_A = ([x]_U, [x]_V)$. Thus $h_0$ is mono, which means that $D_1 = 0$ and $k_1$ is also epi. It is thus an isomorphism

$$H_1(X) \;\cong\; \mathbb{Z}^2. \tag{3.4}$$

For a basis of $H_1(X)$, we choose two simple loops $a, b\colon \mathbb{I} \to X$ around each circle of $X$, as the following ones (assuming that the circles have centre at $(\pm 1, 0)$ and radius 1)

$$a(t) = (-1 + \cos 2\pi t,\ \sin 2\pi t), \qquad b(t) = (1 - \cos 2\pi t,\ -\sin 2\pi t). \tag{3.5}$$

We are using two loops with the same basepoint, at the origin. This is not relevant here, but will be for the fundamental group of $X$, in 7.4.4(b).

The homology classes $[a]_X, [b]_X$ are a basis of $H_1(X)$, since the isomorphism $k_1$ takes them to $([a]_U, 0)$ and $(0, -[b]_V)$, a basis of $H_1(U) \oplus H_1(V)$.

### 3.1.2 Exercises and complements

We can abstract some points from 3.1.1, that will be useful applying the MV-sequence.

(a) If the map $f\colon X \to Y$ is defined on a path connected space, the homomorphism $f_{*0}\colon H_0(X) \to H_0(Y)$ is injective.

(b) If the map $f\colon X \to Y$ takes values in a path connected space, and $X \neq \emptyset$, the homomorphism $f_{*0}\colon H_0(X) \to H_0(Y)$ is surjective.

(c) If, in the MV-sequence (2.83), the space $A = U \cap V$ is path connected, then:

- the homomorphism $h_0 \colon H_0(A) \to H_0(U) \oplus H_0(V)$ is injective,
- the homomorphism $D_1 \colon H_1(X) \to H_0(A)$ is trivial,
- the homomorphism $k_1 \colon H_1(U) \oplus H_1(V) \to H_1(X)$ is surjective.

### 3.1.3 Bouquets of circles

Generalising the figure-eight space, the *bouquet of circles* $C_n$ is a space formed of $n$ circles meeting in a single point (for $n \geqslant 1$)

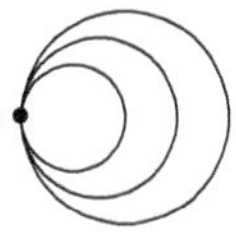

$C_3$ (3.6)

It can be realised as a subspace of $\mathbb{R}^2$ (or of $\mathbb{R}^3$, if we prefer circles with the same radius). But in fact there is no need of a super-space: $C_n$ can be obtained as a sum of pointed circles, a notion already examined in 1.6.2(d) and developed below.

*Exercises and complements.* (a) Compute the homology groups of $C_n$.

(b) Compute the homology groups of the subspace $X$ of $\mathbb{R}^2$ formed of three circles tangent in two points

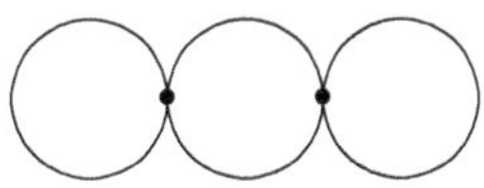
$X$ (3.7)

The reader will note that singular homology does not distinguish this space from $C_3$ (when directly applied to the spaces); but these spaces are not homeomorphic.

### 3.1.4 Sums of pointed spaces

We work in the category $\mathsf{Top}_\bullet$ of pointed spaces; for an object $X$, the basepoint is written as $0_X$, by default.

We want to describe the categorical sum $\bigvee_i X_i$, or *wedge*, of an arbitrary family of pointed spaces, after the binary case examined in 1.6.2(d). In the present section we shall mostly use finite sums, but the infinite ones will be needed later.

By definition, the sum is a pointed space $X$ equipped with a family of pointed maps $u_i\colon X_i \to X$, called injections, that satisfy the universal property:

(*) for every family of pointed maps $f_i\colon X_i \to Y$ $(i \in I)$, there is a unique pointed map $f\colon X \to Y$ such that $fu_i = f_i$ (for all $i \in I$).

The construction used in 1.6.2(d) cannot be extended beyond the finite case (as shown here, in Exercise (d)). We could obtain the sum as the inductive limit of all sums $\bigvee_{i\in J} X_i$ over the finite subsets of $I$ (a general fact proved in 1.7.8(e)), but we prefer to give a more explicit construction.

We begin by forming the sum in $\mathsf{Set}_\bullet$ of the underlying pointed sets, as the quotient $X = S/R$ of the disjoint union of the sets $X_i$

$$S = \bigcup_i X_i \times \{i\}, \tag{3.8}$$

modulo the equivalence relation $R$ that identifies all the points $0_i = (0_{X_i}, i)$, so to give a single basepoint $0_X \in X$. The structural injections are defined as

$$u_i\colon X_i \to X, \qquad u_i(x) = [x, i] \quad (x \in X_i), \tag{3.9}$$

where $[x, i]$ is the equivalence class of $(x, i)$ in $S/R$; all $u_i$ are injective.

The universal property (*) is satisfied (in $\mathsf{Set}_\bullet$): for a family of pointed mappings $f_i\colon X_i \to Y$, we must define $f\colon X \to Y$ letting $f([x, i]) = f_i(x)$, for each $i \in I$ and $x \in X_i$. Proceeding in this way, the mapping $f$ is well defined, and pointed.

Now we put on the pointed set $X$ the final topology for the mappings $u_i$: a subset $U \subset X$ is open if and only if every preimage $u_i^{-1}(U)$ is open in the space $X_i$. The pointed space $X$ is obviously the sum in $\mathsf{Top}_\bullet$.

Equivalently, $X$ is the topological quotient of the sum of the underlying spaces, modulo the equivalence relation $R$ that identifies the basepoints.

*Exercises and complements.* Let $X = \bigvee_i X_i$ be a pointed sum, with injections $u_i\colon X_i \to X$.

(a) An open subset $V$ of $X$ is either a nbd of the basepoint or not. The two forms are characterised as follows:

(i) $V = \bigcup_i u_i U_i$, each $U_i$ is an open nbd of $0_{X_i}$ in $X_i$ $(0_X \in V)$,

(ii) $V = \bigcup_i u_i U_i$, each $U_i$ is open in $X_i$ and $0_{X_i} \notin U_i$ $(0_X \notin V)$.

(b) The injections $u_i\colon X_i \to X$ of a pointed sum are topological embeddings: every $X_i$ is homeomorphic to its image in $X$.

In other words, if $U$ is open in $X_i$, then $u_i(U)$ is open in $u_i(X_i)$. If, moreover, $0_{X_i} \notin U$, then $u_i(U)$ is also open in $X$.

(c) If every pointed space $X_i$ is path connected (resp. connected), so is the pointed sum $X$.

(d) The construction of finite wedges as subspaces of the product, in 1.6.2(d), cannot be extended to the infinite case. *Hint:* use a countable wedge $\bigvee_n \mathbb{S}^0$.

### 3.1.5 Proposition

*Let $X = \bigvee_i X_i$ and $Y = \bigvee_i Y_i$ be pointed sums indexed by the same set $I$, with structural injections $u_i \colon X_i \to X$ and $v_i \colon Y_i \to Y$.*

*(a)* (Functoriality) *A family of pointed maps $f_i \colon X_i \to Y_i$ $(i \in I)$ gives a pointed map*

$$f = \bigvee_i f_i \colon X \to Y, \qquad f([x,i]) = [f_i(x), i] \qquad (x \in X_i), \tag{3.10}$$

*characterised by the relation $fu_i = v_i f_i$ $(i \in I)$.*

*The wedge is thus a functor from the cartesian power $(\mathsf{Top}_\bullet)^I$ to $\mathsf{Top}_\bullet$.*

*(b) A family $\varphi_i \colon f_i \simeq g_i \colon X_i \to Y_i$ $(i \in I)$ of pointed homotopies gives a pointed homotopy*

$$\varphi \colon f \simeq g \colon X \to Y, \qquad \varphi([x,i],t) = [\varphi_i(x,t), i] \quad (x \in X_i,\, t \in \mathbb{I}), \tag{3.11}$$

*the unique map $\varphi \colon X \times \mathbb{I} \to Y$ such that $\varphi(u_i \times \mathbb{I}) = v_i\varphi_i$ (for all $i$).*

*(c)* (Homotopy equivalence) *If $X_i \simeq Y_i$, for all $i \in I$, then $X \simeq Y$.*

**Note.* The cylinder functor of $\mathsf{Top}_\bullet$, introduced in Section 8.6, would reduce point (b) to (a): as a left adjoint, it preserves all colimits.*

*Proof* (a) This is another occurrence of the functoriality of categorical sums, in 1.6.2(g). By the universal property of $\bigvee_i X_i$, $f$ is defined as the pointed map such that $fu_i = v_i f_i$, for all $i \in I$.

(b) The mapping $\varphi$ is well defined in (3.11), because $[\varphi_i(0_{X_i}, t), i] = [0_{Y_i}, i] = 0_Y$ does not depend on the index $i$; this also proves that $\varphi(0_X, t) = 0_Y$, for all $t \in \mathbb{I}$.

To prove the continuity of $\varphi$, we remark that:

- $\bigvee_i X_i = (\Sigma X_i)/R$ is a topological quotient (for the equivalence relation identifying all basepoints $(0_{X_i}, i)$),

- $(\bigvee_i X_i) \times \mathbb{I} = (\Sigma X_i \times \mathbb{I})/(R \times \mathbb{I})$ is also, because $\mathbb{I}$ is locally compact, and therefore the functor $- \times \mathbb{I}$ preserves topological quotients (see 8.3.3).

It is now sufficient to prove that the composite

$$\varphi p \colon \Sigma(X_i \times \mathbb{I}) = (\Sigma X_i) \times \mathbb{I} \longrightarrow (\Sigma X_i \times \mathbb{I})/(R \times \mathbb{I}) \longrightarrow Y$$

is continuous; this is true, because on every summand $X_i \times \mathbb{I}$ the mapping $\varphi p$ coincides with $v_i\varphi_i \colon X_i \times \mathbb{I} \to Y$.

(c) We have a family of pointed homotopy equivalences

$$f_i \colon X_i \rightleftarrows Y_i : g_i, \qquad \varphi_i \colon g_i f_i \simeq \mathrm{id}\, X_i, \quad \psi_i \colon f_i g_i \simeq \mathrm{id}\, Y_i, \tag{3.12}$$

which gives, applying (a) and (b):

$$f \colon X \rightleftarrows Y : g, \qquad \varphi \colon gf \simeq \mathrm{id}\, X, \quad \psi \colon fg \simeq \mathrm{id}\, Y.$$

□

### 3.1.6 Theorem

*Let* $X = \bigvee_{i=1,\dots,r} X_i$ *be a finite sum of pointed spaces; we suppose that the basepoint of each* $X_i$ *has a contractible open nbd* $U_i$ *in* $X_i$. *Then*

$$H_n(X) \;\cong\; \textstyle\bigoplus_i H_n(X_i) \qquad (n > 0), \tag{3.13}$$

*and* $0_X$ *has a contractible open nbd in* $X$.

*Note.* This is an extension of Exercise 3.1.3(a), on a bouquet of circles. As to $H_0(X)$, we already know that, if all $X_i$ are path connected, so is $X$.

*Proof* First, we note that $\bigvee_i u_i U_i$ is a contractible nbd of $0_X$ in $X$. The statement is trivially true for $r = 1$; we suppose it is for $r - 1$ and prove it for $r \geqslant 2$.

We use the MV-sequence of $X$, with the open cover $(U, V)$ defined by the following wedges of pointed subspaces of $X$

$$\begin{aligned} U &= (\textstyle\bigvee_{i<r} u_i X_i) \vee u_r U_r \;\simeq\; \bigvee_{i<r} X_i, \\ V &= (\textstyle\bigvee_{i<r} u_i U_i) \vee u_r X_r \;\simeq\; X_r, \\ A = U \cap V &= \textstyle\bigvee_{i\leqslant r} u_i U_i \;\simeq\; \{*\}. \end{aligned} \tag{3.14}$$

The homotopy equivalences are a consequence of Proposition 3.1.5(c), taking into account that all injections $u_i$ are topological embeddings, so that each $u_i U_i$ is homeomorphic to $U_i$ and contractible. We also note that the contractible space $A$ is path connected (by Exercise 1.4.3(b)).

We want to prove that, for every $n \geqslant 1$, the homomorphism

$$k_n \colon H_n(U) \oplus H_n(V) \to H_n(X) \tag{3.15}$$

is an isomorphism, so that the inductive hypothesis gives:

$$H_n X \;\cong\; H_n U \oplus H_n V \;\cong\; \textstyle\bigoplus_{i<r} H_n X_i \oplus H_n X_r \;\cong\; \bigoplus_i H_n X_i.$$

For $n \geqslant 2$ the claim follows from the following exact sequence, where $H_n(A) = H_{n-1}(A) = 0$

$$H_n(A) \xrightarrow{h_n} H_n(U) \oplus H_n(V) \xrightarrow{k_n} H_n(X) \xrightarrow{D_n} H_{n-1}(A).$$

Finally, for $n = 1$, we have an exact sequence

$$H_1(A) \xrightarrow{h_1} H_1(U) \oplus H_1(V) \xrightarrow{k_1} H_1(X)$$

where $k_1$ is injective, because $H_1(A) = 0$, and is surjective by Exercise 3.1.2(c). □

### 3.1.7 Bouquets of spheres

A *bouquet of spheres* is a finite pointed sum $X = \bigvee_{i=1,...,r} X_i$ where each $X_i$ is a pointed sphere of positive dimension. The space $X$ is path connected.

We recall that all standard spheres are nested, with a common basepoint $e_1$: see (1.46).

*Exercises and complements.* (a) For $n \geqslant 1$, find a bouquet of spheres $X_n$ with the following homology groups

$$H_k(X_n) \cong \mathbb{Z}, \text{ for } 0 \leqslant k \leqslant n, \qquad H_k(X_n) = 0, \text{ for } k > n. \tag{3.16}$$

(b) One can realise a bouquet of spheres $X$ with arbitrary Betti numbers $\beta_1, ..., \beta_n$ and all higher $\beta_k = 0$.

One can add disjoint open points to have an arbitrary $\beta_0 > 0$, if this has an interest.

### 3.1.8 Spaces with non-trivial torsion

So far we have analysed spaces whose homology groups are torsion free. We introduce now a family of path connected spaces $P_n$ $(n \geqslant 2)$ with a single torsion coefficient $n$, in degree 1, and trivial homology groups in higher degree

$$H_1(P_n) \cong \mathbb{Z}_n, \qquad H_k(P_n) = 0, \quad \text{for } k > 1. \tag{3.17}$$

The space $P_n$ is the topological quotient of the standard disc $\mathbb{D}^2$

$$P_n = \mathbb{D}^2/R_n \qquad (n \geqslant 2), \tag{3.18}$$

modulo the equivalence relation $R_n$, which is discrete in the interior of the disc and, on the border $\mathbb{S}^1$, identifies each $n$-tuple of points that are vertices of a regular $n$-sided polygon inscribed in the circle

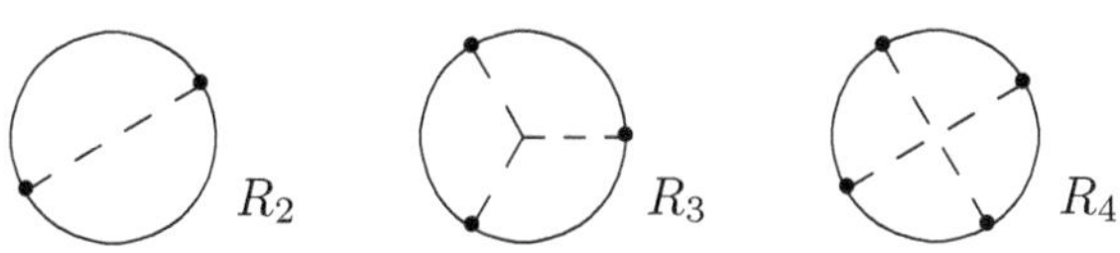


(3.19)

In other words, an equivalence class $\overline{\xi}$ of $R_n$ is either an interior point of the disc or an orbit of the action of the cyclic group $\mathbb{Z}_n$ on $\mathbb{S}^1$, generated by the rotation of $2\pi/n$ radians

$$\overline{\xi} = \{\xi, \xi e^{2\pi i/n}, \xi e^{4\pi i/n}, ..., \xi e^{2(n-1)\pi i/n}\} \subset \mathbb{S}^1 \subset \mathbb{C}.$$

The open disc $\operatorname{int}\mathbb{D}^2$ is topologically embedded in $P_n$, and will be viewed as an open subspace of the latter.

In particular the space $P_2 = \mathbb{D}^2/R_2$ is the quotient of the disc modulo the equivalence relation that identifies the antipodal points of the circle $\mathbb{S}^1$: a well known representation of the real projective plane $\mathbb{P}^2$. (The higher projective spaces $\mathbb{P}^n$ are analysed in Section 3.3.)

The space $\mathbb{P}^2$ is a 2-dimensional manifold. All the other $P_n$ are not locally euclidean: loosely speaking, a typical neighbourhood of a point $\overline{\xi}$ (with $\xi \in \mathbb{S}^1$) is homeomorphic to the union of $n \geqslant 3$ half-discs on the same diameter, and is not homeomorphic to a disc.

*Exercises and complements.* (a) Prove that $H_1(P_n) \cong \mathbb{Z}_n$. This requires some non-trivial work, useful and interesting.

(b) Prove that $H_k(P_n) = 0$, for $k > 1$.

(c) In particular the projective plane has $H_0(\mathbb{P}^2) \cong \mathbb{Z}$, $H_1(P_2) \cong \mathbb{Z}_2$ and trivial higher homology groups.

*(d) Similarly, the *lens spaces* $L_n = \mathbb{S}^3/\mathbb{Z}_n$ are quotients of the sphere $\mathbb{S}^3 \subset \mathbb{C}^2$ modulo the action of $\mathbb{Z}_n$ generated by multiplication by $e^{2\pi i/n}$; this includes the projective space $L_2 = \mathbb{P}^3$. A larger family of lens spaces, including all the odd-dimensional projective spaces, is studied in [Ha], Example 2.43.

### *3.1.9 Exercises* (Torsion coefficients)

The space $P_n = \mathbb{D}^2/R_n$ defined in 3.1.8 for $n \geqslant 2$ will be pointed at the origin $0 \in \operatorname{int}\mathbb{D}^2$.

(a) Construct a path connected space $X$ whose torsion coefficients are all of degree 1, and form an arbitrary sequence of positive integers $\tau_1, ..., \tau_k \in \mathbb{N}^*$, ordered by divisibility.

(b) Construct a path connected space whose torsion coefficients are all of a fixed degree $n \geqslant 1$, and form an arbitrary sequence as in (a).

(c) The spaces $P_4$ and $P_2 \vee P_2$ have different torsion coefficients.

(d) Construct two non-homeomorphic, path connected spaces with a torsion coefficient 6 in degree 1, and no other in any degree.

(e) Putting together point (b) with Exercise 3.1.7(b) and Theorem 3.1.6, every eventually trivial sequence of finitely generated abelian groups can be realised by the homology groups of a space. *(Using Theorem 3.4.7, the restriction to an eventually trivial sequence can be omitted.)*

## 3.2 Compact surfaces

After the 2-sphere and the projective plane, we study some other basic compact surfaces, like the torus, the Klein bottle and the double torus, computing their homology groups.

The exposition will now be less elementary, leaving as understood some topological issues not hard to prove, but requiring analytic work.

### *3.2.1 Quotients of the square*

Many compact surfaces can be obtained as a topological quotient of the standard square $\mathbb{I}^2$, modulo an equivalence relation $R$ which is *trivial out of the border* $\partial\mathbb{I}^2$ (in $\mathbb{R}^2$): each equivalence class of $R$ which is not contained in $\partial\mathbb{I}^2$ is a singleton.

Varying the relation $R$, we can obtain the following instances:

(a) the sphere $\mathbb{S}^2$, if $R$ identifies all the points of $\partial\mathbb{I}^2$,

(b) the projective plane $\mathbb{P}^2$, if $R$ identifies two points of $\partial\mathbb{I}^2$ when they are symmetric with respect to the centre of the square (see 3.1.8(c)),

(c) the torus $\mathbb{T}$, if $R$ identifies two parallel edges 'preserving their direction', and the other two in the same way (see 3.2.2 and 3.2.3),

(d) the Klein bottle $\mathbb{K}$, if $R$ identifies two parallel edges 'preserving direction', and the other two 'reversing direction' (see 3.2.4),

(e) the compact cylinder $\mathbb{S}^1 \times \mathbb{I}$, if $R$ identifies two parallel edges preserving their direction,

(f) the Möbius band $M$, if $R$ identifies two parallel edges reversing their direction (see 3.2.5),

(g) the square $\mathbb{I}^2$ itself, if $R$ is the discrete relation $x = y$.

Below, we describe more precisely the torus, the Klein bottle and the Möbius band, computing their homology groups. The other non-obvious surfaces have already been considered: the sphere in 2.4.2 and the projective plane in 3.1.8.

*Notes.* (i) The term 'surface' is used here in an informal way: the formal definition will be given in Section 4.3.

Essentially, a *surface* (without boundary) is a connected Hausdorff space which is locally homeomorphic to the euclidean plane $\mathbb{R}^2$, while a *surface with boundary* $X$ can also have points with an open nbd homeomorphic to the half-plane $\mathbb{R} \times [0, +\infty[$; these points form the *boundary* of $X$.

In the previous list, the second kind contains: the compact cylinder $\mathbb{S}^1 \times \mathbb{I}$ (with a disconnected boundary, the sum of two circles), the Möbius band $M$ and the disc $\mathbb{D}^2$ (both with boundary a circle).

(ii) The interior $U = ]0, 1[^2$ of the square is homeomorphic to the plane. All the spaces listed above are compactifications of $U$, in which $U$ is embedded as a dense open subspace. (In fact, if $V$ is open in $U$, then it is open in $\mathbb{I}^2$ and saturated for $R$, whence it is open in $\mathbb{I}^2/R$.)

### 3.2.2 The torus

We presented the torus as 'the surface of a lifebuoy', at the beginning of the general Introduction. We are now able to compute its homology groups, and make precise our intuition about its loops and 2-dimensional cavity, in Section 0.1.

(a) One can define the torus as the surface $\mathbb{T} \subset \mathbb{R}^3$ produced by the rotation around the vertical axis $z$ of a circle $C$, which lies in the plane $y = 0$, with centre $(2, 0, 0)$ and radius 1

$$C = \{(x, y, z) \in \mathbb{R}^3 \mid y = 0 \text{ and } (x-2)^2 + z^2 = 1\}, \tag{3.20}$$

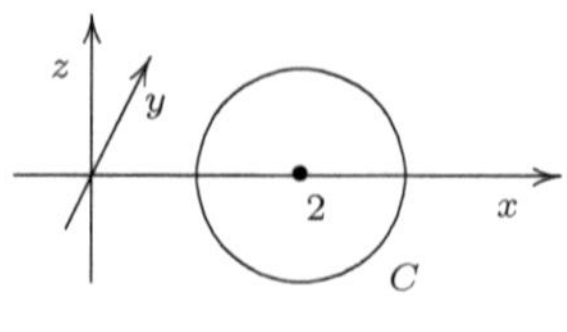


This surface is the image of the following map $f \colon \mathbb{I}^2 \to \mathbb{R}^3$

$$f(s, t) = ((2 + \cos 2\pi t) \cos 2\pi s, \; (2 + \cos 2\pi t) \sin 2\pi s, \; \sin 2\pi t), \tag{3.21}$$

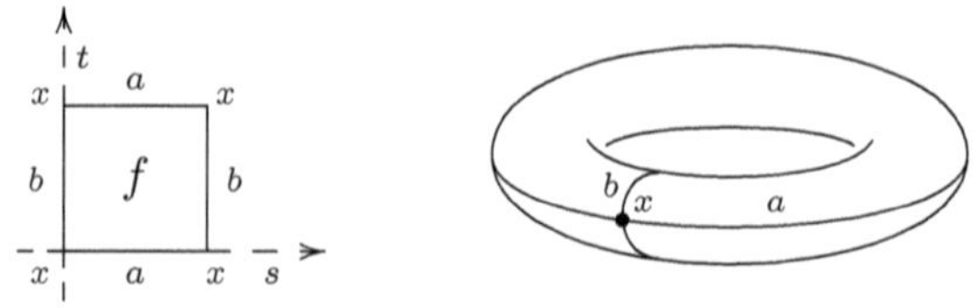


which sends the 1-indexed faces of the square to the meridian $b$, the 2-indexed faces to the equator $a$, and the four vertices of the square to the basepoint $x$ of the loops $a$ and $b$. (We are calling *equator* the circle produced by the rotation of the point $(3, 0, 0)$ of $C$.)

The map $f$ induces a homeomorphism $f'\colon \mathbb{I}^2/R \to \mathbb{T}$ (from a compact to a Hausdorff space).

(b) By this homeomorphism, the torus can also be defined as the topological quotient $\mathbb{I}^2/R$. The equivalence relation $R$ is trivial in the interior of the square, identifies the 1-indexed faces of the square 'preserving their direction', and the 2-indexed faces in the same way:

$$(0,t)\,R\,(1,t), \qquad (s,0)\,R\,(s,1), \qquad\qquad (s,t \in \mathbb{I}). \tag{3.22}$$

(c) Finally, the torus can also be defined as the cartesian product $\mathbb{S}^1 \times \mathbb{S}^1$.

In fact, the relation $R$ is the cartesian power $R_1 \times R_1$ of the equivalence relation on $\mathbb{I}$ that identifies the endpoints. It gives a bijective map $\mathbb{S}^1 \times \mathbb{S}^1 \to \mathbb{T}$, which is a homeomorphism, by the usual argument on compact Hausdorff spaces.

The torus has thus a system of coordinates in $\mathbb{S}^1 \times \mathbb{S}^1$, with latitude on the meridian and longitude on the equator.

The *n-dimensional torus* $\mathbb{T}^n$ is defined as the cartesian power $\mathbb{S}^1 \times ... \times \mathbb{S}^1$ of $n$ factors. It is a compact manifold of dimension $n$, whose homology will be computed in Exercise 6.7.7(e).

### *3.2.3* ***Exercises*** (The homology of the torus)

(a) Prove that

$$H_1(\mathbb{T}) \cong \mathbb{Z}^2, \qquad\qquad H_2(\mathbb{T}) \cong \mathbb{Z}, \tag{3.23}$$

and that all the higher groups are trivial. The Euler–Poincaré characteristic is 0. *Hint:* use an open cover similar to that used for the projective plane, in 3.1.8.

(b) Find generators for $H_1(\mathbb{T})$ and $H_2(\mathbb{T})$. The first point is obvious, the second requires some interesting work.

### *3.2.4* ***The Klein bottle***

The *Klein bottle* $\mathbb{K}$ is a (non-orientable) compact surface.

It can be defined as the quotient $\mathbb{I}^2/R$, modulo the equivalence relation which is trivial in the interior of the square, identifies the 2-indexed faces 'preserving direction', and the 1-indexed faces of the square 'reversing direction':

$$(s,0)\,R\,(s,1), \qquad (0,t)\,R\,(1,1-t) \qquad\qquad (s,t \in \mathbb{I}), \tag{3.24}$$

as represented in the following picture

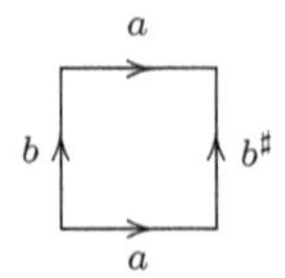

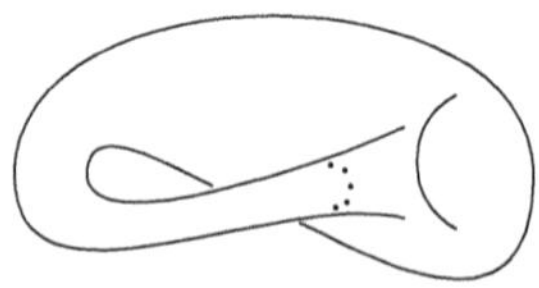

$\mathbb{K}$ cannot be realised in $\mathbb{R}^3$, but has a clear model in $\mathbb{R}^4$, that can be drawn in the plane, as above: the model should be imagined adding a fourth coordinate, to separate the 'intersection circle' where – in $\mathbb{R}^3$ – two portions of the 'tube' intersect each other. The four vertices of the square are identified with the basepoint $x$ of the loops $a$ and $b$ (the images of the 2-indexed and 1-indexed faces of the square).

*An exercise.* (a) Prove that

$$H_1(\mathbb{K}) \cong \mathbb{Z} \oplus \mathbb{Z}_2, \tag{3.25}$$

and that all the higher homology groups are trivial. The Euler–Poincaré characteristic is 0, and there is a torsion coefficient 2 in degree 1.

Find generators for $H_1(\mathbb{K})$.

### 3.2.5 The Möbius band

The *Möbius band* $M$, or *Möbius strip*, is a (non-orientable) compact surface with boundary.

It can be defined as the quotient $\mathbb{I}^2/R$, by the equivalence relation which is trivial in the interior of the square, and identifies two parallel faces (e.g. the 1-indexed ones) 'reversing their direction':

$$(0, t)\, R\, (1, 1-t) \qquad\qquad (s \in \mathbb{I}), \tag{3.26}$$

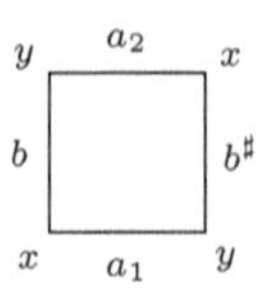

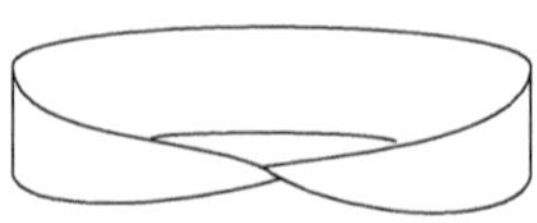

$M$ has as a well known model in $\mathbb{R}^3$, which can be realised with a rectangular strip of paper, gluing the short edges as in the picture above. The 2-indexed faces $a_1$ and $a_2$ are consecutive paths, whose concatenation $a = a_1 * a_2$ is a loop in $M$: a parametrisation of the boundary of $M$, which is homeomorphic to the circle.

Due to the torsion in a strip of paper so glued, the boundary of $M$ and the visible outline, in the picture above, do not agree everywhere.

The following path

$$c \colon \mathbb{I} \to M, \qquad c(s) = [s, 1/2], \tag{3.27}$$

is a simple loop at $[0, 1/2] = [1, 1/2]$, on the 'middle circle' of the strip.

The circle $\mathbb{S}^1$ is thus embedded as a deformation retract of $M$

$$f \colon \mathbb{S}^1 \to M, \qquad f(e^{2\pi i s}) = [s, 1/2] \quad (s \in \mathbb{I}), \tag{3.28}$$

and the homology groups of $M$ are known. The isomorphism $f_* \colon H_1(\mathbb{S}^1) \to H_1(M)$ gives a generator of $H_1(M)$, the homology class $[c]$. It is also evident that the simple loop $a$, around the boundary of $M$, gives $[a] = 2[c]$.

**Exercises and complements.* The following exercises pertain to general topology, and their solution is left to interested readers. But their statement is important, here: the Möbius band allows us to better understand the Klein bottle and (even more) the projective plane – whose models in $\mathbb{R}^4$ cannot be easily visualised.

(a) Pasting two Möbius bands along their boundary we get a Klein bottle

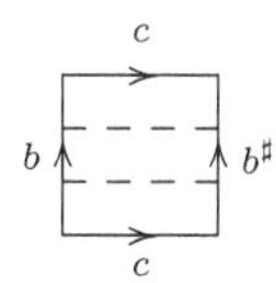


*Hints:* the figure shows the Klein bottle (parametrised on the square); the middle horizontal strip is a Möbius band; the two other strips, pasted along the loop $c$, form another Möbius band; these two bands are pasted on their (dashed) boundary.

(b) Pasting a Möbius band and a disc $\mathbb{D}^2$ along their boundary one gets a projective plane

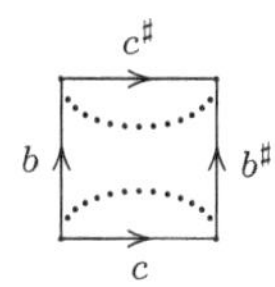


*Hints:* the figure shows the projective plane (parametrised on the square); the middle horizontal strip is a Möbius band; the two other parts, pasted along $c$, form a closed disc; the band and the disc are pasted on their (dotted) boundary.

(c) These pastings can be expressed as pushouts of embeddings, in Top

$$\begin{array}{ccc} \mathbb{S}^1 & \xrightarrow{h} & M \\ {\scriptstyle h}\downarrow & & \downarrow{\scriptstyle u} \\ M & \xrightarrow[v]{} & \mathbb{K} \end{array} \qquad\qquad \begin{array}{ccc} \mathbb{S}^1 & \xrightarrow{h} & M \\ {\scriptstyle k}\downarrow & & \downarrow{\scriptstyle u'} \\ \mathbb{D}^2 & \xrightarrow[v']{} & \mathbb{P}^2 \end{array} \tag{3.29}$$

where $h\colon \mathbb{S}^1 \to M$ is induced by the loop $a\colon \mathbb{I} \to M$ around the boundary and $k\colon \mathbb{S}^1 \to \mathbb{D}^2$ is the inclusion. *The second construction is an instance of a useful procedure, called 'attaching a cell', to the space $M$ (see 4.9.1)*

### *3.2.6 The double torus*

The 'double torus' $X$ is the compact surface represented below

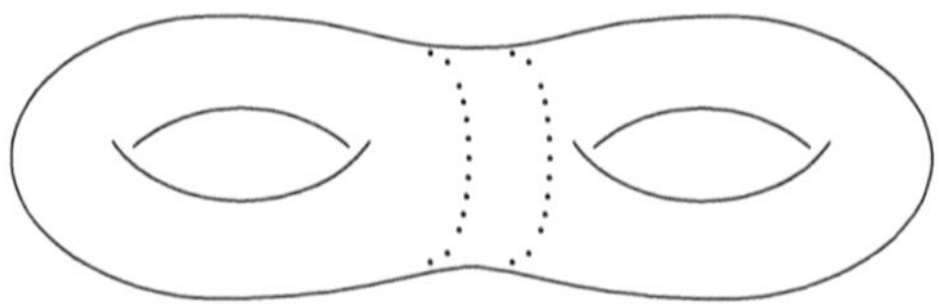

(3.30)

It can be drawn as a 'two-place lifebuoy', as here, or as a sphere with two handles. It can also be obtained as the 'connected sum' $X = \mathbb{T} \sharp \mathbb{T}$ of two tori, as we shall see in 4.3.7.

*An exercise.* (a) Prove that

$$H_1(X) \cong \mathbb{Z}^4, \qquad\qquad H_2(X) \cong \mathbb{Z}, \tag{3.31}$$

and that all the higher groups are trivial. The Euler–Poincaré characteristic is $-2$.

*Hints.* Use the MV-sequence for the open subspace $U$ at the left of the right dotted circle in figure (3.30), and a symmetric $V$ meeting $U$ at a cylinder $A$ open in $X$.

## 3.3 Projective spaces

We analyse now the homology of the real projective spaces $\mathbb{P}^n$, and their interaction with the spheres. The projective plane was already dealt with in 3.1.8(c).

Let us note that the homology of projective spaces can be more easily computed using the theory of CW-complexes: see [Ha], Example 2.42.

### 3.3.1 The topology of projective spaces

We begin by reviewing some well known presentations of the projective spaces. Their equivalence up to homeomorphism is an easy exercise, in point (e).

(a) The (real) projective space $\mathbb{P}^n$ can be defined as

$$\mathbb{P}^n = \mathbb{S}^n/T \qquad (n \geqslant 0), \tag{3.32}$$

the topological quotient of the sphere $\mathbb{S}^n$ modulo the action of the antipodal map $T\colon \mathbb{S}^n \to \mathbb{S}^n$ (see 2.5.3), identifying each $x \in \mathbb{S}^n$ with $-x$.

In other words, it is the orbit space of $\mathbb{S}^n$ for the obvious action of the (finite) multiplicative group $\{1, -1\}$. It is thus a path connected, compact Hausdorff space. In particular, $\mathbb{P}^0$ is a singleton and $\mathbb{P}^1$ is homeomorphic to the circle. The canonical projection

$$p_n\colon \mathbb{S}^n \to \mathbb{P}^n \qquad (n \geqslant 0), \tag{3.33}$$

is open (as any projection onto an orbit space).

(b) The embedding $\mathbb{S}^n \subset \mathbb{R}^{n+1} \setminus \{0\}$ allows us to identify $\mathbb{P}^n$ with the orbit space

$$\mathbb{P}^n = (\mathbb{R}^{n+1} \setminus \{0\})/\mathbb{R}^* \qquad (n \geqslant 0), \tag{3.34}$$

for the action of the multiplicative group of non-zero real numbers: $(x, \lambda) \mapsto \lambda x$.

(c) The map

$$u_n\colon \mathbb{D}^n \to \mathbb{S}^n, \qquad u_n(x) = (x, \sqrt{1 - ||x||^2}), \tag{3.35}$$

embeds the disc as the upper hemisphere $\mathbb{D}^n_+ = \{x \in \mathbb{S}^n \mid x_{n+1} \geqslant 0\}$, as in (1.48). It meets all the orbits of $\mathbb{S}^n$ modulo $T$, and allows us to identify $\mathbb{P}^n$ with the quotient

$$\mathbb{P}^n = \mathbb{D}^n/R \qquad (n \geqslant 0), \tag{3.36}$$

modulo the equivalence relation of antipodism on the sphere $\mathbb{S}^{n-1} = \partial\mathbb{D}^n$ and discrete elsewhere. ($\mathbb{P}^2$ was introduced in this form in 3.1.8.) This gives another quotient map

$$q_n\colon \mathbb{P}^n = \mathbb{D}^n/R \longrightarrow \mathbb{S}^n = \mathbb{D}^n/\mathbb{S}^{n-1} \qquad (n \geqslant 0), \tag{3.37}$$

collapsing $\mathbb{S}^{n-1} = \partial\mathbb{D}^n$ to a point (see 1.4.1(e)); the map $q_n$ will be analysed in Theorem 3.3.2.

(d) The nesting of the spheres (or cartesian spaces, or discs) in 1.4.1 induces a nesting of the projective spaces

$$\{\overline{1}\} = \mathbb{P}^0 \subset \mathbb{P}^1 \subset \mathbb{P}^2 \subset ... \subset \mathbb{P}^n \subset ... \tag{3.38}$$

*An exercise.* (e) Prove the equivalence of the definitions (a), (b), (c), up to homeomorphism.

**3.3.2 Theorem** (Spheres and projective spaces)

For $n > 0$, the projections

$$p_n \colon \mathbb{S}^n \to \mathbb{P}^n = \mathbb{S}^n/T, \quad q_n \colon \mathbb{P}^n = \mathbb{D}^n/R \to \mathbb{S}^n = \mathbb{D}^n/\mathbb{S}^{n-1}, \tag{3.39}$$

give a composed endomap of $\mathbb{S}^n$

$$r_n = q_n p_n \colon \mathbb{S}^n \to \mathbb{P}^n \to \mathbb{S}^n, \qquad \deg r_n = 1 + (-1)^{n+1}, \tag{3.40}$$

whose degree is 2 for $n$ odd and 0 for $n$ even.

*Notes.* For $n$ even this is obvious, because $r_n = r_n T$ and $\deg T = -1$ implies $\deg r_n = -\deg r_n$. The 'odd case' (obvious for $n = 1$) will be used to determine the homology groups of the projective spaces.

*Proof* The analytic description of $r_n$ depends on making precise $q_n$. We identify $\mathbb{P}^n$ with $\mathbb{D}^n_+/R$ and let $q_n$ act on $\mathbb{D}^n_+$ as a restriction of the map $\sigma_n \colon \mathbb{S}^n \to \mathbb{S}^n$ of Exercise 2.5.5(e), that collapses $\mathbb{S}^{n-1}$ to the South pole $\underline{s}$. Noting that $r_n = r_n T$ we have:

$$\begin{gathered} r_n(x) = \sigma_n(x), \text{ if } x \in \mathbb{D}^n_+, \qquad r_n(x) = \sigma_n(Tx), \text{ if } x \in \mathbb{D}^n_-, \\ r_n(\mathbb{S}^{n-1}) = \sigma_n(\mathbb{S}^{n-1}) = \{\underline{s}\}. \end{gathered} \tag{3.41}$$

We use the generator $[z'_n]$ of $H_n(\mathbb{S}^n)$ built in 2.5.5(d) with the cycle $z'_n = a_n - i_{n+1}a_n$, recalling that $a_n$ lives in the upper hemisphere $\mathbb{D}^n_+$ while $i_{n+1}a_n$ lives in the lower one. We have also seen that $b_n = \sigma_n a_n$ is a cycle of $\mathbb{S}^n$ and $[z'_n] = [b_n]$. This proves (3.40)

$$r_n[z'_n] = [\sigma_n a_n] - [\sigma_n T i_{n+1} a_n] = [b_n] - (-1)^n [b_n],$$

taking into account that $\sigma_n T i_{n+1} a_n = \sigma_n i_1 ... i_n a_n = i_1 ... i_n b_n$, because $\sigma_n$ commutes with all the sign changes of the hyperplane $\mathbb{R}^n \subset \mathbb{R}^{n+1}$, and all of them have degree 1. □

***3.3.3 An open cover***

Let $n \geqslant 1$. We shall use the MV-sequence of $\mathbb{P}^n$, with respect to the following open cover $(U_n, V_n)$

$$\begin{aligned} U_n &= (\mathbb{S}^n \setminus \{\underline{n}, \underline{s}\})/T \simeq \mathbb{S}^{n-1}/T \cong \mathbb{P}^{n-1}, \\ V_n &= (\mathbb{S}^n \setminus \mathbb{S}^{n-1})/T \simeq \{*\}, \\ A_n &= U_n \cap V_n \simeq \mathbb{S}^{n-1}. \end{aligned} \tag{3.42}$$

We shall always omit the trivial group $H_i(V_n)$, for $i > 0$. The homomorphism $h_i\colon H_i(A_n) \to H_i(U_n)$ is induced by the inclusion $A_n \subset U_n$, which is homotopic to the canonical projection $p_{n-1}\colon \mathbb{S}^{n-1} \to \mathbb{P}^{n-1}$

$$h_i = (p_{n-1})_{*i}\colon H_i(\mathbb{S}^{n-1}) \to H_i(\mathbb{P}^{n-1}) \qquad (i > 0). \tag{3.43}$$

***3.3.4 Theorem*** (The homology of real projective spaces)

*(i) The homology groups of the projective space $\mathbb{P}^n$ $(n \geqslant 1)$ are determined as:*

$$\begin{aligned} H_i(\mathbb{P}^n) &\cong \mathbb{Z} && \text{for } i = 0 \text{ or } i = n \text{ odd}, \\ H_i(\mathbb{P}^n) &\cong \mathbb{Z}_2 && \text{for } i \text{ odd and } 0 < i < n, \\ H_i(\mathbb{P}^n) &= 0 && \text{for } i \text{ even and } 0 < i \leqslant n, \text{ or } i > n. \end{aligned} \tag{3.44}$$

*(ii) For $i$ odd and $0 < i < n-1$, the canonical embedding $u\colon \mathbb{P}^{n-1} \to \mathbb{P}^n$ gives an isomorphism*

$$u_{*i} = k_i\colon H_i(\mathbb{P}^{n-1}) \to H_i(\mathbb{P}^n). \tag{3.45}$$

*(iii) For $n$ odd, the projections $p_n\colon \mathbb{S}^n \to \mathbb{P}^n$ and $q_n\colon \mathbb{P}^n \to \mathbb{S}^n$ of Theorem 3.3.2 give*

$$\begin{gathered} \operatorname{Im}(p_{n*}\colon H_n(\mathbb{S}^n) \to H_n(\mathbb{P}^n)) = 2H_n(\mathbb{P}^n), \\ q_{n*}\colon H_n(\mathbb{P}^n) \to H_n(\mathbb{S}^n) \ \text{ is an isomorphism.} \end{gathered} \tag{3.46}$$

*Proof* We have already computed the homology groups of $\mathbb{P}^1 \cong \mathbb{S}^1$, in Section 2.4, and of $\mathbb{P}^2$ (as $P_2$), in Section 3.1, so that (i) is verified for $n \leqslant 2$. We extend these results using a part of the MV-sequence associated to the open cover $(U_n, V_n)$ described above, where $U_n \simeq \mathbb{P}^{n-1}$, $A_n \simeq \mathbb{S}^{n-1}$ and $V_n \simeq \{*\}$ is omitted in positive degree

$$H_i(A_n) \xrightarrow{h_i} H_i(U_n) \xrightarrow{k_i} H_i(\mathbb{P}^n) \xrightarrow{D_i} H_{i-1}(A_n). \tag{3.47}$$

(a) First we prove that $H_i(\mathbb{P}^n) = 0$ for $i > n$. We suppose that $H_i(\mathbb{P}^{n-1}) = 0$ for $i > n-1$, and take $i > n$. Then the exact sequence (3.47) has $H_i(U_n) = H_i(\mathbb{P}^{n-1}) = 0$ and $H_{i-1}(A_n) = H_{i-1}(\mathbb{S}^{n-1}) = 0$, whence $H_i(\mathbb{P}^n) = 0$.

(b) We now consider $H_i(\mathbb{P}^n)$ for $0 < i < n-1$. We suppose that what claimed in (3.44) holds for $n-1$ and prove it for $n \geqslant 3$. In the exact sequence (3.47), $k_i$ is mono, because $H_i(A_n) = H_i(\mathbb{S}^{n-1}) = 0$. It is also epi, either by Exercise 3.1.2(c), for $i = 1$, or because $H_{i-1}(A_n) = H_{i-1}(\mathbb{S}^{n-1}) = 0$, for $i > 1$.

Therefore $H_i(\mathbb{P}^n) \cong H_i(\mathbb{P}^{n-1})$ is determined by the inductive hypothesis. We have also proved point (ii).

(c) We still have to compute $H_{n-1}(\mathbb{P}^n)$ and $H_n(\mathbb{P}^n)$, depending on the parity of $n$, and to prove $(\mathrm{iii})_n$, that is, property (iii) in degree $n$.

We want to prove that, for $n$ odd:

$$(*)\qquad \begin{array}{lll} \tilde{H}_{n-1}(\mathbb{P}^n) = 0, & H_n(\mathbb{P}^n) \cong \mathbb{Z}, & \\ H_n(\mathbb{P}^{n+1}) \cong \mathbb{Z}_2, & H_{n+1}(\mathbb{P}^{n+1}) = 0, & (\mathrm{iii})_n \text{ holds true.} \end{array}$$

For $n = 1$ this is known. We prove (*) for $n$ odd $\geqslant 3$, assuming that it holds for $n-2$. Actually, it is sufficient to assume that:

$$(**)\qquad \begin{array}{ll} \tilde{H}_{n-3}(\mathbb{P}^{n-2}) = 0, & H_{n-2}(\mathbb{P}^{n-2}) \cong \mathbb{Z}, \\ H_{n-2}(\mathbb{P}^{n-1}) \cong \mathbb{Z}_2, & H_{n-1}(\mathbb{P}^{n-1}) = 0. \end{array}$$

(d) The first half of (*) follows from the exact sequence

$$0 \to H_n(\mathbb{P}^n) \xrightarrow{D_n} H_{n-1}(A_n) \to 0 \to H_{n-1}(\mathbb{P}^n) \to 0$$

where three groups are already known to be trivial, because of (a) or (**)

$$H_n(U_n) = H_n(\mathbb{P}^{n-1}), \qquad H_{n-1}(U_n) = H_{n-1}(\mathbb{P}^{n-1}),$$
$$H_{n-2}(A_n) = H_{n-2}(\mathbb{S}^{n-1}).$$

Therefore $H_n(\mathbb{P}^n) \cong H_{n-1}(A_n) \cong \mathbb{Z}$ and $H_{n-1}(\mathbb{P}^n) = 0$.

(e) We now prove property $(\mathrm{iii})_n$, using the projection $q_n\colon \mathbb{P}^n \to \mathbb{S}^n$ of Theorem 3.3.2. We recall that $n$ is odd.

The open cover $(U_n, V_n)$ of $\mathbb{P}^n$ is projected by $q_n$ to an open cover $(U'_n, V'_n)$ of $\mathbb{S}^n$

$$\begin{array}{ll} U'_n = q_n(\mathbb{S}^n \setminus \{\underline{n}, \underline{s}\}) \simeq \{*\}, & V'_n = q_n(\mathbb{S}^n \setminus \mathbb{S}^{n-1}) \simeq \{*\}, \\ A'_n = U'_n \cap V'_n = q_n(\mathbb{D}^n_+ \setminus (\mathbb{S}^{n-1} \cup \{\underline{n}\}) \simeq \mathbb{S}^{n-1}, & \end{array} \tag{3.48}$$

which gives the same homology groups as the open cover used for $\mathbb{S}^n$ in Theorem 2.4.2. It is easy to see that the restriction $q'_n\colon A_n \to A'_n$ of $q_n$ is a homeomorphism.

By the naturality of the MV-sequence we have a commutative diagram with exact rows (for $n$ odd)

$$\begin{array}{ccccccc} 0 & \to & H_n(\mathbb{P}^n) & \xrightarrow{D_n} & H_{n-1}(A_n) & \to & 0 \\ \downarrow & & \downarrow{\scriptstyle q_{n*}} & & \downarrow{\scriptstyle q'_{n*}} & & \downarrow \\ 0 & \to & H_n(\mathbb{S}^n) & \xrightarrow[D'_n]{} & H_{n-1}(A'_n) & \to & 0 \end{array} \tag{3.49}$$

where $D_n$, $D'_n$ and $q'_{n*}$ are isomorphisms, so that also $q_{n*}\colon H_n(\mathbb{P}^n) \to H_n(\mathbb{S}^n)$ is an isomorphism: we have proved the second part of $(\mathrm{iii})_n$.

We have seen, in Theorem 3.3.2, that, composing $p_n\colon \mathbb{S}^n \to \mathbb{P}^n$ with the canonical projection $q_n\colon \mathbb{P}^n \to \mathbb{S}^n$, we get an endomap $r_n\colon \mathbb{S}^n \to \mathbb{S}^n$ of degree 2, so that $\mathrm{Im}\,(q_{n*}p_{n*}\colon H_n(\mathbb{S}^n) \to H_n(\mathbb{S}^n)) = 2H_n(\mathbb{S}^n)$ (for $n$ odd).

As $q_{n*}\colon H_n(\mathbb{P}^n) \to H_n(\mathbb{S}^n)$ is an isomorphism, the first part of $(\mathrm{iii})_n$ is also proved:

$$\mathrm{Im}\,(p_{n*}\colon H_n(\mathbb{S}^n) \to H_n(\mathbb{P}^n)) = 2H_n(\mathbb{P}^n).$$

(f) Coming back to the second part of (*), we use an MV-sequence of $\mathbb{P}^{n+1}$

$$0 \to H_{n+1}\mathbb{P}^{n+1} \xrightarrow{D_{n+1}} H_nA_{n+1} \xrightarrow{h_n} H_nU_{n+1} \xrightarrow{k_n} H_n\mathbb{P}^{n+1} \to 0$$

where there are isomorphisms (because $n \geqslant 3$ is odd):

$$\begin{array}{ll} H_{n+1}(U_{n+1}) = H_{n+1}(\mathbb{P}^n) = 0, & H_{n-1}(A_{n+1}) = H_{n-1}(\mathbb{S}^n) = 0, \\ H_n(A_{n+1}) \cong H_n(\mathbb{S}^n) \cong \mathbb{Z}, & H_n(U_{n+1}) \cong H_n(\mathbb{P}^n) \cong \mathbb{Z}. \end{array}$$

Moreover, we have seen in (3.43) that the homomorphism $h_n$ can be identified with $p_{n*}\colon H_n(\mathbb{S}^n) \to H_n(\mathbb{P}^n)$. Applying $(\mathrm{iii})_n$, we get: $\mathrm{Im}\, h_n = 2H_n(U_{n+1})$.

It follows that $H_n(\mathbb{P}^{n+1}) \cong \mathbb{Z}_2$ and $h_n$ is injective (as any non-trivial homomorphism between two infinite cyclic groups). The proof is now complete: $D_{n+1} = 0$ and $H_{n+1}(\mathbb{P}^{n+1}) = 0$. □

## 3.4 Nested inductive limits and homology

Inductive limits of nested sequences of spaces are now used to construct the infinite dimensional spaces $\mathbb{R}^\infty$, $\mathbb{S}^\infty$, $\mathbb{P}^\infty$, and the infinite bouquet of circles $C_\infty$.

Their singular homology is computed, as an inductive limit of already known homology groups.

### *3.4.1 Nested inductive limits of spaces*

A nested sequence of topological spaces

$$X_0 \subset X_1 \subset ... \subset X_n \subset ... \tag{3.50}$$

forms an inductive system, indexed on the totally ordered set $\mathbb{N}$. Its inductive limit is the union of all the underlying sets

$$X = \cup_{n\geqslant 0} X_n, \tag{3.51}$$

equipped with the *final topology* for the inclusions $X_n \subset X$: a subset $A \subset X$ is open (resp. closed) if and only if every trace $A \cap X_n$ is open (resp. closed) in $X_n$. Equivalently, a mapping $f\colon X \to Y$ with values in a topological space is continuous if and only if all its restrictions $X_n \to Y$ are.

The final topology on $X$ could also be called the 'fine topology', as it is the finest inducing the given ones on the subsets $X_n$. Yet, it is often called the 'weak topology' of $X$, in relation with the usual terminology of CW-complexes (see Section 4.9).

*Exercises and complements.* (a) All $X_n$ are subspaces of $X$. If each $X_n$ is open (resp. closed) in $X_{n+1}$, the same holds in $X$.

(b) If all $X_n$ are connected (resp. path connected), so is $X$.

(c) If all $X_n$ are compact, $X$ need not be: see 3.4.3(a).

### *3.4.2 The infinite dimensional cartesian space*

We present, here and in 3.4.3, some interesting examples. Several topological properties of these nested inductive limits will be proved below, in Lemmas 3.4.4–3.4.6.

First, the nested sequence of the euclidean spaces $\mathbb{R}^n$, in (1.45), produces the *infinite-dimensional euclidean space*

$$\mathbb{R}^\infty = \bigcup_{n\geqslant 0} \mathbb{R}^n, \tag{3.52}$$

formed by the essentially finite sequences $t = (t_n)_{n\geqslant 0}$ of real numbers, with the *final topology*, produced by the inductive limit.

It will be useful to compare $\mathbb{R}^\infty$ with the metric space $(\mathbb{R}^\infty, d)$, that is, the same set equipped with the euclidean metric $d$ (the extension of the euclidean metrics of all $\mathbb{R}^n$) and with the *euclidean topology*, defined by the metric. Also this space induces the euclidean topology on all $\mathbb{R}^n$; therefore, the final topology of $\mathbb{R}^\infty$ is finer, and Hausdorff.

In fact, the final topology is strictly finer than the euclidean one. To wit, the following sequence of points of $\mathbb{R}^\infty$

$$x^n = (0, ..., 0, 1/n, 0, ...) \qquad (n > 0), \tag{3.53}$$

(with one non-trivial coordinate $x^n_n$) converges to the origin in the euclidean topology. On the other hand, in the final topology, the sequence forms a closed discrete subspace $C$ (because each trace $C \cap \mathbb{R}^n$ is finite), and does not converge.

Let us note that $C$ is not compact; general results of this kind can be found in Lemma 3.4.6.

*Remarks.* (a) $\mathbb{R}^\infty$ can be viewed as the real vector space $\bigoplus_{n\geqslant 0} \mathbb{R}$, equipped with the *finest topology compatible with the linear structure*. In fact, such a topology would induce on each $\mathbb{R}^n$ a compatible structure, which is necessarily the euclidean one. It is easy to show that $\mathbb{R}^\infty$ is not first countable, and not metrisable.

*(b) $(\mathbb{R}^\infty, d)$ is obviously a normed vector space, with the $l^2$-norm. More generally one can consider the inductive limit of the normed vector spaces $(\mathbb{R}^n, || \, ||_p)$, with the $l^p$-norm (in the category of normed vector spaces and linear contractions).

### 3.4.3 Other examples

(a) Similarly the nested sequence of the euclidean spheres $\mathbb{S}^n$, in (1.47), produces the *infinite-dimensional sphere*

$$\mathbb{S}^\infty = \bigcup_{n\geqslant 0} \mathbb{S}^n = \{t \in \mathbb{R}^\infty \mid d(0, t) = 1\}, \tag{3.54}$$

formed by the essentially finite sequences $t = (t_n)_{n\geqslant 0}$ of real numbers with $\sum_n t_n^2 = 1$, and the inductive limit topology. It is a closed subspace of $\mathbb{R}^\infty$, as proved by Lemma 3.4.4, and a Hausdorff space. It is not compact, as proved by 3.4.6(ii).

We shall see that the space $\mathbb{S}^\infty$ is contractible, in 7.6.5.

(b) The nesting of the projective spaces $\mathbb{P}^n$, in 3.3.1(d), produces the *infinite-dimensional real projective space*

$$\mathbb{P}^\infty = \bigcup_{n\geqslant 0} \mathbb{P}^n, \tag{3.55}$$

It is the quotient of $\mathbb{S}^\infty$ modulo antipodism, as proved by Lemma 3.4.5. It is also a Hausdorff space, and is not compact (by 3.4.6(ii)).

(c) The nesting of the bouquets of circles $C_n$, in 3.1.3, produces the *infinite bouquet of circles*

$$C_\infty = \bigcup_{n\geqslant 1} C_n, \tag{3.56}$$

which can also be obtained as the wedge of a countable family of pointed circles (see 3.1.4).

It is plainly a Hausdorff space, and is not compact (again by 3.4.6(ii)). The same set can be given a coarser, compact topology $\mathbb{H}_1 \subset \mathbb{R}^2$, called the 'Hawaiian earring' (see 4.9.7(b)). (This also shows that $C_\infty$ is Hausdorff.)

It is easy to see that $C_\infty$ is not first countable: the origin has no countable local base. Therefore *it cannot be embedded in the plane*, while all finite bouquets $C_n$ are embedded there (in our construction): an inductive limit of subspaces need not be a subspace.

**3.4.4 Lemma** (Subspaces)

*There are given two nested sequences $(X_n)$, $(Y_n)$ of spaces, with inductive limits $X$ and $Y$. Moreover, for all $n$, $X_n$ is a subspace of $Y_n$ and*

$$(^*) \quad X_{n+1} \cap Y_n = X_n \qquad \text{(the pullback condition).}$$

*If each $X_n$ is open (resp. closed) in $X_{n+1}$* and *in $Y_n$, then $X$ is an open (resp. closed) subspace of $Y$.*

*Note.* Condition (*) cannot be omitted: see 3.4.3(c). *Categorically, the statement lies on the fact that pullbacks commute with filtered colimits.*

*Proof* By construction, $X$ is a sub*set* of $Y$ and the inclusion mapping is continuous. We prove the statement for open subspaces, the other argument is 'the same'. Condition (*) implies that $X_m \cap Y_n = X_n$, for all $m \geqslant n$, and $X \cap Y_n = X_n$.

Let $A$ be an open subspace of $X$. Each trace $A_n = A \cap X_n$ is open in $X_n$. The trace of $A$ on $Y_n$ is the same, by (*):

$$A \cap Y_n = A \cap X \cap Y_n = A \cap X_n = A_n,$$

and is open in $X_n$, which is open in $Y_n$. Finally, $A$ is open in $Y$. □

**3.4.5 Lemma** (Quotients)

*There are given two nested sequences $(X_n)$, $(Y_n)$ of spaces, and a 'consistent' family of topological projections $f_n \colon X_n \to Y_n$ (i.e. each $f_n$ is the restriction of $f_{n+1}$), with inductive limit $f \colon X \to Y$.*

*Then $f$ is also a topological projection. Moreover two points $x, x' \in X$ are identified by $f$ if and only if there is an index $n$ such that $x, x' \in X_n$ and $f_n(x) = f_n(x')$.*

*Proof* The set-theoretical part is obvious. If all $f_n$ are topological projections, take a subset $V \subset Y$ such that $f^{-1}(V)$ is open in $X$. Every trace $V_n = V \cap Y_n$ has an open preimage $f^{-1}(V_n) = f^{-1}(V) \cap X_n$, and is open in the quotient $Y_n$, whence $V$ is open in $Y$. □

**3.4.6 Lemma** (Compactness and separation)

*Let $X$ be the inductive limit of a nested sequence $(X_n)$ of $T_1$-spaces.*

*(i) The space $X$ is $T_1$, and every compact subset $K$ of $X$ is contained in some $X_n$. A sequence converges in $X$ if and only if it is contained in some $(X_n)$ and converges there.*

*(ii) If $(X_n)$ is a strictly increasing sequence of compact $T_1$-spaces, $X$ is not compact.*

*Note.* The $T_1$-condition cannot be omitted: the inductive limit of a system of indiscrete spaces is indiscrete (and compact), as we have seen in 1.7.7(e).

*Proof* It is sufficient to prove (i), as (ii) is a consequence.

The first assertion is obvious: every point of $X$ is closed. As to the second, we suppose that a subset $K \subset X$ is not contained in any $X_n$, and prove that it is not compact.

For every $n$ we can choose a point $x_n \in K \setminus X_n$. All these points form a closed subset $A$ of $X$ (and $K$), since its trace on each $X_n$ is finite, whence closed. The subsets $A_n = \{x_n, x_{n+1}, ...\}$ form a family of closed subsets of $K$ whose intersection is empty (since $A_n \subset K \setminus X_n$), although the intersection of every finite subfamily is not: $K$ is not compact.

Finally, a sequence $(x_n)$ converging to $x_\infty$ in $X$ gives a map $x \colon \mathbb{N}^\bullet \to X$ (see 1.2.2(d)), whose image is compact, and therefore contained in some $X_n$, a subspace of $X$. □

**3.4.7 Theorem** (Singular homology and inductive limits)

*Let $X$ be the inductive limit of a nested sequence $(X_n)$ of $T_1$-spaces. The singular homology group $H_k(X)$ is the inductive limit of the system of abelian groups and (induced) homomorphisms*

$$H_k(X_0) \to H_k(X_1) \to ... \to H_k(X_n) \to ... \tag{3.57}$$

**Note.* There is a similar result for the homotopy groups, in 7.6.5.*

*Proof* It is an easy consequence of the previous lemma, and the compactness of the probe spaces of singular homology: every $k$-cube $a \colon \mathbb{I}^k \to X$ lies in some space $X_n$.

We are working with a fixed integer $k \geqslant 0$. The inductive limit $L = \varinjlim_n H_k(X_n)$ in $\mathsf{Ab}$ is characterised in 1.7.7(b). The inclusions $X_n \to X$ give a canonical homomorphism $L \to H_k(X)$. This is surjective, because every cycle of $X$ lies in some space $X_n$; it is injective, because a cycle of $X_n$ which is the boundary of a chain in $X$, is also a boundary in some $X_{n'}$, for $n' \geqslant n$. □

**3.4.8 Examples**

We can now deduce the homology groups of the spaces introduced above.

All these spaces are path connected, and:

$$H_k(\mathbb{R}^\infty) = H_k(\mathbb{S}^\infty) = 0, \quad \text{for } k > 0, \tag{3.58}$$

$$H_k(\mathbb{P}^\infty) = \mathbb{Z}_2, \ \text{for } k \text{ odd}, \qquad H_k(\mathbb{P}^\infty) = 0, \ \text{for } k \text{ even} > 0, \tag{3.59}$$

$$H_1(C_\infty) = \textstyle\bigoplus_{n\geqslant 1} \mathbb{Z}, \qquad H_k(C_\infty) = 0, \ \text{for } k > 0. \tag{3.60}$$

The countable base of the free abelian group $H_1(C_\infty)$ can be realised choosing a simple loop in each circle. We also note that $\mathbb{R}^\infty$ is easily seen to be contractible, which also determines its homology.

## 3.5 Diagram lemmas in Homological Algebra

The following issues are a basic part of Homological Algebra, and will be of frequent use in the following chapters.

A reader with interests in algebra will find pleasure in proving them, by the technique of diagram chasing introduced in 2.3.3, either independently or following the hints given below. A reader more interested in topology might prefer to skip this section, and come back when necessary.

*These results can be extended to any abelian category, and to various more general contexts: see [G2, G3] and their references.*

### 3.5.1 *The Snake Lemma*

*The two middle squares of the diagram below are given, and supposed to be commutative, with short exact rows. The diagram is completed with the kernels and cokernels of the vertical arrows, and the induced morphisms between them (see* (2.6)*)*

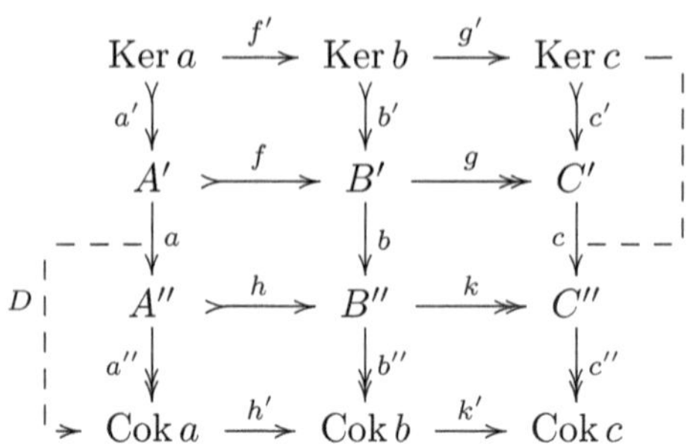


(3.61)

*Then there is a* connecting morphism $D\colon \mathrm{Ker}\, c \to \mathrm{Cok}\, a$, *computed in the proof. (Its dashed arrow in the diagram is the eponymous Snake.)*

*It forms an exact sequence*

$$\operatorname{Ker} a \xrightarrow{f'} \operatorname{Ker} b \xrightarrow{g'} \operatorname{Ker} c \xrightarrow{D} \operatorname{Cok} a \xrightarrow{h'} \operatorname{Cok} b \xrightarrow{k'} \operatorname{Cok} c \qquad (3.62)$$

*which is natural for natural transformations of the two middle squares.*

*Proof* This is a popular lemma in Homological Algebra, closely related to Theorem 2.3.3 on the exact sequence of chain homology.

One can readily deduce the lemma from Theorem 2.3.3, transforming the middle squares into a short exact sequence of chain complexes $A \rightarrowtail B \twoheadrightarrow C$; the complex $A$ is trivial except for $a = \partial_1 \colon A_1 \to A_0$, so that

$$H_1(A) = \operatorname{Ker} a, \qquad H_0(A) = \operatorname{Cok} a.$$

Similarly, $B$ and $C$ consist of the differentials $b \colon B_1 \to B_0$ and $c \colon C_1 \to C_0$, respectively.

The homology sequence of $A \rightarrowtail B \twoheadrightarrow C$ coincides with (3.62) (including an initial and a terminal 0) and tells us how to compute the connecting morphism $D = D_1 \colon H_1(C) \to H_0(A)$ on $z \in \operatorname{Ker} c$, namely $D(z) = [x] = a''(x)$, where $z = g(y)$ and $h(x) = b(y)$ (for $y \in B'$ and $x \in A''$)

$$\begin{array}{ccccc} & & y & \overset{g}{\longmapsto} & z \\ & & {\scriptstyle b}\downarrow & & \downarrow{\scriptstyle c} \\ x & \overset{h}{\longmapsto} & b(y) & \overset{k}{\longmapsto} & 0 \end{array} \qquad (3.63)$$

One can also prove directly the lemma (by the technique of diagram chasing used in the proof of Theorem 2.3.3), and deduce Theorem 2.3.3 from the lemma; but this requires a longer work. □

### 3.5.2 The $3 \times 3$ Lemma

*(a) In the following commutative diagram in* Ab*, we suppose that the three columns and the middle row are short exact*

$$\begin{array}{ccccc} A' & \longrightarrow & B' & \longrightarrow & C' \\ \downarrow & & \downarrow & & \downarrow \\ A & \rightarrowtail & B & \twoheadrightarrow & C \\ \downarrow & & \downarrow & & \downarrow \\ A'' & \longrightarrow & B'' & \longrightarrow & C'' \end{array} \qquad (3.64)$$

*Then the upper row is short exact if and only if the lower row is.*

*(b) The same holds in* $\mathrm{Ch}_+\mathsf{Ab}$.

*Proof* (a) We assume that the upper row is short exact and prove that the lower row is also. It is sufficient to apply the Snake Lemma 3.5.1 to the upper part of diagram (3.64)

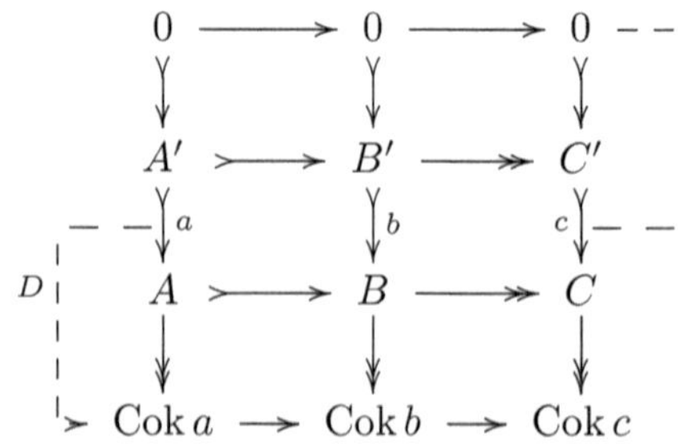


(3.65)

This gives the short exact sequence

$$\mathrm{Cok}\, a \rightarrowtail \mathrm{Cok}\, b \twoheadrightarrow \mathrm{Cok}\, c$$

corresponding to the lower row of diagram (3.64).

Similarly, if we assume that the lower row is short exact, we apply the Snake Lemma to the lower squares of (3.64), and deduce that the upper row is short exact.

(b) It is sufficient to apply (a) in each component. □

### *3.5.3 Exercises* (The Five Lemma)

*There is given a commutative diagram in* Ab, *with exact rows*

$$\begin{array}{ccccccccc} A & \xrightarrow{f} & B & \xrightarrow{g} & C & \xrightarrow{h} & D & \xrightarrow{k} & E \\ {\scriptstyle u}\downarrow & & {\scriptstyle v}\downarrow & & \downarrow{\scriptstyle w} & & \downarrow{\scriptstyle v'} & & \downarrow{\scriptstyle u'} \\ A' & \xrightarrow[f']{} & B' & \xrightarrow[g']{} & C' & \xrightarrow[h']{} & D' & \xrightarrow[k']{} & E' \end{array} \qquad (3.66)$$

*(a) If $u, v, v', u'$ are isomorphisms, $w$ is also.*

*(b) If $v, v'$ are monomorphisms and $u$ is epi, then $w$ is mono.*

*(c) If $v, v'$ are epimorphisms and $u'$ is mono, then $w$ is epi.*

*Hints.* This is easily proved by diagram chasing. Point (a) is the main one, but it is convenient to prove (b) and (c), which imply the former.

*Notes.* The particular case where $A, A', E, E'$ are trivial groups (and the rows are short exact) is called the Short Five Lemma.

*In the theory of abelian categories, (b) and (c) are dual to each other, and it is sufficient to prove one of them. (In this setting, monomorphisms and epimorphisms are characterised by kernels and cokernels.)*

### 3.5.4 Split exact sequences

A pair of consecutive morphisms $(m, q)$ in $\mathsf{Ab}$ is said to be a *split* (short) exact sequence if the following equivalent conditions hold:

$$A \underset{p}{\overset{m}{\rightleftarrows}} B \underset{n}{\overset{q}{\rightleftarrows}} C \qquad (3.67)$$

(i) $(m, q)$ is short exact and there is some $n$ such that $qn = \mathrm{id}\, C$,

(ii) $(m, q)$ is short exact and there is some $p$ such that $pm = \mathrm{id}\, A$,

(iii) there are two morphisms $p, n$ such that: $pm = \mathrm{id}\, A$, $qn = \mathrm{id}\, C$, and $mp + nq = \mathrm{id}\, B$.

*Exercises and complements.* (a) Prove the equivalence of these conditions. Assuming them, $p$ and $n$ determine each other.

(b) If these conditions hold, the morphisms $p, n$ (with $m$ and $q$) make $B$ into the direct sum $A \oplus C$, completing the butterfly diagram (1.101) (with a different notation). This decomposition of $B$ *is not natural* on the initial data $(m, q)$.

(c) If $C$ is a *free* abelian group, the short exact sequence (3.67) necessarily splits. Find an example where this splitting is not determined by $m$ and $q$.

(d) Let $(m, q) = (A \rightarrowtail B \twoheadrightarrow C)$ be a short exact sequence in $\mathrm{Ch}_+\mathsf{Ab}$. We say that it is *componentwise split*, or that it *splits in every degree*, if all its components $(m_k, q_k)$ are split exact sequences of abelian groups. In particular, this certainly holds true if all the components of $C$ are free abelian groups.

A componentwise splitting need not be a splitting in $\mathrm{Ch}_+\mathsf{Ab}$: the partial inverses $p_k \colon B_k \to A_k$ and $n_k \colon C_k \to B_k$ that we can find need not be consistent with the differentials of the complexes. The useful notion is the componentwise one, as we shall see in Chapters 5 and 6.

### 3.5.5 Biproducts revisited

Biproducts in $R\,\mathsf{Mod}$ have been examined in 1.6.4. We add now that a biproduct $C = A \oplus B$ is characterised by a diagram such that:

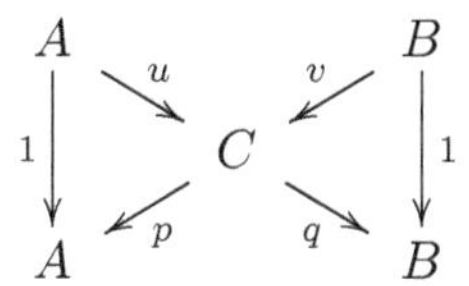


(3.68)

(*) the diagram commutes, and its diagonals are (split) exact sequences.

*Exercises and complements.* (a) By 3.5.4, (*) is equivalent to property 1.6.4(ii), a characterisation of $A \oplus B$ rewritten here as:

$$pu = \mathrm{id}\, A, \qquad qv = \mathrm{id}\, B, \qquad up + vq = \mathrm{id}\, C. \tag{3.69}$$

The same characterisation holds in $\mathrm{Ch}_+\mathsf{Ab}$. *(More generally, it holds in any abelian category.)*

(b) The addition $f+g$ of two homomorphisms $f, g\colon A \to B$ is determined by the categorical structure, as the following composite

$$f+g \;=\; (A \xrightarrow{\Delta} A \oplus A \xrightarrow{f \oplus g} B \oplus B \xrightarrow{\nabla} B), \tag{3.70}$$

- $\Delta = (1, 1)\colon A \to A \oplus A$ is the diagonal of the product,
- $\nabla = [1, 1]\colon B \oplus B \to B$ is the codiagonal of the coproduct,
- $f \oplus g$ is categorically determined as $f \times g$, by the universal property of the categorical product; or, equivalently, by the universal property of the categorical sum.

### 3.5.6 **Exact squares*

Under the hypotheses of the Mayer–Vietoris Theorem 2.3.7, the homology functor $H_n$ takes the left square below to a commutative square of homology groups

$$\begin{array}{ccc} A & \xrightarrow{i} & U \\ {\scriptstyle j}\downarrow & & \downarrow{\scriptstyle u} \\ V & \xrightarrow[v]{} & X \end{array} \qquad\qquad \begin{array}{ccc} H_n(A) & \xrightarrow{i_{*n}} & H_n(U) \\ {\scriptstyle j_{*n}}\downarrow & & \downarrow{\scriptstyle u_{*n}} \\ H_n(V) & \xrightarrow[v_{*n}]{} & H_n(X) \end{array} \tag{3.71}$$

Forgetting the connecting morphism, the theorem gives an exact sequence

$$\begin{gathered} H_n(A) \xrightarrow{h_n} H_n(U) \oplus H_n(V) \xrightarrow{k_n} H_n(X) \\ h_n[z] = (i_{*n}[z], j_{*n}[z]), \qquad k_n([z], [w]) = u_{*n}[z] - v_{*n}[w], \end{gathered} \tag{3.72}$$

which means that the right square (3.71) is *exact*: cf. Hilton [Hi].

The mere commutativity of the square amounts to the condition $k_n h_n = 0$, which means that the sequence (3.71) is of order two, in the sense of 2.1.3(a).

## 3.6 Reduced homology and the simplicial construction

Here we present reduced singular homology, a moderate variation of ordinary singular homology which will be useful in the next chapters.

Then we sketch the simplicial definition of singular homology, based on tetrahedra instead of cubes. The equivalence of these two forms will be proved in Section 6.7.

References are already given in the introduction of this chapter.

### 3.6.1 Reduced singular homology

Reduced singular homology $\tilde{H}_n(X)$ only differs from ordinary homology in degrees 0 and $-1$. We shall see that there are sound reasons to introduce this variation.

First we introduce the category $\mathrm{Ch}_*\mathsf{Ab}$ of *unbounded* chain complexes $A = ((A_n), (\partial_n))$ of abelian groups, indexed by $n \in \mathbb{Z}$

$$... \ A_{n+1} \xrightarrow{\partial_{n+1}} A_n \xrightarrow{\partial_n} A_{n-1} \ ... \qquad \partial_n\partial_{n+1} = 0, \tag{3.73}$$

with obvious morphisms. We have now homology functors $H_n\colon \mathrm{Ch}_*\mathsf{Ab} \to \mathsf{Ab}$, for $n \in \mathbb{Z}$.

The category $\mathrm{Ch}_+\mathsf{Ab}$ of positive chain complexes is a full subcategory of $\mathrm{Ch}_*\mathsf{Ab}$, by adding trivial components in negative degree, as usual.

For a space $X$, the *augmented* (singular) *chain complex* $\tilde{C}(X)$

$$... \ C_2(X) \xrightarrow{\partial_2} C_1(X) \xrightarrow{\partial_1} C_0(X) \xrightarrow{\tilde{\partial}_0} \mathbb{Z} \dashrightarrow 0 \ ... \tag{3.74}$$

adds to the complex $C_+(X)$ of singular chains a single new component $\tilde{C}_{-1}(X) = \mathbb{Z}$, in degree $-1$, and a differential

$$\tilde{\partial}_0\colon C_0(X) \to \tilde{C}_{-1}(X) = \mathbb{Z}, \qquad \tilde{\partial}_0(\Sigma\,\lambda_i x_i) = \Sigma\,\lambda_i, \tag{3.75}$$

which does satisfy the condition $\tilde{\partial}_0\partial_1 = 0$: on each 1-cube $a\colon \mathbb{I} \to X$ we have $\partial_1(a) = a(1) - a(0)$ and $\tilde{\partial}_0\partial_1(a) = 0$. We still write $\tilde{C}_n(X)$ as $C_n(X)$, for $n \geqslant 0$.

The augmented chain complex was used in the solution to Exercise 2.2.5(b), to compute $H_0(X)$ for a path connected space. We can think of $\tilde{C}_{-1}(X)$ as generated by the unique map $\emptyset \to X$, corresponding to an additional standard cube $\mathbb{I}^{-1} = \emptyset$.

The functor $C_+\colon \mathsf{Top} \to \mathrm{Ch}_+\mathsf{Ab}$ is thus extended to a functor

$$\tilde{C}\colon \mathsf{Top} \to \mathrm{Ch}_*\mathsf{Ab}, \tag{3.76}$$

and we have a *reduced singular homology functor* in each degree $n \geqslant -1$ (trivial in lower degrees)

$$\tilde{H}_n\colon \mathsf{Top} \to \mathsf{Ab}, \qquad \tilde{H}_n(X) = H_n(\tilde{C}(X)). \tag{3.77}$$

The obvious natural transformation

$$\rho\colon \tilde{C} \to C_+\colon \mathsf{Top} \to \mathrm{Ch}_*\mathsf{Ab}, \qquad \rho X\colon \tilde{C}(X) \twoheadrightarrow C_+(X), \tag{3.78}$$

$$\begin{array}{ccccccccc}
\dots C_2(X) & \xrightarrow{\partial_2} & C_1(X) & \xrightarrow{\partial_1} & C_0(X) & \xrightarrow{\tilde{\partial}_0} & \tilde{C}_1(X) & \dashrightarrow & 0 \dots \\
\| & & \| & & \| & & \downarrow & & \| \\
\dots C_2(X) & \xrightarrow{\partial_2} & C_1(X) & \xrightarrow{\partial_1} & C_0(X) & \dashrightarrow & 0 & \dashrightarrow & 0 \dots
\end{array}$$

induces a natural transformation in each degree $n \geqslant -1$

$$\rho_{*n} \colon \tilde{H}_n \to H_n \colon \mathsf{Top} \to \mathsf{Ab}, \qquad \rho_{*n}X \colon \tilde{H}_n(X) \to H_n(X), \tag{3.79}$$

which is the identity $\tilde{H}_n(X) = H_n(X)$ for $n > 0$, and has trivial components $\tilde{H}_{-1}(X) \to 0$ in degree $-1$. We are only interested in

$$\rho X = \rho_{*0}X \colon \tilde{H}_0(X) \to H_0(X), \tag{3.80}$$

examined in the exercises below. (Note that we are using the same symbol as in $\rho X \colon \tilde{C}(X) \twoheadrightarrow C_+(X)$.)

### 3.6.2 *Exercises and complements*

We only have to explore reduced homology in degrees $0, -1$.

(a) We begin with two obvious remarks:

- for the empty space, the complex $\tilde{C}(\emptyset)$ has one non-trivial component, in degree $-1$; the only non-trivial group of reduced homology is $\tilde{H}_{-1}(\emptyset) = \mathbb{Z}$ (a canonical identification),

- if $X \neq \emptyset$, the differential $\tilde{\partial}_0 \colon C_0(X) \to \mathbb{Z}$ is surjective and $\tilde{H}_{-1}(X) = 0$; the group $\tilde{H}_0(X)$ is analysed below.

One may prefer to use reduced homology on non-empty spaces, to get rid of $\tilde{H}_{-1}$.

(b) There is a natural exact sequence

$$\begin{gathered}\tilde{H}_0(X) \xrightarrow{\rho X} H_0(X) \xrightarrow{DX} \mathbb{Z} \xrightarrow{pX} \tilde{H}_{-1}(X), \\ (DX)[\Sigma_i\, \lambda_i x_i] = \Sigma_i\, \lambda_i, \qquad (pX)(k) = [k].\end{gathered} \tag{3.81}$$

*Hints.* This can be proved directly, or derived from the short exact sequence of (unbounded) chain complexes $\operatorname{Ker}\rho \rightarrowtail \tilde{C}(X) \twoheadrightarrow C_+(X)$.

(c) For $X = \emptyset$ this exact sequence has: $\tilde{H}_0(\emptyset) = H_0(\emptyset) = 0$ and $p\emptyset = \operatorname{id}\mathbb{Z}$.

(d) For $X \neq \emptyset$, $\tilde{H}_{-1}(X) = 0$ and (3.73) becomes a short exact sequence

$$\tilde{H}_0(X) \rightarrowtail H_0(X) \twoheadrightarrow \mathbb{Z},$$

which necessarily splits (non canonically) in a direct sum (by 3.5.4(c)).

Using the partition $X = \bigcup_{i \in I} X_i$ of the space $X$ in path components (as in (2.54)), the abelian group $H_0(X)$ has a canonical basis $\xi_i = [x_i]$, where

$x_i$ is any point of the path component $X_i$ (Exercise 2.2.5(d)). $\tilde{H}_0(X)$ is thus embedded in $\tilde{H}_0(X)$ as the subgroup

$$\tilde{H}_0(X) = \{\Sigma\, \lambda_i \xi_i \mid \Sigma\, \lambda_i = 0\}. \tag{3.82}$$

Choosing arbitrarily an index $j \in I$, a base of $\tilde{H}_0(X)$ is given by the elements $[x_i - x_j]$, for $i \neq j$, and is in bijective correspondence with the set $I \setminus \{j\}$, namely the set of path components of $X$, one excluded.

(e) The space $X$ is path connected if and only if $\tilde{H}_0(X) = 0$. If $X$ is contractible, *all* its reduced homology groups $\tilde{H}_n(X)$ are trivial.

In several contexts, this property of reduced homology is convenient with respect to ordinary homology. On the other hand, the functors $\tilde{H}_n$ do not preserve categorical sums, for $n \leqslant 0$.

(f) In reduced homology the only non-trivial group of the sphere $\mathbb{S}^n$ is $\tilde{H}_n(\mathbb{S}^n)$, which is infinite cyclic, also for $n = 0$: an advantage in the context of orientable manifolds.

Consider a 0-sphere $S^0 = \{p, q\} \cong \mathbb{S}^0$: then $\tilde{H}_0(S^0)$ is an infinite cyclic group, and admits two elements as generators, either $[q - p]$ or $[p - q]$. Choosing one of them amounts to choose an orientation of $S^0$, from $p$ to $q$ or vice versa.

This will be exploited in Section 4.3: the orientation of an $n$-dimensional manifolds will be based on the orientation of $(n-1)$-spheres in its euclidean neighbourhoods, which leads to $\mathbb{S}^0$ for 1-dimensional manifolds. True, their connected components are lines and circles, and orienting them is not a problem; yet, a sound approach should not make exceptions.

(g) (*The Mayer–Vietoris sequence in reduced homology*). In the same hypotheses of 2.3.7, there is an exact sequence of reduced singular homology groups

$$\dots \to \tilde{H}_n(A) \xrightarrow{h_n} \tilde{H}_n(U) \oplus \tilde{H}_n(V) \xrightarrow{k_n} \tilde{H}_n(X) \xrightarrow{D_n} \tilde{H}_{n-1}(A) \to \dots \tag{3.83}$$

whose homomorphisms are computed as in the non-reduced case.

*(h) Reduced homology can also be defined directly, without going back to chain complexes.

A space $X$ has a unique map $f\colon X \to \{*\}$ with values in the singleton, and:

$$\begin{aligned} \tilde{H}_n(X) &= \mathrm{Ker}\,(f_{*n}\colon H_n(X) \to H_n(\{*\})) \qquad (n \geqslant 0),\\ \tilde{H}_{-1}(X) &= \mathrm{Cok}\,(f_{*0}\colon H_0(X) \to H_0(\{*\})). \end{aligned} \tag{3.84}$$

This approach works for every homology theory; it is used in [EiS], Section I.7 (for $n \geqslant 0$).

### *3.6.3 Suspension and reduced homology*

In reduced homology there is a natural isomorphism *in all degrees*

$$\delta_n X \colon \tilde{H}_{n+1}(\Sigma X) \to \tilde{H}_n(X) \qquad (n \geqslant -1), \tag{3.85}$$

corresponding to the isomorphism of ordinary homology (2.121), which only holds for $n > 0$.

The proof works as in the ordinary case, applying the MV-sequence of reduced homology (in 3.6.2(g)) to the open cover $(U, V)$ of $\Sigma X$ used in 2.4.5(a): the spaces $U, V$ are contractible, which gives $\tilde{H}_n(U) = \tilde{H}_n(V) = 0$, for all $n$.

In particular we now have:

$$\tilde{H}_1(\mathbb{S}^1) \cong \tilde{H}_0(\mathbb{S}^0) \cong \mathbb{Z}, \qquad \tilde{H}_0(\mathbb{S}^0) \cong \tilde{H}_{-1}(\mathbb{S}^{-1}) \cong \mathbb{Z}. \tag{3.86}$$

### *3.6.4 The simplicial form of singular homology*

We end this chapter showing how singular homology is constructed using the tetrahedral models $\Delta^n$, instead of the cubical models $\mathbb{I}^n$: here a space $X$ is explored by studying the probes $\Delta^n \to X$, where $\Delta^0$ is the singleton, $\Delta^1$ is a compact interval, $\Delta^2$ a compact triangle, etc.

The simplicial construction has a formal advantage: it does not need normalisation. On the other hand, it is more complicated in many aspects we have studied or will study below. But we shall see, in Section 6.6, that it gives a simpler theory of cohomology rings.

### *3.6.5 The geometry of tetrahedra*

The *standard tetrahedron* $\Delta^n$ (also called the *standard* $n$*-simplex*) is a subspace of $\mathbb{R}^{n+1}$: the convex hull of the unit points $e_0, ..., e_n$ of the cartesian axes

$$\Delta^n = \{(t_0, ..., t_n) \in \mathbb{R}^{n+1} \mid t_i \geqslant 0,\ \Sigma t_i = 1\} \subset \mathbb{R}^{n+1}, \tag{3.87}$$

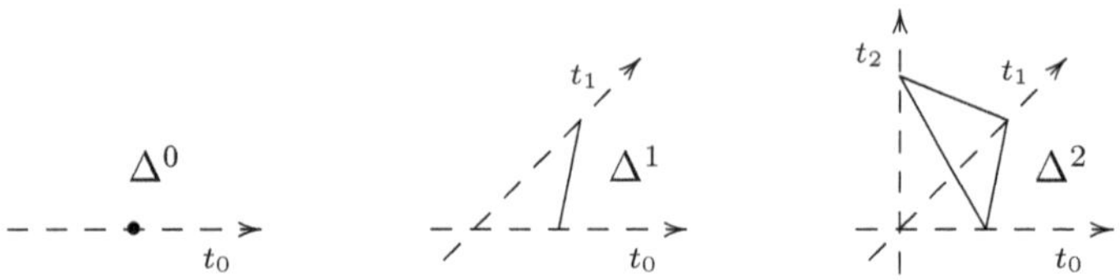

Faces and degeneracies are defined as:

$$\begin{aligned} &\delta_{ni}\colon \Delta^{n-1} \to \Delta^n, \\ &\qquad \delta_{ni}(t_0, ..., t_{n-1}) = (t_0, ..., t_{i-1}, 0, t_i, ..., t_{n-1}) \quad (0 \leqslant i \leqslant n), \\ &\varepsilon_{ni}\colon \Delta^n \to \Delta^{n-1}, \\ &\qquad \varepsilon_{ni}(t_0, ..., t_n) = (t_0, ..., t_i + t_{i+1}, ..., t_n) \quad (0 \leqslant i \leqslant n-1). \end{aligned} \tag{3.88}$$

We generally omit the index $n$. For singular homology we shall only use the faces, and it is important to have a concrete idea of them, in low dimension. (Degeneracies have a role in homotopy theory, similar to that of connections for the cubical setting.)

For $n = 0$, the singleton $\Delta^0 = \{1\}$ will be written as $\{*\}$. The two faces of $\Delta^1$ are its endpoints:

$$\delta_i\colon \{*\} \to \Delta^1, \qquad \delta_0(*) = e_1 \qquad \delta_1(*) = e_0, \tag{3.89}$$

and the three faces of $\Delta^2$ are parametrisations of its edges on $\Delta^1$

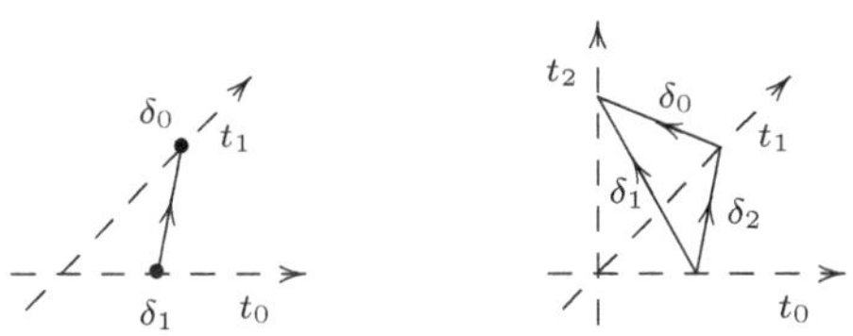


(3.90)

Faces and degeneracies satisfy the *cosimplicial relations*

$$\begin{aligned} &\delta_j\delta_i = \delta_{i+1}\delta_j, && \text{for } j \leqslant i, \\ &\varepsilon_i\varepsilon_j = \varepsilon_j\varepsilon_{i+1}, && \text{for } j \leqslant i, \\ &\varepsilon_j\delta_i = \begin{cases} \delta_{i-1}\varepsilon_j, & \text{for } j < i-1, \\ \text{id}, & \text{for } j = i-1, i, \\ \delta_i\varepsilon_{j-1}, & \text{for } j > i. \end{cases} \end{aligned} \tag{3.91}$$

The system $((\Delta^n), (\delta_i), (\varepsilon_i))$ is called a *cosimplicial object*, in $\mathsf{Top}$.

*Notes.* (a) The relation $\delta_j\delta_i = \delta_{i+1}\delta_j\colon \Delta^0 \to \Delta^2$ expresses the fact that each vertex of $\Delta^2$ belongs to two edges. The proof of the cosimplicial relations of faces is the same as in the cubical case of Exercise 2.2.1(a), for $\alpha = \beta = 0$. The other relations are similarly checked (but are not used to define singular homology).

(b) $\Delta^n$ can also be realised in $\mathbb{R}^n$, as the convex hull of the origin and the unit points $e_1, ..., e_n$. This form, used by some authors, requires disparate formulas for faces (and degeneracies), and makes calculations more complicated.

### 3.6.6 The singular simplicial set

For a space $X$, we apply the contravariant functor

$$\mathsf{Top}(-, X)\colon \mathsf{Top}^{\mathrm{op}} \to \mathsf{Set}$$

to the system $((\Delta^n), (\delta_i), (\varepsilon_i))$, obtaining the *singular simplicial set* $\mathrm{Smp}(X)$ of the space $X$

$$\begin{aligned} &(\mathrm{Smp}(X))_n = \mathsf{Top}(\Delta^n, X), \\ &\partial_i(a) = a\delta_i\colon \Delta^{n-1} \to X, \qquad \text{for } a\colon \Delta^n \to X, \\ &e_i(a) = a\varepsilon_i\colon \Delta^n \to X, \qquad \text{for } a\colon \Delta^{n-1} \to X. \end{aligned} \tag{3.92}$$

The arrows satisfy now the *simplicial relations*, dual to the cosimplicial ones (since we are applying a contravariant functor)

$$\begin{aligned} &\partial_i\partial_j = \partial_j\partial_{i+1}, \qquad \text{for } j \leqslant i, \\ &e_je_i = e_{i+1}e_j, \qquad \text{for } j \leqslant i, \\ &\partial_ie_j = \begin{cases} e_j\partial_{i-1}, & \text{for } j < i-1, \\ \mathrm{id}\,, & \text{for } j = i-1, i, \\ e_{j-1}\partial_i, & \text{for } j > i. \end{cases} \end{aligned} \tag{3.93}$$

In general, a *simplicial set* $K = ((K_n), (\partial_i), (e_i))$ is a sequence of sets $(K_n)_{n\geqslant 0}$ equipped with faces and degeneracies that satisfy the simplicial relations (3.93)

$$\begin{aligned} &\partial_i\colon K_n \to K_{n-1} \qquad (0 \leqslant i \leqslant n), \\ &e_i\colon K_{n-1} \to K_n \qquad (0 \leqslant i \leqslant n-1), \end{aligned} \tag{3.94}$$

$$K_0 \;\leftleftarrows\!\!\!\to\; K_1 \;\Lleftarrow\!\!\!\rightrightarrows\; K_2 \quad \ldots$$

When useful we write the faces and degeneracies as $\partial_{ni}\colon K_n \to K_{n-1}$ and $e_{ni}\colon K_{n-1} \to K_n$, including the degree $n$. The elements of $K_n$ are called *$n$-simplices*, and *vertices* or *edges* when $n$ is 0 or 1, respectively. An $n$-simplex has $n+1$ vertices.

A *morphism* of simplicial sets $f = (f_n)\colon K \to L$ is a sequence of mappings $f_n\colon K_n \to L_n$ that commute with faces and degeneracies. With componentwise composition, they form the category SmpSet.

We have now the *singular simplicial functor*, which acts on a space $X$ as we have seen, and on a map $f\colon X \to Y$ by covariant composition

$$\begin{aligned} &\mathrm{Smp}\colon \mathsf{Top} \to \mathrm{SmpSet}, \qquad X \mapsto \mathrm{Smp}(X), \\ &f_\sharp\colon \mathrm{Smp}(X) \to \mathrm{Smp}(Y), \qquad f_{\sharp n}(a\colon \Delta^n \to X) = fa. \end{aligned} \tag{3.95}$$

*Comments and literature.* (a) Simplicial sets can be defined as presheaves $K\colon \underline{\Delta}^{\mathrm{op}} \to \mathsf{Set}$ on the *simplicial site* $\underline{\Delta}$. The latter is the small category of positive finite ordinals and monotone mappings: see [M2], Section VII.5, or [G5], Section 5.3. There is an excellent, concise book on simplicial objects, by J.P. May [May1].

(b) We already said that the degeneracies have no role in the present context: one could replace simplicial sets with *face-simplicial sets* $K = ((K_n), (\partial_i))$, omitting everything about degeneracies. The latter were introduced by Eilenberg–Zilber [EiZ1] as 'semisimplicial complexes', a name which was later used for simplicial sets and would now lead to confusion. They are also called $\Delta$-*sets*, and can be viewed as presheaves of sets defined on the category of positive finite ordinals and strictly increasing mappings.

### 3.6.7 Homology

We define the *chain functor* of simplicial sets

$$\mathrm{Ch}_+ \colon \mathsf{SmpSet} \to \mathrm{Ch}_+\mathsf{Ab}, \qquad \mathrm{Ch}_+(K) = (C_n(K)), (\partial_n)), \tag{3.96}$$

where, for a simplicial set $K$ the component $C_n(K)$ is the free abelian group $F(K_n) = \mathbb{Z}K_n$. The differential is defined on the basis $K_n$ as a linear combination of the faces $\partial_{ni}\colon K_n \to K_{n-1}$

$$\partial_n(a) = \Sigma_i\, (-1)^i\, \partial_{ni}(a) \in F(K_{n-1}) \qquad (i = 0, 1, ..., n). \tag{3.97}$$

The relation $\partial_n \partial_{n+1} = 0$ is a consequence of the simplicial relations of the faces, as in the cubical case (see 2.2.3(b)).

On a morphism $f\colon K \to L$ of simplicial sets, the components of $\mathrm{Ch}_+(f)$ are the linear extensions of the components $f_n$

$$\begin{gathered} \mathrm{Ch}_+(f) = (C_n(f))_{n \geqslant 0}, \\ C_n(f) = F(f_n)\colon C_n(K) \to C_n(L), \quad \Sigma\, \lambda_i a_i \mapsto \Sigma\, \lambda_i f_n(a_i). \end{gathered} \tag{3.98}$$

Composing with the functor $\mathrm{Smp}\colon \mathsf{Top} \to \mathsf{SmpSet}$ of (3.95), we have the functor *of the singular simplicial chain complex*:

$$\begin{gathered} S_+ \colon \mathsf{Top} \to \mathrm{Ch}_+\mathsf{Ab}, \qquad S_+(X) = \mathrm{Ch}_+(\mathrm{Smp}(X)), \\ f_{\sharp n}(a) = fa \qquad (\text{for } f\colon X \to Y \text{ and } a\colon \Delta^n \to X \text{ in } \mathsf{Top}). \end{gathered} \tag{3.99}$$

Finally, the homology functor $H_n\colon \mathrm{Ch}_+\mathsf{Ab} \to \mathsf{Ab}$ of chain complexes (in (2.24)) gives the $n$-th *singular homology functor*, in simplicial form

$$H_n\colon \mathsf{Top} \to \mathsf{Ab}, \quad H_n(X) = H_n(\mathrm{Ch}_+(\mathrm{Smp}(X))) = H_n(S_+(X)). \tag{3.100}$$

For a topological space $X$, a (formal) linear combination of singular simplices $a_i\colon \Delta^n \to X$, with coefficients in $\mathbb{Z}$

$$c = \Sigma_i\, \lambda_i a_i \in S_n(X) = F(\mathsf{Top}(\Delta^n, X)), \tag{3.101}$$

is called a *singular simplicial n-chain* of the space $X$. Again, it is an *n-cycle* if $\partial_n(c) = 0$, and an *n-boundary* if $c \in \operatorname{Im} \partial_{n+1}$.

Every cycle $c$ has a *homology class* $[c] \in H_n(X)$, which vanishes if and only if $c$ is a boundary.

*Exercises and complements.* (a) A 0-chain is a formal linear combination $\sum \lambda_i x_i$ of points $x_i \colon \{*\} \to X$; it is always a 0-cycle.

(b) A 1-chain is a formal linear combination of 1-simplices $a \colon \Delta^1 \to X$. The boundary of a 1-simplex

$$a \colon \Delta^1 \to X, \quad \partial_1(a) = a(0,1) - a(1,0) = a(e_1) - a(e_0), \qquad (3.102)$$

is again the formal difference of its endpoints; $a$ is a 1-cycle if and only if it is a loop.

(c) Prove that the homology of the singleton is $\mathbb{Z}$ in degree zero and trivial elsewhere, also in the current setting.

This is a consequence of Lemma 3.6.8, below. Nevertheless, a direct proof shows the technical difference with the corresponding cubical issue.

(d) In order to view the maps $\Delta^1 \to X$ as standard paths in the space $X$, we identify $\Delta^1$ with the standard interval $\mathbb{I}$, according to the homeomorphism $\mathbb{I} \to \Delta^1$, $t \mapsto (1-t, t)$, a path from $e_0$ to $e_1$.

This agrees with the differential of a 1-cube

$$a \colon \mathbb{I} \to X, \qquad \partial_1(a) = a(1) - a(0), \qquad (3.103)$$

and with the orientations marked in figure (3.90).

(e) Also here, as in 2.2.5(d), $H_0(X)$ is a free abelian group, with a canonical basis $[x_i]_H$ $(i \in I)$, where $x_i$ is any point of the path component $X_i$. The proof is the same as in 2.2.5(d).

### *3.6.8 Lemma*

*For a non-empty convex euclidean space $X$, $H_0(X) \cong \mathbb{Z}$ and $H_k(X) = 0$ for $k > 0$.*

*Proof* The argument only needs that $X$ be star-shaped, for a point $x_0 \in X$.

The chain complex $J(\mathbb{Z})$ associated to the abelian group $\mathbb{Z}$ (see (2.18)) is a retract of $S_+(X)$, in an obvious way:

$$i \colon J(\mathbb{Z}) \to S_+(X), \qquad\qquad p \colon S_+(X) \to J(\mathbb{Z}),$$
$$i_0(1) = x_0, \qquad\qquad p_0(x) = 1, \quad p_n = 0 \quad (n > 0).$$

Proving that this is a deformation retract will show that $H_+(X)$ is the graded group $J(\mathbb{Z})$.

For simplicity we assume that $x_0$ is the origin; thus, for any $x \in X$, the affine path $a$ from $x_0$ to $x$ can be written as $a(t) = tx$, for $t \in \mathbb{I}$.

We have to build a chain homotopy $\varphi\colon ip \simeq \mathrm{id}$, i.e. a sequence

$$\varphi_n\colon S_n(X) \to S_{n+1}(X), \quad \partial_{n+1}\varphi_n(a) + \varphi_{n-1}\partial_n(a) = a - i_n p_n(a)$$

(for every $n$-simplex $a\colon \Delta^n \to X$).

We define $\varphi_n(a)\colon \Delta^{n+1} \to X$, letting $\varphi_n(a)(1,0,...,0) = 0$ and

$$\varphi_n(a)(t_0,...,t_{n+1}) = (1-t_0)\,a(\lambda t_1,...,\lambda t_{n+1}), \quad \text{for } t_0 < 1, \tag{3.104}$$

where $\lambda = (1-t_0)^{-1}$. Now, for $n = 0$ and $a \in X$ we do have

$$\partial_1\varphi_0(a) = \varphi_n(a)(0,1) - \varphi_n(a)(1,0) = a - 0 = a - i_0 p_0(a).$$

Finally, for $n > 0$ and $a\colon \Delta^n \to X$

$$\begin{aligned} \partial_{n+1}\varphi_n(a) &= \textstyle\sum_{i \leqslant n+1} (-1)^i\, \varphi_n(a)\delta_{n+1,i} \\ &= a - \varphi_{n-1}(\textstyle\sum_{i \leqslant n}(-1)^i a\delta_{ni}) = a - \varphi_{n-1}\partial_n(a). \end{aligned} \tag{3.105}$$

□

### 3.6.9 *Homotopy invariance

The study of the simplicial form of singular homology goes on, much in the same way as for the cubical form previously studied. The proofs and computations of the basic theory are generally more complex; other issues can be simpler (as we shall see much later, in Section 6.6).

Here we hint at diverse ways of proving the homotopy invariance of the simplicial form.

(a) (*A direct, constructive proof*) This is likely the clearest argument, if more complex than in the cubical case. Given a homotopy $\varphi\colon f \simeq g\colon X \to Y$ in Top, we build a chain homotopy

$$\Phi\colon f_\sharp \simeq g_\sharp\colon S_+(X) \to S_+(Y), \tag{3.106}$$

covering the product $\Delta^n \times \mathbb{I}$ with $n+1$ tetrahedra $[v_0,...,v_i,w_i,...,w_n]$ of dimension $n+1$, which only meet at faces. (The square brackets [...] denote the convex hull of the specified points, in $\mathbb{R}^{n+1}$.)

The following figure shows two triangles covering the rectangle $\Delta^1 \times \mathbb{I}$ and meeting at an edge, a diagonal of the square

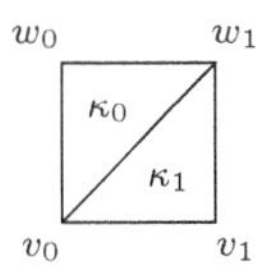


$$\begin{aligned} \Delta^1 \times \mathbb{I} &= [v_0, w_0, w_1] \cup [v_0, v_1, w_1] \\ &= \operatorname{Im}\kappa_0 \cup \operatorname{Im}\kappa_1. \end{aligned} \tag{3.107}$$

In the general case:

- $v_0, ..., v_n$ are the vertices of the $n$-tetrahedron $\Delta^n \times \{0\} \subset \mathbb{R}^{n+1}$,
- $w_0, ..., w_n$ are the vertices of the $n$-tetrahedron $\Delta^n \times \{1\} \subset \mathbb{R}^{n+1}$,
- $\kappa_i \colon \Delta^{n+1} \to \Delta^n \times \mathbb{I}$ is the affine embedding that takes the vertices $e_0, ..., e_{n+1}$ of $\Delta^{n+1}$ to the points $v_0, ..., v_i, w_i, ..., w_n$ of $\Delta^n \times \mathbb{I}$, in this order (for $i = 0, ..., n$).

Finally, for any simplex $a \colon \Delta^n \to X$, we let

$$\Phi_n(a) = \Sigma_{0 \leqslant i \leqslant n} (-1)^i \varphi a \kappa_i,$$

and it is not difficult to verify that $\partial_{n+1}\Phi_n + \Phi_{n-1}\partial_n = ga - fa$. (The details can be found in [Ha], Theorem 2.10.)

In this way Lemma 3.6.8 is pleonastic, since a non-empty convex euclidean space is contractible, and the homology of the singleton is already computed, in Exercise 3.6.7(c).

(b) On the other hand, one can deduce homotopy invariance from Lemma 3.6.8, using the Acyclic Model Theorem of Eilenberg–Mac Lane (see 6.7.3), as in [Sp2], Section 4.4.

Using the terminology of Section 6.7, the point is that the functor $S_+ \colon$ $\mathsf{Top} \to \mathrm{Ch}_+\mathsf{Ab}$ is free on the models $\Delta^n$, while the functor $S_+(- \times \mathbb{I})$ is acyclic on them, because each $\Delta^n \times \mathbb{I}$ is convex.

A similar layout is followed in [Vi], Theorem 1.10, applying the technique of acyclic models in this particular case, rather than the general theorem.

(c) We shall see in Section 6.7 that the Acyclic Model Theorem (with the help of Lemma 3.6.8) proves that the simplicial form of singular homology is naturally isomorphic to the cubical form.

As we already know that the cubical form is homotopy invariant, also the simplicial form is. This approach is followed in [HiW], Theorem 8.4.11.

# 4
# Homology and cohomology theories

Singular homology is generally studied in the extended form of *relative singular homology*, defined on the category $\mathsf{Top}_2$ of *relative pairs of topological spaces*: these are pairs $(X, A)$ formed of a space $X$ and a subspace $A$. This issue is dealt with in the first three sections of this chapter.

In Section 4.4 we list the Eilenberg–Steenrod axioms for homology theories, noting that they hold for (relative) singular homology. We also state the corresponding axioms for the contravariant version, a cohomology theory.

In the next two sections we deal with Alexander–Spanier cohomology and its multiplicative structure: the sequence of all cohomology groups (with coefficients in a ring) becomes a graded ring.

In Sections 4.7–4.9 we briefly introduce Čech cohomology, de Rham cohomology of differentiable manifolds, and CW-complexes. Relative singular homology and cohomology with coefficients in a group are deferred to the next chapter, after some new elements of Homological Algebra.

Alexander–Spanier and Čech cohomologies are closely related, and more important than the singular one, according to some authors. Quoting from the Introduction of Massey's book [Mas2]: "There are two main kinds of homology and cohomology theory: the singular theory and the Čech–Alexander–Spanier type theory. Although the two theories give the same results on topological spaces which are locally nice, it is well known that on general spaces the Čech–Alexander–Spanier type cohomology theory has several advantages over the singular theory."

## 4.1 Relative singular homology

We define relative singular homology, prove its homotopical invariance and the exact homology sequence of a pair of spaces. The term 'pair' has here a special, traditional meaning.

### 4.1.1 Relative pairs of spaces

We begin by introducing the category $\mathsf{Top}_2$ of *pairs of topological spaces.*

An object of this category is a *pair* $(X, A)$ formed of a space $X$ and a *subspace* $A$; more explicitly, it can be called a *relative pair* of topological spaces. A (*relative*) *map* $f\colon (X, A) \to (Y, B)$ is a continuous mapping $f\colon X \to Y$ such that $f(A) \subset B$; they compose as in $\mathsf{Top}$.

$\mathsf{Top}$ is embedded as a full subcategory of $\mathsf{Top}_2$, identifying the space $X$ with the pair $(X, \emptyset)$.

The relative pair $(X, A)$ is usually read as '$X$ modulo $A$' and viewed as a surrogate of a quotient, more 'flexible' than collapsing the subspace $A$ to a point.

A (*relative*) *homotopy* $\varphi\colon f_0 \simeq f_1\colon (X, A) \to (Y, B)$ of relative maps is a map of $\mathsf{Top}_2$ such that:

$$\varphi\colon (X \times \mathbb{I}, A \times \mathbb{I}) \to (Y, B), \qquad \varphi(x, \alpha) = f_\alpha(x) \tag{4.1}$$

(for $x \in X$, $\alpha = 0, 1$). It is the same as an ordinary homotopy $\varphi\colon f_0 \simeq f_1\colon X \to Y$ such that $\varphi(A \times \mathbb{I}) \subset B$.

In this context, all issues related to $\mathsf{Top}$ (objects, maps, homotopies, homology groups...) are called *absolute*, while those related to $\mathsf{Top}_2$ are called *relative*.

*Exercises and complements.* (a) The homotopy relation in $\mathsf{Top}_2$ is a congruence of categories: reflexive, symmetric, transitive and consistent with composition.

(b) *Homotopy equivalence* in $\mathsf{Top}_2$ is defined in the usual way, and amounts to isomorphism in the quotient category $\mathsf{Top}_2/\!\simeq$.

(c) Relative homotopies derive from the *relative cylinder functor*

$$I\colon \mathsf{Top}_2 \to \mathsf{Top}_2, \qquad I(X, A) = (X \times \mathbb{I}, A \times \mathbb{I}), \tag{4.2}$$

with a structure (faces, degeneracy, etc.) similar to that of the absolute cylinder functor, in (1.4.3), and defined by the same formulas; for instance:

$$\partial^\alpha\colon (X, A) \to I(X, A), \qquad \partial^\alpha(x) = (x, \alpha) \qquad (\alpha = 0, 1). \tag{4.3}$$

(d) The category $\mathsf{Top}_\bullet$ is also embedded as a full subcategory of $\mathsf{Top}_2$, identifying the pointed space $(X, x_0)$ with the pair $(X, \{x_0\})$. In this embedding, pointed homotopies are the same as relative homotopies.

(e) The expression 'relative homeomorphism' has a traditional meaning, defined in 4.2.5. An invertible arrow $f\colon (X, A) \to (Y, B)$ in $\mathsf{Top}_2$ will simply be called a *homeomorphism* (of relative pairs). It is the same as a homeomorphism $f\colon X \to Y$ such that $f(A) = B$.

### 4.1.2 Relative chains

For a pair $(X, A)$ in $\mathsf{Top}_2$, we consider the complex $C_+(X) = \mathrm{Ch}_+(\mathrm{Cub}(X))$ of singular chains of the space $X$, and its *subcomplex* $C_+(A)$: we are following the usual abuse of notation, identifying a cube $\mathbb{I}^n \to A$ with a cube of $X$ whose image is contained in $A$ (as in 2.3.4).

The quotient complex, with induced differentials (cf. (2.69))

$$\begin{gathered} C_+(X, A) = C_+(X)/C_+(A), \\ C_n(X, A) = C_n(X)/C_n(A), \qquad \partial_n(\overline{c}) = (\partial_n(c))^-, \end{gathered} \tag{4.4}$$

is called the *complex of relative singular chains* of the pair $(X, A)$, or *of $X$ modulo $A$*. We write as $\overline{c}$ the equivalence class $c + C_n(A)$ of a chain $c \in C_n(X)$.

We have thus a short exact sequence in the category $\mathrm{Ch}_+\mathsf{Ab}$

$$C_+(A) \xrightarrow{m} C_+(X) \xrightarrow{p} C_+(X, A), \tag{4.5}$$

where $m$ is the inclusion and $p$ is the projection on the quotient.

Applying the Induction Lemma for chain complexes (in Exercise 2.3.1(c)), a relative map $f\colon (X, A) \to (Y, B)$ gives a commutative diagram with short exact rows

$$\begin{array}{ccccc} C_+(A) & \rightarrowtail & C_+(X) & \twoheadrightarrow & C_+(X, A) \\ \downarrow{\scriptstyle f_\sharp} & & \downarrow{\scriptstyle f_\sharp} & & \downarrow{\scriptstyle f_\sharp} \\ C_+(B) & \rightarrowtail & C_+(Y) & \twoheadrightarrow & C_+(Y, B) \end{array} \tag{4.6}$$

We are using the same notation $f_\sharp$ for three chain morphisms:

- $f_\sharp\colon C_+(X) \to C_+(Y)$, associated to the absolute map $f\colon X \to Y$,
- $f_\sharp\colon C_+(A) \to C_+(B)$, the restricted chain morphism (associated to the restricted map $A \to B$),
- the induced morphism on quotients, associated to the relative map $f$

$$f_\sharp\colon C_+(X, A) \to C_+(Y, B), \qquad f_\sharp(\overline{c}) = (f_\sharp(c))^-. \tag{4.7}$$

All this defines the functor of *relative* (cubical) *chains*

$$C_+\colon \mathsf{Top}_2 \to \mathrm{Ch}_+\mathsf{Ab}, \qquad C_+(X, A) = C_+(X)/C_+(A), \tag{4.8}$$

which extends the absolute form, as we identify $C_+(X, \emptyset)$ with $C_+(X)$.

*Exercises and complements.* (a) Each component $C_n(X, A)$ is a free abelian group, with a canonical basis $c_n(X, A)$ formed of the non-degenerate $n$-cubes $a\colon \mathbb{I}^n \to X$ whose image is not contained in $A$.

(b) The short exact sequence (4.5) splits in every degree (see 3.5.4(d)).

As usual, the splitting need not be consistent with differentials. In each component there is a canonical splitting, given by the basis described in (a), but this splitting is not respected by relative maps, and is not natural on $\mathsf{Top}_2$.

### 4.1.3 Relative singular homology

Composing the functor $C_+\colon \mathsf{Top}_2 \to \mathrm{Ch}_+\mathsf{Ab}$ of relative chains, with the chain homology functor $H_n\colon \mathrm{Ch}_+\mathsf{Ab} \to \mathsf{Ab}$, we have the *relative singular homology functor* of degree $n \geqslant 0$:

$$\begin{gathered} H_n\colon \mathsf{Top}_2 \to \mathsf{Ab}, \qquad H_n(X,A) = H_n(C_+(X,A)), \\ H_n(f\colon (X,A) \to (Y,B)) = f_{*n}\colon H_n(X,A) \to H_n(Y,B). \end{gathered} \tag{4.9}$$

Relative homology extends the absolute form: $H_n(X) = H_n(X,\emptyset)$ (and similarly on maps). $H_n(X,A)$ is called the *$n$-th group of singular homology of $X$ modulo $A$.*

In particular, a *trivial pair* $(X,X)$ has a trivial chain complex $C_+(X,X)$, and trivial homology $H_n(X,X) = 0$ in every degree. More generally, we shall see that $H_n(X,A) = 0$ (for all $n$) whenever the inclusion $A \to X$ is a homotopy equivalence (in 4.1.6).

*Exercises and complements.* We are given a chain $c \in C_n(X)$ and the associated relative chain $\overline{c} \in C_n(X,A)$. The following (easy) exercises are left to the reader.

(a) $\overline{c}$ is a cycle of $C_n(X,A)$ if and only if: $\partial_n(c) \in C_{n-1}(A)$.

(b) $\overline{c}$ is a boundary of $C_n(X,A)$ if and only if: $c \in B_n(X) + C_n(A)$.

### 4.1.4 Theorem (Homotopy invariance)

*The relative singular homology functor $H_n\colon \mathsf{Top}_2 \to \mathsf{Ab}$ is homotopy invariant. It takes homotopy equivalent relative pairs of spaces to isomorphic groups.*

*Proof* It is an easy consequence of the corresponding Theorem 2.2.6 for the absolute case, and can be written as an exercise. As in the proof of 2.2.6, it is convenient to write the cylinder functor in the form $I(X,A) = (\mathbb{I} \times X, \mathbb{I} \times A)$, to avoid a sign change depending on degree.

We start from a relative homotopy $\varphi\colon f \simeq g\colon (X,A) \to (Y,B)$, as defined in (4.1). The absolute homotopy $\varphi\colon f \simeq g\colon X \to Y$ has an associated chain homotopy (for $a\colon \mathbb{I}^n \to X$):

$$\begin{gathered} \Phi\colon f_\sharp \simeq g_\sharp\colon C_+(X) \to C_+(Y), \\ \Phi_n\colon C_nX \to C_{n+1}Y, \qquad \Phi_n(a) = \varphi(\mathbb{I} \times a)\colon \mathbb{I}^{n+1} \to Y, \\ (\partial_{n+1}\Phi_n + \Phi_{n-1}\partial_n)(a) = ga - fa. \end{gathered} \tag{4.10}$$

Since $\varphi$ is a relative homotopy, $\Phi_n$ restricts to a homomorphism $\Phi'_n\colon C_n(A) \to C_{n+1}(B)$. Applying the Induction Lemma 2.1.4 we have a commutative diagram with short exact rows (for $n \geqslant 0$)

$$\begin{array}{ccccc} C_n(A) & \rightarrowtail & C_n(X) & \twoheadrightarrow & C_n(X,A) \\ \Phi'_n \downarrow & & \downarrow \Phi_n & & \downarrow \Phi''_n \\ C_{n+1}(A) & \rightarrowtail & C_{n+1}(X) & \twoheadrightarrow & C_{n+1}(X,A) \end{array} \qquad (4.11)$$

The induced homomorphisms $\Phi''_n\colon C_n(X,A) \to C_{n+1}(Y,B)$ form a chain homotopy $f_\sharp \simeq g_\sharp\colon C_+(X,A) \to C_+(Y,B)$, so that, for every $n \geqslant 0$, $f_{*n} = g_{*n}\colon H_n(X,A) \to H_n(Y,B)$. □

**4.1.5 Theorem** (The homology sequence of a pair)

*For every pair of topological spaces $(X,A)$, the following sequence is exact and natural (for maps $(X,A) \to (Y,B)$ of $\mathsf{Top}_2$)*

$$\begin{array}{ccccccc} & & \cdots & \longrightarrow & H_2(X,A) & & \\ & & & D_2 & & & \\ H_1(A) & \xrightarrow{u_{*1}} & H_1(X) & \xrightarrow{v_{*1}} & H_1(X,A) & & \\ & & & D_1 & & & \\ H_0(A) & \xrightarrow{u_{*0}} & H_0(X) & \xrightarrow{v_{*0}} & H_0(X,A) & \to & 0. \end{array} \qquad (4.12)$$

*Here $u\colon A \to X$ is the inclusion and $v\colon (X,\emptyset) \to (X,A)$ is defined by the identity of $X$. The connecting homomorphism $D_n$ is calculated as:*

$$D_n\colon H_n(X,A) \to H_{n-1}(A), \qquad D_n[\overline{c}]_{(X,A)} = [\partial_n(c)]_A, \qquad (4.13)$$

*where $\overline{c} \in C_n(X,A)$ is a cycle and $\partial_n(c) \in C_{n-1}(A)$ is a cycle in $A$ (which need not be a boundary in $A$).*

*Proof* In the short exact sequence of chain complexes (4.5)

$$C_+(A) \overset{m}{\rightarrowtail} C_+(X) \overset{p}{\twoheadrightarrow} C_+(X,A), \qquad m = u_\sharp, \;\; p = v_\sharp, \qquad (4.14)$$

the morphisms are induced by the relative maps $u\colon (A,\emptyset) \to (X,\emptyset)$ and $v\colon (X,\emptyset) \to (X,A)$.

The associated homology sequence (in 2.3.3) gives the exact sequence (4.12). Moreover the sequence (4.14) is natural for morphisms of $\mathsf{Top}_2$, and therefore the sequence (4.12) is too.

We only have to check that the morphism $D_n\colon H_n(X,A) \to H_{n-1}(A)$ is indeed calculated as in (4.13).

For a cycle $\overline{c} \in C_n(X, A)$, using the definition of $D_n$ in (2.74)

$$\begin{array}{ccccc} & & c & \longmapsto & \overline{c} & \qquad (n) \\ & & \downarrow & & \downarrow & \\ \partial_n c & \longmapsto & \partial_n c & \longmapsto & 0 & \qquad (n-1) \end{array} \tag{4.15}$$

we lift $\overline{c}$ to a representative chain $c \in C_n(X)$. Its differential $\partial_n(c)$ belongs to $C_{n-1}(A)$ (as remarked in 4.1.3(a)), and $D_n[\overline{c}] = [\partial_n(c)]$. □

### *4.1.6 Exercises and complements*

(a) If $A$ is a deformation retract of $X$, then $H_n(X, A) = 0$ for every $n$. More generally, the same holds if the inclusion $A \to X$ is a homotopy equivalence.

(b) In the general case, $H_0(X, A)$ is the free abelian group generated by the path components of $X$ which do not meet $A$.

(c) Consider the relative pair $(\mathbb{D}^{n+1}, \mathbb{S}^n)$, for $n \geqslant 0$. The connecting morphism of the sequence of relative homology gives a canonical isomorphism

$$D_k \colon H_k(\mathbb{D}^{n+1}, \mathbb{S}^n) \cong \tilde{H}_k(\mathbb{S}^n) \qquad \text{(for all } k). \tag{4.16}$$

All the relative homology groups $H_k(\mathbb{D}^{n+1}, \mathbb{S}^n)$ are thus trivial, except $H_n(\mathbb{D}^{n+1}, \mathbb{S}^n) \cong \mathbb{Z}$.

(d) (*Pointed homology*) For a point $x \in X$, we have identified the pointed space $(X, x)$ with the relative pair $(X, \{x\})$, in 4.1.1(d). The relative homology group $H_n(X, x)$ will be called the *pointed homology* of $X$ at $x$. Prove that it is isomorphic to the reduced homology of $X$

$$H_n(X, x) \cong \tilde{H}_n(X) \qquad (n \geqslant 0), \tag{4.17}$$

in a canonical way, depending on the pair $(X, x)$. (Here there is no need of $\tilde{H}_{-1}$, which is trivial on non-empty spaces.)

In particular $H_0(X, x)$ is canonically isomorphic to the free abelian group generated by the set of the path components of $X$ not containing $x$.

### *4.1.7 Exercises and complements* (Relative and reduced homology)

Reduced homology is well related to relative homology, and better related than ordinary homology (in degree zero, where there is a difference).

(a) First, we remark that relative homology is 'already reduced': for a relative pair $(X, A)$, the quotient chain complex $\tilde{C}(X)/\tilde{C}(A)$ can (and will) be identified with $C_+(X, A) = C(X)/C(A)$.

(b) For a relative pair $(X, A)$, there is an exact sequence in reduced homology which only differs from the ordinary one in degree $\leqslant 0$, and is related to the former by the natural transformation $\rho$, in (3.79), forming a commutative diagram with exact rows

$$\begin{array}{ccccccccccc} \dots H_1(X,A) & \to & \tilde{H}_0A & \to & \tilde{H}_0X & \to & H_0(X,A) & \to & \tilde{H}_{-1}A & \twoheadrightarrow & \tilde{H}_{-1}X \\ \| & & \downarrow{\scriptstyle \rho A} & & \downarrow{\scriptstyle \rho X} & & \| & & \downarrow & & \downarrow \\ \dots H_1(X,A) & \to & H_0A & \to & H_0X & \to & H_0(X,A) & \longrightarrow & 0 & \longrightarrow & 0 \end{array} \qquad (4.18)$$

(c) If $A \neq \emptyset$, then $\tilde{H}_{-1}(A) = \tilde{H}_{-1}(X) = 0$ and we have a commutative diagram with exact rows

$$\begin{array}{ccccccccc} \dots\ H_1(X,A) & \xrightarrow{D_1} & \tilde{H}_0A & \xrightarrow{u_{*0}} & \tilde{H}_0X & \xrightarrow{v_{*0}} & H_0(X,A) & \longrightarrow & 0 \\ \| & & \downarrow{\scriptstyle \rho A} & & \downarrow{\scriptstyle \rho X} & & \| & & \\ \dots\ H_1(X,A) & \xrightarrow[D_1]{} & H_0A & \xrightarrow[u_{*0}]{} & H_0X & \xrightarrow[v_{*0}]{} & H_0(X,A) & \longrightarrow & 0 \end{array} \qquad (4.19)$$

(d) In the pair $(X, A)$, the space $A$ has trivial reduced homology if and only if the canonical map $v \colon X \to (X, A)$ gives a family of isomorphisms $v_{*n} \colon \tilde{H}_n(X) \cong H_n(X, A)$, for $n \geqslant 0$.

(e) In the pair $(X, A)$, the space $X$ has trivial reduced homology if and only if all the connecting homomorphisms $D_n \colon H_n(X, A) \cong \tilde{H}_{n-1}(A)$ of reduced homology are isomorphisms. (This point extends 4.1.6(c).)

(f) The pair $(X, A)$ has $H_n(X, A) = 0$, for all $n$, if and only if the inclusion $u \colon A \to X$ induces isomorphisms $u_{*n} \colon H_n(A) \cong H_n(X)$ (for $n \geqslant 0$), if and only if it induces isomorphisms $u_{*n} \colon \tilde{H}_n(A) \cong \tilde{H}_n(X)$ (for $n \geqslant -1$).

### *4.1.8* **Theorem** (The homology sequence of a triple)

*For a space $X$ and subspaces $B \subset A \subset X$, we have two canonical morphisms in* $\mathsf{Top}_2$

$$(A, B) \xrightarrow{u} (X, B) \xrightarrow{v} (X, A), \qquad (4.20)$$

*given by the inclusion $A \subset X$ and* $\mathrm{id}\, X$, *and an exact homology sequence in* $\mathsf{Ab}$, *called the exact sequence of the (relative)* triple $(X, A, B)$

$$\begin{array}{ccccccc} & & \dots & \longrightarrow & H_2(X,A) & & \\ & & & {\scriptstyle D_2} & & & \\ H_1(A,B) & \xrightarrow{u_{*1}} & H_1(X,B) & \xrightarrow{v_{*1}} & H_1(X,A) & & \\ & & & {\scriptstyle D_1} & & & \\ H_0(A,B) & \xrightarrow{u_{*0}} & H_0(X,B) & \xrightarrow{v_{*0}} & H_0(X,A) & \longrightarrow & 0, \end{array} \qquad (4.21)$$

*where $D_n[\overline{c}] = [\partial_n(c)]$, for every cycle $\overline{c}$ of $C_n(X, A)$.*

*This sequence is natural for morphisms $f\colon (X, A, B) \to (X', A', B')$ of relative triples of spaces; this obviously means a map $f\colon X \to X'$ such that $f(A) \subset A'$ and $f(B) \subset B'$.*

Note. *Theorem 4.1.5 corresponds to the case $B = \emptyset$.*

*Proof* The $3 \times 3$ Lemma, in 3.5.2, proves that the lower row of the following diagram of chain complexes is short exact

$$\begin{array}{ccccc}
C_+(B) & \rightarrowtail & C_+(B) & \twoheadrightarrow & 0 \\
\downarrow & & \downarrow & & \downarrow \\
C_+(A) & \rightarrowtail & C_+(X) & \twoheadrightarrow & C_+(X, A) \\
\downarrow & & \downarrow & & \downarrow \\
C_+(A, B) & \to & C_+(X, B) & \to & C_+(X, A)
\end{array} \qquad (4.22)$$

as we know that the three columns and the other two rows are short exact.

The exact homology sequence of the lower row gives the exact sequence (4.21). Its naturality is obvious.

The connecting morphisms $D_n\colon H_n(X, A) \to H_{n-1}(A, B)$ is indeed computed as stated above: the proof is the same as at the end of 4.1.5. □

### *4.1.9 *Comments*

Theorem 4.1.8 says that relative singular homology, with the connecting homomorphisms $D_n$, forms an *exact connected sequence* of functors from $\mathsf{Top}_2$ to $\mathsf{Ab}$, provided we can interpret the sequence (4.20)

$$(A, B) \xrightarrow{\ u\ } (X, B) \xrightarrow{\ v\ } (X, A),$$

as a short exact sequence in $\mathsf{Top}_2$, and actually as *the* general short exact sequence of topological pairs (up to isomorphism). This layout is developed in [G3], and briefly sketched in [G5], Section 6.7.

Essentially, the category $\mathsf{Top}_2$ is equipped with a class of 'null objects', those of the form $(X, X)$, that is, $X$ *modulo* $X$; a morphism $f\colon (X, A) \to (Y, B)$ is *null* if it factorises through a null object, or equivalently if $f(X) \subset B$. Kernels and cokernels are defined by the usual universal property, with respect to the ideal of null morphisms, and $\mathsf{Top}_2$ satisfies axioms making it a 'homological category' (when equipped with this ideal).

Exact sequences are defined accordingly, and (4.20) is proved to be the general short exact sequence in $\mathsf{Top}_2$. The particular case $A \rightarrowtail X \twoheadrightarrow (X, A)$ shows that $(X, A)$ is indeed the quotient of $X$ modulo $A$, with respect to the ideal of null morphisms.

## 4.2 Excision and compact pairs

We prove the Excision Theorem for singular homology, one of the main properties of relative homology. Loosely speaking, in a relative pair $(X, A)$ we can cut out of each space a subspace $U$ which is 'strongly included' in $A$, in the sense that $\mathrm{cl}_X U \subset \mathrm{int}_X A$.

Then we begin exploring some of its consequences: for a 'strong compact pair' $(X, A)$, the relative homology $H_n(X, A)$ is canonically isomorphic to the reduced homology $\tilde{H}_n(X/A)$ of the quotient of the space $X$ that collapses the subspace $A$ to a point: a case were a relative pair behaves, in homology, as a topological quotient.

We recall that a compact Hausdorff space is normal ([Ke], Chapter 5, Theorem 9).

***4.2.1 Theorem*** (Excision in singular homology)
*If $X$ is a topological space and $U \subset A \subset X$ are subspaces with $\mathrm{cl}_X U \subset \mathrm{int}_X A$, the inclusion map $i\colon (X \setminus U, A \setminus U) \to (X, A)$ of relative pairs induces an isomorphism in homology, in each degree*

$$i_{*n}\colon H_n(X \setminus U, A \setminus U) \cong H_n(X, A), \tag{4.23}$$

*called the* excision isomorphism.

*Proof* It is another consequence of the Subdivision Theorem 2.3.5. By hypothesis, the pair $\mathcal{U} = (X \setminus U, A)$ is a generalised open cover of $X$:

$$\mathrm{int}(X \setminus U) \cup \mathrm{int}\, A = (X \setminus \mathrm{cl}\, U) \cup \mathrm{int}\, A = X,$$

and the inclusion $j\colon C_+(X;\mathcal{U}) \to C_+(X)$ induces isomorphisms in homology. It also gives rise to a commutative diagram in $\mathrm{Ch}_+\mathsf{Ab}$, with short exact rows

$$\begin{array}{ccccc} C_+(A) & \rightarrowtail & C_+(X;\mathcal{U}) & \twoheadrightarrow & E_+ \\ \| & & \downarrow j & & \downarrow i \\ C_+(A) & \rightarrowtail & C_+(X) & \twoheadrightarrow & C_+(X, A) \end{array} \tag{4.24}$$

where $E_+ = C_+(X;\mathcal{U})/C_+(A)$.

Applying Theorem 2.3.3 (on the exact sequence of chain homology, and its naturality), we have a commutative diagram with exact rows

$$\begin{array}{ccccccccc} \dots H_n A & \to & H_n(X;\mathcal{U}) & \to & H_n(E_+) & \to & H_{n-1}A & \to & H_{n-1}(X;\mathcal{U}) \\ \| & & \downarrow j_{*n} & & \downarrow i_{*n} & & \| & & \downarrow j_{*n-1} \\ \dots H_n A & \to & H_n X & \to & H_n(X, A) & \to & H_{n-1}A & \to & H_{n-1}X \end{array} \tag{4.25}$$

Applying the Five Lemma 3.5.3, each $i_{*n}\colon H_n(E_+) \to H_n(X, A)$ is an isomorphism.

It is now sufficient to prove that the chain complex $E_+$ is canonically isomorphic to $C_+(X \setminus U, A \setminus U)$. This follows from a Noether isomorphism of abelian groups, in each component

$$\begin{aligned} E_n &= C_n(X;\mathcal{U})/C_n(A) = (C_n(X \setminus U) + C_n(A))/C_n(A) \\ &\cong C_n(X \setminus U)/(C_n(X \setminus U) \cap C_n(A)) \\ &= C_n(X \setminus U)/(C_n(A \setminus U) = C_n(X \setminus U, A \setminus U). \end{aligned} \tag{4.26}$$

More formally, we are applying a Noether isomorphism in the category of chain complexes. □

### 4.2.2 Compact pairs and other complements

(a) A *compact Hausdorff pair* $(X, A)$ will be a relative pair of compact Hausdorff spaces; equivalently, $X$ is a compact Hausdorff space and $A$ is a closed subspace. We say that $(X, A)$ is a *strong compact pair* if, moreover, $A$ is a strong deformation retract of a compact neighbourhood of $A$ in $X$ (see 1.1.3).

For instance, the pairs $(\mathbb{D}^n, \mathbb{S}^{n-1})$, $(\mathbb{S}^n, \mathbb{S}^{n-1})$ and $(\mathbb{P}^n, \mathbb{P}^{n-1})$ are strong compact pairs.

(b) A relative pair $(X, A)$ has an *associated quotient* $X/A$, where $A$ is collapsed to a single point $x_0$ (as in 1.4.1(e)); the projection $p\colon X \to X/A$ sends all points of $A$ to $x_0$, and leaves the other points unchanged. The quotient $X/A$ is often viewed as a pointed space, with basepoint $x_0$.

The case $A = \emptyset$ requires some attention: here it is convenient to assume that $X/\emptyset$ is the topological sum $X + \{x_0\}$, adding a basepoint $x_0 \notin X$, as made clearer below; note that in this case $p$ is not surjective. (One can avoid considering this case, but we prefer to avoid exceptions.)

In other contexts one may want to form a mere space and take $X/\emptyset = X$. Indeed, we have already seen that the relative pair $(X, \emptyset)$ is identified with the space $X$, and read as $X$ modulo $\emptyset$ (see 4.1.9).

(c) The quotient $X/A$ can be structurally defined as the pushout of the inclusion $i\colon A \to X$ and the (unique) map $A \to \{*\}$

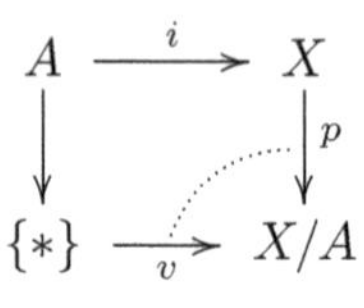


(4.27)

Here $X/A$ is pointed at $x_0 = v(*)$. For $A = \emptyset$, we do get the sum $X + \{x_0\}$.

*(d) Adjoint functors (see Section 8.1) also make the present definition of $X/A$ clearer. The embedding $U\colon \mathsf{Top}_\bullet \to \mathsf{Top}_2$ considered in 4.1.1(d) has a left adjoint $F$

$$\begin{aligned} &U\colon \mathsf{Top}_\bullet \to \mathsf{Top}_2, \qquad && U(Y, y_0) = (Y, \{y_0\}), \\ &F\colon \mathsf{Top}_2 \to \mathsf{Top}_\bullet, \qquad && F(X, A) = X/A, \end{aligned} \tag{4.28}$$

characterised by the natural bijection

$$\mathsf{Top}_2((X, A), (Y, \{y_0\})) \to \mathsf{Top}_\bullet(X/A, (Y, y_0)), \tag{4.29}$$

that takes a relative map $(X, A) \to (Y, \{y_0\})$ to the induced pointed map $X/A \to (Y, y_0)$. If $A = \emptyset$, we must take $F(X, \emptyset) = X + \{x_0\}$.

### 4.2.3 Lemma

*(a) If $(X, A)$ is a compact Hausdorff pair, then $(X/A, x_0)$ is also and the projection $p\colon X \to X/A$ is closed.*

*(b) If $(X, A)$ is a strong compact pair, then $(X/A, x_0)$ is also. More precisely, if $A$ has a compact nbd $U$ in $X$, of which $A$ is a strong deformation retract, then $x_0$ has a compact nbd $U/A$ in $X/A$, of which $\{x_0\}$ is a strong deformation retract.*

*Proof* For $A = \emptyset$ all this is obvious, and we can suppose that $A \neq \emptyset$.

(a) Letting $(X, A)$ be a compact Hausdorff pair, we have to prove that $X/A$ is Hausdorff. Two points $x \neq y$ in the open subspace $X \setminus A$ have disjoint open nbds contained in $X \setminus A$ and saturated for the equivalence relation that collapses $A$. If $x \in A$ and $y \notin A$, there exist two disjoint open sets $U, V$ such that $A \subset U$ and $y \in V$ (because every compact Hausdorff space is regular); again, these open sets are saturated for the equivalence relation.

The projection $p\colon X \to X/A$ is closed, because $X$ is compact and $X/A$ is Hausdorff.

(b) We suppose that $A$ is a strong deformation retract of $U$ (a compact nbd of $A$ in $X$), by a homotopy $\varphi$ (see (1.7))

$$\begin{aligned} &\varphi\colon X \times \mathbb{I} \to X, \quad && \varphi(x, 0) \in A, \\ &\varphi(x, 1) = x, && \varphi(a, t) = a \qquad (\text{for } x \in X,\ a \in A,\ t \in \mathbb{I}). \end{aligned} \tag{4.30}$$

The projection $q\colon U \to U/A$ is closed, as above. There is a (unique) mapping $\psi\colon U/A \times \mathbb{I} \to U/A$ making the following diagram commutative

$$\begin{array}{ccc} U \times \mathbb{I} & \xrightarrow{\ \varphi\ } & U \\ {\scriptstyle q\times\mathbb{I}}\downarrow & & \downarrow{\scriptstyle q} \\ U/A \times \mathbb{I} & \xrightarrow[\ \psi\ ]{} & U/A \end{array} \tag{4.31}$$

because two distinct pairs $(x,t) \neq (x',t')$ are identified by $q \times \mathbb{I}$ if and only if $x, x' \in A$ and $t = t'$; but then $\varphi(x,t) = x$ and $\varphi(x',t') = x'$ are identified by $q$. Moreover $\psi$ is continuous, because so is the composite $\psi(q \times \mathbb{I}) = q\varphi$, and $q \times \mathbb{I}$ is a quotient map (being surjective and closed, as above).

The map $\psi$ satisfies

$$\begin{array}{ll} \psi\colon U/A \times \mathbb{I} \to U/A, & \psi(x,0) = x_0, \\ \psi(x,1) = x, & \psi(x_0,t) = x_0 \quad (\text{for } x \in U/A,\ t \in \mathbb{I}), \end{array} \tag{4.32}$$

so that $\{x_0\}$ is a strong deformation retract of $U/A$, which is indeed a compact nbd of $\{x_0\}$ in $X/A$.

□

### 4.2.4 Theorem

*Let $(X,A)$ be a strong compact pair. The canonical projection $p\colon (X,A) \to (X/A, \{x_0\})$ induces a canonical isomorphism, for every $n \geqslant 0$*

$$p_{*n}\colon H_n(X,A) \longrightarrow H_n(X/A, x_0) \cong \tilde{H}_n(X/A), \tag{4.33}$$

*where the second isomorphism is already known: see* (4.17).

*Proof* We let $X' = X/A$. By hypothesis, $A$ is a strong deformation retract of some compact nbd $U$ (of $A$ in $X$). By Lemma 4.2.3, $\{x_0\}$ is a strong deformation retract of its compact nbd $U' = U/A$.

The compact Hausdorff space $X$ is normal. Since $A$ is a closed subset contained in $\operatorname{int} U$, there is an open subset $V$ of $X$ such that

$$A \subset V \subset \operatorname{cl} V \subset \operatorname{int} U. \tag{4.34}$$

We shall form a commutative diagram, where the solid arrows are isomorphisms, and therefore the dashed ones are also

$$\begin{array}{ccccc} H_n(X,A) & \xrightarrow{\ v_*\ } & H_n(X,U) & \xleftarrow{\ i_*\ } & H_n(X \setminus V,\, U \setminus V) \\ {\scriptstyle p_*}\downarrow & & \downarrow{\scriptstyle p_*} & & \downarrow{\scriptstyle p_*} \\ H_n(X', x_0) & \xrightarrow[\ v_*\ ]{} & H_n(X', U') & \xleftarrow[\ i_*\ ]{} & H_n(X' \setminus pV,\, U' \setminus pV) \end{array} \tag{4.35}$$

(a) First, in the upper row, $A$ is a strong deformation retract of $U$, and the exact sequence of the triple $(X, U, A)$ gives an isomorphism $v_{*n}\colon H_n(X, A) \to H_n(X, U)$ associated to the canonical map $v\colon (X, A) \to (X, U)$.

Similarly, in the lower row, $\{x_0\}$ is a strong deformation retract of $U'$, and we have a canonical isomorphism $v_{*n}\colon H_n(X', x_0) \to H_n(X', U')$.

(b) Secondly, we have two excision isomorphisms

$$\begin{aligned} i_{*n}\colon H_n(X \setminus V,\, U \setminus V) &\longrightarrow H_n(X, U), \\ i_{*n}\colon H_n(X' \setminus pV,\, U' \setminus pV) &\longrightarrow H_n(X', U'), \end{aligned}$$

by excision of $V$ (since $\mathrm{cl}\, V \subset \mathrm{int}\, U$) and of $pV$.

We recall that $p\colon X \to X/A$ is closed and preserves the open sets containing $A$ (which are saturated for $p$).

(c) Thirdly, the projection $p\colon X \to X/A$ restricts to the identity of $X \setminus A$; it gives an identity of relative pairs $(X \setminus V,\, U \setminus V) \to (X' \setminus pV,\, U' \setminus pV)$ and the identity $p_{*n}\colon H_n(X \setminus V,\, U \setminus V) \to H_n(X' \setminus pV,\, U' \setminus pV)$.

(d) We have thus proved that $p_{*n}\colon H_n(X, A) \to H_n(X', x_0)$ is an isomorphism. □

### 4.2.5 Relative homeomorphisms

A relative map $f\colon (X, A) \to (Y, B)$ between strong compact pairs is said to be a *relative homeomorphism* if it has a bijective restriction $X \setminus A \to Y \setminus B$.

This is the traditional terminology: $f$ is not supposed to be an invertible morphism of $\mathsf{Top}_2$.

In this situation $f$ induces a homeomorphism $f'\colon X/A \to X/B$. In fact, we have a commutative diagram in $\mathsf{Top}$, at the left

$$\begin{array}{ccc} X & \xrightarrow{\ f\ } & Y \\ {\scriptstyle p}\downarrow & & \downarrow{\scriptstyle q} \\ X/A & \xrightarrow[f']{} & Y/B \end{array} \qquad \begin{array}{ccc} (X, A) & \xrightarrow{\ f\ } & (Y, B) \\ {\scriptstyle p}\downarrow & & \downarrow{\scriptstyle q} \\ (X/A, x_0) & \xrightarrow[f']{} & (Y/B, y_0) \end{array} \qquad (4.36)$$

where $f'$ is a bijective map between compact Hausdorff spaces (by 4.2.3(a)), and therefore a homeomorphism.

As a consequence, we also have a commutative diagram in $\mathsf{Top}_2$, at the right in (4.36), where $f'$ is a pointed homeomorphism.

### *4.2.6 Relative Homeomorphism Theorem*

*If $f\colon (X, A) \to (Y, B)$ is a relative homeomorphism between strong compact pairs, the induced homomorphism $f_{*n}\colon H_n(X, A) \to H_n(Y, B)$ is an isomorphism, for every $n$.*

*Proof* Applying $H_n$ to the right diagram (4.36), the homeomorphism of relative pairs $f'$ induces an isomorphism; the same is true of the relative maps $p$ and $q$, by Theorem 4.2.4.

It follows that $f_{*n}\colon H_n(X, A) \to H_n(Y, B)$ is an isomorphism. □

### *4.2.7 Exercises*

These exercises are left to the reader. The first is easy, the second much harder.

(a) Taking into account that $\mathbb{D}^n/\mathbb{S}^{n-1} \cong \mathbb{S}^n$, one can apply the exact homology sequence of the strong compact pair $(\mathbb{D}^n, \mathbb{S}^{n-1})$ to determine, by induction, the homology groups of the spheres and their generators, much in the same way as we applied the MV-sequence, in Sections 2.4, 2.5.

*(b) One can make use of the sequences of the strong compact pairs $(\mathbb{P}^n, \mathbb{P}^{n-1})$ and $(\mathbb{S}^n, \mathbb{S}^{n-1})$ to determine the homology groups of the projective spaces, following a pattern similar to that used in Theorem 3.3.4, with the MV-sequence of relative homology.

## 4.3 Local homology and orientable manifolds

The reader likely knows that the orientability of a differentiable manifold is based on the Jacobian determinant of the transition morphisms of an atlas: if these determinants are positive everywhere, the current atlas is oriented, and gives an orientation of the manifold.

For a topological $n$-manifold $M$ the transition morphisms cannot be derived, and we need a different approach. A 'local orientation' at a point $x \in M$ is defined by the choice of a generator $\zeta_x$ of the group $H_n(M, M \setminus \{x\})$: an infinite cyclic group called the *local homology group* of $M$ at $x$; concretely, this amounts to orienting the 'little spheres' centred at $x$. An orientation of $M$ is a consistent choice of local orientations at all points – if such a choice exists.

We only hint here at this classical, non-elementary subject of homology theory. The classification of compact surfaces is briefly described in 4.3.7 and 4.3.8, without proofs. Poincaré duality will be stated in 6.4.8.

For the sake of simplicity, all spaces are here assumed to be Hausdorff;

in particular, their points are closed. On the other hand, manifolds need not be assumed to be paracompact.

*Literature.* Topological manifolds and their homology are studied in [Le] and many other textbooks, like [Mu], Chapter 12, [Mas3], Chapter IX, and [Ha], Section 3.3. The classification of compact surfaces is dealt with in [Mas1] and [Mu]. Differentiable manifolds are the basic topic of Differential Geometry [Sv]. A general theory of 'local structures', from manifolds to fibre bundles, was established by C. Ehresmann, since the 1960's; a new version can be found in [G4]. For more general approaches to 'smooth structures' one can see [IZ] and its references.

The orientation of a topological manifold $M$ can also be based on cohomology ([EiS], Section XI.6), or on a single relative cohomology class of the product $M \times M$ ([Sp2], Sections 6.2 and 6.3).

*Caution.* Isomorphisms should be managed with special care, here: an infinite cyclic group is 'oriented' by the choice of a generator, say $\zeta$ or $-\zeta$; an arbitrary isomorphism need not preserve it.

### 4.3.1 Orienting the spheres

We begin by considering the orientation of spheres; this will be extended to the local orientation of the euclidean spaces, and then to the local and global orientation of manifolds.

Let $C$ be a space homeomorphic to $\mathbb{S}^1$, for instance, a circle in some $\mathbb{R}^k$. Then the group $H_1(C)$ is infinite cyclic. By definition, an *orientation* of $C$ is the choice of a generator $\zeta \in H_1(C)$, or equivalently the choice of an isomorphism $\mathbb{Z} \to H_1(C)$. Concretely, we have seen in 2.4.1(f) that this amounts to choose a way of turning around $C$. The other generator, namely $-\zeta$, gives the *opposite* orientation.

A circle in $\mathbb{R}^2$ has a standard orientation, given by the homology class of the counterclockwise simple loop; but a circle in any higher $\mathbb{R}^k$ has no canonical orientation.

More generally, let $S$ be a space homeomorphic to a standard sphere $\mathbb{S}^n$, for $n \geqslant 1$. Again, the group $H_n(S)$ is infinite cyclic. An *orientation* of $S$ is the choice of a generator $\zeta \in H_n(S)$, or equivalently the choice of an isomorphism $\mathbb{Z} \to H_n(S)$.

For a 0-sphere $S = \{p, q\} \cong \mathbb{S}^0$ we can follow the same pattern, using the reduced homology (as already discussed in 3.6.2(f)): the infinite cyclic group $\tilde{H}_0(\{p, q\})$ is generated by $[q - p]$, and also by $[p - q]$. Choosing one of them amounts to choose an orientation, from $p$ to $q$ or vice versa.

### 4.3.2 Local homology

To extend this notion, we introduce the *local homology of a space $X$ at a point $x \in X$*, as the sequence of groups

$$H_k(X \mid x) = H_k(X, X \setminus \{x\}). \tag{4.37}$$

Its local character is made explicit by the excision isomorphism

$$i_{*k}\colon H_k(U, U \setminus \{x\}) \longrightarrow H_k(X, X \setminus \{x\}),$$

for any open nbd $U$ of $x$ in $X$ (induced by the inclusion $U \subset X$). In fact, we are cutting out the complement $X \setminus U$, which is closed in $X$ and contained in the open subset $X \setminus \{x\}$ (as we are supposing all spaces to be Hausdorff).

The notation $H_k(X \mid x)$ is borrowed from [Ha], Section 3.3. Local homology at $x$ should not be confused with pointed homology at $x$, defined in 4.1.6(d).

More generally we are interested in the *local homology groups* of the space *$X$ at a subspace $A$*

$$H_k(X \mid A) = H_k(X, X \setminus A), \qquad H_k(X \mid A) \cong H_k(U \mid A), \tag{4.38}$$

where $U$ is any open nbd of $\mathrm{cl}\,A$ in $X$: we are using again an excision isomorphism $H_k(U, U \setminus A) \to H_k(X, X \setminus A)$.

### 4.3.3 Local orientations

We now examine the local properties of the euclidean space $X = \mathbb{R}^n$, at a point $x \in X$.

We shall ignore its canonical orientation, because these properties will be used for spaces equipped with a homeomorphism onto $X$.

There are canonical isomorphisms

$$H_k(X, X \setminus \{x\}) \xrightarrow{D_k} \tilde{H}_{k-1}(X \setminus \{x\}) \xrightarrow{N_*} \tilde{H}_{k-1}(S^{n-1}), \tag{4.39}$$

where

- $S^{n-1}$ is an $(n-1)$-sphere of $X$ centred at $x$,

- $D_k$ is the connecting (iso)morphism of the reduced homology sequence of the pair $(X, X \setminus \{x\})$ (with $X$ contractible, as in Exercise 4.1.7(e)),

- the isomorphism $N_*$ is induced by the normalisation map $N\colon X \setminus \{x\} \to S^{n-1}$, for the deformation retract $S^{n-1} \subset X \setminus \{x\}$.

A (local) *orientation* of $X$ *at a point* $x$ is the choice of a generator $\zeta$ of the infinite cyclic group

$$H_n(X \mid x) = H_n(X, X \setminus \{x\}) \cong \tilde{H}_{n-1}(S^{n-1}), \tag{4.40}$$

and amounts to orienting all the $(n-1)$-spheres of $X$ centred at $x$, in a consistent way.

An orientation of $X$ at the point $x$ gives an orientation of $X$ at any point $y$, via the canonical isomorphisms associated to inclusions

$$H_n(X, X \setminus \{x\}) \leftarrow H_n(X, X \setminus B) \rightarrow H_n(X, X \setminus \{y\}), \qquad (4.41)$$

where $B$ is any open ball of $X$ containing $x$ and $y$. (The left isomorphism comes from the homology sequence of the triple $(X, X \setminus \{x\}, X \setminus B)$, in 4.1.8, where $X \setminus B$ is a deformation retract of $X \setminus \{x\}$.)

Globally, this coherent system of local orientations forms an *orientation* of $X$.

*Remarks.* The only non-trivial local homology group $H_k(\mathbb{R}^n \mid x)$ has degree $n$. This is again an intrinsic characterisation of the topological dimension of $\mathbb{R}^n$.

### *4.3.4 Topological manifolds and surfaces*

We briefly review the basic facts of this topic, developed in the literature cited above.

(a) A (topological) *$n$-manifold*, or *manifold of dimension $n$*, is a Hausdorff topological space $X$ locally homeomorphic to $\mathbb{R}^n$, in the sense that every point has an open nbd homeomorphic to $\mathbb{R}^n$. (A non-empty manifold determines its dimension, as we shall see in 4.3.5(a).)

By definition, a manifold $X$ is locally path connected. Therefore every path component of $X$ is open; as they are disjoint, the space $X$ is connected if and only if it is path connected.

Forgetting the empty space, there is, up to homeomorphism, one connected 0-manifold, the singleton. *One can also prove that there is one compact connected 1-manifold, the circle.*

A *surface* is a connected 2-manifold. We have already examined various compact surfaces, like the sphere $\mathbb{S}^2$, the torus, the projective plane $\mathbb{P}^2$, the Klein bottle, etc. A celebrated classification theorem (stated in 4.3.8) says that – up to homeomorphism – they form a countable family.

The higher dimensional manifolds are much more complex.

(b) More generally, an *$n$-manifold with boundary* is a Hausdorff topological space $X$ where every point has an open nbd homeomorphic to $\mathbb{R}^n$ or to the euclidean half-space $H^n = \{x \in \mathbb{R}^n \mid x_1 \geqslant 0\}$.

The (intrinsic) *interior* of $X$ is the set of points which have an open nbd homeomorphic to $\mathbb{R}^n$, and the (intrinsic) *boundary* of $X$ is the complement of the intrinsic interior.

These subspaces of $X$ are intrinsically defined, and should not be confused with relative notions, like the interior or border of $X$ *in* some superspace $X'$. The interior of $X$ is a manifold of dimension $n$, open in $X$; the boundary is a manifold of dimension $n-1$ closed in $X$, usually denoted as $\partial X$.

In particular, a *surface with boundary* is a connected 2-manifold with boundary; the latter is a 1-dimensional manifold, possibly disconnected. Various compact examples were described in 3.2.1.

(c) The cartesian product of two manifolds of dimensions $m, n$ is a manifold of dimension $m+n$. The product $X \times Y$ of a manifold $X$ and a manifold with boundary $Y$ is a manifold with boundary (namely, $X \times \partial Y$).

*(d) A connected manifold $X$ (without boundary) is a *homogeneous space*: for every pair of points $x, y \in X$ there is a homeomorphism $f\colon X \to X$ that takes $x$ to $y$.

This is obvious for $\mathbb{R}^n$, using translations. Therefore, for every $x \in X$, the set $U_x$ of all the points $y \in X$ related in this way to $x$ is open. All $U_x$ form a partition of $X$ in disjoint open sets; by connectedness, there is only one of them.

### *4.3.5 Orientable manifolds*

We can now define orientation for an $n$-manifold $X$ of positive dimension. All manifolds are now assumed to be non-empty.

At any point $x \in X$, the local homology group $H_n(X \mid x) = H_n(X, X \setminus x)$ is isomorphic to $H_n(U \mid x)$, where $U$ is any open nbd of $x$; it is an infinite cyclic group, as shown in (4.39) (taking $U$ homeomorphic to $\mathbb{R}^n$).

A *local orientation* of the $n$-manifold $X$ at a point $x$ is a generator $\zeta_x$ of $H_n(X \mid x)$.

An *orientation* of $X$ is a family $(\zeta_x)_{x\in X}$ of local orientations at each point, which is coherent. This means that any $x \in X$ has a euclidean nbd (homeomorphic to $\mathbb{R}^n$) containing an open $n$-ball $B$ containing $x$, such that there exists a local orientation $\zeta_B \in H_n(X \mid B)$ whose image by the canonical isomorphism $H_n(X \mid B) \to H_n(X \mid y)$ is $\zeta_y$, for all $y \in B$.

*Remarks and complements.* (a) Recalling that $H_k(X \mid x) = 0$ for $k \neq n$ (by (4.39), again), the local homology of $X$ at any point *structurally determines* the dimension of $X$.

(b) If the manifold $X$ has an orientation $(\zeta_x)_{x\in X}$, the family $(-\zeta_x)_{x\in X}$ is also an orientation, the *opposite* one.

(c) A manifold is orientable if and only if all its connected components are.

A connected orientable manifold has precisely two orientations, opposite to each other.

### 4.3.6 *Complements

Further developing the theory of topological manifolds is beyond the bounds of this book. The rest of this section will give further information, without proofs, to awaken interest in reading, or studying, the sources cited above.

The situation is particularly clear in the case of a (non-empty) compact connected $n$-manifold $M$.

(a) $M$ is orientable if and only if $H_n(M) \cong \mathbb{Z}$. In this case, for every $x \in M$, the identity of $M$ induces an isomorphism $i_x \colon H_n(M) \to H_n(M, M \setminus \{x\})$, and a generator $\zeta \in H_n(M)$ produces an orientation of $M$, the family $\zeta_x = i_x(\zeta)$ $(x \in X)$.

(b) $M$ is non-orientable if and only if $H_n(M) = 0$.

(c) $H_k(M) = 0$, for $k > 0$.

(d) Poincaré duality binds the singular homology and cohomology groups of $M$. It will be stated in 6.4.8.

### 4.3.7 *The connected sum of manifolds and surfaces

There is an enticing operation acting on two connected manifolds $X, Y$ of the same dimension $n > 0$. Loosely speaking, the *connected sum* $X \sharp Y$ is obtained by taking out of each of them a 'little' open $n$-disc, and pasting the remaining subspaces $X', Y'$ along their boundaries $S'$ and $S''$, which are spheres of dimension $n - 1$.

The result is a connected manifold *independent* of these choices (up to homeomorphism, of course), essentially because of the homogeneity property of connected manifolds (in 4.3.4(d)). The operation is commutative and associative (up to homeomorphism); the sphere $\mathbb{S}^n$ is a unit element for this operation: in $X \sharp \mathbb{S}^n$ we take out of $X$ an open disc and replace it.

Here we are interested in the connected sum $X \sharp Y$ of connected surfaces; the junction around the dotted circle $S$ looks as follows

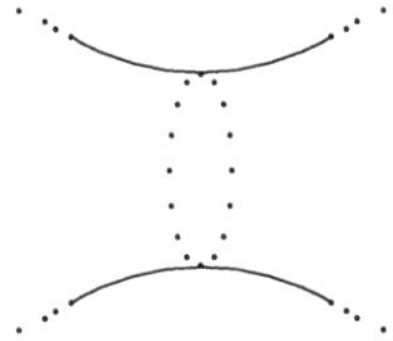

(4.42)

The connected sum $\mathbb{T}^{\sharp 2} = \mathbb{T} \sharp \mathbb{T}$ of two copies of the torus is the double torus drawn and studied in 3.2.6.

Repeating this operation we get an orientable surface $\mathbb{T}^{\sharp p}$ which can be represented as a lifebuoy with $p$ places, or a sphere with $p$ handles. Its homology groups are easily computed, by induction:

$$H_1(\mathbb{T}^{\sharp p}) \cong \mathbb{Z}^{2p}, \qquad H_0(\mathbb{T}^{\sharp p}) \cong H_2(\mathbb{T}^{\sharp p}) \cong \mathbb{Z}. \tag{4.43}$$

#### *4.3.8 *The Classification Theorem of Compact Surfaces*

The compact surfaces form, up to homeomorphism, a countable set with an amazingly simple classification. We only give the statement; the proof, quite complex, can be seen in Massey [Mas1], or in [Mu], Chapter 12.

*(i) Every orientable compact surface is homeomorphic to a connected sum $\mathbb{T}^{\sharp p}$ of $p \geqslant 0$ tori (where $\mathbb{T}^{\sharp 0}$ stands for the 2-sphere, the unit of the connected sum in dimension 2).*

*(ii) Every non-orientable compact surface is homeomorphic to a connected sum $(\mathbb{P}^2)^{\sharp p}$ of $p \geqslant 1$ projective planes.*

*Remarks.* The non-orientable compact surfaces are better viewed considering the following points (not hard to prove).

(a) The connected sum $\mathbb{P}^2 \sharp \mathbb{P}^2$ of two projective planes is the pasting of two Möbius bands along their boundary, that is, a Klein bottle $\mathbb{K}$ (as we have already seen in 3.2.5(b) and 3.2.5(a)).

(b) The connected sums $\mathbb{T} \sharp \mathbb{P}^2$ and $\mathbb{K} \sharp \mathbb{P}^2$ are homeomorphic. *Hint:* it is easy to see that $\mathbb{T} \sharp M \cong \mathbb{K} \sharp M$, where $M$ is a Möbius band without boundary, contained in $\mathbb{P}^2$.

(c) We can now represent any non-orientable compact surface as the connected sum of an orientable surface $\mathbb{T}^{\sharp n}$ with either a projective plane or a Klein bottle:

$$\mathbb{P}^{\sharp 2n+1} \cong \mathbb{T}^{\sharp n} \sharp \mathbb{P}, \qquad \mathbb{P}^{\sharp 2n+2} \cong \mathbb{T}^{\sharp n} \sharp \mathbb{K} \qquad (n \geqslant 0). \tag{4.44}$$

### 4.4 Eilenberg–Steenrod axioms for (co)homology theories

The textbook of Eilenberg and Steenrod [EiS], published in 1952, was a landmark in Algebraic Topology. It establishes a list of axioms sufficiently general to cover most of the theories in use, and sufficiently strong to determine the theory on 'good spaces', for a fixed abelian group 'of coefficients'.

Relative singular homology is a homology theory in this sense, with coefficient group $\mathbb{Z}$. Singular homology and cohomology with an arbitrary coefficient group will be built in Sections 5.3 and 5.5.

### 4.4.1 The axioms

A *homology theory* $H = ((H_n), (D_n))$ on $\mathsf{Top}_2$ consists of the following data.

(a) For every relative pair $(X, A)$ and every $n \geqslant 0$ we have an abelian group, called the *n-th homology group of X modulo A*

$$H_n(X, A). \tag{4.45}$$

The embedding of $\mathsf{Top}$ in $\mathsf{Top}_2$ gives the absolute homology group $H_n(X) = H_n(X, \emptyset)$.

(b) For every relative map $f\colon (X, A) \to (Y, B)$ and every $n \geqslant 0$ we have an *induced homomorphism*

$$H_n(f) = f_{*n}\colon H_n(X, A) \to H_n(Y, B). \tag{4.46}$$

(c) For every relative pair $(X, A)$ and every $n \geqslant 1$ we have a *connecting homomorphism*

$$D_n(X, A)\colon H_n(X, A) \to H_{n-1}(A). \tag{4.47}$$

The following axioms have to be satisfied.

(HT.1) (*Functoriality*) The data (a) and (b) form a functor $H_n\colon \mathsf{Top}_2 \to \mathsf{Ab}$, called the *n-th homology functor* of the theory, for every $n \geqslant 0$.

(HT.2) (*Naturality*) The family of connecting homomorphisms $D_n(X, A)$ forms a natural transformation

$$D_n\colon H_n \to H_{n-1}P''\colon \mathsf{Top}_2 \to \mathsf{Ab} \qquad (n \geqslant 1), \tag{4.48}$$

where $P''\colon \mathsf{Top}_2 \to \mathsf{Top}$ is the second projection: $P''(X, A) = A$.

(HT.3) (*Homotopy invariance*) The functors $H_n\colon \mathsf{Top}_2 \to \mathsf{Ab}$ are invariant under relative homotopies.

(HT.4) (*Exactness*) For every relative pair $(X, A)$, the sequence (4.12) is exact. It is called the *homology sequence* of $(X, A)$.

(HT.5) (*Excision*) If $U \subset A \subset X$ are subspaces and $\mathrm{cl}_X U \subset \mathrm{int}_X A$, the inclusion map $i\colon (X \setminus U, A \setminus U) \to (X, A)$ has an associated isomorphism, for every $n \geqslant 0$, called the *excision isomorphism*

$$i_{*n}\colon H_n(X \setminus U, A \setminus U) \cong H_n(X, A). \tag{4.49}$$

(HT.6) (*Dimension*) The singleton space $\{*\}$ has trivial homology groups in positive degree:

$$H_n(\{*\}) = 0, \quad \text{for } n > 0. \tag{4.50}$$

The group $G = H_0(\{*\})$ of the standard point is called the *coefficient group* of the theory.

*Notes.* The original version of the axioms, in [EiS], is formulated for degrees $n \in \mathbb{Z}$ and an 'allowable' subcategory of $\mathsf{Top}_2$. For instance, the full subcategory formed by the relative pairs of compact Hausdorff spaces is allowable.

Of course, there are diverse singleton spaces $\{x\}$, but all of them are related by a coherent system of homeomorphisms, inducing a coherent system of isomorphisms on the groups $H_0(\{x\})$.

The name of 'Dimension Axiom' is explained by the fact that, without it, one could translate the theory, letting $H'_n = H_{n+h}$, for a fixed integer $h$.

A homology theory in the present sense is said to be *ordinary*. A *generalised homology theory* is not required to satisfy the Dimension Axiom, which allows important theories besides the 'translations' of the ordinary ones; see also 7.6.6(c).

### *4.4.2 The singular theory*

We have already proved that relative singular homology is a homology theory in this sense:

- the data are defined in 4.1.3 and 4.1.5, where the properties of functoriality, exactness and naturality are also established,

- homotopy invariance is proved in Theorem 4.1.4, the excision property in Theorem 4.2.1, and the dimension axiom in (2.56), where we have also seen that $H_0(\{*\}) \cong \mathbb{Z}$, in a canonical way.

Because of this, we say that singular homology is a homology theory *with integral coefficients*, or *with coefficients in* $\mathbb{Z}$.

For a given group $G$, we shall construct, in Chapter 5, the *singular theory with coefficient group* $G$ (up to canonical isomorphism).

Essentially, we construct the singular chain complex $C_+(X; G)$ with coefficients in $G$, letting

$$C_n(X; G) = F_G(\mathrm{Cub}_n(X))/F_G(\mathrm{Deg}_n(X)), \tag{4.51}$$

where, for a set $S$, the direct sum $F_G(S) = \bigoplus_{i \in S} G$ replaces the free abelian group $F(S) = \mathbb{Z}S = \bigoplus_{i \in S} \mathbb{Z}$. An $n$-chain is now an essentially finite formal sum $\sum \lambda_i a_i$ with coefficients $\lambda_i \in G$ and $a_i \colon \mathbb{I}^n \to X$ (up to degenerate chains).

Proceeding in this elementary way would soon become complicated, and a more structural way is usually followed: the chain groups $C_n(X, A; G)$ will be obtained as $C_n(X, A) \otimes G$, using tensor products of abelian groups.

### *4.4.3 Exercises and complements*

A homology theory $H$ is given, with coefficients in $G$. Several issues we have established for the singular theory are a consequence of the previous axioms, and hold for every homology theory, adapting some points related to $G$. We only mention a few cases, but an interested reader can state and work out many more, or look for them in [EiS].

(a) (*Trivial pairs*) Every pair $(X, X)$ has trivial groups $H_n(X, X) = 0$. More generally, if the inclusion $A \to X$ is a homotopy equivalence, then $H_n(X, A) = 0$, for every $n$.

(b) (*The empty space*) In particular, all $H_n(\emptyset)$ are trivial groups.

(c) (*The associated reduced theory*) The theory $H$ has an associated *reduced homology theory* $\tilde{H}$, defined by extending the procedure used in Exercise 3.6.2(h) for the singular theory.

*(d) The exact sequence of a triple, stated and proved for singular homology in Theorem 4.1.8, can be extended to any homology theory. The proof is purely within Homological Algebra, and can be found in [EiS], Section I.10, or in [Sp2], Section 4.8.

*(e) The Mayer–Vietoris Theorem 2.3.7 holds for any homology theory, if restricted to a 'proper triad' (X; U, V): see [EiS], Section I.15.

*(f) The isomorphism $\delta_n X\colon \tilde{H}_{n+1}(\Sigma X) \to \tilde{H}_n(X)$ of reduced singular homology, in (3.85), holds for every (reduced) homology theory.

It is an easy consequence of (e): apply the MV-sequence to the proper triad $(\Sigma X; U, V)$, where $U$ (resp. $V$) is the image in $\Sigma X$ of the subspace $X \times [1/4, 1]$ (resp. $X \times [0, 3/4]$) of the cylinder $X \times \mathbb{I}$.

### 4.4.4 Cohomology theories

Cohomology theories are the contravariant version of homology theories; traditionally, all degrees are written as superscripts.

*Formally, a cohomology theory can be defined as a homology theory with values in $\mathsf{Ab}^{\mathrm{op}}$, the abelian category opposite to $\mathsf{Ab}$.*

A *cohomology theory* $H = ((H^n), (D^n))$ consists of the following data.

(a) For every relative pair $(X, A)$ and every $n \geqslant 0$ we have an abelian group, called the *$n$-th cohomology group of* X *modulo* A

$$H^n(X, A). \tag{4.52}$$

(b) For every relative map $f\colon (X, A) \to (Y, B)$ and every $n \geqslant 0$ we have an *induced homomorphism*

$$H^n(f) = f^{*n}\colon H^n(Y, B) \to H^n(X, A). \tag{4.53}$$

(c) For every relative pair $(X, A)$ and every $n \geqslant 0$ we have a *connecting homomorphism*

$$D^n(X, A)\colon H^n(A) \to H^{n+1}(X, A). \tag{4.54}$$

These data are subject to the following axioms.

(HT*.1) (*Functoriality*) The data (a) and (b) form a contravariant functor $H^n\colon \mathsf{Top}_2 \dashrightarrow \mathsf{Ab}$, called the *n-th cohomology functor* of the theory, for every $n \geqslant 0$.

(HT*.2) (*Naturality*) The family of connecting homomorphisms $D^n(X, A)$ forms a natural transformation of contravariant functors

$$D^n\colon H^nP'' \to H^{n+1}\colon \mathsf{Top}_2 \dashrightarrow \mathsf{Ab} \quad (n \geqslant 0), \tag{4.55}$$

where $P''\colon \mathsf{Top}_2 \to \mathsf{Top}$ is the second projection: $P''(X, A) = A$.

(HT*.3) (*Homotopy invariance*) The contravariant functors $H^n\colon \mathsf{Top}_2 \dashrightarrow \mathsf{Ab}$ are invariant under relative homotopies.

(HT*.4) (*Exactness*) For every relative pair $(X, A)$, the *cohomology sequence*

$$\begin{array}{llllll} 0 \to & H^0(X,A) & \xrightarrow{v^{*0}} & H^0(X) & \xrightarrow{u^{*0}} & H^0(A) \\ & \quad\xleftarrow{\qquad D^0 \qquad} & & & & \\ & H^1(X,A) & \xrightarrow{v^{*1}} & H^1(X) & \xrightarrow{u^{*1}} & H^1(A) \\ & \quad\xleftarrow{\qquad D^1 \qquad} & & & & \\ & H^2(X,A) & \xrightarrow{v^{*2}} & \ldots & & \end{array} \tag{4.56}$$

is exact (for the canonical maps $u\colon A \to X$ and $v\colon (X, \emptyset) \to (X, A)$).

(HT*.5) (*Excision*) If $U \subset A \subset X$ are subspaces and $\mathrm{cl}_X U \subset \mathrm{int}_X A$, the inclusion map $i\colon (X \setminus U, A \setminus U) \to (X, A)$ has an associated *excision isomorphism*, for every $n \geqslant 0$

$$i^{*n}\colon H^n(X, A) \cong H^n(X \setminus U, A \setminus U). \tag{4.57}$$

(HT*.6) (*Dimension*) The singleton space $\{*\}$ has trivial cohomology groups in positive degree:

$$H^n(\{*\}) = 0, \qquad \text{for } n > 0. \tag{4.58}$$

The theory $H$ has *coefficient group* $G = H^0(\{*\})$. Also here we speak of a theory *with integral coefficients* when $G \cong \mathbb{Z}$.

Singular cohomology with an arbitrary coefficient group $G$ will be defined in Chapter 5, using a cochain complex $C^+(X, A; G)$ derived from the singular chain complex

$$C^n(X, A; G) = \mathrm{Hom}(C_n(X, A), G). \tag{4.59}$$

### 4.4.5 Comments

There are different homology theories with the same coefficient group, and different cohomology theories.

In this book we shall construct the cohomology theories of Alexander–Spanier and Čech, and the singular one. Already in degree zero, they give different results on a space $X$, related:

- either to the set $\mathrm{Cc}(X)$ of connected components, in the first two cases (see 4.5.8(d) and 4.7.2(b)),

- or to the set $\Pi_0(X)$ of path components, in the second case (see 5.5.4(d)).

However, all homology theories with the same coefficient group coincide on 'good spaces'. We shall now review, without proof, the original uniqueness theorem of Eilenberg–Steenrod, stated for 'triangulated pairs'.

### 4.4.6 Simplicial complexes

Triangulation of spaces, namely the study of spaces which can be obtained by a 'regular pasting' of tetrahedra, is an old topic, going back to the beginning of Combinatorial and Algebraic Topology.

Following the terminology of [EiS], in Sections II.2 and II.3, triangulation is based on finite simplicial complexes – not to be confused with simplicial sets, defined in 3.6.6.

Loosely speaking, an *n-simplex* $s$ is a formal copy of the standard tetrahedron $\Delta^n$, non-embedded in $\mathbb{R}^{n+1}$. Precisely, $s$ is a (non-empty) finite set of $n+1$ elements, called the *vertices* of $s$. A *point* of $s$ is a real function

$$x\colon s \to \mathbb{R}, \qquad x(v) \geqslant 0, \qquad \Sigma_v\, x(v) = 1 \qquad (\text{for } v \in s), \tag{4.60}$$

whose values are called the *barycentric coordinates* of the point $x$. The *geometric realisation* $|s|$ of $s$ is the space of these points, with the topology defined by the *euclidean distance*

$$d(x,y) = (\Sigma_v\,(x(v) - y(v))^2)^{1/2}.$$

A total ordering $v_0 < v_1 < ... < v_n$ of the set $s$ gives a homeomorphism

$$f\colon |s| \to \Delta^n, \qquad f(x) = (x(v_0), x(v_1), ..., x(v_n)), \tag{4.61}$$

between the space $|s|$ and the $n$-simplex $\Delta^n$, proving that $|s|$ is compact and path connected.

The topology of $|s|$ can also be defined by means of this bijection (independently of the chosen ordering).

A non-empty subset $s' \subset s$ is also called a *face* of $s$, and a *p-face* if it has $p+1$ elements. Plainly, there is a topological embedding $|s'| \to |s|$, where the point $x\colon s' \to \mathbb{R}$ is extended to $s$, letting $x(v) = 0$ for all vertices $v \in s \setminus s'$.

Loosely speaking, again, a *finite simplicial complex* $K$ is union of 'faces' (of any dimension) of the standard tetrahedron $\Delta^n$.

Precisely, $K$ is a simplex $s$ equipped with a set of faces of $s$ (non-empty subsets), which are called *faces* of $K$. The assigned faces are assumed to satisfy the condition:

(*) every face of a face of $K$ is a face of $K$.

The *geometric realisation* $|K|$ is defined to be the subspace of $|s|$ consisting of the points which belong to the realisation of some face of $K$. In other words, $|K| = \bigcup |s'|$, where $s'$ varies in the set of faces of $K$. The *dimension* of $K$ is the maximal dimension of its faces. A simplicial complex with no faces is realised as the empty space.

A *subcomplex* $K'$ of $K$ is a subset $s' \subset s$ equipped with the faces of $K$ which are contained in $s'$. Plainly, there is a topological embedding $|K'| \to |K|$.

*Examples and complements.* (a) Every $n$-simplex $s$ can be viewed as a simplicial complex $K$ of dimension $n$, admitting all its faces. Its geometric realisation is that of $s$, namely $\Delta^n$.

(b) An $n$-simplex $s$ also determines a simplicial complex $K'$, admitting all the faces of $s$ of dimension $< n$ (if any). Its geometric realisation is the border of $\Delta^n$ in the $n$-plane $\sum t_i = 0$ of $\mathbb{R}^{n+1}$. $K'$ is a subcomplex of the previous $K$.

*(c) Finite simplicial complexes form a category, where a map $f\colon K \to K'$ is a mapping $s \to s'$ between the underlying sets that takes every face of $K$ to a face of $K'$. Such a morphism has a geometric realisation as a map $|K| \to |K'|$. (This category will not be used here.)

(d) A general *simplicial complex* is usually defined as a set $K$ equipped with a set of finite non-empty subsets, its *faces*, so that

(i) a face of a face of $K$ is a face of $K$,
(ii) every singleton of $K$ is a face of $K$.

Condition (ii) could be omitted, as in [EiS]: singletons which are not faces are of little relevance.

### *4.4.7 Triangulated pairs*

Finally, a *triangulated pair* $(X, A)$ is a relative pair of topological spaces which has a homeomorphism (i.e. an isomorphism in $\mathsf{Top}_2$)

$$f\colon (X, A) \to (|K|, |K'|), \tag{4.62}$$

where $K$ is a finite simplicial complex and $K'$ is a simplicial subcomplex of $K$.

We are interested in the full subcategory $\mathrm{TrTop}_2$ of $\mathsf{Top}_2$ formed of triangulated pairs.

### 4.4.8 *Uniqueness Theorem [EiS]

*Let $H$ and $H'$ be two homology theories (satisfying Eilenberg–Steenrod's axioms), with coefficient groups $G$ and $G'$, respectively.*

*Restricting these theories to the category $\mathrm{TrTop}_2$ of triangulated pairs, every homomorphism $h\colon G \to G'$ can be uniquely extended to a family of natural transformations*

$$h_n\colon H_n \to H'_n\colon \mathrm{TrTop}_2 \to \mathsf{Ab} \tag{4.63}$$

*(with $h_0(\{*\}) = h$), that commute with the connecting morphisms of $H$ and $H'$, forming commutative diagrams*

$$\begin{array}{ccc} H_n(X,A) & \xrightarrow{h} & H'_n(X,A) \\ {\scriptstyle\partial}\downarrow & & \downarrow{\scriptstyle\partial'} \\ H_{n-1}(A) & \xrightarrow[h]{} & H'_{n-1}(A) \end{array} \tag{4.64}$$

*If $h\colon G \to G'$ is an isomorphism, all the natural transformations $h_n$ are natural isomorphisms.*

*Cohomology theories have a dually-similar uniqueness property.*

*Proof* See [EiS], Theorem III.10.1. □

### 4.4.9 *Simplicial homology

Homology (with a fixed coefficient group) is thus unique on triangulated pairs. It is called *simplicial homology.*

More generally, 'simplicial homology' can also refer to homology on the full subcategory of $\mathsf{Top}_2$ formed of all relative pairs homotopically equivalent to the triangulated ones: the Uniqueness Theorem 4.4.8 still holds in this wider domain. One can thus include all pairs of finite CW-complexes: see Section 4.9.

## 4.5 Alexander–Spanier cohomology

We present now a cohomology theory $\overline{H}^n(-;G)$ with an arbitrary coefficient group $G$, which comes out naturally in contravariant form.

Recalling a remark on contravariant theories, at the end of Section 0.4, we note that the cohomology group $\overline{H}^n(X;G)$ is based on the *functions* $X^{n+1} \to G$ with values in a fixed group, *usually a ring* (see Section 4.6). In particular, $\overline{H}^0(X;\mathbb{Z})$ 'is' the ring $\mathcal{A}(X)$ studied in Section 1.2.

*Literature.* Alexander–Spanier cohomology was introduced by Alexander [Al] for compact metric spaces, in 1935, and extended to general spaces by Spanier [Sp1], in 1948. Spanier's book [Sp2] has an excellent exposition and many results relating this theory to singular and Čech cohomologies (see Section 4.7).

[Mas2] is a more advanced book on Alexander–Spanier cohomology.

### 4.5.1 Cochain complexes

The theory of cochain complexes is quite similar to that of chain complexes, and only needs some adaptation: the degree is written as a superscript, a differential $d^n\colon A^n \to A^{n+1}$ rises the degree by one, while a component of a homotopy lowers it.

A (positive) *cochain complex* of abelian groups $A = ((A^n)_{n\geqslant 0}, (d^n)_{n\geqslant 0}$ is thus a diagram in $\mathsf{Ab}$

$$\dots\ 0 \dashrightarrow A^0 \xrightarrow{d_0} A^1 \quad \dots \quad A^n \xrightarrow{d_n} A^{n+1} \quad \dots \tag{4.65}$$

with *differentials* $d^n\colon A^n \to A^{n+1}$, so that all composites $d^n d^{n-1}$ are trivial. The complex can be extended at the left with trivial groups and differentials.

A morphism $f\colon A \to B$ has components $f^n\colon A^n \to B^n$ that commute with the differentials. They compose componentwise, forming the category $\mathrm{Ch}^+\mathsf{Ab}$ of cochain complexes of abelian groups.

In each component $A^n$ of $A$ we have the subgroups of *n-cocycles* and *n-coboundaries*

$$\begin{aligned} Z^n(A) &= \mathrm{Ker}\,(d^n\colon A^n \to A^{n+1}) \subset A^n, \\ B^n(A) &= \mathrm{Im}\,(d^{n-1}\colon A^{n-1} \to A^n) \subset A^n. \end{aligned} \tag{4.66}$$

In particular $B^0(A) = \mathrm{Im}\,(d^{-1}\colon 0 \to A) = 0$. The condition $d^n d^{n-1} = 0$ is equivalent to saying that $B^n(A) \subset Z^n(A)$ (for $n \geqslant 0$): *every coboundary is a cocycle.*

The *n-th cohomology group* of the complex $A$ is

$$H^n(A) = Z^n(A)/B^n(A) = \mathrm{Ker}\,(d^n)/\mathrm{Im}\,(d^{n-1}) \quad (n \geqslant 0). \tag{4.67}$$

Every cocycle $z \in Z^n(A)$ has a *cohomology class* $[z] \in H^n(A)$, written as $[z]^H$ when useful. Two cocycles $z, z' \in A^n$ are *cohomologous* if $z - z'$ is a coboundary, so that $[z] = [z']$.

This gives a sequence of (covariant!) *cochain cohomology functors*

$$\begin{gathered} H^n \colon \mathrm{Ch}^+\mathsf{Ab} \to \mathsf{Ab}, \qquad A \mapsto H^n(A), \\ H^n(f \colon A \to B) = (f^{*n} \colon H^n(A) \to H^n(B)), \quad f^{*n}[z] = [f^n(z)]. \end{gathered} \tag{4.68}$$

A *homotopy* $\varphi = (\varphi^n) \colon f \simeq g \colon A \to B$ of cochain morphisms is a sequence of homomorphisms $\varphi^n \colon A^n \to B^{n-1}$ such that

$$d^{n-1}\varphi^n + \varphi^{n+1}d^n = g^n - f^n, \tag{4.69}$$

where, of course, $\varphi^0 = 0$. The cohomology functors $H^n \colon \mathrm{Ch}^+\mathsf{Ab} \to \mathsf{Ab}$ are homotopy invariant.

All issues about exactness can be lifted from $\mathsf{Ab}$ to $\mathrm{Ch}^+\mathsf{Ab}$, using the sequence of forgetful functors $U^n \colon \mathrm{Ch}^+\mathsf{Ab} \to \mathsf{Ab}$, $A \mapsto A^n$. The arrows of monomorphisms and epimorphisms of cochain complexes are still written as $\rightarrowtail$ and $\twoheadrightarrow$.

### *4.5.2 Complements*

(a) The relationship between chain and cochain complexes is clearer in the unbounded case: reversing degrees gives an obvious isomorphism that interchanges the two theories

$$\begin{gathered} \mathrm{Ch}_*\mathsf{Ab} \to \mathrm{Ch}^*\mathsf{Ab}, \\ ((A_n), (\partial_n \colon A_n \to A_{n-1})) \;\mapsto\; ((A^n), (d^n \colon A^n \to A^{n+1}), \end{gathered} \tag{4.70}$$

where $A^n = A_{-n}$ and $d^n = \partial_{-n}$, for $n \in \mathbb{Z}$.

*This layout is internal to abelian groups, and makes no use of duality in abelian categories.*

(b) Theorem 2.3.3 has a version for cochain complexes: a short exact sequence $A \rightarrowtail B \twoheadrightarrow C$ in $\mathrm{Ch}^+\mathsf{Ab}$ gives a *long exact cohomology sequence*

$$0 \to H^0(A) \to H^0(B) \to H^0(C) \xrightarrow{D_0} H^1(A) \to \dots \tag{4.71}$$

The connecting homomorphism is computed with the following diagram, an obvious modification of (2.74) for differentials that rise the degree

$$D^n \colon H^n(C) \to H^{n+1}(A), \qquad D^n[c] = [a], \tag{4.72}$$

$$\begin{array}{ccccccc} & & b & \xmapsto{g^n} & c & & (n) \\ & & \downarrow{\scriptstyle d} & & \downarrow{\scriptstyle d} & & \\ a & \xmapsto{f^{n+1}} & db & \xmapsto{g^{n+1}} & 0 & & (n+1) \end{array}$$

(c) A contravariant functor $F\colon \mathsf{Ab} \dashrightarrow \mathsf{Ab}$ that preserves trivial groups takes chain complexes to cochain complexes, and vice versa. The contravariant functor $\mathrm{Hom}(-, G)$ will be used in Section 5.2 to transform the singular chain complex of a space, or a relative pair, into the singular cochain complex with coefficients in $G$.

(d) Kernels, cokernels, exact sequences and (co)chain complexes in the category $R\,\mathsf{Mod}$ of $R$-modules are dealt with as in $\mathsf{Ab}$.

The categories $\mathrm{Ch}_+(R\,\mathsf{Mod})$ and $\mathrm{Ch}^+(R\,\mathsf{Mod})$ of chain and cochain complexes of $R$-modules have homology and cohomology functors

$$\begin{aligned} &H_n\colon \mathrm{Ch}_+(R\,\mathsf{Mod}) \to R\,\mathsf{Mod},\\ &H^n\colon \mathrm{Ch}^+(R\,\mathsf{Mod}) \to R\,\mathsf{Mod}. \end{aligned} \tag{4.73}$$

### *4.5.3 A preliminary cochain complex*

To define Alexander–Spanier cohomology for a space $X$ we begin by considering a cochain complex $C^+(X; G)$ of abelian groups, which does not depend on the topology of $X$

$$\begin{aligned} &C^n(X; G) = \mathsf{Set}(X^{n+1}, G), \qquad d^n\colon C^n(X; G) \to C^{n+1}(X; G),\\ &(d^n\lambda)(x_1, ..., x_{n+2}) = \Sigma_{1\leqslant i\leqslant n+2}\,(-1)^i\,\lambda(x_1, ..., \hat{x}_i, ..., x_{n+2}), \end{aligned} \tag{4.74}$$

where $\lambda\colon X^{n+1} \to G$ is a mapping (of sets). $C^n(X; G)$ is an abelian group for the pointwise sum: $(\lambda + \mu)(x) = \lambda(x) + \mu(x)$. As usual, the symbol $\hat{x}_i$ means that the $i$-th coordinate is omitted.

For a map $f\colon Y \to X$, we have a cartesian power $f^\bullet\colon Y^{n+1} \to X^{n+1}$ and a 'contravariant' cochain morphism

$$\begin{aligned} &f^\sharp\colon C^+(X; G) \to C^+(Y; G),\\ &f^{\sharp n}(\lambda\colon X^{n+1} \to G) = \lambda f^\bullet\colon Y^{n+1} \to G. \end{aligned} \tag{4.75}$$

*Exercises and complements.* These definitions require some simple verifications.

(a) The composite $d^n d^{n-1}\colon C^{n-1}(X; G) \to C^{n+1}(X; G)$ is zero. *Hint:* one can express the differential using cubical degeneracies $\varepsilon_i\colon X^n \to X^{n-1}$, defined as in (2.36).

(b) In (4.75), $f^\sharp$ is indeed a morphism of cochain complexes: $d^n f^{\sharp n} = f^{\sharp n+1} d^n$.

(c) We have defined a functor $C^+(-; G)\colon \mathsf{Top}^{\mathrm{op}} \to \mathrm{Ch}^+\mathsf{Ab}$, contravariant on $\mathsf{Top}$.

*(d) A homomorphism $h\colon G \to G'$ gives an obvious morphism

$$h_\sharp\colon C^+(X;G) \to C^+(X;G'), \qquad h_\sharp(\lambda\colon X^{n+1} \to G) = h\lambda. \tag{4.76}$$

Putting everything together, we have a functor in two variables

$$C^+\colon \mathsf{Top}^{\mathrm{op}} \times \mathsf{Ab} \to \mathrm{Ch}^+\mathsf{Ab}, \qquad (X,G) \mapsto C^+(X;G), \tag{4.77}$$

defined on the product category $\mathsf{Top}^{\mathrm{op}} \times \mathsf{Ab}$.

### 4.5.4 Alexander–Spanier cochains

Now we take the topology of $X$ into account, and define the *complex of Alexander–Spanier cochains of $X$ with coefficient group $G$*, as the quotient

$$\overline{C}^+(X;G) = C^+(X;G)/C_0^+(X;G), \tag{4.78}$$

where $C_0^+(X;G)$ is the subcomplex of $C^+(X;G)$ of *locally zero* functions: a function $\lambda\colon X^{n+1} \to G$ is said to be locally zero (at the diagonal) if it vanishes at a nbd of any point of the diagonal of $X^{n+1}$.

Equivalently, there exists an open cover $\mathcal{U}$ of $X$ such that $\lambda$ vanishes on $U^{n+1}$, for each $U \in \mathcal{U}$. $C_0^n(X;G)$ is the subgroup of $C^n(X;G)$ of locally zero functions; these subgroups are preserved by the differential.

For a map $f\colon Y \to X$, the homomorphism $f^{\sharp n}\colon C^n(X;G) \to C^n(Y;G)$ restricts to a homomorphism $f^{\sharp n}\colon C_0^n(X;G) \to C_0^n(Y;G)$.

In fact, if $\lambda\colon X^{n+1} \to G$ is locally zero with respect to the open cover $\mathcal{U}$ of $X$, then $f^{\sharp n}(\lambda) = \lambda f^\bullet\colon Y^{n+1} \to G$ is locally zero with respect to the open cover $\mathcal{V}$ of $Y$, formed of the open sets $f^{-1}(U) \subset Y$, for $U \in \mathcal{U}$.

The sequence $f^\sharp = (f^{\sharp n})$ is consistent with the differentials, and forms a morphism of cochain complexes.

We have thus two functors:

$$\begin{aligned} &C_0^+(-;G)\colon \mathsf{Top}^{\mathrm{op}} \to \mathrm{Ch}^+\mathsf{Ab}, && X \mapsto C_0^+(X;G),\\ &\overline{C}^+(-;G)\colon \mathsf{Top}^{\mathrm{op}} \to \mathrm{Ch}^+\mathsf{Ab}, && X \mapsto \overline{C}^+(X;G), \end{aligned} \tag{4.79}$$

whose (contravariant) action on morphisms will still be written as $f \mapsto f^\sharp$. Again, the coefficient group can be allowed to vary.

### 4.5.5 Alexander–Spanier cohomology

Composing the functor $\overline{C}^+(-;G)\colon \mathsf{Top}^{\mathrm{op}} \to \mathrm{Ch}^+\mathsf{Ab}$ with the (covariant) functors $H^n$ of cochain cohomology, we get the functors of *Alexander–Spanier cohomology*, with coefficient group $G$

$$\begin{aligned} &\overline{H}^n(-;G)\colon \mathsf{Top}^{\mathrm{op}} \to \mathsf{Ab}, \quad \overline{H}^n(X;G) = H^n(\overline{C}^+(X;G)),\\ &\qquad (f\colon Y \to X) \mapsto (f^{*n}\colon \overline{H}^n(X;G) \to \overline{H}^n(Y;G)). \end{aligned} \tag{4.80}$$

*Exercises.* (a) (*The dimension axiom*) The singleton $X = \{*\}$ gives:

$$\overline{H}^n(X;G) = 0, \text{ for } n > 0, \qquad \overline{H}^0(X;G) \cong G. \tag{4.81}$$

In degree 0, the isomorphism essentially takes a function $\lambda\colon \{*\} \to G$ to its value $\lambda(*)$. *Hint:* compute the cochain complexes $C^+(X;G)$ and $\overline{C}^+(X;G)$.

(b) On the empty space

$$C^n(\emptyset;G) = \overline{C}^n(\emptyset;G) = 0, \qquad \overline{H}^n(\emptyset;G) = 0 \qquad (n \geqslant 0). \tag{4.82}$$

### 4.5.6 Relative Alexander–Spanier cohomology

For a pair $(X, A)$, the inclusion $u\colon A \to X$ gives a morphism of $\mathrm{Ch}^+\mathsf{Ab}$

$$u^\sharp\colon \overline{C}^+(X;G) \to \overline{C}^+(A;G), \qquad u^{\sharp n}(\overline{\lambda}) = (\lambda u^\bullet)^- \tag{4.83}$$

(for $\lambda\colon X^{n+1} \to G$) which is easily seen to be epi.

Here $\lambda u^\bullet\colon A^{n+1} \to G$ is the restriction of $\lambda\colon X^{n+1} \to G$. As to surjectivity, a class $\overline{\mu} \in \overline{C}^n(A;G))$ is represented by a function $\mu\colon A^{n+1} \to G$; we extend the latter to the function $\lambda\colon X^{n+1} \to G$ which vanishes out of $A^{n+1}$, getting $\lambda u^\bullet = \mu$ and $(\lambda u^\bullet)^- = \overline{\mu}$.

We define the *complex of relative Alexander–Spanier cochains* of $(X, A)$, with coefficient group $G$

$$\overline{C}^+(X, A;G) = \mathrm{Ker}\,(u^\sharp\colon \overline{C}^+(X;G) \to \overline{C}^+(A;G)), \tag{4.84}$$

where $\overline{C}^n(X, A;G)$ is the subgroup of all classes $\overline{\lambda} \in \overline{C}^n(X;G)$ such that $\lambda\colon X^{n+1} \to G$ vanishes on $A^{n+1}$.

We have defined a functor

$$\overline{C}^+(-;G)\colon \mathsf{Top_2}^{\mathrm{op}} \to \mathrm{Ch}^+\mathsf{Ab}, \quad (X, A) \mapsto \overline{C}^+(X, A;G), \tag{4.85}$$

which extends the absolute case, because $\overline{C}^+(\emptyset;G) = 0$ (by 4.5.5(b)).

Applying cochain cohomology, we get the functors of *relative Alexander–Spanier cohomology*, with coefficients in $G$

$$\begin{aligned} &\overline{H}^n(-;G)\colon \mathsf{Top_2}^{\mathrm{op}} \to \mathsf{Ab}, \quad \overline{H}^n(X, A;G) = H^n(\overline{C}^+(X, A;G)), \\ &(f\colon (Y, B) \to (X, A)) \mapsto (f^{*n}\colon \overline{H}^n(X, A;G) \to \overline{H}^n(Y, B;G)). \end{aligned} \tag{4.86}$$

### 4.5.7 The exact cohomology sequence

For a relative pair $(X, A)$, the previous definition of $\overline{C}^+(X, A;G)$ gives a (natural) short exact sequence in $\mathrm{Ch}^+\mathsf{Ab}$

$$\overline{C}^+(X, A;G) \xrightarrow{v^\sharp} \overline{C}^+(X;G) \xrightarrow{u^\sharp} \overline{C}^+(A;G), \tag{4.87}$$

where the embedding $v^\sharp$ of $\mathrm{Ker}\, u^\sharp$ is associated to the relative map $v\colon (X, \emptyset) \to (X, A)$ given by $\mathrm{id}\, X$.

The associated exact sequence of relative Alexander–Spanier cohomology

$$\begin{array}{ccccccc} 0 \to \overline{H}^0(X,A;G) & \xrightarrow{v^{*0}} & \overline{H}^0(X;G) & \xrightarrow{u^{*0}} & \overline{H}^0(A;G) & & \\ & & \xleftarrow{\qquad D^0 \qquad} & & & & \\ \overline{H}^1(X,A;G) & \xrightarrow{v^{*1}} & \overline{H}^1(X;G) & \xrightarrow{u^{*1}} & \overline{H}^1(A;G) & & (4.88) \\ & & \xleftarrow{\qquad D^1 \qquad} & & & & \\ \overline{H}^2(X,A;G) & \xrightarrow{v^{*2}} & \dots & & & & \end{array}$$

is natural, with connecting morphism $D^n\colon \overline{H}^n(A;G) \to \overline{H}^{n+1}(X,A;G)$

$$D^n[(u^{\sharp n}\lambda)^-] = [(d^n\lambda)^-] \qquad (\text{for } \lambda\colon X^{n+1} \to G). \tag{4.89}$$

In fact, the restriction $u^{\sharp n}(\lambda) = \lambda u^\bullet$ is a cocycle of $\overline{C}^n(A;G)$; the function $d^n\lambda\colon X^{n+2} \to G$ vanishes on $A^{n+2}$ and is a cocycle in $\overline{C}^{n+1}(X,A;G)$.

### 4.5.8 Exercises and complements (Degree 0)

Alexander–Spanier cohomology in degree zero is related to the ring $\mathcal{A}(X)$ introduced in (1.15), and studied in 1.2.3, 1.2.4.

(a) For a relative pair of spaces $(X, A)$ there is a canonical isomorphism

$$\mathcal{A}(X,A;G) \longrightarrow \overline{H}^0(X,A;G), \qquad \lambda \mapsto [\overline{\lambda}], \tag{4.90}$$

where $\mathcal{A}(X,A;G)$ is the abelian group of locally constant functions $\lambda\colon X \to G$ which vanish on $A$, with pointwise addition.

*Hints.* Build a suitable augmentation $\mathcal{A}(X,A;G) \to \overline{C}^0(X,A;G)$. Note that a function $X \to G$ is locally constant if and only if it is continuous with respect to the discrete topology of $G$.

(b) In particular, for a space $X$, there is a canonical isomorphism

$$\mathcal{A}(X;G) \to \overline{H}^0(X;G), \qquad \lambda \mapsto [\overline{\lambda}], \tag{4.91}$$

where $\mathcal{A}(X;G)$ is the abelian group of locally constant functions $\lambda\colon X \to G$.

(c) With coefficients in a ring $R$, the abelian group $\mathcal{A}(X;R)$ is naturally a ring. In particular, we have the ring $\mathcal{A}(X;\mathbb{Z}) = \mathcal{A}(X)$ previously introduced. The whole graded abelian group $\overline{H}^*(X;R)$ will be made into a graded ring, in the next section.

(d) If the connected components of the space $X$ are open, there is a canonical isomorphism

$$\overline{H}^0(X;G) \cong G^I, \tag{4.92}$$

where $I = \mathrm{Cc}(X)$ is the set of connected components of $X$ (as in (1.14)).

(e) By our results on the ring $\mathcal{A}(X)$ in Section 1.2 we already know that Alexander–Spanier cohomology distinguishes the converging sequence $\mathbb{N}^\bullet$ from any discrete infinite space: $\overline{H}^0(\mathbb{N}^\bullet) \cong \mathcal{A}(\mathbb{N}^\bullet)$ is the countable subring of $\mathbb{Z}^\mathbb{N}$ formed by the eventually constant sequences of $\mathbb{Z}$ (by Exercise 1.2.3(f)), while any discrete infinite space $D$ gives an uncountable ring $\overline{H}^0(D) \cong \mathcal{A}(D) \cong \mathbb{Z}^D$.

On the other hand, $\mathbb{N}^\bullet$ is totally path disconnected and cannot be distinguished from the same set with discrete topology, using either homotopy groups (and sets), or singular homology (or singular cohomology, as we shall see in 5.5.4(c)). This point will become clearer in 7.6.2, using weak homotopy equivalence.

### *4.5.9 *The axioms*

Alexander–Spanier cohomology satisfies the Eilenberg–Steenrod axioms: it is a cohomology theory with coefficient group (isomorphic to) $G$.

We have verified the functoriality axioms in 4.5.5, the naturality and the exactness axioms in 4.5.7, the dimension axiom in Exercise 4.5.5(a).

We refer to Spanier's book [Sp2] for the remaining two. The excision axiom is verified in Section 6.4, Lemma 4. The proof of homotopy invariance is difficult, and takes the whole Section 6.5.

## 4.6 The product in Alexander–Spanier cohomology

Let $R$ be a commutative (unital) ring. For every space $X$, the graded abelian group $\overline{H}^*(X; R)$ of Alexander–Spanier cohomology has a natural structure of graded $R$-algebra, where the product of two elements of degrees $p$ and $q$ has degree $p + q$.

### *4.6.1 Algebras on a commutative ring*

An *R-algebra* $A$ is an $R$-module which is also equipped with an (internal) product $xy$, so that:

(i) the underlying abelian group becomes a ring with this product: the product is associative and distributes over the sum, on both sides,

(ii) the scalar multiplication and the product are consistent:

$$(\lambda x)y = \lambda(xy) = x(\lambda y) \qquad (\text{for } \lambda \in R \text{ and } x, y \in A), \tag{4.93}$$

and we can write $\lambda xy$, without parentheses.

We reserve the name of *multiplication* to the external operation, defined on $R \times A$, and the name of *product* to the internal one, defined on $A \times A$. An $R$-algebra is said to be unital, or commutative, if it is as a ring.

A *homomorphism* of $R$-algebras $f\colon A \to B$ has to preserve all operations: it is both a homomorphism or $R$-modules and of rings. They form the category $R\mathsf{Alg}$ of $R$-algebras.

A *subalgebra* of $A$ is a subset which is both a submodule and a subring. An *$R$-ideal* of $A$ is a sub-$R$-module of $A$ which is an ideal for the ring structure.

*Remarks.* A ring is 'the same' as a $\mathbb{Z}$-algebra, with multiplication given by multiples. A commutative ring is an algebra on itself, with multiplication given by the product.

### 4.6.2 Graded algebras

A *graded $R$-algebra* $A = (A^n)_{n \geqslant 0}$ is a graded $R$-module equipped with a (graded) product

$$A^p \times A^q \to A^{p+q}, \qquad (x, y) \mapsto xy \qquad (p, q \geqslant 0), \tag{4.94}$$

satisfying the following axioms (for $x, x' \in A^p$, $y, y' \in A^q$, $z \in A^r$, $\lambda \in R$)

(i) (*Consistency*) $(\lambda x)y = \lambda(xy) = x(\lambda y)$,

(ii) (*Distributivity*) $(x + x')y = xy + x'y, \qquad x(y + y') = xy + xy'$,

(iii) (*Associativity*) $x(yz) = (xy)z$.

The first two properties say that the mapping (4.94) is $R$-bilinear. The graded $R$-algebra $A$ is said to be *unital* if it has a unit $\underline{1} \in A^0$ (with $\underline{1}x = x = x\underline{1}$, for all $x$). It is said to be *skew-commutative* if

$$xy = (-1)^{pq}\, yx \qquad (\text{for } x \in A^p,\ y \in A^q). \tag{4.95}$$

The 'Koszul sign rule' takes place here: interchanging an element of degree $p$ with an element of degree $q$ comes about with a sign change $(-1)^{pq}$. This behaviour is typical of differential graded structures.

A *differential graded $R$-algebra* $A = ((A^n)_{n \geqslant 0}, (d^n)_{n \geqslant 0})$ is also equipped with $R$-linear differentials, which make it into a cochain complex of $R$-modules, and agree with the product in the following way:

$$\begin{gathered} d^n\colon A^n \to A^{n+1}, \qquad d^n d^{n-1} = 0, \\ d^n(xy) = (d^p x)y + (-1)^p\, x(d^q y) \qquad (\text{for } x \in A^p,\ y \in A^q). \end{gathered} \tag{4.96}$$

It is (and will often be) understood that $n = p+q$. The Koszul rule is applied, viewing $d$ as a generalised morphism of graded modules, of degree 1.

Morphisms of graded $R$-algebras $f = (f^n)\colon A \to B$ are morphisms of graded modules (of degree 0) which preserve the product: $(f^p x)(f^q y) = f^{p+q}(xy)$ (for $x \in A^p$, $y \in A^q$). In the differential case we further require that the components $f^n$ commute with the differentials.

We have thus the categories $R\mathsf{Gra}$ and $R\mathsf{Dga}$ of *graded $R$-algebras* and *differential graded $R$-algebras*. The latter will also be called dg-algebras, leaving the ring $R$ as understood.

From now on, all $R$-algebras are assumed to be unital, with unit $\underline{1}$ (including the graded and differential graded algebras.)

*Remarks.* (a) In a graded $R$-algebra $A$, the component $A^0$ is an $R$-algebra. The ring of scalars $R$ is mapped to $A^0$, by $\lambda \mapsto \lambda\underline{1}$, but this mapping need not be injective. We also note that $d^0(\underline{1}) = 0$ (writing $\underline{1} = \underline{1}\,\underline{1}$), and therefore $d^0(\lambda\underline{1}) = 0$, for all scalars $\lambda \in R$.

(b) A graded $R$-algebra $A$ (resp. a dg-algebra on $R$) can be made into an ordinary $R$-algebra $\hat{A}$ (resp. an ordinary dg-algebra on $R$), letting

$$\hat{A} = \bigoplus A^n, \tag{4.97}$$

and extending sums and products in the obvious way: writing an essentially finite family $(a_n) \in \hat{A}$ as an essentially finite sum $\Sigma_n\, a_n$, the product is

$$(\Sigma_p\, a_p)(\Sigma_q\, b_q) = \Sigma_n\, (\Sigma_{p+q=n}\, a_p\, b_q). \tag{4.98}$$

In the graded algebras associated to cohomology theories (as Alexander–Spanier, or Čech, or de Rham, or the singular theory) this 'transformation' is practically useless: for instance, adding differential forms of different degrees has little use and little sense. But this is not true for a polynomial ring $R[X]$, which has an obvious graduation given by its submodules of monomials.

### *4.6.3 The cohomology algebra*

Let $A = ((A^n)_{n\geqslant 0}, (d^n)_{n\geqslant 0})$ be a differential graded $R$-algebra.

The cohomology graded $R$-module $H^*(A) = (H^n(A))$ inherits a graded product

$$H^p(A) \times H^q(A) \to H^{p+q}(A), \qquad ([x],[y]) \mapsto [xy] \qquad (p, q \geqslant 0), \tag{4.99}$$

that makes it into a graded algebra, with unit $[\underline{1}]$. The following exercises are nearly obvious.

*Exercises and complements.* (a) The product of cohomology classes is well defined.

(b) Cohomology gives a covariant functor

$$H^*\colon R\mathsf{Dga} \to R\mathsf{Gra}. \tag{4.100}$$

### 4.6.4 The Alexander–Spanier graded algebra

For a space $X$, the cochain complex $C^*(X;R)$ has a natural structure of differential graded $R$-algebra

$$\begin{aligned} C^p(X;R) \times C^q(X;R) &\longrightarrow C^{p+q}(X;R), \\ (\lambda\bullet\mu)(x_1,...,x_{n+1}) &= \lambda(x_1,...,x_{p+1})\,\mu(x_{p+1},...,x_{n+1}), \end{aligned} \tag{4.101}$$

as verified in Exercise (a).

Moreover the cochain subcomplex $C_0^+(X;R)$ of locally zero functions is an ideal of this graded algebra, in the obvious sense: if $\lambda$ or $\mu$ is locally zero, then $\lambda\bullet\mu$ is also. Therefore, the quotient $\overline{C}^*(X;R) = C^*(X;R)/C_0^+(X;R)$ inherits a structure of dg-algebra.

The Alexander–Spanier cohomology graded $R$-module is thus a graded $R$-algebra, and we have a functor

$$\overline{H}^*\colon \mathsf{Top} \to R\mathsf{Gra}. \tag{4.102}$$

*Exercises and complements.* (a) The product (4.101) is consistent with the differential, as in (4.96).

(b) (*Functoriality*) Verify that, for a map $f\colon Y \to X$, the cochain morphism $f^\sharp\colon \overline{C}^*(X;R) \to \overline{C}^*(Y;R)$ is a morphism of dg-algebras; therefore it induces a homomorphism of graded algebras, in cohomology.

## 4.7 An elementary approach to Čech cohomology

Čech cohomology is another important theory, naturally arising in contravariant form and well related to Alexander–Spanier theory.

Its construction is more involved, being based on inductive limits and (often) on 'variable coefficients', taking values in a presheaf on the space. We present a more elementary setting, *using a fixed abelian group $G$ of coefficients* (typically, a ring). A reader interested in this discipline should find it easier, after this introduction, to study the usual, more advanced expositions.

*Literature.* Spanier's book [Sp2] studies Čech cohomology and its relationship with Alexander–Spanier's, in Sections 6.7–6.8. One can also see Eilenberg–Steenrod [EiS], Chapters IX and X, and Godement [Go], Chapter 5. Of these books, only [EiS] deals with the functorial aspect of the theory.

### 4.7.1 Main definitions

Let $X$ be a topological space. The Čech cohomology group $\check{H}^n(X;G)$ of $X$, with coefficients in $G$, will be defined as the inductive limit (see Section 1.7) of a preliminary system of cohomology groups $H^n(\mathcal{U};G)$, indexed by the open covers of $X$.

(a) It is convenient to view an open cover $\mathcal{U}$ of $X$ as a *set* (rather than an indexed family) of open subsets of $X$, which cover the latter. These covers form a set $\mathrm{Cov}(X)$, which we equip with a preorder relation $\mathcal{U} \leqslant \mathcal{V}$: namely, $\mathcal{V}$ is *finer* than $\mathcal{U}$ (or a *refinement* of $\mathcal{U}$) if every $V \in \mathcal{V}$ is contained in some $U \in \mathcal{U}$. $\mathrm{Cov}(X)$ is thus a filtered set: two covers $\mathcal{U}$ and $\mathcal{V}$ can be refined by the set of all $U \cap V$, for $U \in \mathcal{U}$ and $V \in \mathcal{V}$.

Of course this is not an order relation. For instance, in the euclidean plane, the open discs *and* the open squares form two covers which refine each other.

(b) We define now a cochain complex $C^+(\mathcal{U};G)$, by a procedure similar to the construction of the complex $C^+(X;G)$, used in 4.5.3 for Alexander–Spanier cohomology.

For $n \geqslant 0$, we let $\mathcal{U}_n$ be the set of $(n+1)$-tuples $(U_0, ..., U_n) \in \mathcal{U}^{n+1}$ such that $\bigcap U_i \neq \emptyset$. In particular, $\mathcal{U}_0 = \{U \in \mathcal{U} \mid U \neq \emptyset\}$.

We define

$$\begin{gathered} C^n(\mathcal{U};G) = \mathsf{Set}(\mathcal{U}_n, G), \quad d^n\colon C^n(\mathcal{U};G) \to C^{n+1}(\mathcal{U};G), \\ (d^n\varphi)(U_0, ..., U_{n+1}) = \textstyle\sum_{0 \leqslant i \leqslant n+1} (-1)^i\, \varphi(U_0, ..., \hat{U}_i, ..., U_{n+1}), \end{gathered} \tag{4.103}$$

where $\varphi\colon \mathcal{U}_n \to G$ is a mapping (of sets), and $C^n(\mathcal{U};G)$ is an abelian group, for the pointwise sum. The property $d^n d^{n-1} = 0$ is readily verified (as in Exercise 4.5.3(a)).

We have thus the *graded cohomology group* $H^*(\mathcal{U};G)$ of the cochain complex $C^+(\mathcal{U};G)$.

(c) If $\mathcal{U} \leqslant \mathcal{V}$, we choose a mapping $\lambda\colon \mathcal{V} \to \mathcal{U}$ such that $V \subset \lambda(V)$, for all $V \in \mathcal{V}$. This defines, contravariantly, a morphism of cochain complexes

$$\lambda^\sharp\colon C^+(\mathcal{U};G) \to C^+(\mathcal{V};G), \quad (\lambda^n\varphi)(V_0, ..., V_n) = \varphi(\lambda V_0, ..., \lambda V_n), \tag{4.104}$$

which depends on the choice of $\lambda$, but induces in cohomology a homomorphism *which only depends on the relation* $\mathcal{U} \leqslant \mathcal{V}$

$$\lambda^n\colon H^n(\mathcal{U};G) \to H^n(\mathcal{V};G), \qquad \lambda^n[\varphi] = [\lambda^n\varphi]. \tag{4.105}$$

In fact, for any similar mapping $\mu\colon \mathcal{V} \to \mathcal{U}$, one can easily verify that there is a cochain homotopy

$$\begin{gathered} \Lambda\colon \lambda^\sharp \simeq \mu^\sharp : C^+(\mathcal{U};G) \longrightarrow C^+(\mathcal{V};G), \\ (\Lambda\varphi)(V_0, ..., V_{n-1}) = \textstyle\sum_{0 \leqslant i < n} (-1)^i\, \varphi(\lambda V_0, ..., \lambda V_i, \mu V_i, ..., \mu V_{n-1}). \end{gathered}$$

(d) We have defined an inductive system of abelian groups $H^n(\mathcal{U};G)$, over the filtered set $\mathrm{Cov}(X)$.

For $\mathcal{U} = \mathcal{V}$ we can take $\lambda = \mathrm{id}\,\mathcal{U}$. If $\mathcal{U} \leqslant \mathcal{V} \leqslant \mathcal{W}$, after choosing appropriate mappings $\lambda\colon \mathcal{V} \to \mathcal{U}$ and $\mu\colon \mathcal{W} \to \mathcal{V}$, we can take $\lambda\mu\colon \mathcal{W} \to \mathcal{U}$ to induce the homomorphism $H^n(\mathcal{U};G) \to H^n(\mathcal{W};G)$.

The *Čech cohomology group* of degree $n$ is now defined as the inductive limit of this system, over the filtered set $\mathrm{Cov}(X)$

$$\check{H}^n(X;G) = \varinjlim{}_{\mathcal{U}}\, H^n(\mathcal{U};G). \tag{4.106}$$

*(e) *Remark.* In (4.103) we can also define $C^n(\mathcal{U};G)$ as the set of mappings $\varphi\colon \mathcal{U}^{n+1} \to G$ such that $\varphi(U_0, ..., U_n) = 0$ when $\bigcap U_i = \emptyset$; but then the definition of $d^n\varphi$ must be corrected, forcing it to vanish in the same way. This correction is automatic in the general case, of coefficients in a presheaf: see (4.110).

### 4.7.2 Exercises (Degree 0)

(a) $H^0(\mathcal{U};G)$ is the abelian group $G(\mathcal{U})$ of *coherent* functions $\mathcal{U}_0 \to G$, which coincide on all non-disjoint pairs $U$, $U' \in \mathcal{U}$.

(b) As in Alexander–Spanier cohomology, $\check{H}^0(X;G)$ is naturally isomorphic to the abelian group $\mathcal{A}(X;G)$ of *locally constant* functions $X \to G$, that is, continuous functions for the discrete topology on $G$ (cf. 4.5.8(a), (b)).

### 4.7.3 Functoriality

For a map $f\colon Y \to X$, we want to define the associated homomorphism $f^{*n}\colon \check{H}^n(X;G) \to \check{H}^n(Y;G)$, producing a sequence of contravariant functors $\check{H}^n\colon \mathsf{Top} \dashrightarrow \mathsf{Ab}$.

(a) We extend the procedure used in 4.7.1(c) for $f = \mathrm{id}\,X$. We take two open covers $\mathcal{U}, \mathcal{V}$ of $X$ and $Y$, respectively, such that $f^*\mathcal{U} \leqslant \mathcal{V}$ (where $f^*\mathcal{U} = \{f^{-1}U \mid U \in \mathcal{U}\}$). Then we choose a mapping $\lambda_f\colon \mathcal{V} \to \mathcal{U}$ such that $V \subset f^{-1}(\lambda_f V)$, for all $V \in \mathcal{V}$ (or, equivalently, $f(V) \subset \lambda_f V$).

Given $\mathcal{U}$, one can always take $\mathcal{V} = f^*\mathcal{U}$; this will be useful below.

This gives a morphism of cochain complexes

$$\lambda_f^\sharp\colon C^+(\mathcal{U};G) \to C^+(\mathcal{V};G), \quad (\lambda_f^n\varphi)(V_0, ..., V_n) = \varphi(\lambda_f V_0, ..., \lambda_f V_n),$$

which depends on the choice of $\lambda_f$, but induces in cohomology a homomorphism

$$\lambda_f^n\colon H^n(\mathcal{U};G) \to H^n(\mathcal{V};G), \qquad \lambda_f^n[\varphi] = [\lambda_f^n\varphi], \tag{4.107}$$

which only depends on $f$, $\mathcal{U}$ and $\mathcal{V}$, as in (4.105).

(b) Suppose now that we have open covers $\mathcal{U} \leqslant \mathcal{U}'$ of $X$ and $\mathcal{V} \leqslant \mathcal{V}'$ of $Y$, with $f^*\mathcal{U} \leqslant \mathcal{V}$ and $f^*\mathcal{U}' \leqslant \mathcal{V}'$. We obtain the left diagram of cochain complexes (with two commutative triangles)

$$\begin{array}{ccc} C^+(\mathcal{U};G) & \xrightarrow{\lambda^\sharp} & C^+(\mathcal{U}';G) \\ {\scriptstyle \lambda_f^\sharp}\downarrow & \searrow\!\!\searrow & \downarrow{\scriptstyle \mu_f^\sharp} \\ C^+(\mathcal{V};G) & \xrightarrow[\mu^\sharp]{} & C^+(\mathcal{V}';G) \end{array} \qquad \begin{array}{ccc} H^n(\mathcal{U};G) & \xrightarrow{\lambda^n} & H^n(\mathcal{U}';G) \\ {\scriptstyle \lambda_f^n}\downarrow & = & \downarrow{\scriptstyle \mu_f^n} \\ H^n(\mathcal{V};G) & \xrightarrow[\mu^n]{} & H^n(\mathcal{V}';G) \end{array} \tag{4.108}$$

induced by four mappings that satisfy the appropriate conditions:

$$\lambda\colon \mathcal{U}' \to \mathcal{U} \quad (U' \subset \lambda U'), \qquad \mu\colon \mathcal{V}' \to \mathcal{V} \quad (V' \subset \mu V'),$$
$$\lambda_f\colon \mathcal{V} \to \mathcal{U} \quad (fV \subset \lambda_f V), \qquad \mu_f\colon \mathcal{V}' \to \mathcal{U}' \quad (fV' \subset \mu_f V').$$

These *two* diagonals are induced by mappings that also satisfy the appropriate conditions, for $V' \in \mathcal{V}'$:

$$\lambda\mu_f\colon \mathcal{V}' \to \mathcal{U}' \to \mathcal{U}, \qquad f(V') \subset \mu_f V' \subset \lambda\mu_f V',$$
$$\lambda_f\mu\colon \mathcal{V}' \to \mathcal{V} \to \mathcal{U}, \qquad V' \subset \mu V' \subset f^{-1}(\lambda_f \mu V').$$

Therefore the diagonals are homotopic and coincide in cohomology, as in (4.107): the right-hand diagram (4.108) commutes.

(c) We can now define the homomorphism induced by $f$ in cohomology

$$f^{*n}\colon \check{H}^n(X;G) \to \check{H}^n(Y;G), \quad f^{*n}[\varphi] = [\lambda_f^n \varphi], \tag{4.109}$$

as the inductive limit of the homomorphisms $\lambda_f^n\colon H^n(\mathcal{U};G) \to H^n(\mathcal{V};G)$.

We are applying a general property of inductive limits (see 1.7.6(c)), using the fact that, for a given cover $\mathcal{U}$ of $X$, we can choose the cover $\mathcal{V} = f^*\mathcal{U}$ of $Y$.

### *4.7.4 *Complements*

(a) To go on, it would now be convenient to deal with the general definition of the Čech cohomology groups $\check{H}^n(X;\Gamma)$, with coefficients in a presheaf $\Gamma$ of abelian groups (see 1.5.6(b)). A good source is Spanier [Sp2], Sections 6.7 and 6.8.

An abelian group $G$ is viewed as a *constant* sheaf on $X$, letting $\Gamma(U) = G$ for $U \neq \emptyset$ (and $\Gamma(\emptyset) = 0$).

In general, $C^n(\mathcal{U};\Gamma)$ is defined as the abelian group of functions $\varphi$ which assign to any $(n+1)$-tuple $(U_0, ..., U_n) \in \mathcal{U}^{n+1}$ an element $\varphi(U_0, ..., U_n) \in \Gamma(\bigcap U_i)$. The differential is defined as

$$\begin{aligned} &(d^n\varphi)(U_0, ..., U_{n+1}) \\ &\quad = \textstyle\sum_{0 \leqslant i \leqslant n+1} (-1)^i\, \varphi(U_0, ..., \hat{U}_i, ..., U_{n+1}) | (U_0 \cap ... \cap U_{n+1}), \end{aligned} \tag{4.110}$$

where a restricted element $\rho^U_V(\gamma)$ is denoted as $\gamma|V$.

In particular, this allows a quick definition of the relative cohomology groups $\check{H}^n(X, A; G)$ and an easy proof of the exact cohomology sequence of the relative pair $(X, A)$ ([Sp2], p. 328).

(b) Alexander–Spanier and Čech cohomology are closely related. Notably, for any *paracompact Hausdorff space* $X$ and any abelian group $G$, there is a natural isomorphism (see [Sp2], Corollary 6.8.8)

$$\check{H}^n(X; G) \;\cong\; \overline{H}^n(X; G). \tag{4.111}$$

(c) Čech cohomology with coefficients in a group satisfies the axioms of Eilenberg–Steenrod: see [EiS], Sections IX.5–IX.7.

There is also a Čech homology theory, which does not satisfy the exactness axiom [EiS].

(d) Eilenberg–Steenrod [EiS] has a Chapter X on 'Special features of the Čech theory', focused on continuity properties and the Excision axiom.

## 4.8 An introduction to de Rham cohomology

In this section we sketch de Rham cohomology for open euclidean spaces. A reader familiar with basic differential geometry will know how this part can be extended to differentiable manifolds. (An automatic extension, from local to global, is also possible, as suggested in 4.8.8.)

*Literature.* Differentiable manifolds, differential forms and de Rham cohomology are studied in Spivak's book [Sv]. De Rham's theorem can be found in the Appendix of [Mas3].

### 4.8.1 Basic aspects

We work in the category $\mathrm{C}^\infty\mathsf{Top}$ of open euclidean spaces (i.e. open subspaces of some $\mathbb{R}^n$) and $\mathrm{C}^\infty$-*maps* between them, that is, maps that have all partial derivatives. A *diffeomorphism* is an isomorphism of $\mathrm{C}^\infty\mathsf{Top}$.

The coordinates of $\mathbb{R}^3$ are written as $x, y, z$; the coordinates of $\mathbb{R}^n$ in the usual form $x^1, x^2, ..., x^n$. The reader is likely acquainted with differential forms $\omega$ on an open subspace $U$ of $\mathbb{R}^3$, written as

$$\begin{aligned}
&\omega = \lambda && \text{(in degree 0)},\\
&\omega = \lambda\, dx + \mu\, dy + \nu\, dz && \text{(in degree 1)},\\
&\omega = \lambda\, dydz + \mu\, dxdz + \nu\, dxdy && \text{(in degree 2)},\\
&\omega = \lambda\, dxdydz && \text{(in degree 3)},
\end{aligned}$$

where $\lambda, \mu, \nu \colon U \to \mathbb{R}$ are $\mathrm{C}^\infty$-functions on U.

A $C^\infty$-function $\lambda\colon U \to \mathbb{R}$ is a form of degree 0, and has a differential: a form of degree 1 whose coefficients are the three partial derivatives of $\lambda$

$$d\lambda = \frac{\partial\lambda}{\partial x}\,dx + \frac{\partial\lambda}{\partial y}\,dy + \frac{\partial\lambda}{\partial z}\,dz. \tag{4.112}$$

A form $\omega = \lambda\,dx + \mu\,dy + \nu\,dz$ of degree 1 has a differential of degree 2

$$d\omega = \Big(\frac{\partial\nu}{\partial y} - \frac{\partial\mu}{\partial z}\Big)\,dydz + \Big(\frac{\partial\nu}{\partial x} - \frac{\partial\lambda}{\partial z}\Big)\,dxdz + \Big(\frac{\partial\mu}{\partial x} - \frac{\partial\lambda}{\partial y}\Big)\,dxdy. \tag{4.113}$$

A form $\omega = \lambda\,dydz + \mu\,dxdz + \nu\,dxdy$ of degree 2 has a differential of degree 3

$$d\omega = \Big(\frac{\partial\lambda}{\partial x} - \frac{\partial\mu}{\partial y} + \frac{\partial\nu}{\partial z}\Big)\,dxdydz. \tag{4.114}$$

We shall now extend differential forms and their calculus to an open subspace $U$ of $\mathbb{R}^n$. The partial derivatives of the $C^\infty$-function $\lambda\colon U \to \mathbb{R}$ will also be written as $D_1\lambda$, ..., $D_n\lambda$.

### *4.8.2 Main definitions*

For an open euclidean space $U \subset \mathbb{R}^n$, we write as

$$\Omega^0(U) = C^\infty\mathsf{Top}(U, \mathbb{R}), \tag{4.115}$$

the commutative algebra of real-valued $C^\infty$-functions on $U$, with the pointwise operations in $\mathbb{R}$. Its unit $\underline{1}$ is the constant function at $1 \in \mathbb{R}$.

A *p-index* $\mathbf{i}$ of dimension $n$ will be a subset of the set $\{1, ..., n\}$ containing $p$ elements; it is empty if $p = 0$ or $p > n$. There are $\binom{n}{p}$ of them, which we rewrite in the form

$$dx^{\mathbf{i}} = dx^{i_1}dx^{i_2} \ldots dx^{i_p}, \qquad dx^{\emptyset} = \underline{1}, \tag{4.116}$$

where it is understood that $\mathbf{i} = \{i_1, i_2, ..., i_p\}$ and $i_1 < i_2 < ... < i_p$, when $\mathbf{i} \neq \emptyset$. Each $dx^{\mathbf{i}}$ will be called a *basic* $p$-form. (As usual, we assume that $\binom{n}{p} = 0$ for $p > n$.)

We write as $\Omega^p(U)$ the free $\Omega^0(U)$-module of dimension $\binom{n}{p}$ generated by the basic $p$-forms. An element of $\Omega^p(U)$, called a *differential form* of degree $p$ on $U$, or a *p-form*, is a formal linear combination of basic forms

$$\omega = \Sigma_{\mathbf{i}}\, \lambda_{\mathbf{i}}\, dx^{\mathbf{i}} = \Sigma_{\mathbf{i}}\, \lambda_{\mathbf{i}}\, dx^{i_1}dx^{i_2} \ldots dx^{i_p}. \tag{4.117}$$

where $\mathbf{i}$ varies in the set of $p$-indices of dimension $n$, and all $\lambda_{\mathbf{i}}$ belong to the algebra $\Omega^0(U)$. $\Omega^p(U)$ will also be considered as a vector space on $\mathbb{R}$.

Thus $\Omega^p(U) = 0$ for $p > n$. For $p = 0$ we find again the algebra of $C^\infty$-functions on $U$, introduced in (4.115).

It is also convenient to extend the meaning of $dx^{i_1}dx^{i_2}...\,dx^{i_p}$ to any $p$-uple $i = (i_1, i_2, ..., i_p)$ of integers in $\{1, ..., n\}$, letting

$$\begin{aligned} &dx^{i_1}dx^{i_2}...\,dx^{i_p} = 0 && \text{if these indices are not distinct,} \\ &dx^{i_1}dx^{i_2}...\,dx^{i_p} = \sigma(i)\,dx^{\mathbf{i}} && \text{otherwise,} \end{aligned} \tag{4.118}$$

where $\sigma(i)$ is the sign of the permutation $i = (i_1, ..., i_p)$ of the ordered set $\mathbf{i}$.

*Remarks.* (a) For $0 \leqslant p \leqslant n$, the module $\Omega^p(U)$ is canonically isomorphic to $\Omega^{n-p}(U)$, sending a $p$-index $\mathbf{i}$ to its complement in $\{1, ..., n\}$.

(b) For the empty set $U = \emptyset$, everything becomes trivial: $\Omega^0(U)$ is the trivial algebra, and $\Omega^p(U) = 0$ for all $p$.

### *4.8.3 The product of differential forms*

We have defined a graded module $\Omega^+(U)$ on the commutative algebra $\Omega^0(U)$; now we make it into a graded algebra.

The *exterior product* of a basic $p$-form and a basic $q$-form is defined as

$$\begin{aligned} &dx^{\mathbf{i}} \wedge dx^{\mathbf{j}} = 0 && \text{if } \mathbf{i} \cap \mathbf{j} \neq \emptyset, \\ &dx^{\mathbf{i}} \wedge dx^{\mathbf{j}} = \sigma(\mathbf{i}, \mathbf{j})\,dx^{\mathbf{k}} && \text{otherwise,} \end{aligned} \tag{4.119}$$

where $\mathbf{k} = \mathbf{i} \cup \mathbf{j}$ (a disjoint union) and $\sigma(\mathbf{i}, \mathbf{j})$ is the sign of the permutation $(i_1, ..., i_p, j_1, ..., j_q)$ of the ordered set $\mathbf{k}$. Using the extended symbols introduced in (4.118), the formulas (4.119) can be rewritten as:

$$(dx^{i_1}...\,dx^{i_p}) \wedge (dx^{j_1}...\,dx^{j_q}) = dx^{i_1}...\,dx^{i_p}dx^{j_1}...\,dx^{j_q}. \tag{4.120}$$

The product is extended to all differential forms, by bilinearity on $\Omega^0(U)$

$$\begin{gathered} \Omega^p(U) \times \Omega^q(U) \to \Omega^{p+q}(U), \\ (\textstyle\sum_{\mathbf{i}} \lambda_{\mathbf{i}}\,dx^{\mathbf{i}}) \wedge (\sum_{\mathbf{j}} \mu_{\mathbf{j}}\,dx^{\mathbf{j}}) = \sum_{\mathbf{i},\mathbf{j}} \lambda_{\mathbf{i}}\mu_{\mathbf{j}}(dx^{\mathbf{i}} \wedge dx^{\mathbf{j}}), \\ (\textstyle\sum_i \lambda_i\,dx^{i_1}...\,dx^{i_p}) \wedge (\sum_j \mu_j\,dx^{j_1}...\,dx^{j_q}) \\ = \textstyle\sum_{i,j} \lambda_i\mu_j\,dx^{i_1}...\,dx^{i_p}dx^{j_1}...\,dx^{j_q}. \end{gathered} \tag{4.121}$$

*Exercises.* All this is simpler than it might initially seem, and does produce a graded algebra.

(a) For an open subspace $U$ of $\mathbb{R}^3$, compute the exterior product of the forms $\omega = \lambda_1\,dx + \lambda_2\,dy + \lambda_3\,dz$ and $\omega' = \mu_1\,dydz + \mu_2\,dxdz + \mu_3\,dxdy$.

(b) The product makes $\Omega^+(U)$ into a skew-commutative graded algebra, as defined in 4.6.2.

### 4.8.4 The differential

We complete the construction, making $\Omega^+(U)$ into a dg-algebra on the real field, called the *exterior algebra of differential forms* on $U$.

The differential of $C^\infty$-functions is well known

$$d^0 \colon \Omega^0(U) \to \Omega^1(U), \qquad d^0(\lambda) = \Sigma_{1 \leqslant k \leqslant n} D_k\lambda \, dx^k. \tag{4.122}$$

We extend it to all degrees, letting

$$\begin{gathered} d^p \colon \Omega^p(U) \to \Omega^{p+1}(U), \qquad d^p(\Sigma_{\mathrm{i}} \lambda_{\mathrm{i}} \, dx^{\mathrm{i}}) = \Sigma_{\mathrm{i}} \, d^0(\lambda_{\mathrm{i}}) \wedge dx^{\mathrm{i}}, \\ d^p(\Sigma_i \lambda_i \, dx^{i_1} ... dx^{i_p}) = \Sigma_{i,k} D_k\lambda_i \, dx^k dx^{i_1} ... \, dx^{i_p}, \end{gathered} \tag{4.123}$$

where $i = (i_1, ..., i_p)$ ranges over the $p$-uples of integers in $\{1, ..., n\}$ and $k = 1, ..., n$. Let us note that this mapping is only $\mathbb{R}$-linear: a homomorphism of real vector spaces.

*Exercises and complements.* (a) For $n = 3$ this definition gives the formulas (4.112)–(4.114).

(b) Prove that the differential acts on a product $\omega \wedge \omega'$ of differential forms (of degrees $p$ and $q$) as in (4.96)

$$d^{p+q}(\omega \wedge \omega') = (d^p\omega) \wedge \omega' + (-1)^p \omega \wedge (d^q\omega'). \tag{4.124}$$

*Hint:* begin with $C^\infty$-functions.

(c) The differential makes $\Omega^+(U)$ into a dg-algebra on the real field.

### 4.8.5 Cohomology

Cocycles and coboundaries form two $\mathbb{R}$-linear subspaces of $\Omega^p(U)$

$$\begin{aligned} Z^p(U) &= \mathrm{Ker}\,(d^p \colon \Omega^p(U) \to \Omega^{p+1}(U)) \subset \Omega^p(U), \\ B^p(U) &= \mathrm{Im}\,(d^{p-1} \colon \Omega^{p-1}(U) \to \Omega^p(U)) \subset Z^p(U). \end{aligned} \tag{4.125}$$

A differential $p$-form $\omega = \Sigma_{\mathrm{i}} \lambda_{\mathrm{i}} \, dx^{\mathrm{i}}$ is said to *closed* if it is a cocycle, i.e. $d^p\omega = 0$. It is said to be *exact* if it is a coboundary: $\omega = d^{p+1}\omega'$, for some form $\omega'$ of degree $p + 1$.

Every exact form is closed; the converse need not be true, and this failure is measured by the quotient

$$H^p_R(U) = Z^p(U)/B^p(U), \tag{4.126}$$

the *de Rham cohomology* vector space of $U$, in degree $p$

*Exercises and complements.* (a) Compute $H^0_R(U)$ and prove that the vector space $H^0_R(U)$ has dimension 1 if and only if $U$ is non-empty and connected.

(b) A 1-form $\omega = \Sigma_i \lambda_i \, dx^i \in \Omega^1(U)$ is closed if and only if

$$\frac{\partial \lambda_i}{\partial x^j} - \frac{\partial \lambda_j}{\partial x^i} = 0, \qquad \text{for all } i, j. \tag{4.127}$$

(c) In Calculus, differential forms on open euclidean subspaces of $\mathbb{R}^3$ are also viewed as vector fields; the differential of a function is then called the *gradient*; saying that $H^1_R(U) = 0$ means that every closed vector field $X = (\lambda_1, \lambda_2, \lambda_3)$ (satisfying the conditions (4.127)) is the gradient of some $C^\infty$-function on $U$. The reader can similarly interpret the conditions $H^2_R(U) = 0$ and $H^3_R(U) = 0$.

### 4.8.6 Smooth maps

A $C^\infty$-map $f = (f^1, ..., f^n)\colon V \to U$ has associated $\mathbb{R}$-linear morphisms

$$\begin{aligned} f^{\sharp p}&\colon \Omega^p(U) \to \Omega^p(V), \\ f^{\sharp p}(\Sigma_i \lambda_i \, dx^{i_1} ... \, dx^{i_p}) &= \Sigma_i \lambda_i{\circ} f \, d^0 f^{i_1} \wedge ... \wedge d^0 f^{i_p}, \end{aligned} \tag{4.128}$$

where the composite $\lambda_i{\circ}f\colon V \to U \to \mathbb{R}$ is not written by juxtaposition, to avoid confusion with pointwise multiplication of real valued functions. In particular, $f^{\sharp 0}(\lambda) = \lambda{\circ}f$.

The following exercises show that $f^\sharp = (f^{\sharp p})\colon \Omega^+(U) \to \Omega^+(V)$ is a morphism of dg-algebras.

*Exercises and complements.* (a) $f^\sharp$ preserves the exterior product of differential forms. *Hint:* begin with $C^\infty$-functions.

(b) The homomorphisms $f^{\sharp p}$ commute with the differentials.

### 4.8.7 Functoriality

We have so forth defined a contravariant functor

$$\Omega^+\colon C^\infty\mathsf{Top} \dashrightarrow \mathbb{R}\mathsf{Dga}. \tag{4.129}$$

The proof of the functorial property $(gf)^\sharp = f^\sharp g^\sharp$ is similar to the previous ones, and is left to an interested reader.

Composing with the cohomology functor $H^*\colon \mathbb{R}\mathsf{Dga} \to \mathbb{R}\mathsf{Gra}$ of (4.100) (for real coefficients), we get the contravariant functor of de Rham cohomology

$$\begin{aligned} H^*_R\colon C^\infty\mathsf{Top} \dashrightarrow \mathbb{R}\mathsf{Dga}, \qquad H^*_R(U) &= H^*(\Omega^+(U)), \\ f^* = H^*_R(f)\colon H^*_R(U) &\to H^*_R(V), \end{aligned} \tag{4.130}$$

on open euclidean spaces (for $f\colon V \to U$ in $C^\infty\mathsf{Top}$).

### *4.8.8 *Complements*

What we have seen can be extended to its natural setting, that is, differentiable manifolds, as we outline now.

(a) A *differentiable manifold* is a topological space $X$ equipped with an *atlas* $(U_i, u_i)_{i \in I}$, where:

- each $U_i$ is a euclidean open set,

- each *chart* $u_i \colon U_i \to X$ is an open embedding, and their images cover $X$,

- each *change of charts* $U_i \nrightarrow U_j$ (induced by $u_i$ and $u_j$) is a partial diffeomorphism between open subsets, and can be realised as a pair of open embeddings $U_i \leftarrow U_{ij} \to U_j$ in $\mathrm{C}^\infty\mathsf{Top}$.

All this would become clearer using the presentation of manifolds by 'intrinsic atlases', as in [G4].

(b) The space $X$ can be viewed as a 'generalised pushout' in $\mathsf{Top}$, the colimit of a diagram $U \colon \hat{I} \to \mathrm{C}^\infty\mathsf{Top}$ consisting of the previous open embeddings, as in the left diagram below (made precise in (d))

$$\begin{array}{ccc} & & U_i \\ & {\scriptstyle u_{ij}}\nearrow & \\ U_{ij} & & \\ & {\scriptstyle u_{ji}}\searrow & \\ & & U_j \end{array} \qquad\qquad \begin{array}{ccccc} & & U_i & & \\ & {\scriptstyle u_{ij}}\nearrow & & \searrow{\scriptstyle u_i} & \\ U_{ij} & & & & X \\ & {\scriptstyle u_{ji}}\searrow & & \nearrow{\scriptstyle u_j} & \\ & & U_j & & \end{array} \tag{4.131}$$

A map $\lambda \colon X \to \mathbb{R}$ amounts thus to a cocone of this system, that is a family of maps $\lambda_i \colon U_i \to \mathbb{R}$ such that $\lambda_i u_{ij} = \lambda_j u_{ji}$ (the coherence condition, for $i, j \in I$); $\lambda$ is $\mathrm{C}^\infty$ on $X$ if all its restrictions $\lambda_i \colon U_i \to \mathbb{R}$ are.

The $\mathbb{R}$-algebra $\Omega^0(X)$ of $\mathrm{C}^\infty$-functions on $X$ can thus be defined as the limit of the (solid) diagram $\Omega^0 U$

$$\begin{array}{ccccc} & & \Omega^0(U_i) & & \\ & {\scriptstyle u_{ij}^*}\swarrow & & \nwarrow{\scriptstyle u_i^*} & \\ \Omega^0(U_{ij}) & & & & \Omega^0(X) \\ & {\scriptstyle u_{ji}^*}\nwarrow & & \swarrow{\scriptstyle u_j^*} & \\ & & \Omega^0(U_j) & & \end{array} \tag{4.132}$$

that is, the composite of the original diagram $U \colon \hat{I} \to \mathrm{C}^\infty\mathsf{Top}$ with the contravariant functor $\Omega^0 \colon \mathrm{C}^\infty\mathsf{Top} \dashrightarrow \mathbb{R}\mathsf{Alg}$.

Similarly, we define the graded differential algebra $\Omega^+(X)$ as a limit

$$\Omega^+(X) = \mathrm{Lim}\, \Omega^+ U, \qquad\qquad H^*_R(X) = H^*(\Omega^+(X)), \tag{4.133}$$

and the de Rham cohomology of $X$ as the cohomology graded algebra of the latter.

(c) For every differentiable manifold $X$, de Rham's theorem establishes a natural isomorphism of graded algebras

$$H^*_R(X) \to H^*(X;\mathbb{R}), \tag{4.134}$$

with the singular cohomology of $X$, with real coefficients, defined in Section 5.5 (see also Section 6.6). The proof can be found in [Mas3], Appendix, Section 3.

(d) A generalised pushout in a category $\mathsf{C}$, used in (b), is a colimit of a particular form, introduced in [CarG].

We start from a non-empty set $I$ and form the set $\hat{I}$ of its subsets $\{i,j\}$ having one or two elements, ordered by the relation $\{i,j\} \leqslant \{i\}$ and viewed as a (connected) category.

A functor $U\colon \hat{I} \to \mathsf{C}$ is represented as the left diagram below

$$\begin{array}{ccc} & \overset{u_{ij}}{\nearrow} & U_i \\ U_{ij} & & \\ & \underset{u_{ji}}{\searrow} & U_j \end{array} \qquad (i,j \in I), \qquad \begin{array}{ccccc} & \overset{u_{ij}}{\nearrow} & U_i & \overset{u_i}{\searrow} & \\ U_{ij} & & & & X \\ & \underset{u_{ji}}{\searrow} & U_j & \underset{u_j}{\nearrow} & \end{array} \tag{4.135}$$

where $u_{ij} = U(\{i,j\} \leqslant i)$ (note that $U_{ij} = U_{ji}$). A cocone of vertex $X$ amounts to a family of morphisms $u_i\colon U_i \to X$ such that all the right-hand squares above commute. The universal cocone, or colimit, is called the *generalised pushout* of the objects $U_i$ *over* the objects $U_{ij}$.

As proved in [CarG], the existence of generalised pushouts in a category amounts to the existence of connected colimits (i.e. colimits based on connected categories); finite generalised pushouts amount to pushouts and coequalisers.

## 4.9 CW-complexes and cellular homology

A CW-complex is a space of an elementary kind, which results from pasting together a family of closed discs of arbitrary dimension, according to specified maps. Their theory is developed in most textbooks on Algebraic Topology, yielding easier computations of many items. It is not used here, as we chose to compute homology groups by a direct use of the MV-sequence; another reason is the interest of peculiar spaces which are not even homotopically equivalent to CW-complexes (see 4.9.6 and 7.6.2).

The relationship between CW-complexes and weak homotopy equivalences will be examined in Section 7.6.

*Literature.* This topic was introduced by J.H.C. Whitehead [W1], in 1949. The second part of this section will be kept at an informative level, with references to Hatcher [Ha], Chapter 0 and Appendix, or Gray [Gr], or a few articles.

### 4.9.1 Attaching spaces and cells

(a) Pushouts in $\mathsf{Top}$ are an important way of constructing spaces, pasting more elementary ones. If, in the following pushout, $j$ is a topological embedding (typically, the inclusion of a subspace)

$$\begin{array}{ccc} A & \xrightarrow{\;j\;} & B \\ {\scriptstyle f}\downarrow & & \downarrow{\scriptstyle v} \\ X & \xrightarrow[u]{} & Y \end{array} \tag{4.136}$$

one says that the pushout space $Y$ is obtained *attaching* $B$ *to* $X$ *along* $f$. It is a quotient of the topological sum $X + B$, denoted as $X +_f B$; its general properties are studied in Lemma 4.9.2.

(b) (*Attaching cells*) Here we are interested in attaching an $n$-disc $\mathbb{D}^n$ (also called an $n$-*cell*) to a topological space $X$, along a map $f\colon \mathbb{S}^{n-1} \to X$ defined on the border of the cell. We have thus a space $Y = X +_f \mathbb{D}^n$ defined by the pushout of the left diagram below

$$\begin{array}{ccc} \mathbb{S}^{n-1} & \hookrightarrow & \mathbb{D}^n \\ {\scriptstyle f}\downarrow & & \downarrow{\scriptstyle v} \\ X & \xrightarrow[u]{} & Y \end{array} \qquad\qquad \begin{array}{ccc} \sum_{i\in I}\mathbb{S}^{n-1} & \hookrightarrow & \sum_{i\in I}\mathbb{D}^n \\ {\scriptstyle [f_i]}\downarrow & & \downarrow{\scriptstyle v} \\ X & \xrightarrow[u]{\qquad} & Y \end{array} \tag{4.137}$$

More generally, we can attach a family of $n$-cells $\mathbb{D}^n_i = \mathbb{D}^n$ $(i \in I)$ along a family of maps $f_i\colon \mathbb{S}^{n-1} \to X$, by the right-hand pushout above. By Lemma 4.9.2, the map $u$ is a *closed embedding*, while $v$ sends homeomorphically each open disc $\mathrm{int}(\mathbb{D}^n_i)$ to an open subset $E^n_i$ of $Y$, also called a *cell* of $Y$.

(c) For instance, one can think of the following two ways of constructing the sphere $Y = \mathbb{S}^2$, either collapsing the border $\mathbb{S}^1$ of the disc $\mathbb{D}^2$ to a point, or gluing two discs by identifying their borders

$$\begin{array}{ccc} \mathbb{S}^1 & \hookrightarrow & \mathbb{D}^2 \\ {\scriptstyle p}\downarrow & & \downarrow{\scriptstyle v} \\ \{*\} & \xrightarrow[u]{} & \mathbb{S}^2 \end{array} \qquad\qquad \begin{array}{ccc} \mathbb{S}^1 + \mathbb{S}^1 & \hookrightarrow & \mathbb{D}^2 + \mathbb{D}^2 \\ {\scriptstyle [\mathrm{id}\,_i]}\downarrow & & \downarrow{\scriptstyle v} \\ \mathbb{S}^1 & \xrightarrow[u]{\qquad} & \mathbb{S}^2 \end{array} \tag{4.138}$$

The same constructions give $\mathbb{S}^n$ out of (one or two) $n$-cells $\mathbb{D}^n$ (for $n \geqslant 0$ and $\mathbb{S}^{-1} = \emptyset$).

***4.9.2 Lemma*** (Attaching spaces)

*There is given a pushout in* Top*, where $j$ is a topological embedding*

$$\begin{array}{ccc} A & \xrightarrow{j} & B \\ {\scriptstyle f}\downarrow & & \downarrow{\scriptstyle v} \\ X & \xrightarrow[u]{} & Y \end{array} \qquad\qquad Y = X +_f B. \tag{4.139}$$

*(i) The space $Y = X +_f B$ is the quotient of $X + B$ where a point $x \in X$ is identified to all the points of the subset $jf^{-1}\{x\} \subset B$, if any. The map $u$ is injective and all identifications of $v$ arise in this way; in particular, $v$ is injective out of $j(A)$.*

*(ii) The square is also a pullback.*

*(iii) The map $u\colon X \to Y$ is a topological embedding. If $j$ is an open (resp. closed) embedding, so is $u$.*

*(iv) Every open set $V$ of $B$ which does not meet $j(A)$ is homeomorphically sent by $v$ to an open set $v(V)$ of $Y$.*

*(v) If $f$ is a topological projection, so is $v$.*

*Proof* The proof is rather easy, and can be carried out as an exercise.

(i) The description of $Y$ as a quotient of $X+B$ comes from the construction of pushouts, for sets and spaces.

(ii) The pullback space $A' = \{(x, b) \in X \times B \mid u(x) = v(b)\}$ has a canonical bijection $w\colon A \to A'$, $w(a) = (f(a), j(a))$, which is an embedding (because $j$ is), whence a homeomorphism.

(iii) To prove that $u$ is also an embedding, we take an open subset $U$ of $X$ and construct an open subset $W$ of $Y$ such that $U = u^{-1}(W)$. The subset $f^{-1}(U)$ is open in $A$; since $j$ is an embedding, there is an open subset $V$ of $B$ such that $f^{-1}(U) = j^{-1}(V)$. We let $W = u(U) \cup v(V) \subset Y$; it is open in $Y$, which has the final topology for $u$ and $v$, because:

$$u^{-1}(W) = u^{-1}u(U) \cup u^{-1}v(V) = U \cup fj^{-1}(V) = U \cup ff^{-1}(U) = U,$$
$$v^{-1}(W) = v^{-1}u(U) \cup v^{-1}v(V) = jf^{-1}(U) \cup V = jj^{-1}(V) \cup V = V.$$

We have used the following facts:

- $u^{-1}u(U) = U$, because $u$ is injective;

- $u^{-1}v(V) = fj^{-1}(V)$ and $v^{-1}u(U) = jf^{-1}(U)$, because the square is a pullback;

- $v^{-1}v(V) = V$, essentially because the equivalence relation of $v$ is determined by that of $f$ and $j^{-1}(V) = f^{-1}(U)$ is saturated for $f$. (Take two *distinct* points $b, b' \in B$, identified by $v$; then $b = j(a)$, $b' = j(a')$ and $f(a) = f(a')$, for some $a, a' \in A$; if $b \in V$, then $a$ belongs to $j^{-1}(V) = f^{-1}(U)$, whence $a' \in f^{-1}(U) = j^{-1}(V)$, and $b' \in jj^{-1}(V) \subset V$.)

Since $u^{-1}(W) = U$, our point is done.

We suppose now that $j(A)$ is open (resp. closed) in $B$, and prove that the same is true of $f(X)$ in $Y$. In fact $f^{-1}f(X) = X$ is open and closed in $X$, and $v^{-1}f(X) = ju^{-1}(X) = j(A)$ is open (resp. closed) in $B$.

(iv) If $V \cap j(A) = \emptyset$, the restriction $v' \colon V \to v(V)$ of $v$ is bijective. Its image $v(V)$ is open in $Y$, because $v^{-1}v(V) = V$ and $u^{-1}v(V) = \emptyset$. Finally, the map $v'$ is open, by the same reason, whence a homeomorphism.

(v) If $f$ is a topological projection, take a subset $W$ of $Y$ such that $v^{-1}(W)$ is open in $B$. Then $f^{-1}u^{-1}(W) = j^{-1}v^{-1}(W)$ is open in $A$, and $u^{-1}(W)$ is open in $X$ (by hypothesis). Finally, $W$ is open in the quotient topology of $Y$. □

### 4.9.3 Main definitions

A *CW-complex* is a topological space $X$ equipped with a *CW-structure* $(X^n)_{n\geqslant 0}$, that is, a nested sequence of subspaces of $X$, where:

(i) $X^0$ is a discrete space, whose singletons $E_1^0$ ($i \in I_0$) are called 0-*cells*,

(ii) for $n \geqslant 1$, the space $X^n$ is obtained from $X^{n-1}$ by attaching a family $\mathbb{D}_i^n = \mathbb{D}^n$ of $n$-*cells*, along maps $f_i \colon \mathbb{S}^{n-1} \to X$, for $i \in I_n$.

(iii) $X$ has the *final* (or inductive) topology defined by these subspaces: a subset $A$ is open (resp. closed) if and only if every trace $A \cap X^n$ is open (resp. closed) in $X^n$.

The index sets $I_n$ are assumed to be disjoint. Letting $X^{-1} = \emptyset$, one can omit (i) and state (ii) for $n \geqslant 0$. As already discussed in 3.4.1, for nested inductive limits, we shall not use the usual term 'weak topology', although Whitehead's acronym 'CW' stems from 'closure-finite' and 'weak topology' [W1].

The CW-complex $X$ is also called a *cellular complex*, or a *cell complex*. It is *finite* if it has a finite number of cells. More generally, it has a *finite dimension* $n$ if there is some $n$-cell and no cell of higher dimension. The *empty CW-complex*, with empty $X^0$ (and all $X^n$) can be given dimension $-\infty$.

Equivalently, one can start with a sequence of spaces $(X^n)_{n\geqslant 0}$ satisfying (i) and (ii), and define the space $X$ as their inductive limit. In fact, $X_n$ 'is' automatically a subspace of $X_{n+1}$, by Lemma 4.9.2, and all of them are subspaces of $X$, by 3.4.1(a).

(b) The subspace $X^n$ is called the $n$-*skeleton* of the system; it is closed in $X^{n+1}$ (by 4.9.2(iii)), and in $X$.

Each $n$-cell $\mathbb{D}^n_i$ is mapped into $X$ (not injectively, in general) by its *characteristic map*

$$v^n_i\colon \mathbb{D}^n_i \longrightarrow \Sigma_{i\in I}\,\mathbb{D}^n \longrightarrow X, \tag{4.140}$$

and one can easily verify that $X$ has the final topology for all $v^n_i$ (applying 4.9.2(v), inductively). Thus, $X$ is a quotient of a sum $\Sigma_{n,i}\,\mathbb{D}^n_i$ of discs.

The euclidean open disc $\mathrm{int}_{\mathbb{R}^n}\mathbb{D}^n_i$ is homeomorphically embedded by each $v^n_i$ as an *open subspace* $E^n_i$ of $X^n$ (by 4.9.2(iv)), which is also called a *cell* of $X$. As a set, $X$ is the disjoint union of its cells $E^n_i$; $X^0$ is the discrete space of all 0-cells $\mathbb{D}^0_i = E^0_i$.

*Let us note that the term 'cell' is ambiguous*; here its meaning is distinguished by the symbols $\mathbb{D}^n_i$ and $E^n_i$. The latter is an open euclidean space, and is open in $X^n$ (but, of course, not in $X$, generally).

(c) A *cellular map* $f\colon X \to Y$ between CW-complexes is a (necessarily continuous) mapping that restricts to continuous mappings $f_n\colon X^n \to Y^n$ between the corresponding skeleta.

### 4.9.4 Examples

(a) The compact interval $\mathbb{I} \cong \mathbb{D}^1$ has an obvious CW-structure, with two 0-cells and one 1-cell. For the line $\mathbb{R}$ one can use the 0-skeleton $\mathbb{Z}$ and countably many 1-cells $\mathbb{D}^1_k = [k, k+1]$, glued to the former by their endpoints. For the cartesian space $\mathbb{R}^n$, one can similarly use a CW-structure with 0-skeleton $\mathbb{Z}^n$ and polyhedra $[k_1, k_1+1] \times ... \times [k_n, k_n+1] \cong \mathbb{D}^n$ as $n$-dimensional cells.

(b) The sphere $\mathbb{S}^n$ has two main CW-structures. The first has one 0-cell $\{*\}$ and one $n$-cell $\mathbb{D}^n$, whose boundary $\mathbb{S}^{n-1}$ is collapsed to $\{*\}$ by the constant map $\mathbb{S}^{n-1} \to \{*\}$. The second has skeleta $\mathbb{S}^0 \subset \mathbb{S}^1 \subset ... \subset \mathbb{S}^n$; there are two 0-cells, forming $\mathbb{S}^0$, then two 1-cells glued to the previous ones, forming $\mathbb{S}^1$; and so on, up to gluing two $n$-discs over their boundary $\mathbb{S}^{n-1}$.

(c) Extending the second construction, the infinite dimensional sphere $\mathbb{S}^\infty$ (see 3.4.3) has a CW-structure with a pair of $n$-cells, for every $n \geqslant 0$.

The first procedure would give $\mathbb{S}^\infty$ as a quotient of the infinite dimensional disc $\mathbb{D}^\infty$. This is out of the present topic, which only uses finite dimensional cells – possibly infinitely many of them.

### 4.9.5 *Topological properties

*The rest of this section is at an informative level*, and most proofs are only referred to other textbooks [Ha, Gr], or articles [Mi, BM].

Let $X$ be a CW-complex, with notation as in 4.9.3.

(a) By Lemma 3.4.6, every compact subset $K$ of $X$ is contained in some skeleton $X^n$.

(b) If $X$ is a finite CW-complex, then it is a quotient of a finite sum of discs: therefore it is compact, and all $X^n$ are.

(c) $X$ is Hausdorff, and even normal (see [Ha], Proposition A.3, in the Appendix). It follows that the image $v_i^n(\mathbb{D}_i^n)$ of any cell $\mathbb{D}_i^n$ is closed in $X$.

(d) $X$ is locally contractible: every point has a local base of contractible open nbds ([Ha], Proposition A.4).

(e) Every space homotopy equivalent to a CW-complex has a *good open cover*, that is, an open cover such that the intersection of every finite subfamily is contractible or empty [Mi].

(f) Every simplicial complex (in the general sense of 4.4.6(d)) is a CW-complex, and every CW-complex is homotopically equivalent to (the geometric realisation of) a simplicial complex: see [Gr], Corollaries 16.44 and 21.15. This relationship can be restricted to finite simplicial complexes (i.e. triangulated spaces) and finite CW-complexes.

### *4.9.6 *Cellular homology*

Applying 44.9 and 49.5(f), all homology theories with a specified coefficient group $G$ coincide on all spaces homotopically equivalent to finite CW-complexes. This theory is called *cellular homology* with coefficients in $G$.

Letting $G = \mathbb{Z}$, for simplicity, we only hint at the basic results (proved in [Ha], Lemma 2.34 and Theorem 2.35). $H_k(-)$ denotes the singular homology groups.

(a) Lemma. *If $X$ is a CW-complex, then:*

*(i) $H_k(X^n, X^{n-1})$ is zero for $k \neq n$, and is free abelian for $k = n$, with a basis in one-to-one correspondence with the set $I_n$ of $n$-cells of $X$;*

*(ii) $H_k(X^n) = 0$ for $k > n$;*

*(iii) The inclusion $X^n \subset X$ induces an isomorphism $H_k(X^n) \to H_k(X)$, for $k < n$.*

Therefore, if $X$ is $n$-dimensional, $H_k(X) = 0$, for $k > n$.

(b) A CW-complex $X$ has a *cellular chain complex* $\mathrm{CW}_+(\mathrm{X})$, constructed from the exact homology sequences of the relative pairs $(X^n, X^{n-1})$ of its skeleton. The $n$-component is the free abelian group:

$$\mathrm{CW_n(X) = H_n(X^n, X^{n-1})} \qquad (\mathrm{n} \geqslant 0). \qquad (4.141)$$

The boundary (for $n \geqslant 1$)

$$\partial_n \colon H_n(X^n, X^{n-1}) \to H_{n-1}(X^{n-1}) \to H_{n-1}(X^{n-1}, X^{n-2}), \qquad (4.142)$$

comes from the boundary of the exact homology sequence of the pair $(X^n, X^{n-1})$, together with the canonical map $X^{n-1} \to (X^{n-1}, X^{n-2})$.

But it can also be described intrinsically, from the structure of the CW-complex.

(c) Theorem. *In each degree $n \geqslant 0$, the homology of the chain complex* $\mathrm{CW}_+(\mathrm{X})$ *is isomorphic to the (singular) homology of $X$.*

(d) Using the first CW-structure of $\mathbb{S}^n$ described in 4.9.4(b), for $n > 0$, we get a cellular chain complex with two non-trivial components, in degree 0 and $n$, isomorphic to $\mathbb{Z}$.

If $n > 1$, all the boundary homomorphisms are trivial, and the homology groups coincide with the components of the cellular chain complex. For $n = 1$, one should use the intrinsic description of the boundary $\partial_1$.

### 4.9.7 *Non-cellular spaces

(a) The rational line $\mathbb{Q}$ and the convergent sequence $\mathbb{N}^\bullet$ are not CW-complexes: in fact, a 0-dimensional CW-complex is discrete, and any higher dimensional CW-complex is at least a continuum.

These spaces are not even homotopy equivalent to a CW-complex, by 4.9.5(e).

On the other hand, $\mathbb{Q}$ and $\mathbb{N}^\bullet$ are weakly homotopy equivalent to the discrete $\mathbb{N}$: see Exercise 7.6.2(a).

(b) The *Hawaiian earring* $\mathbb{H}_1$ is not a CW-complex, because it is not locally contractible. This space is a compact subspace of the cartesian plane, the union of countably many circles tangent at a point, realised here as the union of all the circles with centre $(1/n, 0)$ and radius $1/n$, for $n \geqslant 1$

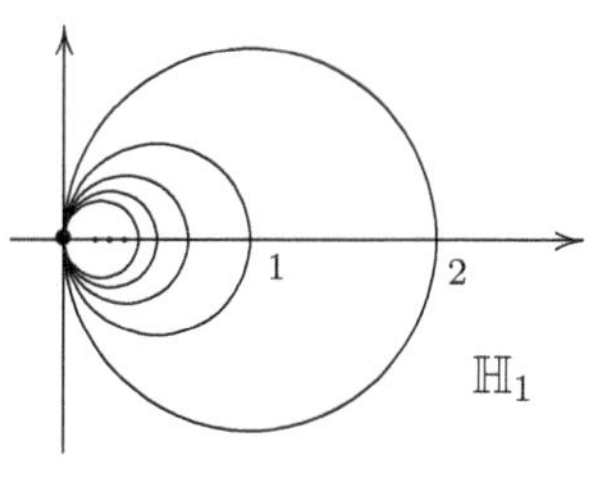


(4.143)

$$\mathbb{H}_1 = \bigcup S_n, \qquad S_n = \{(x, y) \in \mathbb{R}^2 \mid (x - 1/n)^2 + y^2 = 1/n^2\}.$$

Again, this space is not even homotopy equivalent to a CW-complex, by 4.9.5(e): any nbd of the origin contains infinitely many circles.

We have already considered the same *set* with a finer, non-compact topology: the *infinite bouquet of circles* $\mathsf{C}_\infty$, which is the inductive limit of the finite bouquets $C_n$, and a countable wedge of pointed circles (in 3.4.3(c)). Of course, using figure (4.143) for this space, which cannot be embedded in the plane, may be misleading.

(c) One can similarly construct an $n$-dimensional space $\mathbb{H}_n$, the union of countably many $n$-spheres tangent at a point, whose 'anomalous' singular homology is studied in an article of Barrat and Milnor [BM].

In dimension 0, we have already considered $\mathbb{H}_0 = \bigcup_{n\geqslant 1}\{0, 1/n\}$, a subspace of the line homeomorphic to $\mathbb{N}^\bullet$, in (1.20).

# 5
# Singular homology with coefficients

We have already seen cohomology theories that are naturally born with an arbitrary abelian group of coefficients (as Alexander–Spanier and Čech cohomologies), or a precise group of coefficients (as de Rham cohomology, with real coefficients).

In this chapter we introduce singular homology with coefficients in an abelian group $G$, using the tensor product of abelian groups, and the action of the functor $-\otimes G$ on chain complexes.

Similarly, we introduce singular cohomology with coefficients in $G$, using the hom-group $\mathrm{Hom}(A, B)$ and the action of the contravariant functor $\mathrm{Hom}(-, G)$ on chain complexes.

We shall see, in Section 5.4, that an appropriate group of coefficients can detect facts that the integral coefficients do not distinguish.

Moreover, Section 6.6 shows that singular cohomology with coefficients in a ring $R$ forms a graded $R$-algebra, and determines de Rham cohomology when $R$ is the real field – as mentioned in 4.8.8.

## 5.1 Tensor product of modules and Hom

Although we really need the tensor product and Hom functors for abelian groups, we prefer to work in a more natural context: the category $R\,\mathsf{Mod}$ of modules on a commutative ring $R$. The category $\mathsf{Ab}$ is canonically isomorphic to the category $\mathbb{Z}\,\mathsf{Mod}$ of modules on the ring of integers (as recalled in 1.3.6), and we make no distinction between them.

We review the definition and main properties of the tensor product and Hom-functor of $R$-modules, a classical topic dealt with in many books on Homological Algebra, like [M1, CE], or Linear Algebra, as [Bou1]. The exactness properties of these functors are deferred to Section 5.2.

$R$ is a fixed commutative (unital) ring; commutativity is essential, here.

One may prefer to exclude the trivial ring $\{0\}$, on which all modules are trivial. Otherwise, some caution is required in a few points: cf. Remark 5.1.1(a).

### 5.1.1 Linear and multilinear mappings

The hom-set $R\,\mathsf{Mod}(A, B)$ of $R$-homomorphisms $A \to B$, or $R$-linear mappings, has a canonical structure of $R$-module, written as $\mathrm{Hom}_R(A, B)$, or $\mathrm{Hom}(A, B)$ when no confusion may arise.

Its operations are the pointwise addition and pointwise multiplication by scalars $\lambda \in R$

$$(f+g)(a) = f(a) + g(a), \qquad (\lambda f)(a) = \lambda f(a). \tag{5.1}$$

These operations are preserved by compositions: for $h\colon A' \to A$ and $k\colon B \to B'$ we have

$$k(f+g)h = kfh + kgh, \qquad k(\lambda f)h = \lambda(kfh). \tag{5.2}$$

Now, for a product module $A = \prod_i A_i$ $(i = 1, ..., n)$, a mapping $\varphi\colon A \dot{\to} B$ with values in a module $B$ is said to be *multilinear*, or *$n$-linear*, on $R$ if it is $R$-linear in each variable

$$\begin{aligned} &\varphi(a_1, ..., \lambda a_i' + \mu a_i'', ..., a_n) \\ &\quad = \lambda\varphi(a_1, ..., a_i', ..., a_n) + \mu\varphi(a_1, ..., a_i'', ..., a_n). \end{aligned} \tag{5.3}$$

Multilinear mappings $\varphi\colon A \dot{\to} B$ will be denoted by dot-marked arrows. They have a pointwise addition and pointwise scalar multiplication, that extend those of the linear ones (for $n = 1$).

Composing these mappings is less obvious: given a *family* of homomorphisms $h_i\colon A_i' \to A_i$ $(i = 1, ..., n)$ and *one* homomorphism $k\colon B \to B'$, the obvious composite is multilinear

$$k\,\varphi\,(\Pi_i h_i)\colon \Pi_i A_i' \to B', \quad (x_1, ..., x_n) \mapsto k(\varphi(h_1(x_1), ..., h_n(x_n))). \tag{5.4}$$

For abelian groups, a mapping $\varphi\colon \Pi_i A_i \dot{\to} B$ is multilinear on $\mathbb{Z}$ if (and only if) it is *multiadditive*, i.e. additive in each variable.

We have reviewed in 1.2.5 the free abelian group $F(I) = \mathbb{Z}I$ on a set $I$. Here we shall use the *free $R$-module* on a set $I$

$$F(I) = RI = \textstyle\bigoplus_{i \in I} R, \qquad \eta\colon I \to |RI|, \tag{5.5}$$

with the usual mapping $\eta(i) = e_i = (\delta_{ij})_{j \in I}$, the canonical basis. It is characterised by the *universal property*:

(i) for every mapping $f\colon I \to |A|$ with values in (the underlying set of) an $R$-module $A$, there is a unique homomorphism $g\colon RI \to A$ such that $g(e_i) = f(i)$, for all $i \in I$.

By abuse of notation, the index $i$ will also stay for $e_i \in RI$, and we simply say that every mapping $f\colon I \to |A|$ has a unique extension $g\colon RI \to A$ in $R\,\mathsf{Mod}$, with $g(i) = f(i)$, for all $i \in I$. (This is commented in Remark (a), below.)

We have thus the *free $R$-module functor*

$$\begin{aligned} &F\colon \mathsf{Set} \to R\,\mathsf{Mod}, \qquad F(I) = RI = \textstyle\bigoplus_{i\in I} R, \\ &F(f)\colon RI \to RJ, \qquad F(f)(i) = f(i) \qquad (\text{for } f\colon I \to J), \end{aligned} \tag{5.6}$$

which can be written as $F_R$ when useful.

*Remarks.* (a) For the trivial ring $R = \{0\}$, all $R$-modules are trivial and free on any set $I$: the unique map $\eta\colon I \to \{0\}$ trivially satisfies the universal property.

Outside of this case, $0 \neq 1$ in $R$ and the mapping $\eta$ in (5.5) is injective: we can identify $i$ with $e_i$ and $I$ with its image in $RI$. Our abuse of notation is then fully justified.

(b) Being a free object is not a property of the mere category $R\,\mathsf{Mod}$, but of the forgetful functor $U\colon R\,\mathsf{Mod} \to \mathsf{Set}$, as made clear by the universal property (i).

*In fact, the free $R$-module functor is left adjoint to $U$: see Section 8.1.*

### 5.1.2 Tensor product of modules

The *tensor product* of two $R$-modules $A, B$ is defined by a universal property whose goal is to transform bilinear mappings into homomorphism.

Precisely, we look for a module $A \otimes_R B$ *equipped with a bilinear mapping* $\varphi_0\colon A \times B \dot{\to} A \otimes_R B$ such that, for every bilinear mapping $\varphi\colon A \times B \dot{\to} C$

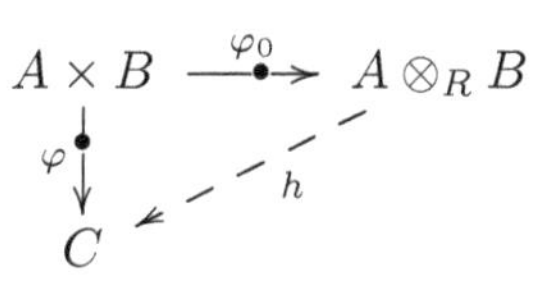


(5.7)

there is a unique $R$-homomorphism $h\colon A \otimes_R B \to C$ such that $\varphi = h\varphi_0$.

The universal property characterises the solution up to a unique isomorphism coherent with the structural bilinear mappings.

In fact, if $\varphi_i\colon A \times B \dot{\to} C_i$ $(i = 1, 2)$ are two solutions of the given problem, there are two (determined) homomorphisms $h, k$ such that

$$h\colon C_1 \to C_2, \qquad k\colon C_2 \to C_1, \qquad \varphi_2 = h\varphi_1, \qquad \varphi_1 = k\varphi_2.$$

Then the homomorphism $kh\colon C_1 \to C_1$ satisfies $(kh)\varphi_1 = \varphi_1 = (\mathrm{id}\,\mathsf{C}_1)\varphi_1$ and is the identity; similarly $hk = \mathrm{id}\,\mathsf{C}_2$, which means that $h$ and $k$ are isomorphisms, inverse to each other.

To prove that the solution exists, we 'construct' one (by a complicated procedure, rarely used in computations)

$$A \otimes_R B = R(A \times B)/H(A, B), \qquad \varphi_0(a, b) = [(a, b)]. \qquad (5.8)$$

Here $R(A \times B)$ is the free $R$-module generated by the *set* $A \times B$, and its elements will be written as essentially finite $R$-linear combinations of pairs $(a, b) \in A \times B$. Secondly, $H(A, B)$ is the submodule of $R(A \times B)$ generated by all the elements of the following types (where $a, a'$ belong to $A$, $b, b'$ to $B$ and $\lambda \in R$)

(i) $(a + a', b) - (a, b) - (a', b)$,
(ii) $(\lambda a, b) - \lambda(a, b)$,
(iii) $(a, b + b') - (a, b) - (a, b')$,
(iv) $(a, \lambda b) - \lambda(a, b)$.

(Taking in (5.8) the quotient modulo $H(A, B)$ is precisely what is required to make the mapping $\varphi_0$ bilinear on $R$.) We write

$$a \otimes b = \varphi_0(a, b) = [(a, b)] \in A \otimes_R B \qquad (\text{for } a \in A,\ b \in B). \qquad (5.9)$$

The generators (i)–(iv) of $H(A, B)$ give the bilinearity of the tensor product of elements:

$$(a + a') \otimes b = a \otimes b + a' \otimes b, \qquad (\lambda a) \otimes b = \lambda(a \otimes b),$$
$$a \otimes (b + b') = a \otimes b + a \otimes b', \qquad a \otimes (\lambda b) = \lambda(a \otimes b).$$

We can write $A \otimes B$ for $A \otimes_R B$, when no ambiguity can arise.

More generally, we define the tensor product $\bigotimes A_i$ of a finite family of $R$-modules, as an $R$-module equipped with a multilinear mapping

$$\varphi_0 \colon \Pi A_i \nrightarrow \bigotimes A_i, \qquad (5.10)$$

universal 'as above', for all $n$-linear mappings $\Pi A_i \nrightarrow C$.

### 5.1.3 Remarks and complements

(a) Every element of $A \otimes B$ can be written as a (finite) sum $\Sigma_i\, a_i \otimes b_i$; obviously, this expression is not unique (generally). The elements of the form $a \otimes b$ form a *canonical system of generators* of $A \otimes_R B$. More generally, if $a$ (resp. $b$) varies in a system of generators of $A$ (resp. $B$), the elements $a \otimes b$ still form a system of generators of $A \otimes B$.

There are cases where all these generators vanish, and $A \otimes_R B = 0$ (the trivial $R$-module).

(b) First, let us note that $a \otimes 0 = 0 \otimes b = 0$ (for $a \in A$ and $b \in B$). Therefore $A \otimes 0 = 0 \otimes B = 0$.

(c) Suppose now that the (commutative) ring $R$ is an integral domain (i.e. has no proper zero-divisors). If $A$ is divisible by a non-zero scalar $\lambda$ (i.e. the equation $\lambda x = a$ has solutions, for every $a \in A$) and $\lambda B = 0$, then $A \otimes B = 0$. In fact, for $a = \lambda a' \in A$ and $b \in B$: $a \otimes b = (\lambda a') \otimes b = a' \otimes (\lambda b) = 0$.

(d) In the same hypothesis on $R$, if $A$ is a *divisible* $R$-module (i.e. the equation $\lambda x = a$ has solutions, for every $a \in A$ and every scalar $\lambda \neq 0$) while $B$ is a *torsion* $R$-module (every $b \in B$ is annihilated by some non-zero scalar), then $A \otimes B = 0$.

(e) Taking $R = \mathbb{Z}$, for every torsion abelian group $B$ we have $\mathbb{Q} \otimes B = 0 = \mathbb{R} \otimes B$. (Any field of characteristic zero is divisible, as an abelian group.)

### *5.1.4 Tensor product of homomorphisms*

Given two homomorphisms $f\colon A \to A'$, $g\colon B \to B'$ there is a homomorphism $f \otimes g$ determined as below on the canonical generators

$$f \otimes g\colon A \otimes_R B \to A' \otimes_R B', \quad (f \otimes g)(a \otimes b) = f(a) \otimes g(b). \tag{5.11}$$

In fact, the bilinear mapping

$$A \times B \to A' \times B' \twoheadrightarrow A' \otimes_R B', \quad (a, b) \mapsto f(a) \otimes g(b),$$

determines such a homomorphism on $A \otimes_R B$.

Plainly, this construction preserves identities and composition:

$$\mathrm{id}\, A \otimes \mathrm{id}\, B = \mathrm{id}\,(A \otimes_R B), \quad (f'f) \otimes (g'g) = (f' \otimes g')(f \otimes g).$$

The tensor product is thus a (covariant) functor in two variables

$$\otimes\colon R\,\mathsf{Mod} \times R\,\mathsf{Mod} \to R\,\mathsf{Mod}. \tag{5.12}$$

This functor is *bilinear*, i.e. additive and homogeneous in each variable, with respect to the $R$-linear structure of morphisms defined above (in 5.1.1):

$$\begin{aligned} (f + f') \otimes g &= f \otimes g + f' \otimes g, \quad & (\lambda f) \otimes g &= \lambda(f \otimes g), \\ f \otimes (g + g') &= f \otimes g + f \otimes g', \quad & f \otimes (\lambda g) &= \lambda(f \otimes g). \end{aligned} \tag{5.13}$$

### *5.1.5 Basic properties of the tensor product*

The ring $R$ is fixed, and we write $A \otimes B$ for $A \otimes_R B$.

(a) The tensor product is *commutative*. More precisely, there is a canonical isomorphism determined as follows on the canonical generators:

$$A \otimes B \to B \otimes A, \quad a \otimes b \mapsto b \otimes a. \tag{5.14}$$

In fact, the obvious bilinear map $A \times B \to B \times A \twoheadrightarrow B \otimes A$ determines a homomorphism $A \otimes B \to B \otimes A$ that acts as above on the canonical generators; symmetrically, we have a homomorphism $B \otimes A \to A \otimes B$ that sends $b \otimes a$ to $a \otimes b$. These homomorphisms are inverse to each other (on generators, hence everywhere).

(b) The tensor product *has a unit*, the $R$-module $R$. More precisely, there is a canonical isomorphism:

$$A \otimes R \to A, \qquad a \otimes \lambda \mapsto \lambda a \quad (a \mapsto a \otimes 1_R). \tag{5.15}$$

In fact the mapping $\varphi_0 \colon A \times R \twoheadrightarrow A$, $(a, \lambda) \mapsto \lambda a$ is bilinear and satisfies the universal property that defines $A \otimes R$.

(c) The tensor product is *associative*, up to a canonical isomorphism:

$$(A \otimes B) \otimes C \to A \otimes (B \otimes C), \quad (a \otimes b) \otimes c \mapsto a \otimes (b \otimes c). \tag{5.16}$$

To give a simple proof, we verify that the trilinear mapping

$$\psi_0 \colon A \times B \times C \twoheadrightarrow (A \otimes B) \otimes C, \qquad (a, b, c) \mapsto (a \otimes b) \otimes c,$$

satisfies the universal property of a ternary tensor product (see (5.10)). Therefore $(A \otimes B) \otimes C$ is a realisation of $A \otimes B \otimes C$. Symmetrically, this is also true of $A \otimes (B \otimes C)$, and the two realisations are isomorphic, as above.

In fact, $\psi_0$ is the composite of the mappings $\varphi_1 \times C$ and $\varphi_2$ defined below (where $\varphi_1$ and $\varphi_2$ are bilinear)

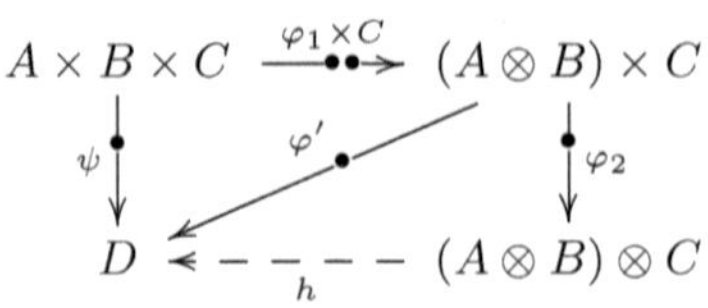


$$\varphi_1(a, b) = a \otimes b, \qquad \varphi_2(z, c) = z \otimes c. \tag{5.17}$$

Now, any trilinear mapping $\psi \colon A \times B \times C \twoheadrightarrow D$ factorises as $\varphi'(\varphi_1 \times C)$ where $\varphi' \colon (A \otimes B) \times C \twoheadrightarrow D$ is bilinear; then we factorise the latter as $\varphi' = h\varphi_2$, by a homomorphism $h$. The uniqueness of $h$ in the factorisation $\psi = h\psi_0$ is obvious, because the elements $(a \otimes b) \otimes c$ generate $(A \otimes B) \otimes C$ (by 5.1.3(a)).

*The properties (a)–(c) say that the tensor product makes $R\,\mathsf{Mod}$ into a symmetric monoidal category, see 8.1.9.*

(d) The tensor product is *distributive* with respect to direct sums, by a canonical isomorphism (an easy consequence of the universal property)

$$(\textstyle\bigoplus_{i \in I} A_i) \otimes B \to \textstyle\bigoplus_{i \in I}(A_i \otimes B), \quad (a_i)_{i \in I} \otimes b \mapsto (a_i \otimes b)_{i \in I}. \tag{5.18}$$

(e) This gives canonical isomorphisms:

$$\begin{aligned} RI \otimes B &\cong \textstyle\bigoplus_{i \in I} B, & R^m \otimes B &\cong B^m, \\ RI \otimes RJ &\cong R(I \times J), & R^m \otimes R^n &\cong R^{mn}. \end{aligned} \tag{5.19}$$

(f) *Tensor products of free modules (and all vector spaces) are thus determined*: if $A, B$ are free on $R$ with bases $(a_i)_{i\in I}$ and $(b_j)_{j\in J}$, then $A \otimes B$ is free with basis $(a_i \otimes b_j)_{(i,j)\in I\times J}$.

(g) For an abelian group $A$, the tensor product $A\otimes_{\mathbb{Z}}\mathbb{Q}$ is made into a vector space on $\mathbb{Q}$, letting $\lambda(a \otimes \mu) = a \otimes \lambda\mu$. The *rank* of the abelian group $A$ is defined as the cardinal

$$\mathrm{rk}(A) = \dim_{\mathbb{Q}}(A \otimes_{\mathbb{Z}} \mathbb{Q}). \tag{5.20}$$

If $A$ is finitely generated, this agrees with the previous definition in 1.2.6(c): the Structure Theorem says that $A$ is isomorphic to a direct sum $\mathbb{Z}^n \oplus \mathrm{t}A$, where $\mathrm{t}A$ is the torsion subgroup of $A$; therefore $A \otimes_{\mathbb{Z}} \mathbb{Q} \cong \mathbb{Q}^n$ (a cartesian power of abelian groups, and also of vector spaces).

(h) More generally, the tensor product $A \otimes_{\mathbb{Z}} R$ of an abelian group with a commutative ring $R$ has a similar structure of $R$-module.

*(i) One could define a tensor product $G\otimes H$ of *groups*, by the universal property of 'bihomomorphisms'. But it is not difficult to show that the solution is just the tensor product $G^{\mathrm{ab}} \otimes_{\mathbb{Z}} H^{\mathrm{ab}}$ of the corresponding abelianised groups (see 7.3.3(e)).

### *5.1.6 The functor Hom*

Another classical issue of linear and homological algebra is the functor Hom (also written as $\mathrm{Hom}_R$).

The set-valued hom-functor $\mathrm{Mor}\colon R\,\mathsf{Mod}^{\mathrm{op}} \times R\,\mathsf{Mod} \to \mathsf{Set}$ (cf. 1.3.8) has an enriched version, recalled in 5.1.1

$$\begin{gathered}\mathrm{Hom}\colon R\,\mathsf{Mod}^{\mathrm{op}} \times R\,\mathsf{Mod} \longrightarrow R\,\mathsf{Mod},\\ (A, B) \mapsto \mathrm{Hom}(A, B), \qquad (h, k) \mapsto k(-)h.\end{gathered} \tag{5.21}$$

$\mathrm{Hom}(A, B)$ is the module of $R$-homomorphisms $A \to B$, with pointwise operations. For two $R$-linear mappings $h\colon A' \to A$ and $k\colon B \to B'$, the mapping $k(-)h\colon \mathrm{Hom}(A, B) \to \mathrm{Hom}(A', B')$ is $R$-linear, as shown in (5.2).

Here we are mostly interested in the covariant functor obtained by fixing the first variable

$$\begin{gathered}\mathrm{Hom}(A, -)\colon R\,\mathsf{Mod} \longrightarrow R\,\mathsf{Mod},\\ (k\colon B \to B') \;\mapsto\; k_\sharp = k(-)\colon \mathrm{Hom}(A, B) \to \mathrm{Hom}(A, B'),\end{gathered} \tag{5.22}$$

but in Section 5.5 we shall use the contravariant functor obtained by fixing the second variable

$$\begin{gathered}\mathrm{Hom}(-, B)\colon R\,\mathsf{Mod} \dashrightarrow R\,\mathsf{Mod},\\ (h\colon A' \to A) \;\mapsto\; h^\sharp = (-)h\colon \mathrm{Hom}(A, B) \to \mathrm{Hom}(A', B).\end{gathered} \tag{5.23}$$

*Exercises and complements.* (a) The functor $\mathrm{Hom}(A,-)$ preserves cartesian products, up to a canonical isomorphism

$$\mathrm{Hom}(A, \Pi_{i\in I} B_i) \to \Pi_{i\in I}\,\mathrm{Hom}(A, B_i), \quad f \mapsto (p_i f)_{i\in I}, \tag{5.24}$$

which takes a homomorphism $f\colon A \to \Pi_{i\in I} B_i$ to the family of its components $p_i f\colon A \to B_i$.

(b) The contravariant functor $\mathrm{Hom}(-,B)$ takes direct sums to cartesian products, up to a canonical isomorphism

$$\mathrm{Hom}(\bigoplus_{i\in I} A_i, B) \to \Pi_{i\in I}\,\mathrm{Hom}(A_i, B), \quad f \mapsto (f u_i)_{i\in I}, \tag{5.25}$$

which sends a homomorphism $f\colon \bigoplus_{i\in I} A_i \to B$ to the family of its co-components $fu_i\colon A_i \to \bigoplus_{i\in I} A_i \to B$.

(c) The $R$-module $R$ gives a functorial isomorphism

$$\mathrm{Hom}(R,-) \cong \mathrm{id}\colon R\,\mathsf{Mod} \to R\,\mathsf{Mod}, \quad (f\colon R \to B) \mapsto f(1_R), \tag{5.26}$$

which is an 'enriched version' of the representability of the forgetful functor $R\,\mathsf{Mod} \to \mathsf{Set}$ by the R-module $R$.

### *5.1.7 The exponential law for sets and modules*

In order to describe the relationship between tensor product and Hom, we begin with a similar, more elementary occurrence in $\mathsf{Set}$.

(a) (*Exponential law for sets*) The hom-functor of $\mathsf{Set}$

$$\mathrm{Mor}\colon \mathsf{Set}^{\mathrm{op}} \times \mathsf{Set} \to \mathsf{Set}, \quad (A,Y) \mapsto \mathsf{Set}(A,Y), \quad (h,k) \mapsto k(-)h,$$

is related to the binary cartesian product by a natural isomorphism in the variables $X, Y$

$$\begin{gathered} \varphi^A_{XY}\colon \mathsf{Set}(X \times A, Y) \longrightarrow \mathsf{Set}(X, \mathsf{Set}(A,Y)), \\ (f\colon X \times A \to Y) \mapsto (g\colon X \to \mathsf{Set}(A,Y)), \\ g(x) = f(x,-)\colon A \to Y. \end{gathered} \tag{5.27}$$

*This is a family of adjunctions, indexed by the 'parameter' $A$: the functor $- \times A\colon \mathsf{Set} \to \mathsf{Set}$ is left adjoint to the functor $\mathsf{Set}(A,-)\colon \mathsf{Set} \to \mathsf{Set}$ (see 8.1.6(a)).*

These hom-sets are often written as 'exponentials' $Y^A = \mathsf{Set}(A,Y)$, because an indexed family $(y_a)_{a\in A}$ of $Y^A$ is the same as a mapping $y\colon A \to Y$. With this notation, the previous family of bijections is rewritten as

$$Y^{X\times A} \cong (Y^A)^X. \tag{5.28}$$

This explains the name of *exponential law*.

(b) (*Exponential law for modules*) Similarly we have a natural family of bijections (for a commutative ring $R$):

$$\begin{aligned} \varphi^A_{XY}\colon R\,\mathsf{Mod}(X\otimes A, Y) &\longrightarrow R\,\mathsf{Mod}(X, \mathrm{Hom}(A,Y)),\\ (f\colon X\otimes A\to Y) &\mapsto (g\colon X\to \mathrm{Hom}(A,Y)),\\ g(x) = f(x\otimes -)&\colon A\to Y. \end{aligned} \tag{5.29}$$

Concretely, a homomorphism $g\colon X\to \mathrm{Hom}(A,Y)$ amounts to a mapping $X\times A \to Y$ which is linear in each variable, and therefore to a homomorphism $f\colon X\otimes A\to Y$.

*For every $R$-module $A$, the endofunctor $-\otimes A$ is left adjoint to the endofunctor $\mathrm{Hom}(A,-)$ (of $R\,\mathsf{Mod}$, see 8.1.9). Moreover (5.29) can be viewed as a natural family of isomorphisms of $R$-modules

$$\varphi^A_{XY}\colon \mathrm{Hom}(X\otimes A, Y) \longrightarrow \mathrm{Hom}(X, \mathrm{Hom}(A,Y)),$$

giving an adjunction enriched over $R\,\mathsf{Mod}$.*

**5.1.8** ***Exercises and complements*** (Tensor product of vector spaces)
The ring of scalars is now a (commutative) field $K$.

Modules on $K$ are called vector spaces and have specific properties, essentially derived from the well-known theory of linear dependence in vector spaces. The category $K\,\mathsf{Mod}$ is usually written as $K\,\mathsf{Vct}$.

(a) In $K\,\mathsf{Vct}$ every monomorphism splits, i.e. has a retraction. In other words, every subspace of a vector space is a retract.

(b) In $K\,\mathsf{Vct}$ every epimorphism splits, i.e. has a section. Every quotient of a vector space is a retract.

(c) All short exact sequences in $K\,\mathsf{Vct}$ split.

(d) For vector spaces $A, B$, there is a canonical homomorphism

$$u_{AB}\colon A\otimes_K B\to \mathrm{Hom}_K(A^*, B),\quad u_{AB}(a\otimes b)(\alpha) = \alpha(a)b, \tag{5.30}$$

where $A^* = \mathrm{Hom}_K(A,K)$ is the dual vector space of $A$ and $\alpha\colon A\to K$.

This family defines a natural transformation

$$u\colon (-\otimes_K -)\to \mathrm{Hom}_K((-)^*, -)\colon K\,\mathsf{Vct}\times K\,\mathsf{Vct}\to K\,\mathsf{Vct}. \tag{5.31}$$

Prove that each component $u_{AB}$ is injective. Moreover, if $A$ and $B$ are *finitely generated*, then $u_{AB}$ is an isomorphism. *Hint:* use a basis of $A$.

(e) It follows that the tensor product of *finitely generated* vector spaces *can be* (and often *is*) defined as: $A\otimes_K B = \mathrm{Hom}_K(A^*, B)$. This shortcut is also used for vector bundles. (Note that it masks commutativity.)

## 5.2 Additive and exact functors

We deal now with additivity and exactness properties of functors between categories of modules, in particular tensor products and Hom-functors.

$R$ and $S$ are always commutative rings.

Kernels, cokernels, images, coimages, exact sequences and split exact sequences are dealt with in any category of modules as we have seen in Ab, in Section 2.1 and in 3.5.4.

### *5.2.1 Additive functors*

A functor $F\colon R\,\mathsf{Mod} \to S\,\mathsf{Mod}$ between categories of modules is *additive* if it preserves the addition of parallel morphisms, which means that, for every pair $A, B$ of $R$-modules, the mapping

$$\mathrm{Hom}_R(A, B) \to \mathrm{Hom}_S(FA, FB), \qquad f \mapsto F(f), \tag{5.32}$$

is a homomorphism of abelian groups.

*The natural framework of this issue is additive categories, see 1.6.5.*

*Exercises and complements.* Let $F\colon R\,\mathsf{Mod} \to S\,\mathsf{Mod}$ be an additive functor.

(a) $F$ preserves the zero object and the zero morphisms (see 1.6.5(a)).

(b) $F$ preserves finite direct sums of modules.

(c) $F$ preserves split exact sequences (see 3.5.4(iii)). In particular, a short exact sequence $A \rightarrowtail B \twoheadrightarrow C$ where $C$ is a free $R$-module is taken to a split exact sequence.

### *5.2.2 Exact functors*

For a functor $F\colon R\,\mathsf{Mod} \to S\,\mathsf{Mod}$ there are diverse forms of preserving exactness, from the strongest (an exact functor) to the weakest (a half exact functor). Many properties will be left as easy exercises; the solution is written below.

(a) The functor $F$ is said to be *exact* if it satisfies the following equivalent conditions:

(i) $F$ preserves exact sequences,

(ii) $F$ preserves short exact sequences,

(iii) $F$ preserves kernels and cokernels.

The equivalence is an exercise, proved in Solution (a). An exact functor also preserves images and coimages, canonical factorisations of homomorphisms, monos and epis (see (2.3)).

*Note.* Condition (iii) means that $F$ takes a (co)kernel of a morphism $f$ to *a* (co)kernel of $F(f)$ (see 2.1.2), without having to preserve distinguished choices: we do not expect $F$ to preserve submodules or quotient modules, 'on the nose': this would rarely happen.

(b) The functor $F$ is *left exact* if it satisfies the following equivalent conditions:

(iv) $F$ preserves kernels,

(v) $F$ takes any short exact sequence $A' \rightarrowtail A \twoheadrightarrow A''$ to an exact sequence $0 \to FA' \to FA \to FA''$.

(b*) The functor $F$ is *right exact* if it satisfies the following equivalent conditions:

(iv*) $F$ preserves cokernels,

(v*) $F$ takes any short exact sequence $A' \rightarrowtail A \twoheadrightarrow A''$ to an exact sequence $FA' \to FA \to FA'' \to 0$.

(c) The functor $F$ is *half exact* if any short exact sequence $A' \rightarrowtail A \twoheadrightarrow A''$ is taken to an exact sequence $FA' \to FA \to FA''$.

Every half exact functor preserves split exact sequences and finite direct sums; moreover, *it is an additive functor.* (Therefore this also holds for left exact, right exact and exact functors.)

(d) Exact functors are plainly closed under composition. The same holds for left exact and right exact functors, but generally fails for the half exact ones.

(e) A contravariant functor $F\colon R\,\mathsf{Mod} \dashrightarrow S\,\mathsf{Mod}$ is the same as a covariant functor $R\,\mathsf{Mod}^{\mathrm{op}} \to S\,\mathsf{Mod}$. It is said to be *left exact* if it satisfies the equivalent properties:

(vi) $F$ takes cokernels to kernels,

(vii) $F$ takes any short exact sequence $A' \rightarrowtail A \twoheadrightarrow A''$ to an exact sequence $0 \to FA'' \to FA \to FA'$.

(f) All these definitions can be extended to categories of chain complexes, which are additive categories where kernels, cokernels and exact sequences have been defined. Thus, the chain homology functor $H_n\colon \mathrm{Ch}_+\mathsf{Ab} \to \mathsf{Ab}$ is half exact, by Theorem 2.3.3.

*The natural framework of these definitions is abelian categories.*

*Solutions.* (a) We have seen in 2.1.3 that exact sequences detect zero objects, kernels and cokernels, and (obviously) short exact sequences. We have also seen that kernels and cokernels detect short exact sequences, which detect the general ones.

(b) The equivalence is readily proved starting from an exact sequence $A' \rightarrowtail A \to A''$ and factorising epi-mono the last morphism.

(c) Let $F$ be half exact. If $(m, p) \colon A' \rightarrowtail A \twoheadrightarrow A''$ is a split exact sequence, $m$ is a section and $p$ is a retraction, which are preserved by any functor. This shows that $(Fm, Fp)$ is still a split exact sequence.

The other preservation properties are a consequence:

- the zero object $Z$ is characterised by $(\mathrm{id}\, Z, \mathrm{id}\, Z)$ being a split exact sequence,
- a biproduct $A \oplus B$ is characterised, in 3.5.5, by a commutative butterfly diagram with split exact sequences,
- an addition $f + g$ of parallel homomorphism is characterised by binary biproducts, in (3.70).

**5.2.3 Lemma** (Extending additive functors)

*(a) An additive functor* $F \colon R\,\mathsf{Mod} \to S\,\mathsf{Mod}$ *has an obvious extension to an additive functor* $F_+ \colon \mathrm{Ch}_+ R\,\mathsf{Mod} \to \mathrm{Ch}_+ S\,\mathsf{Mod}$ *that preserves homotopies. (One can say that the extension is* homotopy equivariant.*) The extension will often be written as* $F$*.*

*(b) For an exact functor* $F \colon R\,\mathsf{Mod} \to S\,\mathsf{Mod}$*, the extension* $F_+$ *is exact and commutes with homology. More precisely, on every chain complex of* $R$*-modules* $A$ *we have a natural isomorphism (for every* $n \geqslant 0$*)*

$$\varphi_n \colon H_n(F_+A) \to F(H_n(A)), \tag{5.33}$$

*induced by the natural isomorphism* $Z_n(FA) \to F(Z_n(A))$*.*

*Proof* Point (a) is obvious. As to (b), the homology of $A$ is characterised by the short exact sequence

$$B_n(A) \rightarrowtail Z_n(A) \xrightarrow{p} H_n(A) \tag{5.34}$$

preserved by the exact functor $F$.

Similarly, the homology of the complex $FA$ is characterised by the short exact sequence in the upper row of the following diagram

$$\begin{array}{ccccc} B_n(FA) & \rightarrowtail & Z_n(FA) & \xrightarrow{p} & H_n(FA) \\ \downarrow{\scriptstyle \chi_n} & & \downarrow{\scriptstyle \psi_n} & & \downarrow{\scriptstyle \varphi_n} \\ F(B_nA) & \rightarrowtail & F(Z_nA) & \xrightarrow[Fp]{} & F(H_n(A)) \end{array} \tag{5.35}$$

Since $F$ preserves images and kernels, we have natural isomorphisms $\psi_n$ and $\chi_n$, which induce a natural isomorphism $\varphi_n\colon H_n(FA) \to F(H_nA)$, by the Induction Lemma 2.1.4. □

***5.2.4 Theorem*** (Exactness properties of the tensor product)
*For every $R$-module $A$, the tensor functor $F = - \otimes_R A\colon R\,\mathsf{Mod} \to R\,\mathsf{Mod}$ is right exact.*

*Proof* *Category Theory will give this for free, as we shall see in 8.1.4(d): every left adjoint functor preserves categorical colimits, in particular cokernels (and sums, as we have already verified for $F$, in 5.1.5(d)).*

We verify directly our claim, following nevertheless the outline suggested by the adjunction: the exponential law in 5.1.7(b). We already know that $F$ preserves zero objects and zero morphisms.

We have an $R$-homomorphism $f\colon B \to C$, with cokernel $p\colon C \to C'$ (in the left diagram below), and we want to prove that $F(p) = p \otimes A\colon C \otimes A \to C' \otimes A$ is a cokernel of $F(f) = f \otimes A\colon B \otimes A \to C \otimes A$ (in the right diagram below), verifying the universal property of cokernels (in 2.1.2(i*))

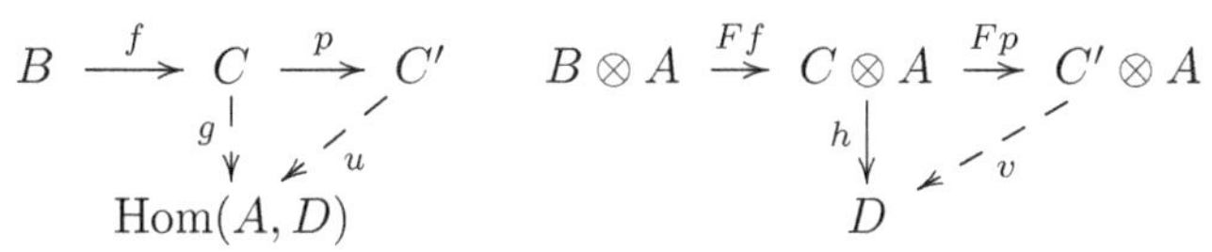

(5.36)

First, $F(p)F(f) = F(pf) = 0$. Secondly, we suppose that $h\colon C \otimes A \to D$ also gives $h.F(f) = 0$.

By the exponential law, $h\colon C \otimes A \to D$ corresponds to a homomorphism in the left diagram

$$g\colon C \to \mathrm{Hom}(A, D), \qquad h(c \otimes a) = g(c)(a) \qquad (\text{for } c \in C,\ a \in A),$$

and $gf = 0$, because $gf(b)(a) = h(f(b) \otimes a) = h.Ff(b \otimes a) = 0$ on every $b \in B$ and $a \in A$.

Similarly, a homomorphism $u\colon C' \to \mathrm{Hom}(A, D)$ in the left diagram corresponds to a homomorphism $v$ in the right one

$$v\colon C' \otimes A \to D, \qquad v(c' \otimes a) = u(c')(a) \qquad (\text{for } c' \in C',\ a \in A).$$

Finally, the condition $g = up$ is equivalent to $h = v.Fp$. By hypothesis, there is a unique $u\colon C' \to \mathrm{Hom}(A, D)$ such that $g = up$; therefore there is also a unique $v\colon C' \otimes A \to D$ such that $h = v.Fp$. □

**5.2.5 Theorem** (Exactness properties of Hom)

*For every R-module X*

*(i) the functor* $G = \mathrm{Hom}_R(X, -)\colon R\,\mathsf{Mod} \to R\,\mathsf{Mod}$ *is left exact,*

*(ii) the functor* $H = \mathrm{Hom}_R(-, X)\colon R\,\mathsf{Mod}^{\mathrm{op}} \to R\,\mathsf{Mod}$ *is left exact.*

*Proof* Both points can easily be verified directly. *In Category Theory one proves that a representable functor preserves categorical limits, in particular kernels (and products). Point (i) also follows from the fact that $G$ is a right adjoint.*

(i) If $k\colon K \to A$ is a kernel of $f\colon A \to B$, we prove that $G(k) = k_\sharp$ is a kernel of $G(f) = f_\sharp$. In the right diagram below we take a morphism $u\colon D \to Hom(X, A)$ such that $G(f).u = 0$

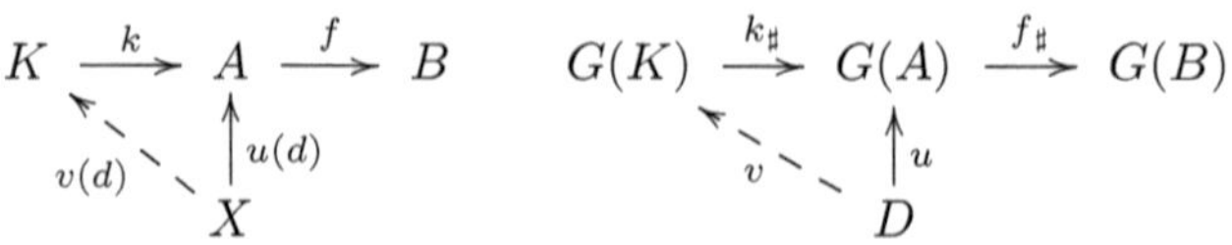


(5.37)

Therefore, for every $d \in D$, the homomorphism $u(d)\colon X \to A$ in the left diagram is annihilated by $f$ and factorises as $k.v(d)$, by a unique homomorphism $v(d)\colon X \to K$. Coming back to the right diagram, we have defined the unique mapping $v\colon D \to G(K) = \mathrm{Hom}(X, K)$ such that $G(k).v = u$, and it is a homomorphism because $k\colon K \to A$ is injective.

(ii) According to 5.2.2(e) we prove that, if $c\colon B \to C$ is a cokernel of $f\colon A \to B$, then $H(c) = c^\sharp$ is a kernel of $H(f) = f^\sharp$. In the right diagram below we take a morphism $u$ such that $H(f).u = 0$

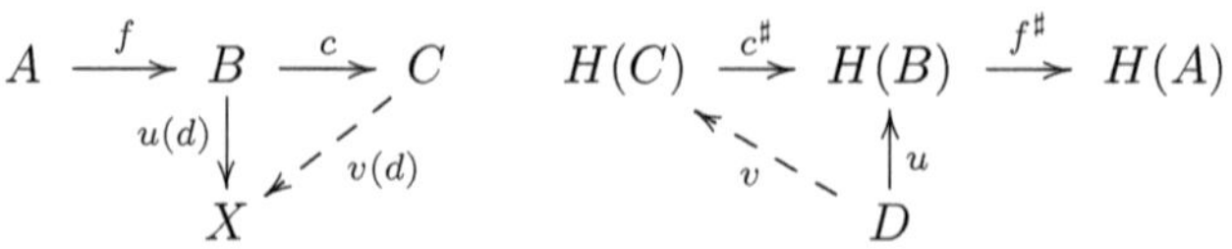


(5.38)

For every $d \in D$, the homomorphism $u(d)\colon B \to X$ gives $u(d).f = H(f)(u(d)) = 0$ and factorises in the left diagram as $v(d).c$, by a unique homomorphism $v(d)\colon C \to X$. In the right diagram we have defined the unique mapping $v\colon D \to \mathrm{Hom}(C, X)$ such that $H(c).v = u$, and it is a homomorphism because $c\colon B \to C$ is an epimorphism. □

**5.2.6 *Flat, projective and injective modules***

(a) An $R$-module $A$ is said to be *flat* if the functor $F = - \otimes_R A$ is exact; this is equivalent to saying that $F$ preserves monomorphisms.

(b) An $R$-module $P$ is said to be *projective* if the functor $G = \mathrm{Hom}_R(P, -)\colon R\,\mathsf{Mod} \to R\,\mathsf{Mod}$ is exact. This is equivalent to saying that $G$ preserves epimorphisms, and can be expressed by the following property:

(i) if $p\colon A \to B$ is an epimorphism, every homomorphism $h\colon P \to B$ can be lifted to $A$, i.e. there is some homomorphism $k\colon P \to A$ such that $pk = h$

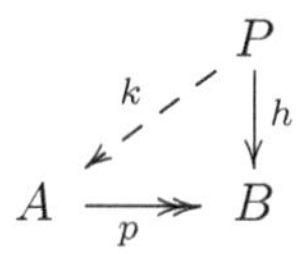


(5.39)

(c) An $R$-module $Q$ is said to be *injective* if the contravariant Hom-functor $H = \mathrm{Hom}_R(-, B)\colon R\,\mathsf{Mod} \dashrightarrow R\,\mathsf{Mod}$ is exact, that is, preserves exact sequences. This is equivalent to saying that $H$ takes monomorphisms to epimorphisms, and can be expressed by the following property:

(i*) if $m\colon B \to A$ is a monomorphism, every homomorphism $h\colon B \to Q$ can be extended to $A$, i.e. there exists $k\colon A \to Q$ such that $km = h$

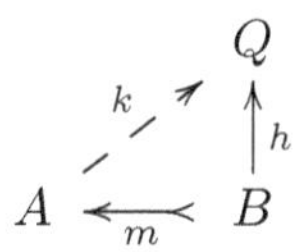


(5.40)

### *5.2.7 Exercises and complements*

(a) A free $R$-module $A$ is flat and projective: the functors $- \otimes_R A$ and $\mathrm{Hom}_R(A, -)$ are exact. (Of course, $A \otimes_R -$ is also.)

(b) Projective modules are closed under direct sums and direct summands (see 1.6.3(b)). Injective modules are closed under products and summands.

(c) An $R$-module is projective if and only if it is a direct summand of a free module.

*(d) Every projective module is flat. This will be proved in Exercise 6.1.6(b).

*(e) The properties (i) and (i*) of 5.2.6 are related by reversing all arrows and interchanging epi with mono: projective and injective objects are dual issues, if we state them in the selfdual framework of abelian categories.

### *5.2.8 Exercises and complements on abelian groups, I*

The rest of this section is about abelian groups.

(a) Prove that $\mathbb{Z}_m \otimes \mathbb{Z}_n \cong \mathbb{Z}_d$, where $d = \gcd(m, n)$ is the greatest common divisor of $m, n \geqslant 2$. In particular, if $m$ and $n$ are coprime integers, we get $\mathbb{Z}_m \otimes \mathbb{Z}_n = 0$. *Hint:* use the exactness properties of the tensor product.

(b) Compute $\mathrm{Hom}(\mathbb{Z}_m, A)$, for $n \geqslant 2$. Prove that $\mathrm{Hom}(\mathbb{Z}_m, \mathbb{Z}_n) \cong \mathbb{Z}_d$, where $d = \gcd(m, n)$.

(c) A flat abelian group $A$ is always torsion free (see 1.2.6). *The converse is also true, but less easy to prove: see 6.2.8(a).*

(d) In particular, a finite non-trivial abelian group is not flat. For instance, $- \otimes \mathbb{Z}_2 \colon \mathsf{Ab} \to \mathsf{Ab}$ takes the monomorphism $2 \colon \mathbb{Z} \to \mathbb{Z}$ (multiplication by 2) to $0 \colon \mathbb{Z}_2 \to \mathbb{Z}_2$.

### *5.2.9 Exercises and complements on abelian groups, II*

(a) We shall use, without proof, a well known theorem about abelian groups: *every subgroup of a free abelian group is free.*

The proof is not easy, and depends on the axiom of choice. It can be found, for instance, in a classical textbook on abelian groups [Fu], in Theorem II.12.1.

(b) An abelian group is projective if and only if it is free.

(c) An additive functor $F \colon \mathsf{Ab} \to \mathsf{Ab}$ preserves all *upper unbounded* exact sequences of *free abelian groups.* (Note that upper unbounded sequences include all sequences ending with a trivial group in the last position – which can be repeated indefinitely.)

This cannot be extended to upper bounded sequences. We already remarked that the additive functor $- \otimes \mathbb{Z}_2 \colon \mathsf{Ab} \to \mathsf{Ab}$ annihilates the monomorphism $2 \colon \mathbb{Z} \to \mathbb{Z}$ (multiplication by 2). In other words, it does not preserve the exactness of a sequence $0 \to \mathbb{Z} \to \mathbb{Z}$ of free abelian groups.

*(d) An abelian group is injective if and only if it is divisible (see 5.1.3(d)). We shall not prove this result, which again depends on the axiom of choice. One can see [CE], Theorem I.3.2, or [M1], Corollary III.7.3.

## 5.3 Relative singular homology with coefficients

We have defined relative singular homology as a homology theory with coefficients in the group $\mathbb{Z}$ of integers. Working with the tensor functor $- \otimes G$, we can modify it into a homology theory with coefficients in an arbitrary abelian group $G$. The new theory may be able to detect facts that the original one cannot see.

For instance, *homology modulo* 2 (i.e. with coefficients in $\mathbb{Z}_2$) is able to prove that the canonical projection $\mathbb{P}^2 \to \mathbb{S}^2$ is not homotopically trivial (see Exercise 5.4.2(c)). One cannot get this result working with integral coefficients: for $k > 0$, one at least of the groups $H_k(\mathbb{P}^2)$ or $H_k(\mathbb{S}^2)$ is trivial.

$G$ will denote a fixed abelian group.

### 5.3.1 Extending tensor products, I

The additive functor $- \otimes G \colon \mathsf{Ab} \to \mathsf{Ab}$ has an obvious componentwise extension to chain complexes (as in 5.2.3(a))

$$\begin{gathered} - \otimes G \colon \mathrm{Ch}_+\mathsf{Ab} \to \mathrm{Ch}_+\mathsf{Ab}, \quad A \otimes G = ((A_n \otimes G), (\partial_n \otimes G)), \\ (f \otimes G)_n = f_n \otimes G \colon A_n \otimes G \to B_n \otimes G, \end{gathered} \tag{5.41}$$

where $f \colon A \to B$ is a morphism of chain complexes.

This extension preserves homotopies: a homotopy $\varphi \colon f \simeq g \colon A \to B$ in $\mathrm{Ch}_+\mathsf{Ab}$ gives a homotopy

$$\begin{gathered} \varphi \otimes G \colon f \otimes G \simeq g \otimes G \colon A \otimes G \to B \otimes G, \\ (\varphi \otimes G)_n = \varphi_n \otimes G \colon A_n \otimes G \to B_{n+1} \otimes G. \end{gathered} \tag{5.42}$$

*Remarks.* Letting $G$ vary in $\mathsf{Ab}$, the extension (5.41) becomes a functor $\mathrm{Ch}_+\mathsf{Ab} \times \mathsf{Ab} \to \mathrm{Ch}_+\mathsf{Ab}$, that will be used in Section 5.6. In Section 6.5 we shall introduce a further extension, the tensor product of chain complexes, defined on $\mathrm{Ch}_+\mathsf{Ab} \times \mathrm{Ch}_+\mathsf{Ab}$.

### 5.3.2 Chains with coefficients

We now apply this functor $-\otimes G \colon \mathrm{Ch}_+\mathsf{Ab} \to \mathrm{Ch}_+\mathsf{Ab}$ to the singular chains of a space.

We have thus the *singular chain complex* of a space $X$, *with coefficients in* $G$, forming a functor

$$\begin{gathered} C_+(-;G) \colon \mathsf{Top} \to \mathrm{Ch}_+\mathsf{Ab}, \qquad C_+(X;G) = C_+(X) \otimes G, \\ C_n(X;G) = C_n(X) \otimes G, \\ f_{\sharp n}((\Sigma\, \lambda_i a_i) \otimes g) = (\Sigma\, \lambda_i f a_i) \otimes g, \end{gathered} \tag{5.43}$$

where $f \colon X \to Y$ and $a_i \colon \mathbb{I}^n \to X$ are in $\mathsf{Top}$, while $g \in G$.

Similarly, we have the *singular chain complex* of relative pairs of spaces, *with coefficients in* $G$

$$C_+(-;G) \colon \mathsf{Top}_2 \to \mathrm{Ch}_+\mathsf{Ab}, \qquad C_+(X, A; G) = C_+(X, A) \otimes G. \tag{5.44}$$

It is important to note that, for a relative pair $(X, A)$, the short exact sequence

$$C_+(A) \xrightarrow{u_\sharp} C_+(X) \xrightarrow{v_\sharp} C_+(X, A) \qquad (5.45)$$

splits in every degree (because all $C_n(X, A)$ are free, see Exercise 4.1.2(b)), and is preserved by the additive functor $- \otimes G$ (by 5.2.1(c)): we have a componentwise split sequence of chain complexes

$$C_+(A; G) \xrightarrow{u_\sharp} C_+(X; G) \xrightarrow{v_\sharp} C_+(X, A; G). \qquad (5.46)$$

*Note.* We already remarked, in (4.51), that $C_n(X; G) \cong \bigoplus_a G$, where $a$ varies in $\mathrm{Cub}_n X \setminus \mathrm{Deg}_n X$. Defining $C_n(X; G)$ in this way, without using tensor products, is a shortcut which would soon prove ineffective.

### *5.3.3 Singular homology with coefficients*

Applying chain homology to the previous complexes, we have now the absolute and relative *singular homology groups with coefficients in G*

$$H_n(-; G)\colon \mathsf{Top} \to \mathsf{Ab}, \qquad H_n(X; G) = H_n(C_+(X; G)), \qquad (5.47)$$

$$H_n(-; G)\colon \mathsf{Top}_2 \to \mathsf{Ab}, \qquad H_n(X, A; G) = H_n(C_+(X, A; G)). \qquad (5.48)$$

With $G = \mathbb{Z}$ we find the previous chain complexes and homology with integral coefficients, up to canonical isomorphism (and generally identified)

$$\begin{array}{ll} C_+(X; \mathbb{Z}) \cong C_+(X), & C_+(X, A; \mathbb{Z}) \cong C_+(X, A), \\ H_n(X; \mathbb{Z}) \cong H_n(X), & H_n(X, A; \mathbb{Z}) \cong H_n(X, A). \end{array} \qquad (5.49)$$

*Remarks.* For a commutative ring $R$, the chain complex $C_+(X, A; R)$ is obviously a complex of $R$-modules, and the homology groups $H_n(X, A; R)$ are $R$-modules (cf. 5.1.5(h)).

The important cases, after the ring $\mathbb{Z}$, are the fields $\mathbb{R}, \mathbb{Q}$ and the ring $\mathbb{Z}_n$. The last case gives the modules $H_n(X, A; \mathbb{Z}_n)$ of singular homology *modulo n* (vector spaces, for $n$ prime).

*Exercises and complements.* The elementary computations of 2.2.5 can be easily extended to arbitrary coefficients.

(a) Compute $H_n(\{*\}; G)$.

(b) Compute $H_0(X; G)$, where $X$ is 0-connected (i.e. path connected and non-empty).

(c) The partition $X = \cup_{i\in I} X_i$ of a space in path components gives canonical isomorphisms

$$\bigoplus_{i\in I} C_+(X_i;G) \to C_+(X;G), \quad (c_i)_{i\in I} \mapsto \Sigma_{i\in I}\, c_i, \tag{5.50}$$

$$\bigoplus_{i\in I} H_n(X_i;G) \to H_n(X;G), \quad ([z_i])_{i\in I} \mapsto \Sigma_{i\in I}\, [z_i]. \tag{5.51}$$

(d) If the map $f\colon X \to Y$ is defined on a path connected space, the homomorphism $f_{*0}\colon H_0(X;G) \to H_0(Y;G)$ is injective.

### 5.3.4 Subdivision Theorem

*For a space $X$ and a generalised open cover $\mathcal{U} = (U_i)$, the inclusion $j\colon C_+(X;\mathcal{U}) \to C_+(X)$ gives a canonical morphism $j\otimes G\colon C_+(X;\mathcal{U})\otimes G \to C_+(X;G)$, which is a monomorphism and induces isomorphism in homology, in every degree*

$$(j \otimes G)_{*n}\colon H_n(X;\mathcal{U};G) \to H_n(X;G). \tag{5.52}$$

*Proof* We deduce this from the original Subdivision Theorem 2.3.5, with integral coefficients. We start from the short exact sequence of chain complexes

$$C_+(X;\mathcal{U}) \xrightarrow{\;j\;} C_+(X) \xrightarrow{\;p\;} E_+ \tag{5.53}$$

produced by the inclusion $j$.

This sequence is componentwise split, because the canonical basis of $C_n(X;\mathcal{U})$ is a subset of the canonical basis of $C_n(X)$. Applying $-\otimes G$ we get a short exact sequence of chain complexes

$$C_+(X;\mathcal{U}) \otimes G \xrightarrow{\;j\otimes G\;} C_+(X;G) \xrightarrow{\;p\otimes G\;} E_+ \otimes G \tag{5.54}$$

In the long exact homology sequence of (5.53) all $j_{*n}$ are isomorphisms (by Theorem 2.3.5), and therefore $H_n(E_+) = 0$, for all $n$.

The complex $E_+$ is thus an exact sequence of free abelian groups, and an unbounded one (when extended with zeros). Applying Exercise 5.2.9(c), also $E_+ \otimes G$ is an exact sequence and $H_n(E_+ \otimes G) = 0$, for all $n$.

Finally, by the exactness of the homology sequence of (5.54), all the homomorphisms $(j \otimes G)_{*n}$ are isomorphisms. □

**5.3.5 Theorem**

*Relative singular homology with coefficients in $G$ satisfies the Eilenberg–Steenrod axioms for a homology theory with coefficient group $G$.*

*Proof* The Functoriality axiom was proved from the start, in 5.3.3. The others are proved as follows.

(a) *Exactness and Naturality.* For a relative pair $(X, A)$, the natural short exact sequence of chain complexes (5.46)

$$C_+(A; G) \xrightarrow{u_\sharp} C_+(X; G) \xrightarrow{v_\sharp} C_+(X, A; G) \tag{5.55}$$

gives a natural, exact homology sequence

$$\begin{array}{l} \dots\ H_n(A; G) \xrightarrow{u_{*n}} H_n(X; G) \xrightarrow{v_{*n}} H_n(X, A; G) \xrightarrow{D_n} \\ H_{n-1}(A; G) \to \qquad \dots \qquad \to H_0(X, A; G) \to 0 \end{array} \tag{5.56}$$

where $D_n[z] = [\partial_n z]$, for $z \in Z_n(X, A; G)$.

(b) *Homotopy invariance.* Let $\varphi\colon f \simeq g\colon (X, A) \to (Y, B)$ be a relative homotopy. In the proof of Theorem 4.1.4 we built a homotopy of chain complexes $\Phi''\colon f_\sharp \simeq g_\sharp\colon C_+(X, A) \to C_+(Y, B)$.

Applying the additive functor $- \otimes G$ we get a homotopy

$$\Phi'' \otimes G\colon f_\sharp \simeq g_\sharp\colon C_+(X, A; G) \to C_+(Y, B; G).$$

(c) *Excision.* As in Theorem 4.2.1, using now the Subdivision Theorem 5.3.4, with coefficients in $G$.

(d) *Dimension.* By Exercise 5.3.3(a), which also determines the coefficient group. □

**5.3.6 Theorem** (The Mayer–Vietoris sequence)

*There are given a topological space $X$, two subspaces $U$ and $V$ such that $X = \operatorname{int} U \cup \operatorname{int} V$, and an abelian group $G$. We let $A = U \cap V$.*

*There is an exact sequence of singular homology groups*

$$\begin{array}{l} \dots\ H_n(A; G) \xrightarrow{h_n} H_n(U; G) \oplus H_n(V; G) \xrightarrow{k_n} H_n(X; G) \xrightarrow{D_n} \\ H_{n-1}(A; G) \to \ \dots \end{array} \tag{5.57}$$

*called the* exact sequence of Mayer–Vietoris, *or* MV-sequence, *of the space $X$ with coefficients in $G$, with respect to the generalised open cover $(U, V)$.*

*These homomorphisms are computed as in* (2.84).

*The sequence is natural for continuous mappings $f\colon X \to X'$, where*

$$X' = \operatorname{int} U' \cup \operatorname{int} V', \qquad f(U) \subset U', \qquad f(V) \subset V'. \tag{5.58}$$

*Proof* We apply the same argument as in 2.3.7, using now the Subdivision Theorem 5.3.4. □

## 5.4 Acyclic spaces and essential maps

Acyclic spaces and acyclic maps, with respect to singular homology, are characterised by the vanishing of all functors $H_n(-;G)$, for $n > 0$. It will be interesting to see that integral coefficients are sufficient to determine the question for spaces, but are not for maps, where a suitable coefficient group can prove the opposite property, the 'homological essentiality' of a map.

We examine various instances, related to spheres and projective spaces. Homology modulo 2 will be of help, to take on a study begun in Section 3.3.

All this is considered here in singular homology: the converging sequence $\mathbb{N}^\bullet$ is acyclic in singular homology, but has non-trivial Alexander–Spanier cohomology groups: see 4.5.8(e).

### *5.4.1 Acyclic properties*

(a) A space $X$ is said to be *acyclic* (in singular homology) if $H_n(X) = 0$ for all $n > 0$. (This is certainly true if its path components are contractible.)

Then the same holds with coefficients in any group $G$. In fact, the following augmented chain complex of an arbitrary space X

$$... \ C_2(X) \ \xrightarrow{\partial_2} \ C_1(X) \ \xrightarrow{\partial_1} \ C_0(X) \ \xrightarrow{\operatorname{cok}\partial_1} \ H_0(X) \ \to \ 0 \tag{5.59}$$

is always exact at $C_0(X)$ and $H_0(X)$. Therefore, the space $X$ is acyclic if and only if the previous sequence is exact. In this case, as the sequence is unbounded and all objects are free abelian groups, tensoring by $G$ gives an exact sequence (by 5.2.9(c)), and $H_n(X;G) = 0$ for all $n > 0$.

(b) We say that a map $f\colon X \to Y$ is *acyclic* (in singular homology) if $H_n(f;G) = 0$, for any group $G$ and any $n > 0$.

Otherwise, we say that $f$ is *homologically essential* (in singular homology): this means that $H_n(f;G) \neq 0$, for some group $G$ and some degree $n > 0$; as a consequence, $f$ cannot be homotopic to a constant map, and cannot factorise through an acyclic space, nor a contractible one. In particular, $X$ and $Y$ are not acyclic.

A homologically essential map with values in a sphere must be surjective: taking out a point gives a contractible space.

Integral coefficients are not sufficient to settle this point, as shown by Exercise 5.4.2(c).

(c) The same terminology can be applied to relative pairs and their maps.

### *5.4.2 Exercises and complements*

The following exercises show part of the interest of introducing coefficient groups.

(a) We already know, from the Exercises of 5.3.3, that $H_0(\mathbb{S}^0; G) \cong G^2$, with trivial higher groups. Prove that, for $n > 0$:

$$H_0(\mathbb{S}^n; G) \cong H_n(\mathbb{S}^n; G) \cong G, \qquad H_k(\mathbb{S}^n; G) = 0 \ \ (0 \neq k \neq n). \tag{5.60}$$

(b) For $n \geqslant 2$, the space $P_n = \mathbb{D}^2/R_n$ defined in 3.1.8 gives:

$$H_1(P_n; G) \cong G/nG, \qquad H_2(P_n; G) \cong \{g \in G \mid ng = 0\}, \tag{5.61}$$

and the higher groups are trivial.

(c) The projection $q\colon P_n \to \mathbb{S}^2$ (viewing both spaces as quotients of the disc $\mathbb{D}^2$) is homologically essential. In particular, this holds for the projection $\mathbb{P}^2 \to \mathbb{S}^2$ of the projective plane. (Higher projective spaces will be dealt with in Section 6.3.)

*Hints.* Points (a) and (b) suggest to use homology modulo $n$, so that both $H_2(P_n; \mathbb{Z}_n)$ and $H_2(\mathbb{S}^2; \mathbb{Z}_n)$ are non-trivial. In fact, one can prove that $q$ induces an isomorphism $H_2(P_n; \mathbb{Z}_n) \to H_2(\mathbb{S}^2; \mathbb{Z}_n)$ via suitable MV-sequences. Integral coefficients cannot see this fact, because $H_1(\mathbb{S}^2) = 0$ and $H_2(P_n) = 0$: in each positive degree, one of the homology functors is 'blind' on our spaces.

*(d) The Universal Coefficient Theorem, in 6.2.5–6.2.6, will show that singular homology with integral coefficients determines the *groups* of singular homology with any coefficients. (More precisely, we need both groups $H_n(X)$ and $H_{n-1}(X)$ to determine $H_n(X; G)$.)

Exercise (c) shows that *this fails for the morphisms.*

## 5.5 Relative singular cohomology with coefficients

Transforming the singular chain complex of a space with the contravariant functor $\mathrm{Hom}(-, G)$, we obtain a cohomology theory with coefficients in an abelian group $G$, which is still based on the singular probes.

### 5.5.1 Extending the Hom functor, I

Extending the contravariant (additive) functor $\mathrm{Hom}(-, G)\colon \mathsf{Ab} \dashrightarrow \mathsf{Ab}$ to chain complexes, we get a contravariant functor with values in cochain complexes (where $f\colon A \to B$ is a morphism of chain complexes)

$$\begin{aligned} &\mathrm{Hom}(-, G)\colon \mathrm{Ch}_+\mathsf{Ab} \dashrightarrow \mathrm{Ch}^+\mathsf{Ab}, \ (\mathrm{Hom}(A, G))^n = \mathrm{Hom}(A_n, G), \\ &\quad d^n = \mathrm{Hom}(\partial_{n+1}, G)\colon \mathrm{Hom}(A_n, G) \to \mathrm{Hom}(A_{n+1}, G), \\ &\quad f^n = \mathrm{Hom}(f_n, G)\colon \mathrm{Hom}(B_n, G) \to \mathrm{Hom}(A_n, G). \end{aligned} \tag{5.62}$$

This functor preserves homotopies: a homotopy in $\mathrm{Ch}_+\mathsf{Ab}$

$$\varphi\colon f \simeq g\colon A \to B, \quad \varphi_n\colon A_n \to B_{n+1}, \quad \partial_{n+1}\varphi_n + \varphi_{n-1}\partial_n = g_n - f_n,$$

gives a homotopy of cochain complexes (see (4.69))

$$\begin{aligned} \psi &= \mathrm{Hom}(\varphi, G)\colon \mathrm{Hom}(f, G) \simeq \mathrm{Hom}(g, G), \\ \psi^n &= \mathrm{Hom}(\varphi_{n-1}, G)\colon \mathrm{Hom}(B_n, G) \to \mathrm{Hom}(A_{n-1}, G), \\ d^{n-1}\psi^n + \psi^{n+1}d^n &= \mathrm{Hom}(\varphi_{n-1}\partial_n + \partial_{n+1}\varphi_n, G) = g^n - f^n. \end{aligned} \tag{5.63}$$

### 5.5.2 Singular cochains

We now apply the functor $\mathrm{Hom}(-, G)$ to the singular chains of a space, with integral coefficients.

We have thus the *singular cochain complex* of a space $X$, *with coefficients in* $G$, a contravariant functor which is the composite of $C_+\colon \mathsf{Top} \to \mathrm{Ch}_+\mathsf{Ab}$ and $\mathrm{Hom}(-, G)$

$$\begin{aligned} &C^+(-; G)\colon \mathsf{Top} \dashrightarrow \mathrm{Ch}^+\mathsf{Ab}, \qquad C^+(X; G) = \mathrm{Hom}(C_+(X), G), \\ &\quad C^n(X; G) = \mathrm{Hom}(C_n(X), G), \qquad d^n = \mathrm{Hom}(\partial_{n+1}, G), \\ &\quad f^{\sharp n}\colon \mathrm{Hom}(C_n(Y), G) \longrightarrow \mathrm{Hom}(C_n(X), G), \\ &\quad f^{\sharp n}(\mu\colon C_n(Y) \to G) = \mu f_{\sharp n}\colon C_n(X) \to G, \end{aligned} \tag{5.64}$$

where $f\colon X \to Y$ is a map of topological spaces.

More generally, we have the *singular cochain complex* of relative pairs, *with coefficients in* $G$

$$\begin{aligned} &C^+(-; G)\colon \mathsf{Top}_2 \dashrightarrow \mathrm{Ch}^+\mathsf{Ab}, \\ &C^+(X, A; G) = \mathrm{Hom}(C_+(X, A), G). \end{aligned} \tag{5.65}$$

The (componentwise split) short exact sequence of the relative pair $(X, A)$

(in (5.45)) gives here a (componentwise split) short exact sequence of cochain complexes

$$C^+(X,A;G) \xrightarrow{v^\sharp} C_+(X;G) \xrightarrow{u^\sharp} C_+(A;G). \tag{5.66}$$

*Comments and complements.* (a) A cochain $\lambda\colon C_n(X) \to G$ can be viewed as a mapping on $n$-cubes which vanishes on the degenerate ones (and is linearly extended on cubical chains)

$$\begin{gathered}\lambda\colon \mathrm{Cub}_n X \to |G|, \qquad \lambda(\mathrm{Deg}_n X) = \{0\}, \\ f^{\sharp n}(\mu)(a) = \mu(fa) \qquad (\text{for } \mu\colon \mathrm{Cub}_n Y \to G,\ a\colon \mathbb{I}^n \to X).\end{gathered} \tag{5.67}$$

(b) A relative cochain in $C^n(X,A;G)$ is thus the same as an absolute cochain $\lambda\colon C_n(X) \to G$ which vanishes on degenerate cubes and all the cubes of $A$

$$\lambda\colon \mathrm{Cub}_n X \to |G|, \qquad \lambda(\mathrm{Deg}_n X \cup \mathrm{Cub}_n A) = \{0\}, \tag{5.68}$$

(c) For a relative map $f\colon (X,A) \to (Y,B)$, the associated homomorphism of relative cochains is described as in (5.67)

$$f^{\sharp n}\colon C^n(Y,B;G) \to C^n(X,A;G), \qquad f^{\sharp n}(\mu)(a) = \mu(fa), \tag{5.69}$$

where $\mu\colon \mathrm{Cub}_n Y \to G$ vanishes on $\mathrm{Deg}_n Y \cup \mathrm{Cub}_n B$, and $a\colon \mathbb{I}^n \to X$.

(d) (*Restrictions and extensions*) The embedding $u\colon A \to X$ gives a *restriction*

$$u^{\sharp n}\colon C^n(X;G) \twoheadrightarrow C^n(A;G), \qquad u^{\sharp n}(\lambda) = \lambda u_{\sharp n} = \lambda|_A, \tag{5.70}$$

whose value on the cochain $\lambda\colon \mathrm{Cub}_n X \to |G|$ will also be written as $\lambda|_A$.

This epimorphism has a canonical section

$$C^n(A;G) \rightarrowtail C^n(X;G), \qquad \xi \mapsto \xi_X\colon C_n(X) \to G, \tag{5.71}$$

which will be called *extension* (in degree $n$): the cochain $\xi\colon C_n(A) \to G$ is extended to $X$ letting $\xi_X(a) = 0$ for every cube $a\colon \mathbb{I}^n \to X$ whose image is not contained in $A$.

In this way we have constructed a *componentwise* splitting of the short exact sequence (5.66): the extension homomorphisms do not commute with the differentials.

### 5.5.3 Singular cohomology

Applying the (covariant) functor of cochain cohomology to the previous complexes, we have now the absolute and relative *singular cohomology groups with coefficients in G*

$$H^n(-;G)\colon \mathsf{Top} \dashrightarrow \mathsf{Ab}, \qquad H^n(X;G) = H^n(C^+(X;G)), \tag{5.72}$$

$$H^n(-;G)\colon \mathsf{Top}_2 \dashrightarrow \mathsf{Ab}, \quad H^n(X,A;G) = H^n(C^+(X,A;G)). \tag{5.73}$$

A cocycle $\lambda \in C^n(X,A;G)$ is a relative cochain $\lambda\colon \mathsf{C}_n(X) \to G$ which vanishes on all $n$-boundaries

$$\lambda(\mathrm{Deg}_n X \cup \mathrm{Cub}_n A \cup \partial_{n+1}(\mathrm{Cub}_{n+1} X)) = \{0\}. \tag{5.74}$$

With $G = \mathbb{Z}$ we get singular cohomology *with integral coefficients.* With the ring $\mathbb{Z}_n$ we get singular cohomology *modulo n.*

### 5.5.4 Exercises and complements

The elementary computations of 2.2.5 can be adapted to cohomology with coefficients in $G$.

(a) Compute $H^n(\{*\};G)$.

(b) Compute $H^0(X;G)$, where $X$ is 0-connected (i.e. path connected and non-empty).

(c) For a space $X$ with path components $X_i$ and inclusions $u_i\colon X_i \to X$ $(i \in I)$, there is a canonical isomorphism

$$H^n(X;G) \to \textstyle\prod_{i\in I} H^n(X_i;G), \qquad [\lambda] \mapsto ([\lambda_i])_{i\in I}, \tag{5.75}$$

where $\lambda_i = u_i^{\sharp n}(\lambda)$: the value $\lambda_i(a)$ on an $n$-cube $a\colon \mathbb{I}^n \to X_i$ is $\lambda(u_i a)$; or simply $\lambda(a)$ when we view $a$ as a map $\mathbb{I}^n \to X$ with image in $X_i$.

More formally, the homomorphisms $u_i^{*n}\colon H^n(X;G) \to H^n(X_i;G)$ are the projections of a cartesian product.

(d) $H^0(X;G)$ is (literally) the group of 0-cochains $\lambda\colon X \to G$ which are constant on path components. It is isomorphic to $G^I$ (with the previous notation).

(e) *Reduced* singular cohomology is defined with the augmented chain complex $\tilde{C}(X)$ in (3.74), letting

$$\tilde{H}^n(X;G) = H^n(\mathrm{Hom}(\tilde{C}(X);G)) \qquad (n \geqslant -1). \tag{5.76}$$

It coincides with singular cohomology in degree $n \geqslant 1$, and can be easily determined in lower degree: $\tilde{H}^{-1}(X;G)$ 'is' $G$ for $X = \emptyset$ and trivial otherwise, while $\tilde{H}^0(X;G)$ is the group of 0-cochains $\lambda\colon X \to G$ which are constant on path components, modulo the subgroup of the constant ones.

### 5.5.5 Subdivision Theorem

*For a topological space $X$ and a generalised open cover $\mathcal{U} = (U_i)$, the inclusion $j\colon C_+(X;\mathcal{U}) \to C_+(X)$ gives a canonical morphism* $\mathrm{Hom}(j,G)\colon C^+(X;G) \to \mathrm{Hom}(C_+(X;\mathcal{U}),G)$, *which is an epimorphism and induces isomorphism in cohomology, in every degree*

$$\mathrm{Hom}(j,G)^{*n}\colon H^n(X;G) \longrightarrow H^n(X;\mathcal{U};G). \tag{5.77}$$

*Proof* As in Theorem 5.3.4, we start from the short exact sequence (5.53), of chain complexes

$$C_+(X;\mathcal{U}) \xrightarrow{j} C_+(X) \xrightarrow{p} E_+ \tag{5.78}$$

which we know to be componentwise split.

Applying the additive functor $\mathrm{Hom}(-,G)$ we get a short exact sequence of cochain complexes

$$\mathrm{Hom}(E_+,G) \xrightarrow{p^\sharp} C^+(X;G) \xrightarrow{j^\sharp} \mathrm{Hom}(C_+(X;\mathcal{U}),G). \tag{5.79}$$

Applying the original Subdivision Theorem 2.3.5 we have seen that the complex $E_+$ is an exact sequence of free abelian groups, and an unbounded one (extended with zeros). Applying Exercise 5.2.9(c), in contravariant form, the cochain complex $\mathrm{Hom}(E_+,G)$ is also an exact sequence and $H^n(\mathrm{Hom}(E_+,G)) = 0$ for all $n$.

Finally, by the exactness of the cohomology sequence of (5.79), all the homomorphisms $\mathrm{Hom}(j,G)^{*n}$ are isomorphisms. □

### 5.5.6 Theorem

*Relative singular cohomology with coefficients in $G$ satisfies the Eilenberg–Steenrod axioms for a cohomology theory with coefficients in $G$.*

*Proof* The Functoriality axiom is proved in 5.5.3.

(a) *Exactness and Naturality.* A relative pair $(X,A)$ has a natural short exact sequence of cochain complexes (5.66). Therefore its cohomology sequence (4.56) is exact and natural.

(b) *Homotopy invariance.* A relative homotopy $\varphi\colon f \simeq g\colon (X,Y) \to (Y,B)$ gives a homotopy of chain complexes $\Phi''\colon f_\sharp \simeq g_\sharp\colon C_+(X,A) \to C_+(Y,B)$ (cf. 4.1.4), turned by the additive functor $\mathrm{Hom}(-,G)$ into a homotopy

$$\mathrm{Hom}(\Phi'',G)\colon f^\sharp \simeq g^\sharp\colon C^+(Y,B;G) \longrightarrow C^+(X,A;G).$$

(c) *Excision.* As in Theorem 4.2.1, using the Subdivision Theorem 5.5.5.

(d) *Dimension* and *coefficient group.* By Exercise 5.5.3(a). □

**5.5.7 Theorem** (The Mayer–Vietoris sequence)

*Let $X$ be a topological space, $U$ and $V$ subspaces of $X$ such that $X = \operatorname{int} U \cup \operatorname{int} V$, and $A = U \cap V$. Let $G$ be an abelian group.*

*There is a long exact sequence of singular cohomology groups*

$$\begin{aligned} \ldots\; H^n(X;G) &\xrightarrow{k^n} H^n(U;G) \oplus H^n(V;G) \xrightarrow{h^n} H^n(A;G) \xrightarrow{D_n} \\ H^{n+1}(X;G) &\to \ldots \end{aligned} \tag{5.80}$$

*called the* cohomology sequence of Mayer–Vietoris, *or* cohomology MV-sequence *of the space $X$ with coefficients in $G$, with respect to the generalised open cover $(U, V)$.*

*With the notation of* (2.82) *for the inclusions $i\colon A \to U$, $u\colon U \to X$, ...*

$$\begin{aligned} k^n([\lambda]) &= (u^{*n}[\lambda], -v^{*n}[\mu]) = ([\lambda|_U], -[\lambda|_V]), \\ h^n([\mu],[\nu]) &= i^{*n}[\mu] + j^{*n}[\nu] = [\mu|_A] + [\nu|_A], \\ D^n([\xi]) &= [(d^n \xi_U)_X], \end{aligned} \tag{5.81}$$

*where $\lambda \in Z^n(X;G)$, $\mu \in Z^n(U;G)$, $\nu \in Z^n(V;G)$, and $\xi \in Z^n(A;G)$. The restrictions $\lambda|_U, \lambda|_V$, ... are defined in* (5.70)*; the extensions $\xi_U$ and $(d^n \xi_U)_X$ in* (5.71).

*The sequence is natural for continuous mappings $f\colon X \to X'$, where*

$$X' = \operatorname{int} U' \cup \operatorname{int} V', \qquad f(U) \subset U', \qquad f(V) \subset V'.$$

*Proof* We start from the short exact sequence of chain complexes (2.86)

$$C_+(A) \overset{h'}{\rightarrowtail} C_+(U) \oplus C_+(V) \overset{k'}{\twoheadrightarrow} C_+(X; \mathcal{U}) \tag{5.82}$$

$$\begin{aligned} h'_n(c) &= (i_{\sharp n}(c), j_{\sharp n}(c)) = (c, c), \\ k'_n(c, c') &= u_{\sharp n}(c) - v_{\sharp n}(c) = c - c'. \end{aligned}$$

The sequence splits componentwise, and is taken by $\operatorname{Hom}(-, G)$ to a short exact sequence of cochain complexes

$$C^+(X;G;\mathcal{U}) \overset{k''}{\rightarrowtail} C^+(U;G) \oplus C^+(V;G) \overset{h''}{\twoheadrightarrow} C^+(A;G) \tag{5.83}$$

$$\begin{aligned} k''_n(\lambda) &= (u^{\sharp n}(\lambda), -v^{\sharp n}(\lambda)) = (\lambda|_U, -\lambda|_V), \\ h''_n(\mu, \nu) &= i^{\sharp n}(\mu) + j^{\sharp n}(\nu) = \mu|_A + \nu|_A. \end{aligned}$$

The cohomology of $C^+(X;G;\mathcal{U})$ is identified to that of $C^+(X;G)$, by the (current) Subdivision Theorem 5.5.5. This gives the cohomology sequence (5.80).

The induced homomorphisms $h^n$ and $k^n$ are computed as stated in (5.81). The differential acts indeed as $D^n([\xi]) = [(d^n\xi_U)_X]$

$$\begin{array}{ccccccc} & & (\xi_U, 0) & \overset{h''}{\longmapsto} & \xi & & (n) \\ & & \downarrow & & \downarrow & & \\ (d^n\xi_U)_X & \overset{k''}{\longmapsto} & (d\xi_U, 0) & \overset{h''}{\longmapsto} & 0 & & (n+1) \end{array} \tag{5.84}$$

- the cocycle $\xi \in Z^n(A;G)$ is extended to a cochain $\xi_U \in C^n(U;G)$, letting $\xi_U(a) = 0$ for every $n$-cube $a\colon \mathbb{I}^n \to U$ whose image is not contained in $A$,

- the cochain $d^n\xi_U$ is similarly extended to the cochain $(d^n\xi_U)_X$, which is a cocycle of $X$.

The restriction of $\lambda = (d^n\xi_U)_X$ to $V$ is indeed zero: for any cube $a\colon \mathbb{I}^{n+1} \to X$ with image contained in $V$:

- if $\mathrm{Im}\,(a) \subset A$, then

$$\lambda(a) = (d^n\xi_U)(a) = \xi_U(\partial^{n+1}a) = \xi(\partial^{n+1}a) = d^n\xi(a) = 0,$$

- otherwise, $\mathrm{Im}\,(a)$ is not contained in $U$, and $\lambda(a) = 0$. □

## 5.6 *Changing the coefficient group

This brief section offers a glimpse on the variation of the coefficient group. It will not be used elsewhere.

### 5.6.1 *Extending tensor products, II*

As already remarked in 5.3.1, the tensor product of chain complexes and abelian groups is a functor in two variables

$$\begin{gathered} -\otimes -\colon \mathrm{Ch}_+\mathsf{Ab} \times \mathsf{Ab} \longrightarrow \mathrm{Ch}_+\mathsf{Ab}, \\ A \otimes G = ((A_n \otimes G), (\partial_n \otimes G)), \\ (f \otimes h)_n = f_n \otimes h\colon A_n \otimes G \to B_n \otimes G', \end{gathered} \tag{5.85}$$

where $f\colon A \to B$ is a morphism of chain complexes and $h\colon G \to G'$ a homomorphism of abelian groups. In fact

$$(\partial_n \otimes G')(f_n \otimes h) = \partial_n f_n \otimes h = f_n\partial_{n-1} \otimes h = (f_{n-1} \otimes h)(\partial_n \otimes G).$$

Here we are interested in keeping the chain complex $A$ fixed, and letting the coefficient group vary, giving a functor

$$\begin{gathered} A\otimes -\colon \mathsf{Ab}\to \mathrm{Ch}_+\mathsf{Ab}, \qquad G\mapsto A\otimes G, \qquad h\mapsto h_\sharp = A\otimes h, \\ H_n(A\otimes -)\colon \mathsf{Ab}\to\mathsf{Ab}, \qquad G\mapsto H_n(A\otimes G), \\ (h\colon G\to G')\mapsto h_{*n} = (A\otimes h)_{*n}\colon H_n(A\otimes G)\to H_n(A\otimes G'). \end{gathered} \tag{5.86}$$

### 5.6.2 The Bockstein operator in homology

Suppose that $A$ is a chain complex of *free* abelian groups. A short exact sequence of abelian groups

$$(m,p)\colon G'\rightarrowtail G\twoheadrightarrow G'' \tag{5.87}$$

gives raise to a sequence of chain complexes

$$A\otimes G' \xrightarrow{A\otimes m} A\otimes G \xrightarrow{A\otimes p} A\otimes G'' \tag{5.88}$$

that is still short exact, since each functor $A_n\otimes -\colon \mathsf{Ab}\to\mathsf{Ab}$ is exact (by Exercise 5.2.7(a)).

We have thus a long exact homology sequence

$$\begin{gathered} \ldots\ H_n(A\otimes G') \xrightarrow{m_{*n}} H_n(A\otimes G) \xrightarrow{p_{*n}} H_n(A\otimes G'') \xrightarrow{\beta_n} \\ H_{n-1}(A\otimes G') \to \ \ldots \end{gathered} \tag{5.89}$$

natural for morphisms $f\colon A\to B$ of chain complexes.

The connecting homomorphism $\beta_n\colon H_n(A\otimes G'')\to H_{n-1}(A\otimes G')$ is called the *Bockstein operator*, for the short exact sequence (5.87) and the chain complex $A$ (with free components).

In particular we can apply this to the chain complex $C_+(X,A)$ of relative chains of a pair of topological spaces, getting an exact homology sequence

$$\begin{gathered} \ldots\ H_n(X,A;G') \xrightarrow{m_{*n}} H_n(X,A;G) \xrightarrow{p_{*n}} H_n(X,A;G'') \xrightarrow{\beta_n} \\ H_{n-1}(X,A;G') \to \ \ldots \end{gathered} \tag{5.90}$$

natural for relative maps $f\colon (X,A)\to(Y,B)$ in $\mathsf{Top}_2$. The connecting homomorphism $\beta_n$ is called the (homology) *Bockstein operator*, for the short exact sequence (5.87) and the relative pair $(X,A)$.

### *5.6.3 Extending the Hom functor, II*

Similarly, the Hom functor of chain complexes and abelian groups (in 5.5.1) can be viewed as a functor in two variables, contravariant in the first

$$\begin{gathered} \mathrm{Hom}(-,-)\colon \mathrm{Ch}_+\mathsf{Ab}^{\mathrm{op}} \times \mathsf{Ab} \longrightarrow \mathrm{Ch}^+\mathsf{Ab}, \\ (\mathrm{Hom}(A,G))^n = \mathrm{Hom}(A_n,G), \\ d^n = \mathrm{Hom}(\partial_{n+1},G)\colon \mathrm{Hom}(A_n,G) \to \mathrm{Hom}(A_{n+1},G), \\ (\mathrm{Hom}(f,h))^n = \mathrm{Hom}(f_n,h)\colon \mathrm{Hom}(B_n,G) \to \mathrm{Hom}(A_n,G'), \end{gathered} \tag{5.91}$$

where $f\colon A \to B$ is a morphism of chain complexes and $h\colon G \to G'$ a homomorphism of abelian groups.

Keeping the chain complex $A$ fixed, and letting the coefficient group vary, we have covariant functors

$$\begin{gathered} \mathrm{Hom}(A,-)\colon \mathsf{Ab} \to \mathrm{Ch}_+\mathsf{Ab}, \qquad G \mapsto \mathrm{Hom}(A,G), \\ H^n(\mathrm{Hom}(A,-))\colon \mathsf{Ab} \to \mathsf{Ab}, \qquad G \mapsto H^n(\mathrm{Hom}(A,G)), \\ (h\colon G \to G') \mapsto h^{*n}\colon H^n(\mathrm{Hom}(A,G)) \to H^n(\mathrm{Hom}(A,G')). \end{gathered} \tag{5.92}$$

### *5.6.4 The Bockstein operator in cohomology*

Working as in 5.6.2, let $A$ be a chain complex of free abelian groups and $(m,p)\colon G' \rightarrowtail G \twoheadrightarrow G''$ a short exact sequence in $\mathsf{Ab}$.

The exact functor $\mathrm{Hom}(A,-)$ gives a short exact sequence in $\mathrm{Ch}^+\mathsf{Ab}$

$$\mathrm{Hom}(A,G') \xrightarrow{m^\sharp} \mathrm{Hom}(A,G) \xrightarrow{p^\sharp} \mathrm{Hom}(A,G'') \tag{5.93}$$

and a long exact homology sequence

$$\begin{gathered} \dots\ H^n\mathrm{Hom}(A,G') \xrightarrow{m^{*n}} H^n\mathrm{Hom}(A,G) \xrightarrow{p^{*n}} H^n\mathrm{Hom}(A,G'') \xrightarrow{\beta^n} \\ H^{n+1}\mathrm{Hom}(A,G') \to \dots \end{gathered}$$

contravariantly natural for morphisms $f\colon A \to B$ of chain complexes.

Applying this to the chain complex $C_+(X,A)$ of a relative pair, we get a long exact cohomology sequence

$$\begin{gathered} \dots\ H^n(X,A;G') \xrightarrow{m^{*n}} H^n(X,A;G) \xrightarrow{p^{*n}} H^n(X,A;G'') \xrightarrow{\beta^n} \\ H^{n+1}(X,A;G') \to \dots \end{gathered} \tag{5.94}$$

contravariantly natural for relative maps $f\colon (X,A) \to (Y,B)$ in $\mathsf{Top}_2$; $\beta^n$ is the (cohomology) *Bockstein operator* for the short exact sequence $(m,p)$ of coefficient groups and the relative pair $(X,A)$.

### 5.6.5 Remarks

The usual short exact sequence $\mathbb{Z} \rightarrowtail \mathbb{Z} \twoheadrightarrow \mathbb{Z}_n$ can be used to link homology and cohomology groups with coefficients in $\mathbb{Z}_n$ and the ordinary homology groups with integral coefficients – although this is more effectively done with the Universal Coefficient Theorems of Sections 6.2 and 6.4.

*An exercise.* (a) One can find again, in this way, the results of Exercises 5.4.2(c), (e).

# 6
# Universal coefficients and products

We show now that the singular theory with integral coefficients determines the singular homology *groups* and singular cohomology *groups* for any coefficient group $G$. (We already now that this fails on maps, from Section 5.4.)

To prove this we have to introduce new tools of Homological Algebra. We begin with a brief exposition of derived functors in categories of modules; this gives the Tor and Ext functors, derived – respectively – of the tensor product and Hom-functor. We are then ready for the Universal Coefficient Theorem, for singular homology and cohomology.

We end by sketching higher results, on the multiplicative structure of singular cohomology and the homology of a product of spaces. The latter is studied by the method of 'acyclic models', which also yields the equivalence of the cubical and simplicial forms of singular homology.

This chapter is less elementary than the previous ones. A reader arrived at this point should be able to follow these developments.

## 6.1 Derived functors

Derived functors are a powerful topic of Homological Algebra, with diverse applications in Algebraic Topology, Algebraic Geometry, Commutative Algebra, and other fields.

We work in categories of modules. In this book we shall only use two particular cases, the torsion product (a derived functor of the tensor product) and the Ext functor (a derived functor of Hom), for abelian groups. Yet, the pattern is better perceived in the general construction, rather than following the shortcuts of particular cases.

*Literature.* The classical text on derived functors (and satellites) for categories of modules is Cartan–Eilenberg [CE].

Its extension to abelian categories is exposed in Grothendieck's article [Gt] and Mac Lane's book [M1], Chapter XII.

### *6.1.1 Introduction*

We have a right exact functor (which is additive, by 5.2.2(c))

$$F\colon R\,\mathsf{Mod} \to S\,\mathsf{Mod}, \tag{6.1}$$

and we want to construct the sequence of its left derived functors

$$L_nF = F_n\colon R\,\mathsf{Mod} \to S\,\mathsf{Mod} \qquad (n \geqslant 0), \tag{6.2}$$

starting with $F_0 = F$. This sequence is meant to cope with the failure of exactness of $F$, in the sense that every short exact sequence of $R$-modules

$$(m, p)\colon A \rightarrowtail B \twoheadrightarrow C \tag{6.3}$$

will have an associated exact sequence of left derived functors

$$\begin{array}{ccccc} & & \cdots & \longrightarrow & F_2(C) \\ & \overset{\partial_2}{\swarrow} & & & \\ F_1(A) & \xrightarrow{F_1 m} & F_1(B) & \xrightarrow{F_1 p} & F_1(C) \\ & \overset{\partial_1}{\swarrow} & & & \\ F(A) & \xrightarrow{Fm} & F(B) & \xrightarrow{Fp} & F(C) \longrightarrow 0. \end{array} \tag{6.4}$$

The *connecting morphisms* $\partial_n\colon F_n(C) \to F_{n-1}(A)$ form a natural transformation

$$\partial_n\colon F_nP''' \to F_{n-1}P'\colon \mathrm{Sh}R\,\mathsf{Mod} \to S\,\mathsf{Mod}, \tag{6.5}$$

where $P'''$ and $P'$ are projections $\mathrm{Sh}R\,\mathsf{Mod} \to R\,\mathsf{Mod}$ of the category of short exact sequences of $R$-modules, as in (2.71).

All this forms an *exact connected sequence* of functors $((F_n), (\partial_n))$. (The sequence $((H_n), (D_n))$ of homology functors of chain complexes is also such a sequence, by Theorem 2.3.3.

### *6.1.2 Projective resolutions*

We start from a *projective resolution* of an $R$-module $A$. This is a chain complex $A_+ = ((A_n), (\partial_n))$ equipped with an augmentation $\varepsilon\colon A_0 \to A$, such that:

(i) all components $A_n$ are projective $R$-modules (see 5.2.6(b)),

(ii) the following sequence is exact

$$\ldots\ A_2 \xrightarrow{\partial_2} A_1 \xrightarrow{\partial_1} A_0 \xrightarrow{\varepsilon} A \to 0. \tag{6.6}$$

The resolution will be written in the form $\varepsilon\colon A_+ \to A$, as a morphism of chain complexes with values in the chain complex $A = J(A)$, where $A$ has degree zero and the rest is trivial (see (2.18)).

We prove below that every module has projective resolutions, determined up to homotopy equivalence of chain complexes.

*Exercises and complements.* These exercises are the base of the theory of derived functors and a significant piece of Homological Algebra. Solutions are given below, but the reader is invited to solve them as independently as possible.

(a) A module $A$ has a *standard free resolution*, with free components. It is constructed using the fact that $A$ is a quotient of the free module $P(A) = F(|A|)$ generated by its underlying set, via the canonical epimorphism $\varepsilon\colon P(A) \twoheadrightarrow A$ that sends any generator to the same element of $A$.

(b) For an $R$-homomorphism $f\colon A \to B$ and two projective resolutions $\varepsilon\colon A_+ \to A$ and $\eta\colon B_+ \to B$, there is a chain morphism $f_+\colon A_+ \to B_+$ that *lifts* $f$, forming a commutative diagram with exact rows

$$\begin{array}{ccccccccc}
\dots A_2 & \xrightarrow{\partial_2} & A_1 & \xrightarrow{\partial_1} & A_0 & \xrightarrow{\varepsilon} & A & \longrightarrow & 0 \\
\downarrow{\scriptstyle f_2} & & \downarrow{\scriptstyle f_1} & & \downarrow{\scriptstyle f_0} & & \downarrow{\scriptstyle f} & & \\
\dots B_2 & \xrightarrow[\partial_2]{} & B_1 & \xrightarrow[\partial_1]{} & B_0 & \xrightarrow[\eta]{} & B & \longrightarrow & 0
\end{array} \qquad (6.7)$$

*Hint:* prove the existence of $f_0\colon A_0 \to B_0$ and proceed upwards, using the fact that $\varepsilon\colon A_+ \to A$ satisfies (i) and $\eta\colon B_+ \to B$ satisfies (ii).

(c) If the chain morphisms $f_+, g_+\colon A_+ \to B_+$ lift the same homomorphism $f\colon A \to B$ to projective resolutions of $A$ and $B$, there exists a chain homotopy $\varphi\colon f_+ \simeq g_+$. *Hint:* by additivity, it is sufficient to work on a chain morphism $h_+\colon A_+ \to B_+$ that lifts the zero morphism $0\colon A \to B$.

(d) If $\varepsilon\colon A_+ \to A$ and $\eta\colon B_+ \to A$ are projective resolutions of the same module, the chain complexes $A_+$ and $B_+$ are homotopy equivalent. *Hint:* use two liftings $f_+\colon A_+ \to B_+$ and $g_+\colon B_+ \to A_+$ of $\mathrm{id}\, A$.

(e) For a composite $h = gf\colon A \to B \to C$, any three liftings $f_+\colon A_+ \to B_+$, $g_+\colon B_+ \to C_+$ and $h_+\colon A_+ \to C_+$ to projective resolutions have a chain homotopy $h_+ \simeq g_+ f_+$.

Similarly, any commutative diagram in $R\,\mathsf{Mod}$ can be lifted to a *homotopically commutative* diagram in $\mathrm{Ch}_+R\,\mathsf{Mod}$: i.e. commutative up to chain homotopy – and properly commutative in the homotopy category $\mathrm{ho}(\mathrm{Ch}_+R\,\mathsf{Mod})$.

*Solutions.* (a) We form the standard free resolution of $A$ letting:

- $A_0 = P(A)$, with the canonical epimorphism $\varepsilon \colon A_0 \twoheadrightarrow A$,

- $A_1 = P(\operatorname{Ker}\varepsilon)$, with $\partial_1 = (A_1 \twoheadrightarrow \operatorname{Ker}\varepsilon \rightarrowtail A_0)$ and $\operatorname{Im}\partial_1 = \operatorname{Ker}\varepsilon$,

- $A_n = P(\operatorname{Ker}\partial_{n-1})$, with $\partial_n = (A_n \twoheadrightarrow \operatorname{Ker}\partial_{n-1} \rightarrowtail A_{n-1})$ and $\operatorname{Im}\partial_n = \operatorname{Ker}\partial_{n-1}$ (for $n \geqslant 2$).

In a particular case, there are often more 'economical' resolutions.

(b) It is an easy consequence of the projective property of all $A_n$, combined with the exactness of the lower sequence.

In diagram (6.7), the homomorphism $f\varepsilon \colon A_0 \to B$ starts from a projective module and can be lifted to $f_0 \colon A_0 \to B_0$ such that $\eta f_0 = f\varepsilon$

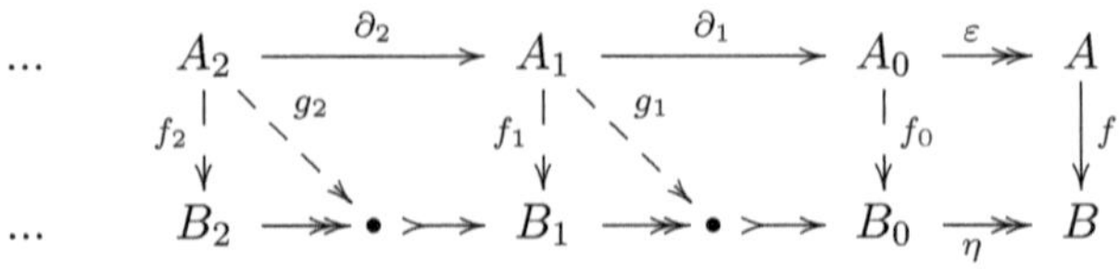


(6.8)

Now, after factorising $\partial_1 \colon B_1 \to B_0$ through $\operatorname{Im}\partial_1 = \operatorname{Ker}\eta$, the composite $f_0\partial_1 \colon A_1 \to B_0$ factorises through $\operatorname{Ker}\eta$, because $\eta f_0 \partial_1 = f\varepsilon\partial_1 = 0$.

The homomorphism $g_1$ obtained in this way can be lifted to $f_1 \colon A_1 \to B_1$, making a commutative diagram. We go on in the same way.

(c) The difference $h_+ = f_+ - g_+ \colon A_+ \to B_+$ lifts $0 \colon A \to B$; we prove that it is homotopic to $0 \colon A_+ \to B_+$, working as previously

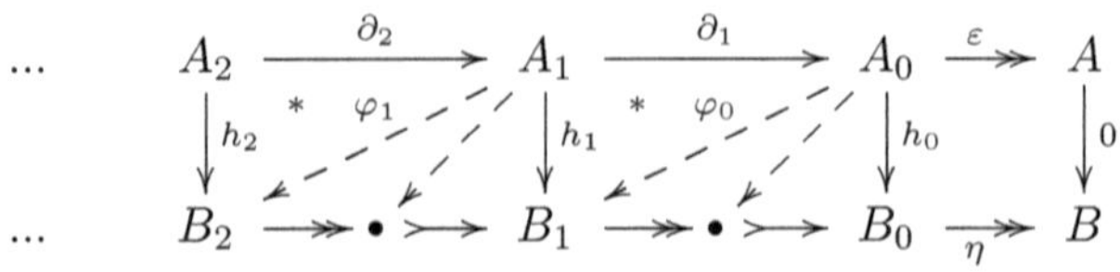


(6.9)

The homomorphism $h_0 \colon A_0 \to B_0$ gives $\eta h_0 = 0$, and can be lifted to $\operatorname{Im}\partial_1 = \operatorname{Ker}\eta$, and then to $\varphi_0 \colon A_0 \to B_1$ (because $A_0$ is projective), giving $\partial_1\varphi_0 = h_0$. The *-marked triangle need not commute, but the difference $h_1 - \varphi_0\partial_1 \colon A_1 \to B_1$ is annihilated by $\partial_1$ (because $\partial_1 h_1 - \partial_1\varphi_0\partial_1 = \partial_1 h_1 - h_0\partial_1 = 0$) and can similarly be lifted, first to $\operatorname{Im}\partial_2 = \operatorname{Ker}\partial_1$ and then to $\varphi_1 \colon A_1 \to B_2$, giving $\partial_2\varphi_1 = h_1 - \varphi_0\partial_1$. And so on.

(d) Take two liftings $f_+ \colon A_+ \to B_+$ and $g_+ \colon B_+ \to A_+$ of $\operatorname{id} A$. Then $g_+f_+ \colon A_+ \to A_+$ and $\operatorname{id} A_+$ lift $\operatorname{id} A$, and are homotopic. Similarly, $f_+g_+ \simeq \operatorname{id} B_+$.

(e) Both morphisms $h_+$ and $g_+f_+ \colon A_+ \to C_+$ lift $h = gf$.

### 6.1.3 Left derived functors

We are now ready to define the *left derived functors* $L_nF = F_n \colon R\,\mathsf{Mod} \to S\,\mathsf{Mod}$ of the right exact functor $F$ (for $n \geqslant 0$).

For every $R$-module $A$ we choose a projective resolution $\varepsilon \colon A_+ \to A$. Using the obvious extension $F \colon \mathrm{Ch}_+R\,\mathsf{Mod} \to \mathrm{Ch}_+S\,\mathsf{Mod}$ of the additive functor $F$ to chain complexes (see 5.2.3(a)), we define

$$F_n(A) = H_n(FA_+). \tag{6.10}$$

Every $R$-homomorphism $f \colon A \to B$ has a lifting $f_+ \colon A_+ \to B_+$ with respect to the chosen projective resolutions $\varepsilon \colon A_+ \to A$ and $\eta \colon B_+ \to B$; all of them give the same homomorphism (by Exercises 6.1.2(b), (c))

$$F_n(f) = H_n(Ff_+) \colon F_n(A) \to F_n(B). \tag{6.11}$$

*Exercises and complements.* (a) The mappings $F_n$ defined above are additive functors.

(b) The functor $F_0(A) = H_0(FA_+)$ is canonically isomorphic to the original functor $F$, and we shall identify them.

*More generally, one can define as above the left derived functors of any additive functor; then the sequence (6.4) would end with the derived $F_0$, instead of the original $F$.*

(c) Another choice of projective resolutions gives a sequence of functors $F'_n$ linked to the previous ones by natural isomorphisms $F_n \to F'_n$.

(d) Without using the axiom of choice, one can define the functors $F_n$ using the standard free resolutions of 6.1.2(a).

### 6.1.4 Theorem (Lifting short exact sequences)

*A short exact sequence of $R$-modules $(m,p) \colon A \rightarrowtail B \twoheadrightarrow C$ has a* projective resolution. *This means a componentwise split exact sequence of chain complexes $(m_+, p_+) \colon A_+ \rightarrowtail B_+ \twoheadrightarrow C_+$ with liftings of $m$ and $p$ to projective resolutions of $A, B$ and $C$.*

*One can start with two arbitrary projective resolutions $\varepsilon \colon A_+ \twoheadrightarrow A$ and $\vartheta_+ \colon C_+ \twoheadrightarrow C$, and complete the rest*

$$\begin{array}{ccccccc}
\ldots A_2 & \xrightarrow{\partial_2} & A_1 & \xrightarrow{\partial_1} & A_0 & \xrightarrow{\varepsilon} & A \\
\downarrow m_2 & & \downarrow m_1 & & \downarrow m_0 & & \downarrow m \\
\ldots B_2 & \dashrightarrow & B_1 & \dashrightarrow & B_0 & \dashrightarrow & B \\
\downarrow p_2 & & \downarrow p_1 & & \downarrow p_0 & & \downarrow p \\
\ldots C_2 & \xrightarrow[\partial_2]{} & C_1 & \xrightarrow[\partial_1]{} & C_0 & \xrightarrow[\vartheta]{} & C
\end{array} \tag{6.12}$$

*Proof* (a) For every $n \geqslant 0$, we take $B_n = A_n \oplus C_n$ (a projective module, by Exercise 5.2.7(b)). We let $m_n \colon A_n \to B_n$ be the first injection and $p_n \colon B_n \to C_n$ the second projection.

We have thus a commutative solid diagram

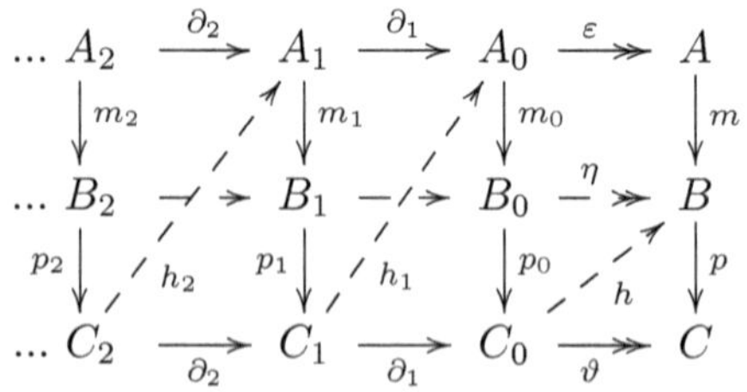


(6.13)

where the upper and lower rows are projective resolutions. All the columns are split exact sequences, and we still have to define the homomorphisms of the middle row.

(b) We define the homomorphism

$$\eta \colon A_0 \oplus C_0 \to B, \qquad \eta(a, c) = m\varepsilon(a) + h(c),$$

for some homomorphism $h \colon C_0 \to B$ such that $ph = \vartheta$ ($C_0$ is projective). This makes the two rightmost squares commute, in diagram (6.13):

$$\eta m_0(a) = \eta(a, 0) = m\varepsilon(a), \quad p\eta(a, c) = pm\varepsilon(a) + ph(c) = \vartheta p_0(a, c).$$

(d) Now we define

$$\partial_1 \colon A_1 \oplus C_1 \to A_0 \oplus C_0, \quad \partial_1(a, c) = (\partial_1(a) + h_1(c), \, \partial_1(c)),$$

for some $h_1 \colon C_1 \to A_0$ such that $m\varepsilon h_1 = -h\partial_1 \colon C_1 \to B$.

(Indeed $-h\partial_1 \colon C_1 \to B$ can first be lifted to $A = \operatorname{Ker} p$, by exactness of the rightmost column, and then to $A_0$, because $\varepsilon$ is epi and $C_1$ is projective.)

This makes two more commutative squares:

$$\partial_1 m_1(a) = (\partial_1(a), 0) = m_0\partial_1(a), \quad p_0\partial_1(a, c) = \partial_1(c) = \partial_1 p_1(a, c).$$

Moreover $\operatorname{Im} \partial_1 = \operatorname{Ker} \eta$, in the middle row. First

$$\eta\partial_1(a, c) = \eta(\partial_1(a) + h_1(c), \partial_1(c)) = m\varepsilon\partial_1(a) + m\varepsilon h_1(c) + h\partial_1(c) = 0.$$

Secondly, take an element $(a, c) \in \operatorname{Ker} \eta$: thus $c \in \operatorname{Ker} \vartheta \colon C_0 \to C$ and there is some $c' \in C_1$ such that $\partial_1(c') = c$. Now $m\varepsilon(a) = -h(c) = -h\partial_1(c') = m\varepsilon h_1(c')$, and there is some $a' \in A_1$ such that $\partial_1(a') = a - h_1(c')$. Finally we have found a pair $(a', c') \in A_1 \oplus C_1$ such that:

$$\partial_1(a', c') = (\partial_1(a') + h_1(c'), \, \partial_1(c')) = (a, c).$$

(e) We let

$$\partial_2 \colon A_2 \oplus C_2 \to A_1 \oplus C_1, \qquad \partial_2(a,c) = (\partial_2(a) + h_2(c),\, \partial_2(c)),$$

for some $h_2 \colon C_2 \to A_1$ such that

$$m_0\partial_1 h_2 = -h_1\partial_2 \colon C_2 \to A_0.$$

One goes on in the same way. □

### 6.1.5 The sequence of derived functors

Let $(m,p)\colon A \rightarrowtail B \twoheadrightarrow C$ be a short exact sequence of $R$-modules, with a (componentwise split) projective resolution $(m_+, p_+)\colon A_+ \rightarrowtail B_+ \twoheadrightarrow C_+$.

The right exact functor $F\colon R\,\mathsf{Mod} \to S\,\mathsf{Mod}$ gives a short exact sequence of chain complexes, which splits componentwise

$$(Fm_+, Fp_+)\colon FA_+ \rightarrowtail FB_+ \twoheadrightarrow FC_+. \tag{6.14}$$

We have thus proved our goal: the short exact sequence $(m,p)$ gives raise to the long exact homology sequence (6.4); this is the *exact sequence of left derived functors* of $F$, with connecting morphisms

$$\partial_n \colon F_n(C) \to F_{n-1}(A) \qquad (n \geqslant 1). \tag{6.15}$$

The whole process is natural: a morphism $(u,v,w)\colon (m,p) \to (m',p')$ of short exact sequences (as in (2.70)) has a lifting

$$(u_+, v_+, w_+)\colon (m_+, p_+) \nrightarrow (m'_+, p'_+)$$

which is a *weak morphism* of short exact sequences of chain complexes: the two squares below commute up to chain homotopy (by 6.1.2(e))

$$\begin{array}{ccccc}
A_+ & \xrightarrow{m_+} & B_+ & \xrightarrow{p_+} & C_+ \\
{\scriptstyle u_+}\downarrow & \simeq & \downarrow{\scriptstyle v_+} & \simeq & \downarrow{\scriptstyle w_+} \\
A'_+ & \xrightarrow[m'_+]{} & B'_+ & \xrightarrow[p'_+]{} & C'_+
\end{array} \tag{6.16}$$

Applying the functor $F$ and chain homology we get a morphism of exact sequences of $S$-modules, with components

$$\begin{array}{ccccccc}
\ldots\ F_n(A) & \xrightarrow{F_n m} & F_n(B) & \xrightarrow{F_n p} & F_n(C) & \xrightarrow{\partial_n} & F_{n-1}(A)\ \ldots \\
{\scriptstyle F_n(u)}\downarrow & & {\scriptstyle F_n(v)}\downarrow & & {\scriptstyle F_n(w)}\downarrow & & {\scriptstyle F_{n-1}(u)}\downarrow \\
\ldots\ F_n(A') & \xrightarrow[F_n m']{} & F_n(B') & \xrightarrow[F_n p']{} & F_n(C') & \xrightarrow[\partial_n]{} & F_{n-1}(A')\ \ldots
\end{array} \tag{6.17}$$

### *6.1.6 Exercises and complements*

(a) A right exact functor $F\colon R\,\mathsf{Mod} \to S\,\mathsf{Mod}$ is exact if and only if all its left derived functors (in positive degree) are trivial ($F_n = 0$, for $n > 0$), if and only if $F_1 = 0$.

(b) For a right exact functor $F\colon R\,\mathsf{Mod} \to S\,\mathsf{Mod}$, the derived functors $F_n$ with $n > 0$ vanish on all projective $R$-modules. In particular, every projective module is flat.

*(c) The theory of left satellites, developed in [CE], characterises the left derived functor $L_nF$ as the $n$-th left satellite $S_nF$. It is important to note that $S_{m+n}F = S_m(S_nF)$.

### *6.1.7 The case of abelian groups*

Letting $R = \mathbb{Z}$, an abelian group $A$ has always a *short* free resolution

$$0 \to A_1 \xrightarrow{\partial} A_0 \xrightarrow{\varepsilon} A \to 0 \tag{6.18}$$

where $\varepsilon\colon A_0 \to A$ is the canonical epimorphism defined on the free abelian group $\mathbb{Z}|A|$ and $\partial\colon A_1 \to A$ is the kernel of $\varepsilon$ (recalling that a subgroup of a free abelian group is free).

Therefore, for a right exact functor $F\colon \mathsf{Ab} \to S\,\mathsf{Mod}$, all left derived functors vanish in degree $n \geqslant 2$, and there is a single left derived functor $LF = F_1\colon \mathsf{Ab} \to S\,\mathsf{Mod}$ of interest:

$$F_1(A) = H_1(A_+) = \mathrm{Ker}\,(F\partial\colon FA_1 \to FA_0). \tag{6.19}$$

A short exact sequence $(m,p)\colon A \rightarrowtail B \twoheadrightarrow C$ gives an exact sequence

$$0 \to F_1A \xrightarrow{F_1m} F_1B \xrightarrow{F_1p} F_1C \xrightarrow{\partial} FA \xrightarrow{Fm} FB \xrightarrow{Fp} FC \to 0 \tag{6.20}$$

showing that $F_1$ is left exact.

In the category $\mathsf{Ab}$ one can develop the theory of left derived functors using short projective resolutions, which vanish in degree $\geqslant 2$. Yet, this reduced version can mask the nature and power of the process, resulting less clear than the complete picture.

### *6.1.8 Right derived functors*

Loosely speaking, we get the dual notion of right derived functors of a left exact functor, reversing all arrows and replacing projective objects with the injective ones.

*This duality has a precise sense if we work in the selfdual context of abelian categories.*

Concretely, we are given a right exact additive functor

$$F\colon R\,\mathsf{Mod} \to S\,\mathsf{Mod}, \tag{6.21}$$

and we want to construct the sequence of its right derived functors

$$R^nF = F^n\colon R\,\mathsf{Mod} \to S\,\mathsf{Mod} \qquad (n \geqslant 0), \tag{6.22}$$

starting with $F^0 = F$.

Again, this sequence corrects the failure of exactness of $F$, in the sense that every short exact sequence of $R$-modules $(m, p)\colon A \rightarrowtail B \twoheadrightarrow C$ has an associated *exact sequence of right derived functors*

$$\begin{array}{ccccccc} 0 & \to & F(A) & \xrightarrow{Fm} & F(B) & \xrightarrow{Fp} & F(C) \\ & & & & \xleftarrow{d^0} & & \\ & & F^1(A) & \xrightarrow{F^1m} & F^1(B) & \xrightarrow{F^1p} & F^1(C) \\ & & & & \xleftarrow{d^1} & & \\ & & F^2(A) & \longrightarrow & \ldots & & \end{array} \tag{6.23}$$

The *connecting morphisms* $d^n\colon F^n(C) \to F^{n+1}(A)$ are the components of a natural transformation

$$d^n\colon F^nP''' \to F^{n+1}P'\colon \mathrm{Sh}R\,\mathsf{Mod} \to S\,\mathsf{Mod}, \qquad (n \geqslant 0), \tag{6.24}$$

forming an exact connected sequence of functors $((F^n), (d^n))$ (with a connecting morphism of degree 1).

The theory is developed in a similar way to what we have done above, using for every $R$-module $A$ an *injective resolution*. This is a cochain complex $A^+ = ((A^n), (d^n))$ equipped with an augmentation $\varepsilon\colon A \rightarrowtail A^0$, such that:

(i) all components $A^n$ are injective $R$-modules (defined in 5.2.6(c)),

(ii) the following sequence is exact

$$0 \to A \xrightarrow{\varepsilon} A^0 \xrightarrow{d^0} A^1 \xrightarrow{d^1} A^2 \to \ldots \tag{6.25}$$

The resolution can be written in the form $\varepsilon\colon A \rightarrowtail A^+$, as a monomorphism of cochain complexes.

Showing that any $R$-module has an injective resolution is less easy than in the projective case. The basic step, every $R$-module is a submodule of an injective one, is proved in [M1], Theorem III.7.4.

The right derived functor $F^n$ is now defined by the cohomology of the cochain complex $FA^+$:

$$F^n(A) = H^n(FA^+). \tag{6.26}$$

*Exercises and complements.* (a) A left exact functor $F\colon R\,\mathsf{Mod} \to S\,\mathsf{Mod}$

is exact if and only if all its right derived functors (in positive degree) are trivial ($F^n = 0$, for $n > 0$), if and only if $F^1 = 0$. *Hint:* the solution is similar to that of Exercise 6.1.6(a).

*(b) The theory of right satellites, developed in [CE], characterises the right derived functor $R^n F$ as the $n$-th right satellite $S^n F$. Again, $S^{m+n}F = S^m(S^n F)$.

(c) (*Deriving contravariant functors*) Let $F\colon R\,\mathsf{Mod} \dashrightarrow S\,\mathsf{Mod}$ be a contravariant left exact functor (see 5.2.2(e)). Its *right derived functors* $F^n$ are constructed as in 6.1.8, starting from a projective resolution $\varepsilon\colon A_+ \to A$ of the $R$-module $A$ and taking the cohomology of the *cochain* complex $FA_+$.

Similarly, if $F$ is right exact, one can obtain its *left derived functors* as in 6.1.3, using an injective resolution $\varepsilon\colon A \to A^+$ and taking the homology of the chain complex $FA^+$.

*Finally, we have considered four 'different' situations, which can be unified in a single one in the theory of abelian categories [Gt, M1].*

### 6.1.9 *The case of abelian groups

We have recalled, in 5.2.9(d), that an abelian group is injective if and only if it is divisible. But a quotient of a divisible abelian group is plainly divisible. It follows that in $\mathsf{Ab}$ every object $A$ has a *short* injective resolution $\varepsilon\colon A \rightarrowtail A^+$ with $A^n = 0$ for $n \geqslant 2$

$$0 \to A \xrightarrow{\varepsilon} A^0 \xrightarrow{d} A^1 \to 0. \tag{6.27}$$

For a left exact functor $F\colon \mathsf{Ab} \to S\,\mathsf{Mod}$, all right derived functors vanish in degree $\geqslant 2$, and we only consider the right derived functor $RF = F^1\colon \mathsf{Ab} \to S\,\mathsf{Mod}$.

A short exact sequence $(m,p)\colon A \rightarrowtail B \twoheadrightarrow C$ gives an exact sequence

$$0 \to FA \xrightarrow{Fm} FB \xrightarrow{Fp} FC \xrightarrow{d} F^1A \xrightarrow{F^1m} F^1B \xrightarrow{F^1p} F^1C \to 0 \tag{6.28}$$

showing that $F^1$ is right exact.

## 6.2 Torsion product and universal coefficients

It is now convenient to work in $\mathsf{Ab}$. We introduce the torsion product $\mathrm{Tor}(A,B)$ as the left derived functor of $A \otimes B$, working equivalently on the variable $A$ or $B$. A concrete construction of $\mathrm{Tor}(A,B)$ will be briefly presented in 6.2.8.

$G$ is always an abelian group, used as a coefficient group for homology.

**Literature.* For $R$-modules, the torsion product $\mathrm{Tor}_n^R(A, B)$ is the $n$-th left derived functor of $A \otimes B$, on one of the variables. The theory, much more complex than in $\mathsf{Ab}$, is fully developed in [CE].

### 6.2.1 The torsion product

We begin by fixing the second variable. The functor $- \otimes B \colon \mathsf{Ab} \to \mathsf{Ab}$ is right exact. Its left derived functor, *the* Tor *product*

$$\mathrm{Tor}(-, B) \colon \mathsf{Ab} \to \mathsf{Ab}, \tag{6.29}$$

is left exact, and can be computed on a free resolution $(\partial, \varepsilon) \colon A_1 \rightarrowtail A_0 \twoheadrightarrow A$

$$\mathrm{Tor}(A, B) = H_1(A_+ \otimes B) = \mathrm{Ker}\,(\partial \otimes B \colon A_1 \otimes B \to A_0 \otimes B). \tag{6.30}$$

A short exact sequence $A' \rightarrowtail A \twoheadrightarrow A''$ of abelian groups gives an exact sequence

$$\begin{aligned} 0 \to \mathrm{Tor}(A', B) \to \mathrm{Tor}(A, B) \to \mathrm{Tor}(A'', B)& \\ \xrightarrow{\partial} A' \otimes B \to A \otimes B \to A'' \otimes B \to 0& \end{aligned} \tag{6.31}$$

which is natural on $\mathsf{ShAb}$. The Tor product is also denoted as $A * B$.

We prove below that, if we work on the left variable, we get the same result; as a consequence, $\mathrm{Tor}(A, B) \cong \mathrm{Tor}(B, A)$.

*Exercises and complements.* (a) The abelian group $B$ is flat if and only if $\mathrm{Tor}(-, B) = 0$. This is certainly the case if $B$ is a free abelian group.

(More precisely, we shall see in 6.2.8(a) that $B$ is flat if and only if it is torsion free.)

(b) Prove that $\mathrm{Tor}(\mathbb{Z}_m, \mathbb{Z}_n) \cong \mathbb{Z}_d \cong \mathbb{Z}_m \otimes \mathbb{Z}_n$, where $d = \gcd(m, n)$.

### 6.2.2 Proposition

*Fixing the first variable, we get a functor* $\underline{\mathrm{Tor}}(A, -) \colon \mathsf{Ab} \to \mathsf{Ab}$, *left derived of* $A \otimes - \colon \mathsf{Ab} \to \mathsf{Ab}$.

*Then there is a natural isomorphism* $\mathrm{Tor}(A, B) \cong \underline{\mathrm{Tor}}(A, B)$, *for all pairs* $(A, B)$ *of abelian groups.*

*Proof* We choose two free resolutions

$$A_1 \rightarrowtail A_0 \twoheadrightarrow A, \qquad\qquad B_1 \rightarrowtail B_0 \twoheadrightarrow B.$$

This gives a commutative diagram, where

$$\begin{array}{ccccc}
0 & \longrightarrow & 0 & \longrightarrow & \mathrm{Tor}(A,B) \\
\downarrow & & \downarrow & & \downarrow \\
A_1 \otimes B_1 & \rightarrowtail & A_1 \otimes B_0 & \twoheadrightarrow & A_1 \otimes B \\
\downarrow & & \downarrow & & \downarrow \\
A_0 \otimes B_1 & \rightarrowtail & A_0 \otimes B_0 & \twoheadrightarrow & A_0 \otimes B \\
\downarrow & & \downarrow & & \downarrow \\
A \otimes B_1 & \longrightarrow & A \otimes B_0 & \twoheadrightarrow & A \otimes B
\end{array} \qquad (6.32)$$

- the two middle rows are short exact, because $A_1$ and $A_0$ are free,
- the right column is exact, by definition of $\mathrm{Tor}(A, B)$,
- the other two columns are exact, because $B_1$ and $B_0$ are free.

Applying the Snake Lemma 3.5.1 to the two middle rows we get a natural exact sequence linking the kernels and the cokernels of the three vertical arrows

$$0 \to \mathrm{Tor}(A,B) \to A \otimes B_1 \to A \otimes B_0 \to A \otimes B \to 0.$$

This shows that the connecting homomorphism $\mathrm{Tor}(A,B) \to A \otimes B_1$ is a kernel of $A \otimes B_1 \to A \otimes B_0$; but this 'is' $\underline{\mathrm{Tor}}(A, B)$, computed by a free resolution of $B$. □

### 6.2.3 Corollary

*(a) For all pairs $(A, B)$ of abelian groups, there is a natural isomorphism* $\mathrm{Tor}(A,B) \cong \mathrm{Tor}(B,A)$.

*(b) Every subgroup of a flat abelian group is flat.*

Note. Point (b) is proved here without using the characterisation of flat abelian groups as the torsion-free ones.

*Proof* (a) $\mathrm{Tor}(A, B)$ is the left derived functor of $- \otimes B$ computed in $A$. By Proposition 6.2.2, it is also the left derived functor of $A \otimes -$ computed in $B$. By the commutativity of the tensor product, $\mathrm{Tor}(A, B)$ is also the left derived functor of $- \otimes A$ computed in $B$, i.e. $\mathrm{Tor}(B, A)$, by definition.

(b) If $H$ is a subgroup of the flat abelian group $A$, the short exact sequence $H \rightarrowtail A \twoheadrightarrow A/H$ gives, for every abelian group $B$, an exact sequence $0 \to \mathrm{Tor}(H,B) \to \mathrm{Tor}(A,B) = 0$, showing that $\mathrm{Tor}(H,-) = 0$. By (a), we also have $\mathrm{Tor}(-,H) = 0$. □

### 6.2.4 A natural morphism

Let $A$ be a chain complex of abelian groups. We begin to explore the homology group $H_n(A \otimes G)$, introducing a canonical morphism

$$\alpha_n\colon H_n(A) \otimes G \to H_n(A \otimes G), \qquad \alpha_n([z] \otimes g) = [z \otimes g]. \tag{6.33}$$

This is well defined. If $z$ is an $n$-cycle in $A$, $\partial_n(z \otimes g) = \partial_n z \otimes g = 0$. If $z$ is an $n$-boundary, $z \otimes g$ is also. Moreover the expression $z \otimes g$ is linear in each variable.

We have defined a natural transformation

$$\alpha_n\colon H_n(-) \otimes - \longrightarrow H_n(- \otimes -)\colon \mathrm{Ch}_+\mathsf{Ab} \times \mathsf{Ab} \to \mathsf{Ab}. \tag{6.34}$$

*Remarks.* (a) If $G$ is a flat abelian group, then $\alpha_n\colon H_n(A) \otimes G \to H_n(A \otimes G)$ is an isomorphism.

In fact, the functor $- \otimes G\colon \mathsf{Ab} \to \mathsf{Ab}$ is exact, and 'preserves homology', as we have seen in 5.2.3(b).

(b) This becomes obvious for a free abelian group $G \cong \bigoplus_{i\in I} \mathbb{Z}$.

Then $A \otimes G \cong \bigoplus_{i\in I} A$ is a direct sum of chain complexes (see 2.1.8), and we have seen that its $n$-th homology group is canonically isomorphic to $\bigoplus_{i\in I} H_n(A) \cong H_n(A) \otimes G$.

### 6.2.5 Universal Coefficient Theorem (for chain homology)

*For every chain complex $A$ of* flat *abelian groups, and every abelian group $G$, there is a split exact sequence*

$$H_n(A) \otimes G \overset{\alpha_n}{\rightarrowtail} H_n(A \otimes G) \overset{\beta_n}{\twoheadrightarrow} \mathrm{Tor}(H_{n-1}(A), G) \tag{6.35}$$

*which is natural on morphisms $f\colon A \to B$ of chain complexes and morphisms $h\colon G \to G'$ of abelian groups.*

*Note.* The ensuing isomorphism

$$H_n(A \otimes G) \cong (H_n(A) \otimes G) \oplus (\mathrm{Tor}(H_{n-1}(A), G)) \tag{6.36}$$

is *not* natural on chain morphisms, and should be used with prudence (see 6.2.7(b)).

*Proof* It is a particular case of the Künneth Theorem 6.5.4, replacing the chain complex $B$ with the complex $J(G)$. □

### 6.2.6 Universal coefficients for singular homology

For every relative pair $(X, A)$ of spaces and every abelian group $G$, the previous theorem gives a split exact sequence

$$H_n(X,A) \otimes G \xrightarrow{\alpha_n} H_n(X,A;G) \xrightarrow{\beta_n} \mathrm{Tor}(H_{n-1}(X,A),G) \qquad (6.37)$$

which is natural on relative maps $f\colon (X,A) \to (Y,B)$ and morphisms $h\colon G \to G'$ of abelian groups.

### 6.2.7 Comments and exercises

There are given two relative maps $f, g\colon (X,A) \to (Y,B)$.

(a) If $f_{*n}\colon H_n(X,A) \to H_n(Y,B)$ is an isomorphism for $n \geqslant 0$ (with integral coefficients), the same holds with any coefficient group. More generally, if $H_n(X,A)$ and $H_n(Y,B)$ are isomorphic for $n \geqslant 0$, the same holds with coefficients in any abelian group.

(b) On the other hand, if $f_{*n} = g_{*n}\colon H_n(X,A) \to H_n(Y,B)$ for $n \geqslant 0$ (with integral coefficients), one cannot generally deduce that the same holds with every coefficient group: the isomorphism (6.36) is not natural.

For instance, we have already seen, in Exercise 5.4.2(c), that the projection $q\colon \mathbb{P}^2 \to \mathbb{S}^2$ and any constant map $\mathbb{P}^2 \to \mathbb{S}^2$ have the same (trivial) homology homomorphisms with integral coefficients, but different ones with coefficients in $\mathbb{Z}_2$. (By Theorem 6.3.2, the same holds more generally for the projection $\mathbb{P}^n \to \mathbb{S}^n$, for $n > 0$.)

(c) However, if the abelian group $G$ is flat, we have a natural isomorphism

$$\alpha_n\colon H_n(X,A) \otimes G \to H_n(X,A;G), \qquad \alpha_n([z] \otimes g) = [z \otimes g], \qquad (6.38)$$

so that the induced homomorphism $f_{*n}$ with integral coefficients determines $H_n(f;G)$ 'as' $f_{*n} \otimes G$.

In particular, it is the case with coefficients in $\mathbb{R}$, or $\mathbb{Q}$, or any field of characteristic 0.

(d) Similarly, if all the groups $H_k(X,A)$ and $H_k(Y,B)$ are flat, we have a natural isomorphism (6.38), and the same conclusion as above. In particular, if $f_{*n} = 0$ (with integral coefficients) for all $n > 0$, the map $f$ is acyclic, as defined in 5.4.1(c).

(e) Using Universal Coefficients, in 6.2.6, prove that the torus and the Klein bottle have the following homology groups with coefficients in $\mathbb{Z}_2$, and trivial higher groups:

$$\begin{aligned} H_i(\mathbb{T};\mathbb{Z}_2) &\cong H_i(\mathbb{K};\mathbb{Z}_2) \cong \mathbb{Z}_2 \qquad \text{for } i = 0, 2, \\ H_1(\mathbb{T};\mathbb{Z}_2) &\cong H_1(\mathbb{K};\mathbb{Z}_2) \cong \mathbb{Z}_2 \oplus \mathbb{Z}_2. \end{aligned} \qquad (6.39)$$

(f) One can easily extend (e) to an arbitrary coefficient group $G$. The homology groups of $\mathbb{P}^n$ modulo 2 will be dealt with in Section 6.3. Other calculations can be worked out on the homology groups determined in Chapter 3.

### 6.2.8 *Complements

We end this section with a concrete construction of the group $\mathrm{Tor}(A, B)$, based on the 'common torsion' of the abelian groups $A$ and $B$. The proof can be found in [M1], Section V.6.

The construction is similar to that of the tensor product, in 5.1.2. We let

$$\mathrm{Tor}(A, B) = T(A, B)/H(A, B), \tag{6.40}$$

where $T(A, B)$ is the free abelian group generated by the following triples

$$(a, m, b) \qquad (a \in A,\ m \in \mathbb{Z},\ b \in B;\ ma = 0 = mb), \tag{6.41}$$

and $H(A, B)$ is the subgroup generated by the elements of the following kinds

$$\begin{aligned}
&(a + a', m, b) - (a, m, b) - (a', m, b) && (\text{for } ma = ma' = mb = 0),\\
&(a, m, b + b') - (a, m, b) - (a, m, b') && (\text{for } ma = mb = mb' = 0),\\
&(a, mn, b) - (na, m, b) && (\text{for } mna = mb = 0),\\
&(a, mn, b) - (a, m, nb) && (\text{for } ma = mnb = 0).
\end{aligned}$$

From this construction one can deduce two important facts.

(a) A torsion-free abelian group is flat (and conversely, by 5.2.8(c)).

If $B$ is torsion free, this construction shows that $\mathrm{Tor}(A, B) = 0$ for every $A$. Indeed, consider a generator $(a, m, b) \in T(A, B)$. If $m \neq 0$, then $b = 0$ and $[a, m, 0] = 0$, by linearity in the last variable. Otherwise $[a, 0, b] = [0a, 0, b] = 0$.

(b) $\mathrm{Tor}(A, B)$ is always a torsion abelian group.

In fact, any generator $[a, m, b]$, with $ma = 0 = mb$, is annihilated by $m$:

$$m[a, m, b] = [ma, m, b] = [0, m, b] = 0.$$

## 6.3 Projective spaces and homology modulo 2

Homology modulo 2 of the projective spaces is important, and can be easily determined, using the Universal Coefficient Theorem. We apply these results to complete a study of the homological essentiality of maps between $\mathbb{P}^n$ and $\mathbb{S}^n$.

***6.3.1 Theorem*** (Projective spaces and homology modulo 2)
*The projective space $\mathbb{P}^n$ has the following homology groups modulo 2, and trivial higher groups*

$$H_i(\mathbb{P}^n; \mathbb{Z}_2) \cong \mathbb{Z}_2, \qquad \text{for } 0 \leqslant i \leqslant n. \tag{6.42}$$

*Proof* It follows from Universal Coefficients (see 6.2.6), using the homology groups $H_i(\mathbb{P}^n)$ with integral coefficients (see Theorem 3.3.4), and some algebraic results: $\mathbb{Z}_2 \otimes \mathbb{Z}_2 \cong \mathrm{Tor}(\mathbb{Z}_2, \mathbb{Z}_2) \cong \mathbb{Z}_2$ (in 5.2.8 and 6.2.1). □

***6.3.2 Theorem***

*The canonical embedding $u\colon \mathbb{P}^{n-1} \to \mathbb{P}^n$ gives an isomorphism*

$$u_{*i}\colon H_i(\mathbb{P}^{n-1}; \mathbb{Z}_2) \cong H_i(\mathbb{P}^n; \mathbb{Z}_2), \qquad \text{for } 0 \leqslant i < n. \tag{6.43}$$

*The projections $p_n\colon \mathbb{S}^n \to \mathbb{P}^n$ and $q_n\colon \mathbb{P}^n \to \mathbb{S}^n$ of Theorem 3.3.2 give, respectively, a trivial homomorphism and an isomorphism*

$$p_{n*} = 0\colon H_n(\mathbb{S}^n; \mathbb{Z}_2) \longrightarrow H_n(\mathbb{P}^n; \mathbb{Z}_2), \qquad \text{for } n > 0, \tag{6.44}$$

$$q_{n*}\colon H_n(\mathbb{P}^n; \mathbb{Z}_2) \cong H_n(\mathbb{S}^n; \mathbb{Z}_2), \qquad \text{for } n > 0. \tag{6.45}$$

*Proof* The argument is simplified by the fact that we already know the groups $H_i(\mathbb{P}^n; \mathbb{Z}_2)$. On maps, not everything comes out of Universal Coefficients: we shall also use the MV-sequence associated to the open cover $(U_n, V_n)$ described in 3.3.3, where $U_n \simeq \mathbb{P}^{n-1}$, $A_n \simeq \mathbb{S}^{n-1}$ and $V_n \simeq \{*\}$ is omitted *in positive degree.* To simplify the notation, $H_i$ will denote homology with coefficients in $\mathbb{Z}_2$.

(a) We prove (6.43) and (6.44). The exact MV-sequence (for $i > 0$)

$$H_i(\mathbb{S}^{n-1}) \xrightarrow{h_i} H_i(\mathbb{P}^{n-1}) \xrightarrow{k_i} H_i(\mathbb{P}^n) \xrightarrow{D_i} H_{i-1}(\mathbb{S}^{n-1})$$

proves that $k_i = u_{*i}$ is an isomorphism, for $0 < i < n-1$. For $i = 0$, this is obvious.

For $i = n-1 > 0$, we have the exact sequence

$$H_{n-1}(\mathbb{S}^{n-1}) \xrightarrow{h_{n-1}} H_{n-1}(\mathbb{P}^{n-1}) \xrightarrow{k_{n-1}} H_{n-1}(\mathbb{P}^n) \to 0$$

where we already know that all groups are isomorphic to $\mathbb{Z}_2$. Therefore the epimorphism $u_{*n-1} = k_{n-1}$ is an isomorphism, and $h_{n-1}$ is trivial. But $h_{n-1}\colon H_{n-1}(A_n) \to H_{n-1}(U_n)$ is induced by the inclusion $A_n \subset U_n$, which is homotopic to the canonical projection $p_{n-1}\colon \mathbb{S}^{n-1} \to \mathbb{P}^{n-1}$: we have proved that $p_{*n} = 0$, for $n > 0$.

(b) Finally we prove (6.45), for $n > 0$. We already know, from the proof of Theorem 3.3.4, point (e), that the open cover $(U_n, V_n)$ of $\mathbb{P}^n$ is projected by $q_n \colon \mathbb{P}^n \to \mathbb{S}^n$ to an open cover $(U'_n, V'_n)$ of $\mathbb{S}^n$, with the same homology groups as the open cover used for $\mathbb{S}^n$ in Theorem 2.4.2. Moreover, the restriction $q'_n \colon A_n \to A'_n$ of $q_n$ is a homeomorphism.

By the naturality of the MV-sequence we have a commutative diagram with exact rows

$$\begin{array}{ccccc} H_n(\mathbb{P}^{n-1}) & \longrightarrow & H_n(\mathbb{P}^n) & \longrightarrow & H_{n-1}(A_n) \\ \| & & \downarrow{\scriptstyle q_{*n}} & & \downarrow{\scriptstyle \cong} \\ 0 & \longrightarrow & H_n(\mathbb{S}^n) & \longrightarrow & H_{n-1}(A'_n) \end{array}$$

where $q_{n*} \colon H_n(\mathbb{P}^n) \to H_n(\mathbb{S}^2)$ is a monomorphism between groups isomorphic to $\mathbb{Z}_2$, hence an isomorphism. □

### 6.3.3 Complements on essential maps

We end this section collecting the results we have seen on the homological essentiality of the maps $p_n$, $q_n$, $r_n$ of 3.3.2 (for $n > 0$)

$$r_n = q_n p_n \colon \mathbb{S}^n \to \mathbb{P}^n \to \mathbb{S}^n, \qquad \deg r_n = 1 + (-1)^{n+1}. \tag{6.46}$$

(a) *Case $n$ odd.* $H_n(r_n; \mathbb{Z})$ is the multiplication by 2 in an infinite cyclic group: integral coefficients are thus sufficient to prove that $r_n$ is homologically essential. Plainly, the same is true of $p_n$ and $q_n$.

(b) *Case $n$ even.* $H_i(p_n; \mathbb{Z})$, $H_i(q_n; \mathbb{Z})$ and $H_i(r_n; \mathbb{Z})$ are trivial for all $i > 0$ (because $H_i(\mathbb{S}^n) = 0$ for $i \neq n$ and $H_n(\mathbb{P}^n) = 0$ for $n$ even). Since all $H_i(\mathbb{S}^n)$ are free abelian groups, $r_n$ is acyclic, by 6.2.7(d).

(c) Theorem 6.3.2 says that the map $q_n \colon \mathbb{P}^n \to \mathbb{S}^n$ is always homologically essential, since $H_n(q_n; \mathbb{Z}_2)$ is an isomorphism of non-trivial groups. It also says that $p_n \colon \mathbb{S}^n \to \mathbb{P}^n$ is 'acyclic modulo 2', for $n > 0$.

(d) Finally, we have also seen that all the embeddings $\mathbb{P}^{n-1} \to \mathbb{P}^n$ $(n > 0)$ are homologically essential (in 6.3.2).

## 6.4 Ext functor and universal coefficients

We go on working in $\mathsf{Ab}$. We introduce the functor $\mathrm{Ext}(A, B)$ as the derived functor of $\mathrm{Hom}(A, B)$ with respect to the covariant variable $B$ or, equivalently, the contravariant variable $A$.

$G$ is always an abelian group, used as a coefficient group for cohomology.

**Literature.* For $R$-modules, the derived functors $\mathrm{Ext}^n_R(A, B)$ are dealt with in [CE] and most books on Homological Algebra. A construction of $\mathrm{Ext}^1_R(A, B)$, by 'extensions' of $B$ by $A$, can be found in [M1], Section III.1.

### *6.4.1 The Ext functor*

Fixing the first variable, the covariant functor $\mathrm{Hom}(A, -)\colon \mathsf{Ab} \to \mathsf{Ab}$ is left exact. Its right derived functor

$$\mathrm{Ext}(A, -)\colon \mathsf{Ab} \to \mathsf{Ab}, \tag{6.47}$$

is right exact, and can be computed on an injective resolution $(\varepsilon, d)\colon B \rightarrowtail B^0 \twoheadrightarrow B^1$ as

$$\begin{aligned} \mathrm{Ext}(A, B) &= H^1(\mathrm{Hom}(A, B^+)) \\ &= \mathrm{Cok}\,(\mathrm{Hom}(A, d)\colon \mathrm{Hom}(A, B^0) \to \mathrm{Hom}(A, B^1)). \end{aligned} \tag{6.48}$$

For every short exact sequence $(m, p)\colon B' \rightarrowtail B \twoheadrightarrow B''$ of abelian groups, we have an exact sequence

$$\begin{aligned} 0 \to \mathrm{Hom}(A, B') &\to \mathrm{Hom}(A, B) \to \mathrm{Hom}(A, B'') \\ \xrightarrow{d} \mathrm{Ext}(A, B') &\to \mathrm{Ext}(A, B) \to \mathrm{Ext}(A, B'') \to 0 \end{aligned} \tag{6.49}$$

which is natural on $\mathsf{Ab}$ (contravariantly) and $\mathsf{ShAb}$ (covariantly).

*Exercises and complements.* (a) $A$ is a free abelian group if and only if $\mathrm{Ext}(A, -) = 0$.

(b) Prove that $\mathrm{Ext}(\mathbb{Z}_m, \mathbb{Z}_n) \cong \mathbb{Z}_d \cong \mathrm{Hom}(\mathbb{Z}_m, \mathbb{Z}_n)$, where $d = \gcd(m, n)$. *Hint:* it is convenient to use the next proposition.

(c) Prove that $\mathrm{Ext}(\mathbb{Z}_m, \mathbb{Z}) \cong \mathbb{Z}_m$. *Hint:* as in (b).

### *6.4.2 Proposition*

*Fixing the second variable $B$, every short exact sequence $A' \rightarrowtail A \twoheadrightarrow A''$ of abelian groups gives an exact sequence*

$$\begin{aligned} 0 \to \mathrm{Hom}(A'', B) &\to \mathrm{Hom}(A, B) \to \mathrm{Hom}(A', B) \\ \xrightarrow{d} \mathrm{Ext}(A'', B) &\to \mathrm{Ext}(A, B) \to \mathrm{Ext}(A', B) \to 0 \end{aligned} \tag{6.50}$$

*natural on* $\mathsf{ShAb}$ *(contravariantly) and* $\mathsf{Ab}$ *(covariantly).*

*Proof* It is again an application of the Snake Lemma 3.5.1. Taking an injective resolution $(\varepsilon, d)\colon B \rightarrowtail B^0 \twoheadrightarrow B^1$ of $B$, we have a commutative diagram where

$$\begin{array}{ccccc}
\mathrm{Hom}(A'',B) & \rightarrowtail & \mathrm{Hom}(A,B) & \longrightarrow & \mathrm{Hom}(A',B) \\
\downarrow & & \downarrow & & \downarrow \\
\mathrm{Hom}(A'',B^0) & \rightarrowtail & \mathrm{Hom}(A,B^0) & \twoheadrightarrow & \mathrm{Hom}(A',B^0) \\
\downarrow & & \downarrow & & \downarrow \\
\mathrm{Hom}(A'',B^1) & \rightarrowtail & \mathrm{Hom}(A,B^1) & \twoheadrightarrow & \mathrm{Hom}(A',B^1) \\
\downarrow & & \downarrow & & \downarrow \\
\mathrm{Ext}(A'',B) & \longrightarrow & \mathrm{Ext}(A,B) & \twoheadrightarrow & \mathrm{Ext}(A',B)
\end{array} \qquad (6.51)$$

- the columns are exact, by definition of Ext,
- the two middle rows are short exact, because $B^0$ and $B^1$ are injective.

This gives the exact sequence (6.50), with its naturality properties. □

### *6.4.3 Corollary*

*Defining* $\mathrm{Ext}(A, B)$ *as the right derived functor of the left exact functor*

$$\mathrm{Hom}(-, B)\colon \mathsf{Ab}^{\mathrm{op}} \to \mathsf{Ab},$$

*we get the same result, up to natural isomorphism on* $\mathsf{Ab}^{\mathrm{op}} \times \mathsf{Ab}$.

*Proof* The right derived functor of the contravariant left exact functor $F = \mathrm{Hom}(-, B)$ is defined in 6.1.8(c): $F^1(A) = H^1(FA_+)$, where $A_+$ is a projective resolution of the abelian group $A$.

We apply Proposition 6.4.2 to a free resolution $A_1 \rightarrowtail A_0 \twoheadrightarrow A$ of the abelian group $A$; the chain complex $A_+$ is reduced to $m\colon A_1 \to A_0$.

$\mathrm{Ext}(A_0, B)$ and $\mathrm{Ext}(A_1, B)$ vanish, because $A_0$ and $A_1$ are projective, and the exact sequence

$$0 \to \mathrm{Hom}(A, B) \to \mathrm{Hom}(A_0, B) \to \mathrm{Hom}(A_1, B) \to \mathrm{Ext}(A, B) \to 0$$

shows that $\mathrm{Ext}(A, B)$ is the cokernel of $m^\sharp\colon \mathrm{Hom}(A_0, B) \to \mathrm{Hom}(A_1, B)$. But this is precisely the alternative form $\underline{\mathrm{Ext}}(A, B) = H^1(\mathrm{Hom}(A_+, B))$, computed on the projective resolution of $A$. □

### *6.4.4 A natural morphism*

Let $A$ be a chain complex. We begin to explore the cohomology group $H^n(\mathrm{Hom}(A, G))$ by introducing a canonical morphism

$$\alpha^n\colon H^n(\mathrm{Hom}(A, G)) \to \mathrm{Hom}(H_n(A), G), \qquad \alpha^n[\lambda][z] = \lambda(z). \qquad (6.52)$$

Here $z \in Z_n(A)$ and $\lambda\colon A_n \to G$ is a cocycle of $\mathrm{Hom}(A_n, G)$, which means that it vanishes on any boundary of $A_n$. Therefore $\lambda(z)$ only depends on the homology class $[z]$.

We have defined a natural transformation

$$\alpha^n\colon H^n(\mathrm{Hom}(-,-)) \to \mathrm{Hom}(H_n(-),-)\colon \mathrm{Ch}_+\mathsf{Ab}^{\mathrm{op}} \times \mathsf{Ab} \to \mathsf{Ab}. \quad (6.53)$$

*Remarks.* (a) If $G$ is an injective group (i.e. a divisible one), the homomorphism $\alpha^n\colon H^n(\mathrm{Hom}(A,G)) \to \mathrm{Hom}(H_n(A),G)$ is an isomorphism. In fact, the functor $\mathrm{Hom}(-,G)\colon \mathsf{Ab}^{\mathrm{op}} \to \mathsf{Ab}$ is exact, and 'preserves homology'.

### *6.4.5 Universal Coefficient Theorem* (for chain cohomology)

*For every chain complex $A$ of* free *abelian groups, and every abelian group $G$, there is a split exact sequence*

$$\mathrm{Ext}(H_{n-1}A, G) \xrightarrow{\beta^n} H^n(\mathrm{Hom}(A,G)) \xrightarrow{\alpha^n} \mathrm{Hom}(H_nA, G) \quad (6.54)$$

*which is natural on morphisms $f\colon A \to B$ of chain complexes (contravariantly), and on morphisms $h\colon G \to G'$ of abelian groups (covariantly).*

*Note.* Again, we have an isomorphism which is not natural on $\mathrm{Ch}_+\mathsf{Ab}$

$$H^n(\mathrm{Hom}(A,G)) \cong (\mathrm{Hom}(H_n(A),G)) \oplus (\mathrm{Ext}(H_{n-1}(A),G)). \quad (6.55)$$

*Proof* It is a particular case of the Künneth Theorem 6.5.5, replacing the cochain complex $B$ with $J(G)$. (A direct proof can be found in [M1], Theorem III.4.1, or [Mas3], Theorem VII.4.3.) □

### *6.4.6 Universal coefficients for singular cohomology*

For every relative pair $(X, A)$ of spaces, and every abelian group $G$, the previous theorem gives a split exact sequence

$$\mathrm{Ext}(H_{n-1}(X,A),G) \xrightarrow{\beta^n} H^n(X,A;G) \xrightarrow{\alpha^n} \mathrm{Hom}(H_n(X,A),G) \quad (6.56)$$

which is natural on relative maps $f\colon (X,A) \to (Y,B)$ (contravariantly), and on morphisms $h\colon G \to G'$ of abelian groups (covariantly).

*Exercises.* (a) If $H_{n-1}(X,A)$ is a free abelian group, then $\alpha^\sharp\colon H^n(X,A;G) \to \mathrm{Hom}(H_n(X,A),G)$ is an isomorphism.

In particular, the sphere $\mathbb{S}^n$ $(n > 0)$ has the following cohomology groups, with coefficients in $G$:

$$H^k(\mathbb{S}^n; G) \cong G, \ \text{for } k = 0, n; \qquad H^k(\mathbb{S}^n; G) = 0, \ \text{otherwise}. \quad (6.57)$$

(b) The projective space $\mathbb{P}^n$ has the following cohomology groups, with coefficients in $\mathbb{Z}$ and $\mathbb{Z}_2$ (and trivial ones in degree $> n$)

$$\begin{aligned} H^k(\mathbb{P}^n) &\cong \mathbb{Z}_2, &&\text{for } k \text{ even, } 0 < k < n, \\ H^k(\mathbb{P}^n) &\cong \mathbb{Z}, &&\text{for } k = 0 \text{ or } k = n \text{ odd.} \end{aligned} \tag{6.58}$$

$$H^k(\mathbb{P}^n; \mathbb{Z}_2) \cong \mathbb{Z}_2, \quad \text{for } 0 \leqslant k \leqslant n. \tag{6.59}$$

### 6.4.7 Betti numbers in cohomology

The abelian group $\mathbb{Q}$ is both torsion free and divisible. We have thus isomorphisms of vector spaces on $\mathbb{Q}$ (by 6.2.4(a) and 6.4.4(a))

$$\alpha_n \colon H_n(X) \otimes \mathbb{Q} \longrightarrow H_n(X; \mathbb{Q}), \tag{6.60}$$

$$\alpha^n \colon H^n(X; \mathbb{Q}) \longrightarrow \mathrm{Hom}(H_n(X), \mathbb{Q}). \tag{6.61}$$

If $H_n(X)$ is finitely generated, $H_n(X) \cong \mathbb{Z}^{\beta_n(X)} \oplus \mathrm{t}(H_n(X))$ and

$$\begin{aligned} &H_n(X) \otimes \mathbb{Q} \cong \mathbb{Q}^{\beta_n(X)} \cong \mathrm{Hom}(H_n(X), \mathbb{Q}), \\ &\beta_n(X) = \dim_{\mathbb{Q}} H_n(X; \mathbb{Q}) = \dim_{\mathbb{Q}} H^n(X; \mathbb{Q}). \end{aligned} \tag{6.62}$$

The Betti numbers of $X$ are thus equivalently determined by homology and cohomology.

### 6.4.8 *Manifolds and Poincaré duality

The homology and cohomology groups of orientable topological manifolds are closely related, by 'Poincaré duality'. Here we only hint at these celebrated results; an interested reader is referred to [Vi], Chapter 6, or [Mas3], Chapter IX, or [Ha], Section 3.3.

(a) The basic theorem is concerned with a compact orientable $n$-manifold $M$. In this case, all the homology and cohomology groups of $M$ are finitely generated, and there is a canonical isomorphism

$$H^k(M) \to H_{n-k}(M), \qquad \text{for } 0 \leqslant k \leqslant n, \tag{6.63}$$

all the higher homology and cohomology groups being trivial. One can see the proof in [Ha], Theorem 3.30.

Projective spaces yield a typical example. For $n$ odd, $\mathbb{P}^n$ is orientable, and the groups we have computed agree with Poincaré duality (see 3.4.3 and 6.4.6(b)). For $n$ even $> 0$, $\mathbb{P}^n$ is not orientable; the groups $H^k(\mathbb{P}^n)$ and $H_{n-k}(\mathbb{P}^n)$ are never isomorphic, for $k \leqslant n$.

(b) Recalling that $\beta_k(X) = \dim_{\mathbb{Q}} H_k(X;\mathbb{Q}) = \dim_{\mathbb{Q}} H^k(X;\mathbb{Q})$, a compact orientable $n$-manifold $M$ has

$$\beta_k(M) = \beta_{n-k}(M), \qquad \text{for } 0 \leqslant k \leqslant n, \tag{6.64}$$

as we have already seen for the $n$-sphere and the projective space $\mathbb{P}^n$ (in the orientable case, that is, for $n$ odd), and will see for the $n$-torus in Exercise 6.7.7(e).

(c) More generally, an orientable $n$-manifold $M$ has a canonical isomorphism (for every abelian group $G$)

$$H^k_c(M;G) \to H_{n-k}(M;G), \tag{6.65}$$

where $H^k_c$ denotes *singular cohomology with compact supports.* This can be found in [Mas3], Sections IX.3–4, or in [Ha], Section 3.3. Various cohomology and homology theories with compact supports are studied in [Mas2].

## 6.5 Tensor product of chain complexes

The Universal Coefficient Theorem, in 6.2.5, can be extended to the homology group $H_n(A \otimes B)$ of a tensor product of chain complexes, proving that it is determined by the homology groups $H_k(A)$ and $H_k(B)$, for $0 \leqslant k \leqslant n$.

(Again, the morphisms induced in homology are not determined in this way.)

This is made precise in the Künneth Theorem 6.5.4.

*Literature.* There are various versions of the Künneth theorem, under different hypotheses. Theorem V.10.4 in [M1] is the same as the present statement (up to allowing unbounded chain complexes). Theorem VI.3 in [CE] and Theorem VI.9.13 in [Do] are much more general.

### *6.5.1 Extending the tensor product, III*

There is a tensor product $A \otimes B$ of chain complexes

$$\begin{aligned} (A \otimes B)_n &= \textstyle\bigoplus_{p+q=n} (A_p \otimes B_q), \\ \partial_n(a \otimes b) &= \partial_p a \otimes b + (-1)^p\, a \otimes \partial_q b \quad (a \in A_p,\, b \in B_q), \end{aligned} \tag{6.66}$$

which extends the tensor product of abelian groups via the canonical embedding $J\colon \mathsf{Ab} \to \mathrm{Ch}_+\mathsf{Ab}$, in (2.18). (The Koszul sign rule applies, as in (4.96).)

The soundness of this definition is verified in the easy exercises below. The homology of $A \otimes B$ is determined by the Künneth Theorem 6.5.4.

*Exercises and complements.* (a) The differential $\partial_n$ of $A \otimes B$ is well defined on canonical generators, by the formula above. Moreover, $\partial_{n-1}\partial_n = 0$.

(b) Up to canonical isomorphisms, the tensor product of chain complexes is commutative, associative and admits a unit: the chain complex $J(\mathbb{Z})$ associated to the additive group $\mathbb{Z}$.

### 6.5.2 The tensor functor

The tensor product $f \otimes g$ of two morphisms $f\colon A \to A'$ and $g\colon B \to B'$ of chain complexes has the following components

$$\begin{gathered}(f \otimes g)_n = \textstyle\bigoplus_{p+q=n} (f_p \otimes g_q)\colon (A \otimes B)_n \to (A' \otimes B')_n, \\ (f_p \otimes g_q)(a \otimes b) = f_p(a) \otimes g_q(b),\end{gathered} \tag{6.67}$$

for $a \in A_p$, $b \in B_q$. Commutativity with differentials is obvious.

This defines the tensor functor of chain complexes

$$- \otimes -\colon \mathrm{Ch}_+\mathsf{Ab} \times \mathrm{Ch}_+\mathsf{Ab} \to \mathrm{Ch}_+\mathsf{Ab}. \tag{6.68}$$

*Exercises and complements.* (a) A sequence of chain complexes

$$A \otimes B \xrightarrow{f \otimes g} A' \otimes B' \xrightarrow{f' \otimes g'} A'' \otimes B''$$

is exact if and only if every sequence $(f_p \otimes g_q, f'_p \otimes g'_q)$ is exact in $\mathsf{Ab}$. In particular, the functor $- \otimes B\colon \mathrm{Ch}_+\mathsf{Ab} \to \mathrm{Ch}_+\mathsf{Ab}$ is exact if and only if all the components of $B$ are flat.

(b) Let $(f, g)\colon A' \rightarrowtail A \twoheadrightarrow A''$ be a short exact sequence of chain complexes, and $B$ a chain complex. If $A''$ has flat components, tensoring by $B$ gives a short exact sequence $(f \otimes B, g \otimes B)\colon A' \otimes B \rightarrowtail A \otimes B \twoheadrightarrow A\text{"} \otimes B$.

(c) *Remarks.* In this section we often use chain complexes of flat abelian groups. We know that an abelian group is flat if (and only if) it is torsion free, as stated in 6.2.8(a) (the proof is only referred to).

Most of the time one could use the stronger hypothesis of *free* components (obviously flat): we only need the more general case in a few marginal points.

### 6.5.3 A natural morphism

Extending 6.2.4, let $A$ and $B$ be chain complexes. We begin to explore the homology group $H_n(A \otimes B)$ introducing a canonical homomorphism

$$\alpha_{pq}\colon H_p(A) \otimes H_q(B) \to H_{p+q}(A \otimes B), \quad \alpha_{pq}([a] \otimes [b]) = [a \otimes b], \tag{6.69}$$

whose result is written as $[a] \times [b]$, and called the *homology external product*.

The homomorphism is well defined. If $a$ and $b$ are cycles in $A$ and $B$, $a \otimes b$ is a cycle of $A \otimes B$, by (6.66). If $a = \partial a'$ and $b$ is a cycle, then $\partial(a' \otimes b) = a \otimes b$; and symmetrically.

This defines a natural transformation

$$\alpha_n \colon \bigoplus_{p+q=n} H_p \otimes H_q \to H_n(- \otimes -) \colon \mathrm{Ch}_+\mathsf{Ab} \times \mathrm{Ch}_+\mathsf{Ab} \to \mathsf{Ab},$$
$$\alpha_n \colon \bigoplus_{p+q=n} H_p(A) \otimes H_q(B) \to H_n(A \otimes B), \tag{6.70}$$

where $\alpha_n$ has co-components $\alpha_{pq}$.

*Exercises and complements.* (a) If $G$ is a flat abelian group (viewed as a complex), we have seen in 6.2.4 that $\alpha$ gives a natural isomorphism

$$\alpha_n = \alpha_{n0} \colon H_n(A) \otimes G \to H_n(A \otimes G), \quad \alpha_n([a] \otimes g) = [a \otimes g]. \tag{6.71}$$

(b) More generally, if $B$ is a complex with zero differentials and flat components, the exact functor $- \otimes B \colon \mathrm{Ch}_+\mathsf{Ab} \to \mathrm{Ch}_+\mathsf{Ab}$ gives a natural isomorphism

$$\alpha_n \colon \bigoplus_{p+q=n} H_p(A) \otimes B_q \to H_n(A \otimes B), \quad \alpha_n([a] \otimes b) = [a \otimes b]. \tag{6.72}$$

**6.5.4 Künneth Theorem** (for chain complexes)
*For a pair $A, B$ of chain complexes of abelian groups, where $A$ (or $B$) has* flat *components, there is a short exact sequence*

$$\bigoplus_{p+q=n} H_pA \otimes H_qB \overset{\alpha_n}{\rightarrowtail} H_n(A \otimes B) \overset{\beta_n}{\twoheadrightarrow} \bigoplus_{p+q=n-1} \mathrm{Tor}(H_pA, H_qB) \tag{6.73}$$

*which is natural on* $\mathrm{Ch}_+\mathsf{Ab} \times \mathrm{Ch}_+\mathsf{Ab}$, *and splits (not in a natural way).*
*In particular, in degree 0, we have an isomorphism*

$$\alpha_0 \colon H_0(A) \otimes H_0(B) \to H_0(A \otimes B), \quad \alpha_0([a] \otimes [b]) = [a \otimes b]. \tag{6.74}$$

*Proof* The proof is long and complex: see 6.5.6. □

**6.5.5 Künneth Theorem** (for cochain complexes)
*For a pair $A, B$ of cochain complexes of abelian groups, where $A$ has* free *components, there is a short exact sequence*

$$\bigoplus_{p+q=n-1} \mathrm{Ext}(H^pA, H^qB) \overset{\beta^n}{\rightarrowtail} H^n(A \otimes B) \overset{\alpha^n}{\twoheadrightarrow} \bigoplus_{p+q=n} H^pA \otimes H^qB \tag{6.75}$$

*which is natural on* $\mathrm{Ch}_+\mathsf{Ab}^{\mathrm{op}} \times \mathrm{Ch}_+\mathsf{Ab}$, *and splits (not in a natural way).*

*Proof* The proof is similar to that of Theorem 6.5.4. A more general, more difficult result is given in [CE], Theorem VI.3a, under somewhat symmetric hypotheses on $A$ and $B$. □

### 6.5.6 Proof of Theorem 6.5.4

We assume that $A$ has flat components. The Tor product $\mathrm{Tor}(X, Y)$ is written as $X * Y$.

(a) We form two chain complexes with zero differentials

$$Z = (Z_n A)_{n \geqslant 0}, \qquad Z' = (Z'_n A)_{n \geqslant 0},$$

where $Z'_n(A) = A_n/Z_n(A) \cong B_{n-1}(A) \subset A_{n-1}$. They have flat components, by 6.2.3(b).

Applying $- \otimes B$ to the short exact sequence $Z \rightarrowtail A \twoheadrightarrow Z'$ of chain complexes, we get a short exact sequence $Z \otimes B \rightarrowtail A \otimes B \twoheadrightarrow Z' \otimes B$, by Exercise 6.5.3(b).

(b) We have thus an exact homology sequence around the group $H_n = H_n(A \otimes B)$

$$H_{n+1}(Z' \otimes B) \xrightarrow{D_{n+1}} H_n(Z \otimes B) \xrightarrow{\sigma_n} H_n \xrightarrow{\tau_n} H_n(Z' \otimes B) \xrightarrow{D_n} H_{n-1}(Z \otimes B).$$

We rewrite the sequence in the upper part of the following diagram, using the natural isomorphism (6.72)

$$\textstyle\bigoplus_{p+q=n} C_p \otimes H_q B \to H_n(C \otimes B), \qquad c \otimes [b] \mapsto [c \otimes b],$$

for the complexes $C = Z'$ or $C = Z$ (with zero differentials and flat components). The direct sums range over all pairs $(p, q)$ of integers with $p + q = n$

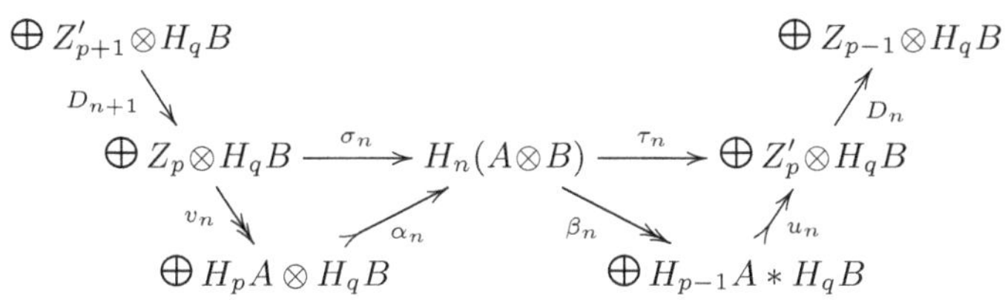


(6.76)

(c) We begin now to form the lower part of the previous diagram, as in 2.1.3(c). The homology group $H_p A$ is characterised by the short exact sequence

$$Z'_{p+1} \overset{h_p}{\rightarrowtail} Z_p \overset{v_p}{\twoheadrightarrow} H_p A \qquad h_p(\overline{a}) = \partial_{p+1}(a) \quad (a \in A_{p+1}),$$

where we use the canonical isomorphism $Z'_{p+1} \to B_p(A)$ induced by the differential $\partial_{p+1} \colon A_{p+1} \to A_p$.

Tensoring by $H_q B$, and recalling that the abelian group $Z_p$ is flat, we have a natural exact sequence

$$H_p A * H_q B \overset{u_{pq}}{\rightarrowtail} Z'_{p+1} \otimes H_q B \xrightarrow{h_{pq}} Z_p \otimes H_q B \overset{v_{pq}}{\twoheadrightarrow} H_p A \otimes H_q B$$

from which we deduce two consequences:

- the relation $\operatorname{Im} h_{pq} = \operatorname{Ker} v_{pq}$ (added for $p + q = n$), gives in (6.76): $v_n = \operatorname{cok} D_{n+1} = \operatorname{coim} \sigma_n$,

- the relation $\operatorname{Im} u_{p-1,q} = \operatorname{Ker} h_{p-1,q}$ (added again), gives in (6.76): $u_n = \ker D_n = \operatorname{im} \tau_n$.

(d) The epi-mono factorisations $\sigma_n = \alpha_n v_n$ and $\tau_n = u_n \beta_n$ give thus the short exact sequence $(\alpha_n, \beta_n)$.

The monomorphism $\alpha_n$ we have obtained is indeed computed as in 6.5.3, because $\sigma_n$ is induced by the inclusion $Z \rightarrowtail A$

$$\sigma_n(a \otimes [b]) = [a \otimes b], \qquad \alpha_n([a] \otimes [b]) = [a \otimes b] \quad (a \in Z_pA,\, b \in Z_qB).$$

(e) Finally, we have to show that the sequence (6.73) splits. (Here we follow Dold's proof in [Do], Theorem VI.9.13.)

First, we suppose that $A$ and $B$ have free components. Then the short exact sequence $Z_pA \rightarrowtail A_p \twoheadrightarrow Z'_pA$ splits (because $Z'_pA$ is free), and $Z_pA$ is a direct summand of $A_p$; the projection $v_p\colon Z_pA \to H_pA$ can thus be extended to a homomorphism $\varphi_p\colon A_p \to H_pA$, which vanishes on boundaries; similarly, by our additional hypothesis on $B$, the projection $Z_qB \to H_qB$ is extended to a homomorphism $\psi_q\colon B_q \to H_qB$.

Their tensor products give a homomorphism

$$(\varphi \otimes \psi)_n\colon (A \otimes B)_n \longrightarrow \bigoplus H_pA \otimes H_qB,$$

which vanishes on boundaries and induces a homomorphism

$$(\varphi \otimes \psi)_{*n}\colon H_n(A \otimes B) \longrightarrow \bigoplus H_pA \otimes H_qB. \tag{6.77}$$

This is a retraction of $\alpha_n\colon \bigoplus H_pA \otimes H_qB \to H_n(A \otimes B)$: for every $a \in Z_pA$ and $b \in Z_qB$:

$$(\varphi \otimes \psi)_{*n}\alpha_n([a] \otimes [b]) = (\varphi \otimes \psi)_{*n}[a \otimes b] = \varphi_p(a) \otimes \psi_q(b) = [a] \otimes [b].$$

(f) Coming back to the general case, we apply Lemma 6.5.7 (written below): there exist two complexes $A', B'$ of free abelian groups and two chain morphisms $f\colon A' \to A$, $g\colon B' \to B$ that induce homology isomorphisms $f_{*n}$ and $g_{*n}$, in every degree.

The naturality of the sequence (6.73) gives a commutative diagram with short exact rows and $n' = n - 1$

$$\begin{array}{ccccc}
\bigoplus_{p+q=n} H_pA' \otimes H_qB' & \overset{u^\sharp}{\rightarrowtail} & H_n(A' \otimes B') & \overset{v^\sharp}{\twoheadrightarrow} & \bigoplus_{p+q=n'} H_pA' * H_qB' \\
\downarrow {\scriptstyle f_* \otimes g_*} & & \downarrow {\scriptstyle (f \otimes g)_*} & & \downarrow {\scriptstyle f_* * g_*} \\
\bigoplus_{p+q=n} H_pA \otimes H_qB & \underset{u^\sharp}{\rightarrowtail} & H_n(A \otimes B) & \underset{v^\sharp}{\twoheadrightarrow} & \bigoplus_{p+q=n'} H_pA * H_qB
\end{array}$$

The left and the right vertical arrows are isomorphisms, and $(f \otimes g)_{*n}$ is also, by the Five Lemma 3.5.3. (We are actually applying the 'Short Five Lemma', with trivial groups at the left and the right hand.)

As we have proved that the upper row splits, the lower also does. □

### 6.5.7 Lemma

*For every chain complex A of abelian groups there exists a complex $A'$ with free components and a chain morphism $f\colon A' \to A$ that induces isomorphisms $f_{*n}\colon H_n(A') \to H_n(A)$.*

*Note.* This is Lemma V.10.5 in [M1]. ([Do] has similar results in II.4.)

*Proof* For a fixed $n \geqslant 0$, we begin by building a complex $F$ with free components, the correct homology in degree $n$ and trivial homology elsewhere.

$F$ is the upper row of the following commutative diagram, where

$$\begin{array}{ccccccccc} \cdots \longrightarrow & 0 & \longrightarrow & F_{n+1} & \xrightarrow{\partial_{n+1}} & & F_n & \longrightarrow & 0 & \longrightarrow \cdots \\ & \downarrow & & \downarrow u_{n+1} & \searrow & & \downarrow u_n & & \downarrow & \\ \cdots \to & A_{n+2} & \to & A_{n+1} & \twoheadrightarrow & \bullet \rightarrowtail & A_n & \underset{\partial_n}{\to} & A_{n-1} & \to \cdots \end{array} \tag{6.78}$$

- $F_n$ is a free abelian group with an epimorphism $F_n \twoheadrightarrow Z_nA$,

- $u_n$ is the composite $F_n \twoheadrightarrow Z_nA \rightarrowtail A_n$ (annihilated by $\partial_n\colon A_n \to A_{n-1}$),

- $F_{n+1} = u_n^{-1}(B_nA)$ is a (free) subgroup of $F_n$, and $\partial_{n+1}\colon F_{n+1} \to F_n$ is the inclusion,

- for $u_{n+1}$ we first lift $u_n\partial_{n+1}\colon F_{n+1} \to A_n$ to $B_nA$, and then to $A_{n+1}$ (as $F_{n+1}$ is free).

The chain complex $F$ has trivial homology groups in degree $\neq n$, while $u_n\colon F_n \to A_n$ induces an isomorphism

$$H_nF = F_n/F_{n+1} \cong Z_nA/B_nA = H_nA.$$

Now we rename the chain complex $F$ as $A^{(n)}$ (it depends on $n \geqslant 0$) and form the required complex $A'$ as the direct sum $A' = \bigoplus_n A^{(n)}$ of chain complexes. □

## 6.6 Products in singular cohomology

We show now that singular cohomology with coefficients in a ring $R$ forms a graded ring, as we have already seen for the cohomology theories of

Alexander–Spanier and de Rham. The multiplicative structure is provided by the 'cup product'.

*We use here the simplicial form of singular cohomology*, where the cup product has a simpler definition. In 6.6.9 we investigate 'why' a multiplicative structure should arise in cohomology.

$R$ is a fixed coefficient ring, which is commutative; the rings commonly used are $\mathbb{Z}$, $\mathbb{Z}_n$, $\mathbb{Q}$ and $\mathbb{R}$.

*Literature.* Hatcher's book [Ha] has many computations of graded algebras of singular cohomology, and a rich information on the difficult problem of realising a graded algebra in such a way: cf. Section 3.2.

*The cup product can also be defined in cubical form, but this has to go through heavier computations, by a more complex diagonal approximation $C_+(X) \to C_+(X) \otimes C_+(X)$, as in [HiW], Section 9.3, or by the chain homotopy equivalence $C_+(X \times X) \to C_+(X) \otimes C_+(X)$ (dealt with in Section 6.7), as in [Mas3], Section VIII.3.*

### 6.6.1 Singular cochains with coefficients in a ring

The chain complex $S_+(X)$ of simplicial singular chains of the space $X$ was defined in 3.6.6

$$\begin{aligned} &S_+(X) = \mathrm{Ch}_+\mathrm{Smp}(X) = ((F(\mathsf{Top}(\Delta^n, X))_n, (\partial_n)), \\ &\partial_n(a) = \Sigma_{0 \leqslant i \leqslant n} (-1)^i \, \partial_{ni}(a) = \Sigma_{0 \leqslant i \leqslant n} (-1)^i \, a\delta_{ni}, \end{aligned} \tag{6.79}$$

where $a \colon \Delta^n \to X$ is a simplex of $X$.

We recall that the tetrahedron $\Delta^n$ is the convex hull of the unit points $e_0, ..., e_n$ of $\mathbb{R}^{n+1}$; in particular, $\Delta^0 = \{1\}$. The differential is based on their faces:

$$\begin{aligned} &\delta_{ni} \colon \Delta^{n-1} \to \Delta^n \qquad (0 \leqslant i \leqslant n), \\ &\delta_{ni}(t_0, ..., t_{n-1}) = (t_0, ..., t_{i-1}, 0, t_i, ..., t_{n-1}). \end{aligned} \tag{6.80}$$

From this complex, we form the complex of simplicial singular cochains with coefficients in the abelian group $G$:

$$\begin{aligned} &S^+(-; G) \colon \mathsf{Top}^{\mathrm{op}} \to \mathrm{Ch}^+\mathsf{Ab}, \\ &S^n(X; G) = \mathrm{Hom}(S_n(X), G) = \mathsf{Set}(\mathrm{Smp}_n(X), G), \\ &(d^n\lambda)(a) = \Sigma_{0 \leqslant i \leqslant n+1} (-1)^i \, \lambda(a\delta_i), \\ &f^{\sharp n} \colon S^n(X; G) \to S^n(Y; G), \qquad (f^{\sharp n}\lambda)(b) = \lambda(fb), \end{aligned} \tag{6.81}$$

for $a \colon \Delta^{n+1} \to X$, $\lambda \colon \mathrm{Smp}_n(X) \to G$, $f \colon Y \to X$ and $b \colon \Delta^n \to Y$.

The simplicial definition of the singular cohomology groups is thus:

$$H^n(X;G) = H^n(S^+(X;G)). \tag{6.82}$$

From now on, we work with coefficients in the ring $R$.

$S^+(X;R)$ is thus a cochain complex of $R$-modules. An $n$-cochain is simply a mapping $\lambda\colon \mathrm{Smp}_n(X) \to R$ defined on the $n$-simplices of $X$, and a 0-cochain is a mapping $\lambda\colon X \to R$. Their $R$-linear combinations are computed pointwise.

### 6.6.2 Cup product

These cochains have a natural product, called the cup product $\lambda \smile \mu$.

It is based on the product of the ring $R$, together with the geometry of tetrahedra, which steps in by means of new higher faces.

Namely, the *back p-face* adds $q$ zeros at the right, while the *front q-face* adds $p$ zeros at the left (for $p + q = n$)

$$\begin{aligned} &\tau_p^- : \Delta^p \to \Delta^n, && \tau_p^-(t_0, ..., t_p) = (t_0, ..., t_p, 0..., 0),\\ &\tau_q^+ : \Delta^q \to \Delta^n, && \tau_q^+(t_0, ..., t_q) \mapsto (0..., 0, t_0, ..., t_q). \end{aligned} \tag{6.83}$$

In particular:

$$\begin{gathered} \tau_0^-(1) = e_0 \in \Delta^n, \qquad \tau_0^+(1) = e_n \in \Delta^n,\\ \tau_{n-1}^- = \delta_n \colon \Delta^{n-1} \to \Delta^n, \qquad \tau_{n-1}^+ = \delta_0 \colon \Delta^{n-1} \to \Delta^n,\\ \tau_n^- = \tau_n^+ = \mathrm{id}\,\Delta^n. \end{gathered}$$

Now we define the *cup product* of cochains with coefficients in $R$

$$\begin{aligned} &- \smile - \colon S^p(X;R) \times S^q(X;R) \to S^n(X;R) && (p+q=n),\\ &(\lambda \smile \mu)(a) \;=\; \lambda(a\tau_p^-)\,\mu(a\tau_q^+) && (a\colon \Delta^n \to X). \end{aligned} \tag{6.84}$$

In this way, the complex $S^+(X;R)$ becomes a differential graded $R$-algebra (see 4.6.2), as we verify below. The unit

$$1 \in S^0(X;R) = \mathsf{Set}(|X|, R), \qquad 1\colon X \to R, \tag{6.85}$$

is the constant function at the unit $1_R$ of $R$, defined on the set $|X| = \mathrm{Smp}_0(X)$.

For $\lambda \in S^n(X;R)$ and $\rho \in S^0(X;R)$ (i.e. a function $\rho\colon X \to R$) we simply have:

$$(\rho \smile \lambda)(a) = \rho(a(e_0))\,\lambda(a), \qquad (\lambda \smile \rho)(a) = \lambda(a)\,\rho(a(e_n)). \tag{6.86}$$

The cup product is related to the Alexander–Whitney diagonal approximation map $\tau_{pq}\colon S_{p+q}(X) \to S_p(X) \otimes S_q(X)$ (for simplicial singular chains), as we shall see in 6.6.6.

*Exercises and complements.* We have to check the properties listed below. The verifications are obvious, except the last, which works by commuting ordinary faces and higher faces of tetrahedra; it is written down in the solutions of Chapter 9.

We assume that $\lambda, \lambda' \in S^p(X; R)$, $\mu, \mu' \in S^q(X; R)$, $\nu \in S^r(X; R)$ and $\rho \in R$.

(a) $(\rho\lambda) \smile \mu = \rho(\lambda \smile \mu) = \lambda \smile (\rho\mu)$,

(b) $\lambda \smile (\mu \smile \nu) = (\lambda \smile \mu) \smile \nu$,

(c) $(\lambda + \lambda') \smile \mu = \lambda \smile \mu + \lambda' \smile \mu, \qquad \lambda \smile (\mu + \mu') = \lambda \smile \mu + \lambda \smile \mu'$,

(d) $d^n(\lambda \smile \mu) = d^p\lambda \smile \mu + (-1)^p\, \lambda \smile d^q\mu \qquad (n = p + q)$.

### 6.6.3 The cohomology algebra

The cohomology graded $R$-module of $X$, with coefficients in the ring $R$

$$H^*(X; R) = H^*(S^+(X; R)), \tag{6.87}$$

becomes a graded $R$-algebra, with the induced *cup product* of cohomology classes

$$H^p(X; R) \times H^q(X; R) \longrightarrow H^{p+q}(X; R), \qquad [\lambda] \smile [\mu] = [\lambda \smile \mu], \tag{6.88}$$

and unit $1 \in H^0(X; R) = Z^0(X; R)$, the cohomology class of the constant function $1\colon X \to R$.

The general case was dealt with in 4.6.3. We recall that, because of formula 6.6.2(d) on the differential of a product, a product of cocycles is a cocycle; the coboundaries form a bilateral graded ideal within cocycles.

A map $f\colon X \to Y$ gives a cochain morphism $f^\sharp\colon S^+(Y; G) \to S^+(X; G)$ which preserves the graded product and the unit

$$\begin{aligned} &(f^{\sharp n}(\lambda \smile \mu))(a) = (\lambda \smile \mu)(fa) = (\lambda(fa\tau_p^-)\, \mu(fa\tau_q^+)) \\ &\qquad = (f^{\sharp p}\lambda) \smile (f^{\sharp q}\mu)(a) \qquad\qquad (a\colon \Delta^n \to X), \\ &(f^{\sharp 0}(1))(x) = 1(fx) = 1 \qquad\qquad (x \in X). \end{aligned} \tag{6.89}$$

We have thus defined two contravariant functors, with values in dg-algebras and graded algebras:

$$S^+(-; R)\colon \mathsf{Top}^{\mathrm{op}} \to R\mathsf{Dga}, \qquad H^*(-; R)\colon \mathsf{Top}^{\mathrm{op}} \to R\mathsf{Gra}. \tag{6.90}$$

**Complements.* (a) The algebra $H^*(X;R)$ is skew-commutative:

$$[\lambda] \smile [\mu] = (-1)^{pq} \, [\mu] \smile [\lambda], \tag{6.91}$$

where $\lambda$ and $\mu$ have degree $p$ and $q$. One can see a proof in [Vi], Theorem 5.11, or [Sp2], Subsection 5.6.11.

(b) For a relative pair $(X, A)$, there are relative forms of the cup product of cochains, and therefore of cohomology classes

$$\begin{aligned} H^p(X,A;R) \times H^q(X;R) &\longrightarrow H^{p+q}(X,A;R), \\ H^p(X;R) \times H^q(X,A;R) &\longrightarrow H^{p+q}(X,A;R), \\ H^p(X,A;R) \times H^q(X,A;R) &\longrightarrow H^{p+q}(X,A;R), \end{aligned} \tag{6.92}$$

because, if the cochain $\lambda$ or the cochain $\mu$ vanishes on the subspace $A$, then the cochain $\lambda \smile \mu$ also does.

### 6.6.4 Path components and cup product

Let $X$ be a space and $(X_i)_{i \in I}$ its partition in path components, with inclusions $u_i \colon X_i \to X$.

(a) $H^0(X;R) = Z^0(X;R)$ is the $R$-algebra of the functions $\rho \colon X \to R$ constant on path components; the cohomology class of this function can be equivalently written as $\rho$ or $[\rho]$.

There is thus a natural isomorphism

$$H^0(X;R) \longrightarrow \textstyle\prod_{i \in I} H^0(X_i;R), \qquad \rho \mapsto (\rho_i)_{i \in I}, \tag{6.93}$$

where $\rho_i = \rho u_i \colon X_i \to R$ is the restriction of $\rho$. In particular, if $X$ is 0-connected, $H^0(X;R)$ will be identified with $R$.

This isomorphism will now be extended to the whole graded algebra $H^*(X;R)$.

(b) The homomorphisms of graded $R$-algebras (for $i \in I$)

$$p_i = u_i^* \colon H^*(X;R) \longrightarrow H^*(X_i;R), \qquad [\lambda] \mapsto [\lambda_i] = [{u_i}^\sharp \lambda], \tag{6.94}$$

are computed in the obvious way: in degree $n$, the value $\lambda_i(a)$ on a simplex $a \colon \Delta^n \to X_i$ is $\lambda(u_i a)$; or simply $\lambda(a)$ if we view $a$ as a map $\Delta^n \to X$ with image in $X_i$.

The consistency with the cup product is based on the fact that, on any simplex $a$, the restrictions $a\tau_p^-$ and $a\tau_q^+$ stay in the same path component.

As in (5.5.4)(c), these homomorphisms are the projections of a cartesian product, and there is a canonical isomorphism of graded $R$-algebras

$$H^*(X;R) \longrightarrow \textstyle\prod_{i \in I} H^*(X_i;R), \qquad [\lambda] \mapsto ([\lambda_i])_{i \in I}. \tag{6.95}$$

(c) For a 0-cocycle $\rho = (\rho_i) \in Z^0(X;R)$, the cup product with any cochain $\lambda \in S^n(X;R)$ is computed as in (6.86), on a simplex $a\colon \Delta^n \to X$ with image in $X_i$

$$(\rho \smile \lambda)(a) = \rho_i\, \lambda(a), \qquad (\lambda \smile \rho)(a) = \lambda(a)\, \rho_i. \tag{6.96}$$

These products are thus determined by the $R$-linear structure of the algebra $Z^0(X;R) = H^0(X;R)$.

(d) (*Basic products*) In $H^*(X;R)$, the cup products $[\lambda] \smile [\mu]$ where one of the factors has degree zero will be called (in this book) *basic products*: they are easily calculated, by the previous formula, and cannot completely vanish (unless the ring $R$ is trivial, or the space $X$ is empty).

We say that the graded algebra $H^*(X;R)$ has a *basic cup product* if all the other cup products vanish: $[\lambda] \smile [\mu] = 0$ whenever both factors have a positive degree. We shall see in 6.6.5 that a space can have a basic cup product with integral coefficients, and an interesting cup product modulo 2.

(e) The *characteristic function* of the component $X_i$ in the ring $R$

$$\begin{gathered}\eta_i\colon X \to R,\\ \eta_i(x) = 1_R \text{ if } x \in X_i, \qquad \eta_i(x) = 0_R \text{ if } x \notin X_i,\end{gathered} \tag{6.97}$$

is a 0-cocycle of $X$. Now $p_i(\eta_i) = 1_i$ and $p_j(\eta_i) = 0$ for $j \neq i$.

For every function $\lambda\colon \mathrm{Smp}_n(X) \to R$

$$(\eta_i \smile \lambda)(a) = (\lambda \smile \eta_i)(a) = \lambda(a), \quad \text{if } \mathrm{Im}\,(a) \subset X_i, \tag{6.98}$$

and is 0 otherwise.

The projection $p_i\colon H^*(X;R) \to H^*(X_i;R)$ restricts thus to an isomorphism

$$\eta_i(H^*(X;R)) \to H^*(X_i;R), \tag{6.99}$$

defined on the principal ideal of $H^*(X;R)$ generated by the cohomology class $\eta_i \in H^0(X;R)$.

*(f) In any product $R = \prod_{i\in I} R_i$ of unital rings, possibly non commutative, the unit of $R_i$ produces as above an idempotent element $\eta_i$ in the centre of $R$, which generates a principal ideal isomorphic to $R_i$.

### *6.6.5* ***Exercises and complements*** (Cohomology algebras)

These exercises are solved below (or a reference is given).

(a) (*The spheres*) All the graded algebras $H^*(\mathbb{S}^n;R)$ have a basic cup product.

(b) The same is true of $H^*(\mathbb{P}^2; \mathbb{Z})$, but coefficients in $\mathbb{Z}^2$ give a different result: see (e).

(c) (*The torus*) Compute the graded algebra $H^*(\mathbb{T}) = H^*(\mathbb{T}; \mathbb{Z})$ of the torus, showing that $H^1(\mathbb{T})$ has two generators whose cup product generates $H^2(\mathbb{T})$. *Hint:* use the Universal Coefficient Theorem for cohomology, with suitable generators of $H_1(\mathbb{T})$ and $H_2(\mathbb{T})$, in simplicial form.

*The graded algebra $H^*(\mathbb{T}^n)$ of the $n$-dimensional torus is a (rather obvious) extension of this case: see [Ha], Example 3.11.*

(d) The previous relationship between $H^1$ and $H^2$ fails for the pointed sum $X = \mathbb{S}^1 \vee \mathbb{S}^1 \vee \mathbb{S}^2$, which has the same cohomology groups as the torus, but a basic cup product. (Of course there is no need of the cohomology ring to show that the torus is not homeomorphic to $X$: the latter is not even a manifold, and can be disconnected taking out the basepoint.)

*(e) We already know that $H^k(\mathbb{P}^n; \mathbb{Z}_2) \cong \mathbb{Z}_2$, for $0 \leqslant k \leqslant n$ (by Exercise 6.4.6(b)). In fact, these groups form an interesting graded $\mathbb{Z}_2$-algebra: $H^*(\mathbb{P}^n; \mathbb{Z}_2)$ 'is' the quotient $\mathbb{Z}_2[t]/(t^{n+1})$ of a polynomial algebra in one indeterminate, generated (as a graded algebra) by the non-trivial element of $H^1(\mathbb{P}^n; \mathbb{Z}_2) \cong \mathbb{Z}_2$. For a proof, see [Sp2], Section 5.8, or [Ha], Theorem 3.12.

*Solutions.* (a) For $n = 0$, this is obvious. For $n > 0$, the graded $R$-module $H^*(\mathbb{S}^n; R)$ has non-trivial components in degree 0 and $n$. In positive degree, $[\lambda] \smile [\mu] = 0$, simply because $H^{2n}(\mathbb{S}^n) = 0$.

(b) Obvious, as the only non trivial cohomology group in positive degree is $H^2(\mathbb{P}^2; \mathbb{Z}) = \mathbb{Z}_2$ (by Exercise 6.4.6(b)).

(c) For every $n \geqslant 0$, $H_n(\mathbb{T})$ is a free abelian group of finite rank. The Universal Coefficient Theorem for cohomology gives a canonical isomorphism

$$\alpha^n \colon H^n(\mathbb{T}) \to \mathrm{Hom}(H_n(\mathbb{T}), \mathbb{Z}), \qquad \alpha^n[\lambda][z] = \lambda(z), \tag{6.100}$$

for $\lambda \in Z^n(\mathbb{T})$ and $z \in Z_n(\mathbb{T})$.

The standard generators $[a], [b]$ of $H_1(\mathbb{T})$, realised by two loops around the equator and a meridian (as in 3.2.3(b)), give two (free) generators of the abelian group $H^1(\mathbb{T})$

$$\begin{aligned} &[\lambda] \in H^1(\mathbb{T}), \quad \lambda(a) = 1, \quad \lambda(b) = 0, \\ &[\mu] \in H^1(\mathbb{T}), \quad \mu(a) = 0, \quad \mu(b) = 1. \end{aligned} \tag{6.101}$$

The cup product will be determined proving that, at the level of cochains

$$\lambda \smile \lambda = 0 = \mu \smile \mu, \qquad \lambda \smile \mu = \omega, \tag{6.102}$$

where $[\omega]$ is a generator of $H^2(\mathbb{T})$.

We use as generator $[z]$ of $H_2(\mathbb{T})$ the chain $z = f - g$ produced by two simplices $f, g\colon \Delta^2 \to \mathbb{T}$ meeting at an edge; letting $p\colon \mathbb{I}^2 \to \mathbb{T}$ be the projection, we define $f = pf'$ and $g = pg'$, where $f', g'\colon \Delta^2 \to \mathbb{I}^2$ are *affine* simplices of the square, defined as follows on the vertices

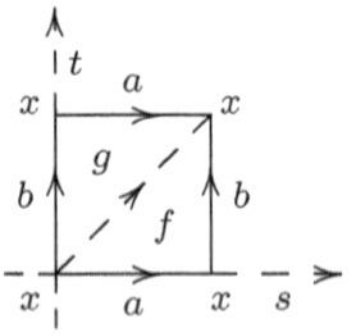


$$\begin{aligned} &f'(e_0)=(0,0), \quad f'(e_1)=(1,0), \quad f'(e_2)=(1,1), \\ &g'(e_0)=(0,0), \quad g'(e_1)=(0,1), \quad g'(e_2)=(1,1). \end{aligned} \tag{6.103}$$

The fact that $z$ is a cycle is obvious. The fact that $[z]$ generates $H_2(\mathbb{T})$ would be proved in the initial study of singular homology in simplicial form, and is left as understood.

We can now compute $\lambda \smile \mu$, $\lambda \smile \lambda$ and $\mu \smile \mu$ on the cycle $z$

$$\begin{aligned} (\lambda \smile \mu)(f - g) &= \lambda(f\tau_1^-)\, \mu(f\tau_q^+) - \lambda(g\tau_1^-)\, \mu(g\tau_q^+) \\ &= \lambda(f\delta_2)\, \mu(f\delta_0) - \lambda(g\delta_2)\, \mu(g\delta_0) \\ &= \lambda(a)\mu(b) - \lambda(b)\, \mu(a) = 1, \end{aligned}$$

$$\begin{aligned} (\lambda \smile \lambda)(f - g) &= \lambda(f\tau_1^-)\, \lambda(f\tau_q^+) - \lambda(g\tau_1^-)\, \lambda(g\tau_q^+) \\ &= \lambda(f\delta_2)\, \lambda(f\delta_0) - \lambda(g\delta_2)\, \lambda(g\delta_0) \\ &= \lambda(a)\lambda(b) - \lambda(b)\, \lambda(a) = 0, \end{aligned}$$

and similarly $\mu \smile \mu = 0$.

(d) The cohomology groups of $X$ are easily determined, as above.

The homology group $H_2(X)$ is freely generated by the homology class of a map $f\colon \Delta^2 \to \mathbb{S}^2 \subset X$ that collapses the border of $\Delta^2$ to a point, and is an embedding on the interior part. Now, any product $\lambda \smile \mu$ of 1-cochains of $X$ vanishes on $[f]$, because all the faces of $f$ are constant loops.

### 6.6.6 The Alexander–Whitney diagonal approximation

There is a natural transformation

$$\begin{aligned} &\tau\colon S_+ \to S_+ \otimes S_+ \colon \mathsf{Top} \to \mathrm{Ch}_+\mathsf{Ab}, \\ &\tau_{pq}X\colon S_{p+q}(X) \longrightarrow S_p(X) \otimes S_q(X), \\ &\tau_{pq}(a) = a\tau_p^- \otimes a\tau_q^+ \qquad (a\colon \Delta^{p+q} \to X), \end{aligned} \tag{6.104}$$

called the *Alexander–Whitney diagonal approximation.* It is constructed by means of the higher faces introduced in (6.83), for $p+q=n$

$$\begin{aligned} \tau_p^- &: \Delta^p \to \Delta^n, & \tau_p^-(t_0, ..., t_p) &= (t_0, ..., t_p, 0..., 0), \\ \tau_q^+ &: \Delta^q \to \Delta^n, & \tau_q^+(t_0, ..., t_q) &\mapsto (0..., 0, t_0, ..., t_q). \end{aligned} \tag{6.105}$$

*Exercises and complements.* (a) Prove that the sequence of homomorphisms

$$\tau_n X \colon S_n(X) \to \textstyle\bigoplus_{p+q=n} S_p(X) \otimes S_q(X)$$

of co-components $\tau_{pq}X$ commutes with the differentials. (Naturality is obvious.)

(b) In degree 0, we simply have the diagonal homomorphism

$$\tau_{00}X \colon S_0(X) \to S_0(X) \otimes S_0(X), \qquad \tau_{00}(x) = x \otimes x \quad (x \in X) \tag{6.106}$$

which induces in homology a natural homomorphism $[x] \mapsto [x \otimes x]$.

(c) The cup product of cohomology arises in a natural way from the natural transformation (6.104).

### 6.6.7 Cross product

For two spaces $X, Y$ there is also a *cross product,* or *external product,* of singular cochains

$$\begin{aligned} S^p(X;R) \times S^q(Y;R) &\longrightarrow S^{p+q}(X \times Y; R), \\ (\lambda, \mu) &\mapsto \lambda \times \mu = p^{\sharp 1}(\lambda) \smile p^{\sharp 2}(\mu), \\ (\lambda \times \mu)(a) = \lambda(p_1 a \tau_p^-)\, \mu(p_2 a \tau_q^+) & \quad (\text{for } a \colon \Delta^n \to X \times Y), \end{aligned} \tag{6.107}$$

where $p_1 \colon X \times Y \to X$ and $p_2 \colon X \times Y \to Y$ are the cartesian projections.

This product is $R$-bilinear and consistent with the differential, in the sense that (as proved below):

$$d^n(\lambda \times \mu) = d^p(\lambda) \times \mu + (-1)^p \lambda \times d^q(\mu) \qquad (n = p+q). \tag{6.108}$$

There is thus a *cross product,* or *external product,* in cohomology:

$$\begin{aligned} H^p(X;R) \times H^q(Y;R) &\longrightarrow H^{p+q}(X \times Y, A; R), \\ [\lambda] \times [\mu] &= [\lambda \times \mu]. \end{aligned} \tag{6.109}$$

*Exercises and complements.* (a) Prove formula (6.108).

(b) We have defined the cross product in terms of the cup product. The converse can also be done, taking $X = Y$.

### 6.6.8 Complements

The diagonal approximation $\tau\colon S_+ \to S_+ \otimes S_+$ has also an equivalent 'external form', working on two spaces $X, Y$. This is a natural transformation on $\mathsf{Top} \times \mathsf{Top}$

$$\begin{aligned} \vartheta\colon S_+(X \times Y) &\longrightarrow S_+(X) \otimes S_+(Y), \\ (a, b) \mapsto a\tau_p^- \otimes b\tau_q^+ &\qquad (a\colon \Delta^{p+q} \to X,\, b\colon \Delta^{p+q} \to Y), \end{aligned} \tag{6.110}$$

whose components $\vartheta(X, Y)$ are chain homotopy equivalences, as we shall see in Section 6.7.

Given the Alexander–Whitney diagonal approximation, we define $\vartheta_{pq}$ as the composite $\vartheta_{pq} = (p_{1\sharp} \otimes p_{2\sharp})\tau_{pq}$

$$S_{p+q}(X \times Y) \to S_p(X \times Y) \otimes S_q(X \times Y) \to S_p(X) \otimes S_q(Y), \tag{6.111}$$

where $p_1\colon X \times Y \to X$ and $p_2\colon X \times Y \to Y$ are the cartesian projections.

Conversely, given all $\vartheta_{pq}$, we define $\tau_{pq}$ by the diagonal map $\Delta$

$$\tau_{pq} = \vartheta_{pq}\Delta_\sharp\colon S_{p+q}(X) \to S_{p+q}(X \times X) \to S_p(X) \otimes S_q(X). \tag{6.112}$$

### 6.6.9 *The dual affinity of spaces and rings

There are geometrical reasons why a multiplication appears in *co*homology theories: every space has a diagonal map $\Delta\colon X \to X \times X$, which – in the theory of coalgebras – is a *co*monoid structure (satisfying diagrammatic properties of coassociativity and counitarity).

In fact, we have seen that the diagonal, combined with the chain homotopy equivalence $S_+(X \times X) \to S_+(X) \otimes S_+(X)$ recalled above, gives the Alexander–Whitney diagonal approximation $\tau\colon S_+ \to S_+ \otimes S_+$, which supplies the cup product in a natural way (worked out in Exercise 6.6.6(c)).

More essentially, there seems to be a dual affinity between topological spaces and rings, according to which rings (or commutative rings) are able to give a 'mirror image' of topological spaces, rather than a covariant translation.

(a) As a concrete instance of this affinity, Gelfand duality says that the category of compact Hausdorff spaces is equivalent to the opposite of the category of commutative (unital) $C^*$-algebras, by the contravariant functor taking a compact Hausdorff space $X$ to the $C^*$-algebra $C(X) = \mathsf{Top}(X, \mathbb{C})$ of complex continuous functions on $X$.

This leads to the domain of Noncommutative Geometry, introduced by A. Connes [Co]: a general $C^*$-algebra is viewed as a (possibly) noncommutative space. Directed and weighted algebraic topology have a strong relationship with this topic [G1].

(b) Pointless Topology also simulates spaces by the opposite of a category of an algebraic character.

In fact, a space $X$ determines the ordered set $\mathcal{O}(X)$ of its open subsets; this is a *frame*, that is, a complete lattice where arbitrary joins distribute over binary meets. Contravariantly, a map $f\colon X \to Y$ gives the preimage homomorphism $f^*\colon \mathcal{O}(Y) \to \mathcal{O}(X)$, that preserves arbitrary joins and binary meets. Pointless Topology studies the category $\mathsf{Loc} = \mathsf{Frm}^{\mathrm{op}}$ of 'locales', the formal opposite of the category of frames [Jo1, Jo2, Bo3].

(c) This dual affinity also appears in elementary categorical facts.

The terminal object $\{*\}$ of $\mathsf{Top}$ determines the category $\mathsf{Top}_\bullet$ of pointed spaces as the *slice category* $\mathsf{Top}\backslash\{*\}$: an object $(X, x_0)$ is viewed as a map $x_0\colon \{*\} \to X$.

In a dually similar way, the initial object $\mathbb{Z}$ of $\mathsf{Rng}$ determines the slice category $\mathsf{Rng}/\mathbb{Z}$, of *copointed rings* $p\colon R \to \mathbb{Z}$. This is equivalent to the category $\mathsf{Rng}'$ *of possibly-non-unital rings*, sending the (unital) homomorphism $p$ to the ideal $\operatorname{Ker} p$. Details can be found in [G5], Exercise 3.3.2(b).

*(d) More specifically, $\mathsf{Top}$ and the dual of the category of commutative rings share similar distributive properties, of product over sum [CaLW].

## 6.7 Acyclic models and products of spaces

This last section of Chapter 6 establishes some important results, among which:

(i) the equivalence of the simplicial and cubical forms of singular homology,

(ii) the Eilenberg–Zilber Theorem on the singular homology of a product of spaces.

These results are proved as applications of a powerful technique, the Acyclic Models Theorem.

*Literature.* Acyclic models were introduced by Eilenberg and Mac Lane [EiM5] in 1953, formalising the role of the 'simplicial models' $\Delta^n$ and the 'cubical models' $\mathbb{I}^n$ in the corresponding constructions of singular homology; their equivalence is a consequence of the Acyclic Model Theorem. Application (ii) was exposed in the subsequent article of the same journal, by Eilenberg and Zilber [EiZ2].

The main theorem is also exposed in [Sp2], Section 4.2, Theorem 8, in a basic form that covers application (ii), but does not cover cubical chains (because of their procedure of normalisation). [HiW], in Sections 8.4 and 8.7, uses for both applications the method of Acyclic Models, rather than the theorem.

Here we prove the Acyclic Models Theorem in Theorem 6.7.3, beginning with the simplified version and extending the proof to the general form.

### *6.7.1 Acyclic models, the basic version*

We consider a category $\mathsf{C}$ equipped with a set $\mathcal{M}$ of objects, called *models*. We begin by exposing the simplified version used in [Sp2, Vi]. The reader should think of the simplicial case, made explicit below, in Example (a).

First, a functor $F\colon \mathsf{C} \to \mathsf{Ab}$ is said to be *free on the models* of $\mathcal{M}$ if there is a family of *distinguished elements* $\xi_j \in F(M_j)$ (with $j \in J$ and $M_j \in \mathcal{M}$) such that, for every object $X$ in $\mathsf{C}$, $F(X)$ is the free abelian group generated by all the elements $a_j = (Fa)(\xi_j)$, for $j \in J$ and $a\colon M_j \to X$ in $\mathsf{C}$; these elements $a_j$ can be called the *$F$-models of $X$*.

Now we consider a functor $F\colon \mathsf{C} \to \mathrm{Ch}_+\mathsf{Ab}$ with values in chain complexes. We say that $F$ is *free on the models* if each component $F_n\colon \mathsf{C} \to \mathsf{Ab}$ is free, with distinguished elements $\xi_j \in F_n(M_j)$ (for $j \in J_n$ and $M_j \in \mathcal{M}$). Every abelian group $F_n(X)$ has thus a corresponding basis of $F_n$-models

$$a_j = (F_n a)(\xi_j), \qquad \text{for } j \in J_n \text{ and } a \in \mathsf{C}(M_j, X). \tag{6.113}$$

We are supposing that all the sets $J_n$ are disjoint, so that we need not specify the degree $n$ in $\xi_j$ and $M_j$.

The functor $F\colon \mathsf{C} \to \mathrm{Ch}_+\mathsf{Ab}$ is *acyclic on the models* if $H_n(F(M)) = 0$, for all $n > 0$ and $M \in \mathcal{M}$.

*Examples.* (a) The basic example is the category $\mathsf{Top}$, equipped with the *simplicial models*, i.e. the set of all tetrahedra $\Delta^n$, for $n \geqslant 0$. The simplicial singular complex functor

$$S_+\colon \mathsf{Top} \to \mathrm{Ch}_+\mathsf{Ab}, \qquad S_n(X) = \mathbb{Z}(\mathsf{Top}(\Delta^n, X)), \tag{6.114}$$

is free and acyclic on the models $\Delta^n$ ($n \geqslant 0$).

In fact, for every $n \geqslant 0$, $S_n$ is free on the single model $\Delta^n$, with $\xi_n = \mathrm{id}\,\Delta^n$, because the abelian group $S_n(X)$ is freely generated by the $n$-simplices $(S_n a)(\xi_n) = a$, for all $a\colon \Delta^n \to X$. Acyclicity on these models was verified in Lemma 3.6.8.

(b) To study cartesian products $X \times Y$ (in 6.7.5), we shall use the category $\mathsf{Top} \times \mathsf{Top}$, with models consisting of all cartesian products $\Delta^p \times \Delta^q$, for $p, q \geqslant 0$.

(c) This approach does not cover the cubical singular complex functor

$$C_+\colon \mathsf{Top} \to \mathrm{Ch}_+\mathsf{Ab}, \qquad C_n(X) = \mathbb{Z}(\mathsf{Top}(\mathbb{I}^n, X))/\mathbb{Z}\mathrm{Deg}_n(X), \tag{6.115}$$

which is acyclic on the models $\mathbb{I}^n$ $(n \geqslant 0)$, but is not free on them, because of normalisation.

Let us note that the non-normalised cubical functor is not even acyclic on the singleton $\mathbb{I}^0$, as shown in the solution of Exercise 2.2.5(a).

### 6.7.2 A more complex version

To deal with the cubical layout we resort to the original setting of [EiM5]. We are still examining a functor $F\colon \mathsf{C} \to \mathsf{Ab}$ defined on a category $\mathsf{C}$ with a set $\mathcal{M}$ of models, and we want to extend the property of being free on them.

First we build a new functor $\tilde{F}\colon \mathsf{C} \to \mathsf{Ab}$, where

(i) for an object $X$ of $\mathsf{C}$, $\tilde{F}(X)$ is the free abelian group generated by all pairs $(a, \xi)$, where $a\colon M_a \to X$ is a morphism of $\mathsf{C}$ defined on a model $M_a \in \mathcal{M}$, and $\xi \in F(M_a)$,

(ii) for a morphism $f\colon X \to Y$ in $\mathsf{C}$, the homomorphism $\tilde{F}f\colon \tilde{F}X \to \tilde{F}Y$ is defined on the previous basis, as $\tilde{F}(f)(a, \xi) = (fa\colon M_a \to Y, \xi)$.

There is a natural transformation

$$\pi\colon \tilde{F} \to F, \qquad (\pi X)(a, \xi) = F(a)(\xi) \in F(X), \tag{6.116}$$

for $a\colon M_a \to X$, $\xi \in F(M)$.

We say that $F$ is *essentially free on the models* if there is also a natural transformation $\rho\colon F \to \tilde{F}$ such that $\pi\rho = \mathrm{id}$ (a section of $\pi$).

Such a functor is called 'representable' in the original paper [EiM5]; but this term has now a different, well-established meaning in category theory: see 1.3.8.

*Exercises and complements.* The solutions are below.

(a) Every functor $F\colon \mathsf{C} \to \mathsf{Ab}$ free on the models of $\mathcal{M}$ is also essentially free on them.

(b) The cubical singular complex functor (6.114) is essentially free (and acyclic) on the models $\mathbb{I}^n$ $(n \geqslant 0)$.

*Solutions.* (a) We define $\rho\colon F \to \tilde{F}$ taking the basis element $a_j = (Fa)(\xi_j)$ (indexed by the pair $(j, a)$) to the pair $(a, \xi_j)$. Then $\pi\rho(a_j) = (Fa)(\xi_j) = a_j$.

(b) Each component $C_n(X)$ of cubical chains is freely generated by the family of non-degenerate cubes $a\colon \mathbb{I}^n \to X$.

Letting $\mathcal{M}_n = \{\mathbb{I}^n \mid n \geqslant 0\}$ and $\xi_n = \mathrm{id}\,\mathbb{I}^n$, we can define

$$\rho_n X\colon C_n(X) \to \tilde{C}_n(X), \qquad \rho_n(a) = (a, \xi_n) \tag{6.117}$$

(for $a\colon \mathbb{I}^n \to X$ non-degenerate): this gives $\pi_n\rho_n(a) = (C_n a)(\xi_n) = a$.

**6.7.3 Acyclic Models Theorem**

*Let* $\mathsf{C}$ *be a category with models* $\mathcal{M}$ *and* $F, G\colon \mathsf{C} \to \mathrm{Ch}_+\mathsf{Ab}$ *two functors.*

*(i) Suppose that* $F$ *is free on these models and* $G$ *is acyclic on them.*

*Then every natural transformation* $\Phi\colon H_0F \to H_0G\colon \mathsf{C} \to \mathsf{Ab}$ *can be lifted to a natural transformation* $\varphi\colon F \to G$ *such that* $H_0\varphi = \Phi$*, and any two of them* $\varphi, \psi$ *are naturally chain homotopic: there exists a family of natural transformations* $\tau_n\colon F_n \to G_{n+1}$ *forming a homotopy* $\tau X = (\tau_n X)_n\colon \varphi X \simeq \psi X\colon FX \to GX$.

*(ii) More generally, the same conclusion holds if each* $F_n$ *is essentially free on a subset* $\mathcal{M}_n \subset \mathcal{M}$*, and* $G$ *is acyclic on* $\mathcal{M}$.

*Proof* See 6.7.9. □

**6.7.4 Theorem** (Equivalence of the simplicial and cubical form)

*The simplicial and the cubical functors of singular chains*

$$S_+\colon \mathsf{Top} \to \mathrm{Ch}_+\mathsf{Ab}, \qquad C_+\colon \mathsf{Top} \to \mathrm{Ch}_+\mathsf{Ab}, \tag{6.118}$$

*are linked by natural transformations* $\varphi\colon S_+ \rightleftarrows C_+ : \psi$*, whose composites are naturally chain-homotopic to identities.*

*The corresponding homology functors are naturally isomorphic.*

*Proof* (a) First we take in $\mathsf{Top}$ the set of simplicial models $\Delta^n$, for $n \geqslant 0$.

We have already seen, in 6.7.1(a), that the simplicial functor $S_+$ is free on these models. On the other hand, the cubical functor $C_+$ is acyclic on them (and all contractible spaces).

We have thus a natural transformation $\varphi\colon S_+ \to C_+$ that lifts the obvious isomorphism $\Phi\colon H_0S_+ \to H_0C_+$ that takes the simplicial homology class $[x]_\vartriangle$ of a point to the cubical homology class $[x]_\Box$ of the same point (see 3.6.7(e)); $\varphi$ is determined up to natural chain homotopy.

(b) Secondly, we take in $\mathsf{Top}$ the set of cubical models $\mathbb{I}^n$, for $n \geqslant 0$.

On these models, the cubical functor $C_+$ is essentially free, as we have seen in 6.7.2(b), and the simplicial functor $S_+\colon \mathsf{Top} \to \mathrm{Ch}_+\mathsf{Ab}$ is acyclic, by Lemma 3.6.8.

We have thus a natural transformation $\psi\colon C_+ \to S_+$ that lifts the inverse isomorphism $\Phi^{-1}$, and is determined up to natural chain homotopy.

(c) The composite $\psi\varphi\colon S_+ \to S_+$ lifts $\mathrm{id}\, H_0S_+$, and is chain homotopic to the identity. The composite $\varphi\psi$ is also. □

### 6.7.5 Eilenberg–Zilber Theorem

*The functors*

$$\begin{aligned}
&F^{\triangle}\colon \mathsf{Top}\times\mathsf{Top}\to \mathrm{Ch}_+\mathsf{Ab}, && F^{\triangle}(X,Y)=S_+(X)\otimes S_+(Y),\\
&G^{\triangle}\colon \mathsf{Top}\times\mathsf{Top}\to \mathrm{Ch}_+\mathsf{Ab}, && G^{\triangle}(X,Y)=S_+(X\times Y),\\
&F^{\square}\colon \mathsf{Top}\times\mathsf{Top}\to \mathrm{Ch}_+\mathsf{Ab}, && F^{\square}(X,Y)=C_+(X)\otimes C_+(Y),\\
&G^{\square}\colon \mathsf{Top}\times\mathsf{Top}\to \mathrm{Ch}_+\mathsf{Ab}, && G^{\square}(X,Y)=C_+(X\times Y),
\end{aligned} \tag{6.119}$$

*are linked by natural chain homotopy equivalences, forming the left square below, commutative up to chain homotopy*

$$\begin{array}{ccccccc}
F^{\triangle}(X,Y) & \xrightarrow{f} & G^{\triangle}(X,Y) & \qquad & H_nF^{\triangle}(X,Y) & \xrightarrow{f_{*n}} & H_nG^{\triangle}(X,Y)\\
{\scriptstyle h}\downarrow & \simeq & \downarrow{\scriptstyle k} & & {\scriptstyle h_{*n}}\downarrow & = & \downarrow{\scriptstyle k_{*n}}\\
F^{\square}(X,Y) & \xrightarrow[g]{} & G^{\square}(X,Y) & & H_nF^{\square}(X,Y) & \xrightarrow[g_{*n}]{} & H_nG^{\square}(X,Y)
\end{array} \tag{6.120}$$

*These natural transformations induce in homology a commutative diagram of natural isomorphisms, in the right square. Moreover, $f$ is determined up to chain homotopy by $f_{*0}$, and the same holds for $g, h, k$.*

*Proof* It is a consequence of the Acyclic Model Theorem, as proved in [EiZ2].

(a) First, we take as models of the category $\mathsf{Top}\times\mathsf{Top}$ the pairs $(\Delta^p, \Delta^q)$, for $p, q \geqslant 0$; in dimension $n$ there are $n+1$ models.

The functors $F^{\square}$ and $G^{\square}$ are acyclic on these models, while:

(i) $F^{\triangle}$ is acyclic and free on these models, with distinguished elements $\eta_{pq} = \mathrm{id}\,(\Delta^p)\otimes \mathrm{id}\,(\Delta^q)$,

(ii) $G^{\triangle}$ is acyclic and free on these models, with distinguished elements $\xi_{pq} = \mathrm{id}\,(\Delta^p\times\Delta^q)$.

(b) Secondly, we take as models of $\mathsf{Top}\times\mathsf{Top}$ the pairs $(\mathbb{I}^p, \mathbb{I}^q)$ for $p, q \geqslant 0$. Now, the functors $F^{\triangle}$ and $G^{\triangle}$ are acyclic on these models, while:

(iii) $F^{\square}$ is acyclic and essentially free on these models, with

$$\rho_nX\colon F^{\square}_n(X,Y)\to \tilde{F}^{\square}_n(X,Y), \qquad \rho_{pq}(a\otimes b) = (a\otimes b, \eta_{pq}),$$

where $a\colon \mathbb{I}^p\to X$ and $b\colon \mathbb{I}^q\to Y$ are non-degenerate cubes, and $\eta_{pq} = \mathrm{id}\,(\mathbb{I}^p)\otimes\mathrm{id}\,(\mathbb{I}^q)$,

(iv) $G^{\square}$ is acyclic and essentially free on these models, with

$$\rho_nX\colon G^{\square}_n(X,Y)\to \tilde{G}^{\square}_n(X,Y), \qquad \rho_{pq}(a\times b) = (a\times b, \xi_{pq}),$$

where $a\colon \mathbb{I}^p \to X$ and $b\colon \mathbb{I}^q \to Y$ are non-degenerate cubes, and $\xi_{pq} = \mathrm{id}\,(\mathbb{I}^p \times \mathbb{I}^q)$.

(c) The bijective correspondence $\Pi_0(X) \times \Pi_0(Y) \to \Pi_0(X \times Y)$ (between the bases of $H_0$) gives the right commutative square in (6.120), for $n = 0$ (natural on $\mathsf{Top}^2$).

The Acyclic Model Theorem leads to the conclusion. □

### *6.7.6 Corollary* (Eilenberg–Zilber)

*For topological spaces $X, Y$ there is a short exact sequence*

$$\bigoplus_{p+q=n} H_pX \otimes H_qY \overset{\alpha_n}{\rightarrowtail} H_n(X \times Y) \overset{\beta_n}{\twoheadrightarrow} \bigoplus_{p+q=n-1} \mathrm{Tor}(H_pX, H_qY)$$

*which is natural on* $\mathsf{Top} \times \mathsf{Top}$, *and splits in a non-natural way.*

*Proof* It is a straightforward consequence of Theorem 6.7.5 (using either the cubical or the simplicial chains) and the Künneth Theorem 6.5.4. □

### *6.7.7 Exercises and complements*

There are given two spaces $X, Y$.

(a) If $H_n(X) = 0$ for $n > n'$ and $H_n(Y) = 0$ for $n > n''$, then $H_n(X \times Y) = 0$ for $n > n' + n''$.

(b) If the homology groups of the space $X$ are flat, there is a canonical isomorphism

$$\alpha_n \colon \bigoplus_{p+q=n} H_p(X) \otimes H_q(Y) \longrightarrow H_n(X \times Y), \tag{6.121}$$

which is natural on $\mathsf{Top} \times \mathsf{Top}$. (We only need that $H_p(X)$ be flat for $p < n$.)

(c) (*Poincaré polynomial*) Suppose that each homology group $H_k(X)$ has a finite rank $\beta_k(X)$, eventually zero (see 2.2.7). It is convenient to write the sequence of Betti numbers as a polynomial, in the polynomial ring $\mathbb{Z}[t]$ (with a formal indeterminate $t$)

$$\beta X(t) = \Sigma_{k \geqslant 0}\, \beta_k(X)\, t^k. \tag{6.122}$$

It is called the *Poincaré polynomial* of $X$; it determines the graded group $H_*(X)$, if all $H_k(X)$ are free abelian groups.

If $Y$ satisfies the same hypotheses, this is also true of $X \times Y$ (by Corollary 6.7.6), and we can express $\beta(X \times Y)$ as a product of polynomials

$$\beta(X \times Y)(t) = \Sigma_{k \geqslant 0}\, (\Sigma_{p+q=k}\, \beta_p(X)\, \beta_q(Y)) t^k = \beta X(t)\, \beta Y(t). \tag{6.123}$$

Without assuming the Betti numbers to be eventually zero, we have a formal series in the ring $\mathbb{Z}[[t]]$, with the same result as above: note that every coefficient $\sum_{p+q=n} \beta_p(X)\, \beta_q(Y)$ is still given by a finite sum.

(d) (*Euler–Poincaré characteristic*) In the same hypotheses on $X$ and $Y$ (their Betti numbers are finite and eventually zero), we have

$$\chi(X \times Y) = \chi(X)\, \chi(Y). \tag{6.124}$$

*Hint:* $\chi(X)$ can be expressed as a value of the polynomial $\beta X$.

(e) Compute the homology groups of the $n$-torus $\mathbb{T}^n = \mathbb{S}^1 \times ... \times \mathbb{S}^1$ (with $n$ factors).

### 6.7.8 Complements

The chain homotopy equivalences 6.7.5 between the functors $F^\Box$ and $G^\Box$, or $F^\triangle$ and $G^\triangle$, were obtained by applying the Acyclic Models Theorem. Yet, there are specific candidates for some of these chain maps (that are determined up to chain homotopy by the homomorphisms induced on $H_0$).

In fact, there are *easy* constructions $F^\Box \to G^\Box$ and $G^\triangle \to F^\triangle$, but not the other way round.

(a) (*Product of spaces by cubical chains*) We can build a natural transformation $\zeta \colon F^\Box \to G^\Box$, exploiting the fact that a product of standard cubes $\mathbb{I}^p \times \mathbb{I}^q$ 'is' the standard cube $\mathbb{I}^{p+q}$.

This gives a chain map $\zeta \colon C_+(X) \otimes C_+(Y) \to C_+(X \times Y)$, where $\zeta_n$ has co-components

$$\begin{aligned} &\zeta_{pq} \colon C_p(X) \otimes C_q(Y) \longrightarrow C_n(X \times Y) \qquad (n = p+q), \\ &\zeta_{pq}(a \otimes b) = a \times b \colon \mathbb{I}^n \to X \times Y \qquad (a \colon \mathbb{I}^p \to X,\ b \colon \mathbb{I}^q \to Y). \end{aligned} \tag{6.125}$$

A (more complex) chain map $\eta \colon C_+(X \times Y) \to C_+(X) \otimes C_+(Y)$ such that $\eta\zeta$ is the identity is constructed in [Mas3], Section VI.5; then, a homotopy equivalence $\zeta\eta \simeq \mathrm{id}$ is proved to *exist*, by the method of Acyclic Models.

(b) (*Product of spaces by simplicial chains*) We have already built, in (6.110), a natural transformation $\vartheta \colon G^\triangle \to F^\triangle$, related to the Alexander–Whitney diagonal approximation.

(c) Singular homology can equivalently be built with the *normalised* simplicial chain complex, namely the quotient of the ordinary complex modulo degenerate simplicial chains.

This can be seen in [HiW], Section 8.2; the equivalence is proved in Section 8.4, by the method of Acyclic Models (and could be proved here, as in Theorem 6.7.4).

### 6.7.9 *Proof of Theorem 6.7.3

(a) We begin to prove (i).

For a space $X$, consider the following solid diagram

$$\begin{array}{ccccccccc} \dots & \xrightarrow{\partial_2} & F_1(X) & \xrightarrow{\partial_1} & F_0(X) & \xrightarrow{\partial_0} & (H_0F)X & = & F_{-1}(X) \\ & & \downarrow{\scriptstyle\varphi_1} & & \downarrow{\scriptstyle\varphi_0} & & \downarrow{\scriptstyle\Phi} & & \downarrow{\scriptstyle\varphi_{-1}} \\ \dots & \xrightarrow[\partial_2]{} & G_1(X) & \xrightarrow[\partial_1]{} & G_0(X) & \xrightarrow[\partial_0]{} & (H_0G)X & = & G_{-1}(X) \end{array} \qquad (6.126)$$

where we have augmented the chain complexes $FX$ and $GX$ so that, in both rows, $\partial_0(z) = [z]$ is the natural projection on $H_0$, and $\operatorname{Im}\partial_1 = \operatorname{Ker}\partial_0$. We also rename $\Phi = \varphi_{-1}$ for convenience in the induction step, and remark that the lower row is exact (in every degree) when $X$ is a model.

We want to lift the natural transformation $\Phi\colon H_0F \to H_0G$ to a natural transformation $\varphi\colon F \to G$. The argument is a modified version of the lifting procedure in 6.1.2(b) (on projective resolutions): here we are bound by the naturality requirement; moreover, we only know that the complex $FX$ is free on the given models and the complex $GX$ is acyclic on models.

(b) *Case* $n = 0$. The first lifting, namely $\varphi_0X\colon F_0(X) \to G_0(X)$, already has to pass through the models. First, for every distinguished element $\xi_j \in F_0(M_j)$, with $j \in J_0$, we choose an element $(\varphi_0M_j)(\xi_j) \in G_0(M_j)$ such that $\partial_0(\varphi_0M_j)(\xi_j) = (\Phi M_j)\partial_0(\xi_j)$

$$\begin{array}{ccccccc} F_0(X) & \xleftarrow{F_0(a)} & F_0(M_j) & \xrightarrow{\partial_0} & (H_0F)M_j & \longrightarrow & 0 \\ \downarrow{\scriptstyle\varphi_0} & & \downarrow{\scriptstyle\varphi_0} & & \downarrow{\scriptstyle\Phi} & & \\ G_0(X) & \xleftarrow[G_0(a)]{} & G_0(M_j) & \xrightarrow[\partial_0]{} & (H_0G)M_j & \longrightarrow & 0 \end{array} \qquad (6.127)$$

Note that we have not (yet) defined $\varphi_0M_j$ on the whole $F_0(M_j)$. Rather, we define $\varphi_0X$ for an arbitrary space, on the basis of $F_0(X)$ formed by the elements $a_j = (F_0a)(\xi_j)$ (for $j \in J_0$ and all maps $a\colon M_j \to X$), letting

$$\varphi_0X\colon F_0(X) \to G_0(X), \qquad (\varphi_0X)(a_j) = (G_0a)(\varphi_0M_j)(\xi_j), \qquad (6.128)$$

as required by the left square above.

The family of these components is natural: for a map $f\colon X \to Y$ we have $(G_0f)(\varphi_0X) = (\varphi_0Y)(F_0f)$, as we verify on the basis $a_j = (F_0a)(\xi_j)$

$$\begin{aligned} (G_0f)(\varphi_0X)(a_j) &= (G_0f)(G_0a)(\varphi_0M_j)(\xi_j) \\ &= (G_0(fa))(\varphi_0M_j)(\xi_j) = (\varphi_0Y)(F_0(fa))(\xi_j) = (\varphi_0Y)(F_0f)(a_j). \end{aligned}$$

Moreover $H_0\varphi_0 = \Phi$ because the rightmost square in (6.126) commutes, by the naturality of $\Phi$

$$\begin{aligned}(\Phi X)\partial_0(a_j) &= (\Phi X)\partial_0(F_0 a)(\xi_j) = \Phi(H_0F(a))\partial_0(\xi_j)\\ &= (H_0G(a))(\Phi M_j)\partial_0(\xi_j) = (H_0G(a))\partial_0(\varphi_0 M_j)(\xi_j)\\ &= \partial_0(G_0 a)(\varphi_0 M_j)(\xi_j) = \partial_0(\varphi_0 X)(a_j).\end{aligned}$$

(c) *Inductive step.* Suppose that, for $n \geqslant 1$, we have defined the natural transformations $\varphi_0, \varphi_1, \dots \varphi_{n-1}\colon F_{n-1} \to G_{n-1}$, commuting so far with the differentials.

We want to lift $\varphi_n X\colon F_n(X) \to G_n(X)$. For each distinguished element $\xi_j \in F_n(M_j)$ ($j \in J_n$) we choose an element $(\varphi_n M_j)(\xi_j) \in G_n(M_j)$ such that $\partial_n(\varphi_n M_j)(\xi_j) = (\varphi_{n-1}M_j)(\partial_n \xi_j)$

$$\begin{array}{ccccccc} F_n(X) & \xleftarrow{F_n(a)} & F_n(M_j) & \xrightarrow{\partial_n} & F_{n-1}(M_j) & \xrightarrow{\partial_{n-1}} & F_{n-2}(M_j) \\ \downarrow{\scriptstyle \varphi_n} & & \downarrow{\scriptstyle \varphi_n} & & \downarrow{\scriptstyle \varphi_{n-1}} & & \downarrow{\scriptstyle \varphi_{n-2}} \\ G_n(X) & \xleftarrow[G_n(a)]{} & G_n(M_j) & \xrightarrow[\partial_n]{} & G_{n-1}(M_j) & \xrightarrow[\partial_{n-1}]{} & G_{n-2}(M_j) \end{array} \qquad (6.129)$$

This is possible, because $\partial_{n-1}(\varphi_{n-1}M_j)(\partial_n\xi_j) = (\varphi_{n-2}M_j)\partial_{n-1}(\xi_j) = 0$, and $(\varphi_{n-1}M_j)(\partial_n\xi_j)$ is a boundary in $G_{n-1}(M_j)$ (by exactness).

Now we define the component $\varphi_n X$, on the basis formed by the elements $a_j = (F_n a)(\xi_j)$ (for $j \in J_n$ and all maps $a\colon M_j \to X$), letting

$$\varphi_n X\colon F_n(X) \to G_n(X), \qquad (\varphi_n X)(a_j) = (G_n a)(\varphi_n M_j)(\xi_j), \qquad (6.130)$$

and we verify as above that this is a natural transformation and preserves commutativity, in diagram (6.126).

(d) *Homotopy.* We suppose that $\varphi X\colon FX \to GX$ lifts the trivial homomorphism $(H_0F)X \to (H_0G)X$, and we want to prove that it is homotopic to the zero transformation $0\colon FX \to GX$, by a family of natural transformations $\tau_n\colon F_n \to G_{n+1}$ such that $\partial_{n+1}\tau_n + \tau_{n-1}\partial_n = \varphi_n$

$$\begin{array}{ccccccc} \dots F_2(X) & \xrightarrow{\partial_2} & F_1(X) & \xrightarrow{\partial_1} & F_0(X) & \xrightarrow{\partial_0} & (H_0F)X \\ \downarrow{\scriptstyle \varphi_2} & \swarrow{\scriptstyle \tau_1} & \downarrow{\scriptstyle \varphi_1} & \swarrow{\scriptstyle \tau_0} & \downarrow{\scriptstyle \varphi_0} & & \downarrow{\scriptstyle 0} \\ \dots G_2(X) & \xrightarrow[\partial_2]{} & G_1(X) & \xrightarrow[\partial_1]{} & G_0(X) & \xrightarrow[\partial_0]{} & (H_0G)X \end{array} \qquad (6.131)$$

Again, the lifting is an enriched version of the corresponding argument for projective resolutions, in Exercise 6.1.2(c).

(e) *Case $n = 0$.* We have to build a natural transformation $\tau_0 X\colon F_0(X) \to G_1(X)$ such that $\partial_1\tau_0 = \varphi_0$.

First, for every distinguished element $\xi_j \in F_0(M_j)$ $(j \in J_0)$, we choose an element $(\tau_0 M_j)(\xi_j) \in G_1(M_j)$ such that $\partial_1(\tau_0 M_j)(\xi_j) = (\varphi_0 M_j)(\xi_j)$

$$\begin{array}{ccccccc} & & F_0(X) & \xleftarrow{F_0(a)} & F_0(M_j) & \xrightarrow{\partial_0} & (H_0F)M_j \\ & \swarrow{\scriptstyle \tau_0} & & \swarrow{\scriptstyle \tau_0} & \downarrow{\scriptstyle \varphi_0} & & \downarrow{\scriptstyle 0} \\ G_1(X) & \xleftarrow[G_0(a)]{} & G_1(M_j) & \xleftarrow[G_0(a)]{} & G_0(M_j) & \xrightarrow[\partial_0]{} & (H_0G)M_j \end{array} \tag{6.132}$$

This is legitimate: $(\varphi_0 M_j)(\xi_j)$ is a boundary in $G_0(M_j)$, by the right square above. We define the component $\tau_0 X$ on the usual basis formed by the elements $a_j = (F_0 a)(\xi_j)$ (for $j \in J_0$ and all maps $a\colon M_j \to X$), letting

$$\tau_0 X\colon F_0(X) \to G_1(X), \qquad (\tau_0 X)(a_j) = (G_1 a)(\tau_0 M_j)(\xi_j). \tag{6.133}$$

We verify as in (b) the naturality of $\tau_0$ and the identity $\partial_1\tau_0 = \varphi_0$.

(f) The inductive step, defining the natural transformation $\tau_n X\colon F_n(X) \to G_{n+1}(X)$ after the lower ones, is performed as in (d).

(g) We end by showing how the previous part is modified, to prove (ii).

By hypothesis, every functor $F_n\colon \mathsf{C} \to \mathsf{Ab}$ has a natural retraction $\pi\rho\colon F_n \to \tilde{F}_n \to F_n$. We recall that, for an object $X$ of $\mathsf{C}$, $\tilde{F}_n(X)$ is the free abelian group generated by all pairs $(a, \xi)$, where $a\colon M \to X$ is a morphism of $\mathsf{C}$ defined on a model $M \in \mathcal{M}_n$ and $\xi \in F_n(M)$.

We have the solid diagram:

$$\begin{array}{ccccccc} \dots\ F_2(X) & \xrightarrow{\partial_2} & F_1(X) & \xrightarrow{\partial_1} & F_0(X) & \xrightarrow{\partial_0} & (H_0F)X \\ \downarrow{\scriptstyle \rho_2} & & \downarrow{\scriptstyle \rho_1} & & \downarrow{\scriptstyle \rho_0} & & \\ \dots\ \tilde{F}_2(X) & & \tilde{F}_1(X) & & \tilde{F}_0(X) & & \downarrow{\scriptstyle \Phi} \\ \downarrow{\scriptstyle \tilde{\varphi}_2} & & \downarrow{\scriptstyle \tilde{\varphi}_1} & & \downarrow{\scriptstyle \tilde{\varphi}_0} & & \\ \dots\ G_2(X) & \xrightarrow[\partial_2]{} & G_1(X) & \xrightarrow[\partial_1]{} & G_0(X) & \xrightarrow[\partial_0]{} & (H_0G)X \end{array} \tag{6.134}$$

For every $\xi \in F_0(M)$, with $M \in \mathcal{M}_0$, we choose an element $(\varphi_0 M)(\xi) \in G_0(M)$ such that $\partial_0(\varphi_0 M)(\xi) = (\Phi M)\partial_0(\xi)$. Then we define the homomorphism $\tilde{\varphi}_0 X\colon \tilde{F}_0(X) \to G_0(X)$ letting, for $a\colon M \to X$

$$\tilde{\varphi}_0(a, \xi) = (G_0 a)(\varphi_0 M)(\xi),$$

and we let $\varphi_0 X = (\tilde{\varphi}_0 X)(\rho_0 X)\colon F_0(X) \to G_0(X)$.

The rest is similarly modified. □

# 7
# Fundamental groupoid and homotopy groups

Homotopy groups explore $n$-dimensional properties of a space $X$, by *probes* $\mathbb{S}^n \to X$. Their study, quite complex, forms a large domain of Algebraic Topology. Here we only want to present the basic ideas and the relationship between these groups and singular homology.

After the definition and basic properties of the fundamental groupoid and fundamental group, we prove the main computation tool, the van Kampen theorem. Then we briefly introduce the (absolute and relative) higher homotopy groups (Section 7.5), and the Whitehead theorems on weak homotopy equivalences (Section 7.6).

This chapter can be read together with Chapter 2, and is written in an elementary way, up to the first half of Section 7.5.

*Literature.* The fundamental group of a pointed space, also called the *Poincaré group*, was introduced in 1895 by H. Poincaré in a celebrated article called 'Analysis situs' [Po]. It was the first algebraic structure used to study spaces, and the beginning of Algebraic Topology – before topology had this name. The higher homotopy groups where introduced by Čech, in 1931, and studied by Hurewicz since 1935.

Many books on Algebraic Topology deal with the fundamental group, like [HiW, Sp2, Mu, Ha]. For the fundamental groupoid we refer to R. Brown's article and books [Br1]–[Br3]. Higher homotopy groups can be approached with [Sp2], Chapter 7, [Ha], Chapter 4, and the book [Hu], by S.T. Hu, a classical reference on Homotopy Theory.

## 7.1 The fundamental groupoid

The fundamental groupoid $\Pi_1 X$ of a space is a global structure, that contains the fundamental group $\pi_1(X, x)$ at each point $x \in X$, all of them nicely organised inside the groupoid – a (small) category where all arrows are invertible.

This structure gives raise to a functor $\Pi_1 \colon \mathsf{Top} \to \mathsf{Gpd}$ that transforms topological spaces into small groupoids, their maps into functors, and their homotopies into natural isomorphisms, making the analogy remarked in 1.5.5 concrete.

### 7.1.1 Reviewing paths

The basic terminology on paths, loops and homotopies was fixed in Section 1.4.

A path $a \colon \mathbb{I} \to X$ from $x$ to $x'$ (its endpoints) is viewed as a homotopy $a \colon x \simeq x' \colon \{*\} \to X$. If $x = x'$, the path is called a loop at $x$, its basepoint.

We write as $P(X)$ the set of paths of the space $X$, as $P(X, x, x')$ the set of paths from $x$ to $x'$, as $\Omega(X, x) = P(X, x, x)$ the set of loops at $x$.

Consecutive paths in $P(X)$ have a concatenation. The *regular concatenation* of a sequence of consecutive paths is written as $a_1 * a_2 * \ldots * a_n$ (cf. (1.76)), the reversed path as $a^\sharp$.

*Exercises.* There are given a path $a \colon x \simeq y$, a loop $b \colon y \simeq y$, and a path $c \colon y \simeq z$ in a space $X$.

(a) Write the parametrisation of the paths $a * b * c$ and $a * b * b * c$.

(b) Write the parametrisation of the loop $a * b * a^\sharp \in \Omega(X, x)$.

### 7.1.2 The fundamental groupoid

We associate to a space $X$ a small groupoid $\Pi_1 X$, called the *fundamental groupoid* of $X$.

An object (or *vertex*) of $\Pi_1 X$ is any point $x \in X$; an arrow $[a] \colon x \to x'$ is a class of paths in $X$, from $x$ to $x'$, up to homotopy with fixed endpoints.

In other words (as already said in 1.4.7), two paths $a_0, a_1 \colon \mathbb{I} \to X$ from $x$ to $x'$ are identified if there exists a continuous mapping $h \colon \mathbb{I} \times \mathbb{I} \to X$ which deforms one into the other, without moving the endpoints

$$h(s, i) = a_i(s), \quad h(0, t) = x, \quad h(1, t) = x' \qquad (s, t \in \mathbb{I},\ i = 0, 1), \qquad (7.1)$$

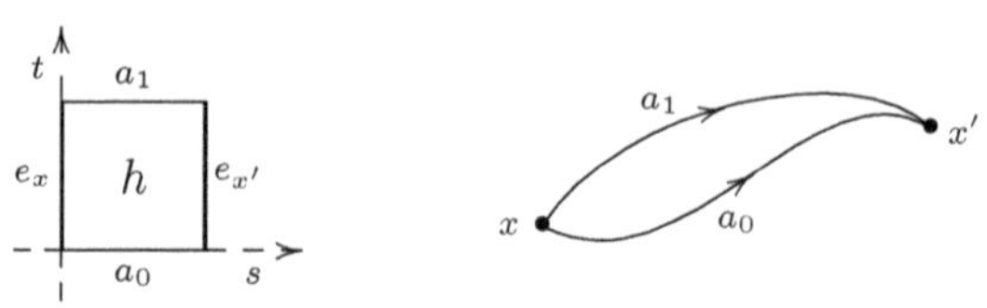


The map $h$ is also called a 2-homotopy from $a_0$ to $a_1$ and denoted as $h \colon a_0 \simeq_2 a_1 \colon x \simeq x'$. The 2-homotopy relation $a_0 \simeq_2 a_1$ is an equivalence relation in the set of paths $P(X, x, x')$, as verified in Exercise 1.4.7(b).

In $\Pi_1 X$ consecutive arrows $[a]\colon x \to x'$ and $[b]\colon x' \to x''$ are composed by concatenation of paths (written in the natural order):

$$[a]\,[b] = [a * b]\colon x \to x''. \tag{7.2}$$

As verified in the exercises below this operation is well defined and gives a groupoid, with identities represented by constant loops and inverses represented by reversed paths.

At each point $x \in X$ the endoarrows $[a]\colon x \to x$ of $\Pi_1 X$ form the fundamental group studied in the next section

$$\pi_1(X, x) = (\Pi_1 X)(x, x). \tag{7.3}$$

*Exercises and complements.* These exercises are an easy extension of the corresponding issues for the fundamental group – in case this is already known to the reader.

(a) In $\Pi_1 X$ the partial operation of composition is well defined.

(b) It is associative. This is a particular case of 1.4.7(d), but – owing to its importance – it can be useful to give a direct proof. *Hint:* the concatenations $a * (b * c)$ and $(a * b) * c$ of three consecutive paths in $X$ are interchanged by an invertible reparametrisation of $\mathbb{I}$.

(c) Every vertex $x \in X$ has an identity, namely the class $[e_x]$ of the constant loop at $x$.

(d) The inverse of $[a]\colon x \to y$ is the class $[a]^{-1} = [a^\sharp]$ of the reversed path.

(e) One can give an equivalent construction by 'Moore paths', parametrised on intervals $[0, r]$, so that concatenation is associative from the start; [HiW] follows this way for the fundamental group.

### 7.1.3 The fundamental-groupoid functor

The previous construction gives a functor with values in the category of small groupoids

$$\begin{gathered} \Pi_1\colon \mathsf{Top} \to \mathsf{Gpd}, \qquad X \mapsto \Pi_1 X, \\ f_*\colon \Pi_1 X \to \Pi_1 Y, \qquad f_*(x) = f(x), \quad f_*([a]) = [fa]. \end{gathered} \tag{7.4}$$

A map $f\colon X \to Y$ is sent to the functor $\Pi_1 f = f_*$ which takes the class of a path $a\colon x \simeq x'$ in $X$ to the class of the path $fa\colon fx \simeq fx'$ in $Y$.

This requires some straightforward verifications, listed in the exercises below.

*Exercises and complements.* (a) For any map $f\colon X \to Y$ and all $x, x' \in X$, the mapping $f_*\colon \Pi_1 X(x, x') \to \Pi_1 Y(fx, fx')$ is well defined.

(b) (*Local functoriality*) The transformation $f_*\colon \Pi_1 X \to \Pi_1 Y$ is a functor between groupoids.

(c) (*Global functoriality*) $\Pi_1\colon \mathsf{Top} \to \mathsf{Gpd}$ is a functor:

- given two consecutive maps $f\colon X \to Y$ and $g\colon Y \to Z$, $(gf)_* = g_* f_*$,
- given a space $X$, the functor $(\mathrm{id}\, X)_*$ is the identity of the groupoid $\Pi_1 X$.

(d) In degree zero, we have already seen in (1.71) the homotopy invariant functor $\Pi_0\colon \mathsf{Top} \to \mathsf{Set}$, that takes every space to the set of its path components

$$\begin{aligned} &\Pi_0\colon \mathsf{Top} \to \mathsf{Set}, && \Pi_0(X) = |X|/\!\simeq, \\ f_* = {}&\Pi_0(f)\colon \Pi_0(X) \to \Pi_0(Y), && f_*([x]_P) = [f(x)]_P, \end{aligned} \tag{7.5}$$

where $[x]_P$ is the path component of $x$ in $X$, and $f\colon X \to Y$ is a map.

This functor can be seen as a composite $F\Pi_1\colon \mathsf{Top} \to \mathsf{Gpd} \to \mathsf{Set}$, for a suitable functor $F$.

### *7.1.4* **Theorem** (Homotopy equivariance)

*(a) The functor $\Pi_1$ has a canonical enrichment: a homotopy $\varphi\colon f \simeq g\colon X \to Y$ has an associated* homotopy of groupoids, *i.e. a natural isomorphism*

$$\begin{aligned} &\varphi_*\colon f_* \to g_*\colon \Pi_1 X \to \Pi_1 Y, \\ &\varphi_*(x) = [\varphi(x,-)] \in \Pi_1 Y(f(x), g(x)) \qquad (x \in X), \end{aligned} \tag{7.6}$$

*where $\varphi(x,-)\colon \mathbb{I} \to Y$ is a path from $f(x)$ to $g(x)$ in $Y$.*

*The naturality condition on an arrow $[a]\colon x \to x'$ in $\Pi_1 X$ means the commutativity of the following square, in the groupoid $\Pi_1 Y$*

$$\begin{array}{ccc} f(x) & \xrightarrow{\varphi_*(x)} & g(x) \\ {\scriptstyle [fa]}\downarrow & & \downarrow{\scriptstyle [ga]} \\ f'(x) & \xrightarrow[\varphi_*(x')]{} & g'(x) \end{array} \qquad [fa]\,\varphi_*(x') = \varphi_*(x)\,[ga]. \tag{7.7}$$

*(b) A homotopy equivalence of spaces is taken by $\Pi_1$ to a categorical equivalence of groupoids.*

*Proof* We define $\varphi_*(x) = [\varphi(x,-)]$, as in (7.6): this makes sense, because $\varphi_x = \varphi(x,-)\colon \mathbb{I} \to Y$ is a path from $f(x)$ to $g(x)$, in $Y$, and its class $[\varphi_x]$ is an arrow in $\Pi_1 Y$.

The commutativity of the square (7.7) follows from the continuous mapping $\Phi = \varphi(a \times \mathrm{id}\,\mathbb{I})\colon \mathbb{I}^2 \to Y$, whose boundary is represented below

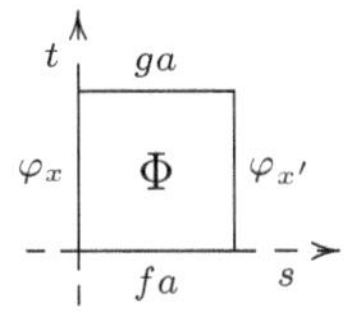


(7.8)

Applying Lemma 7.1.8, we get a 2-homotopy $fa * \varphi_{x'} \simeq_2 \varphi_x * ga$.

(b) Given a homotopy equivalence $f\colon X \rightleftarrows Y : g$, with homotopies $gf \simeq 1_X$ and $fg \simeq 1_Y$, we get two functors $f_*\colon \Pi_1 X \rightleftarrows \Pi_1 Y : g_*$ whose composites are isomorphic to identity functors. □

***7.1.5 Exercises and complements*** (Path connectedness)

(a) As in 1.4.6, we say that a space $X$ is 0-connected if it is non-empty and path connected.

Equivalently, it has precisely one path component: $\Pi_0(X)$ is a singleton. Equivalently again, $\Pi_1 X$ is a *connected* groupoid: it is not empty and for any two points $x, x' \in X$ there is some arrow $x \to x'$ in $\Pi_1 X$.

(b) The following conditions on a space $X$ are equivalent:

(i) any two paths $a, b\colon x \simeq y$ in $X$ are 2-homotopic,
(ii) any loop $a\colon x \simeq x$ in $X$ is 2-homotopic to the constant loop $e_x$,
(iii) for any $x, x' \in X$ there is at most one arrow $x \to x'$ in $\Pi_1 X$,
(iv) for any $x \in X$ there is a unique arrow $x \to x$ in $\Pi_1 X$, its identity.

(c) The following conditions are equivalent, and define a *1-connected* space (as anticipated in 1.4.7):

(i′) $X$ is 0-connected and satisfies the equivalent properties above,
(ii′) the groupoid $\Pi_1 X$ is non-empty and *indiscrete*: for any $x, x' \in X$ there is a unique arrow $x \to x'$ in $\Pi_1 X$,
(iii′) $\Pi_1 X$ is equivalent to the category **1**.

*A further equivalent condition can be added, using a result that will be proved in 7.3.1:

(iv′) $X$ is non-empty, path connected, and $\pi_1(X, x) = 0$ for some $x \in X$.*

(d) The property of being 0-connected, or 1-connected, is invariant up to homotopy equivalence. Every contractible space is 1-connected. $\mathbb{S}^0$ and $\mathbb{S}^1$ are not, but we shall see in 7.4.4(a) that all higher spheres $\mathbb{S}^n$ are (for $n \geqslant 2$).

(e) A space $X$ is said to be *totally path disconnected* if all its path components are singletons. This is equivalent to saying that the groupoid $\Pi_1 X$

is discrete, i.e. its only arrows are the identities. This property is also invariant up to homotopy equivalence.

Every discrete space and the rational line $\mathbb{Q}$ (with euclidean topology) are of this kind. (We may require that there are at least two points!)

**7.1.6 Proposition** (Cartesian products)

*The functor* $\Pi_1\colon \mathsf{Top} \to \mathsf{Gpd}$ *preserves cartesian products, as made precise in the following points.*

*(a) For a product* $X = \prod_{i\in I} X_i$ *of topological spaces (with projections* $p_i\colon X \to X_i$*), there is a canonical isomorphism of groupoids*

$$\varphi\colon \Pi_1 X \longrightarrow \textstyle\prod_{i\in I} \Pi_1 X_i, \qquad \varphi([a]) = ([p_i a])_{i\in I}. \tag{7.9}$$

*(b)* (Naturality) *Every family of maps* $f_i\colon X_i \to Y_i$ *gives a commutative square in* $\mathsf{Gpd}$ *(with* $f = \prod_i f_i$*)*

$$\begin{array}{ccc} \Pi_1(\prod_i X_i) & \xrightarrow{\ \varphi\ } & \prod_i(\Pi_1 X_i) \\ {\scriptstyle f_*}\big\downarrow & & \big\downarrow{\scriptstyle \prod_i f_{i*}} \\ \Pi_1(\prod_i Y_i) & \xrightarrow[\ \varphi\ ]{} & \prod_i(\Pi_1 Y_i) \end{array} \tag{7.10}$$

**Note.* The commutative square (7.10) says that (7.9) is the component on the family $(X_i)_{i\in I}$ of an isomorphism of functors defined on the cartesian-power category $\mathsf{Top}^I = \prod_{i\in I} \mathsf{Top}$

$$\begin{gathered} \varphi\colon F \to G\colon \mathsf{Top}^I \to \mathsf{Gpd}, \\ F((X_i)_{i\in I}) = \Pi_1(\textstyle\prod_{i\in I} X_i), \qquad G((X_i)_{i\in I}) = \prod_{i\in I}(\Pi_1 X_i). \end{gathered} \tag{7.11}$$

*Proof* (a) A path $a\colon x \simeq x'$ in $X = \prod_{i\in I} X_i$ gives a path $p_i a\colon p_i x \simeq p_i x'$ in each $X_i$, and amounts to their family $(p_i a)_{i\in I}$, by the universal property of the product (in 1.6.1(i)). Similarly a 2-homotopy $h\colon a \simeq_2 b\colon x \simeq x'$ in $X$ amounts to the family of 2-homotopies $p_i h\colon p_i a \simeq_2 p_i b$ $(i \in I)$.

The mapping $a \mapsto (p_i a)_{i\in I}$ is consistent with path concatenation, and induces an isomorphism of groupoids $\varphi\colon [a] \mapsto ([p_i a])_{i\in I}$.

(b) In the square (7.10), the two composed diagonal functors send an arrow $[a] \in \Pi_1 X$ to the same $I$-indexed family in $\prod_i(\Pi_1 Y_i)$

$$\begin{gathered} \varphi f_*([a]) = \varphi([fa]) = ([p_i fa])_{i\in I}, \\ (\textstyle\prod_i f_{i*})\,\varphi([a]) = (\prod_i f_{i*})([p_i a])_{i\in I}) = ([f_i p_i a])_{i\in I}. \end{gathered}$$

□

### 7.1.7 Double homotopies and 2-homotopies

A 2-cube $A\colon \mathbb{I}^2 \to X$ will be seen as a *double homotopy* among four paths, its faces

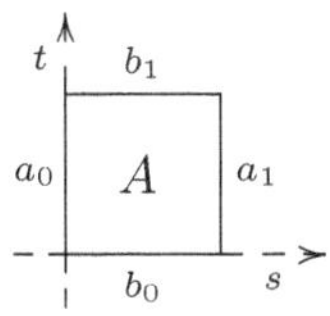


$$\partial_1^\alpha A = a_\alpha, \qquad \partial_2^\alpha A = b_\alpha. \tag{7.12}$$

It is a 2-homotopy $b_0 \simeq_2 b_1$ when both paths $a_\alpha = \partial_1^\alpha A$ are constant.

Two double homotopies $A, B$ have a *horizontal concatenation* $A *_1 B$, in the first variable, provided they are horizontally consecutive, that is, $\partial_1^+ A = \partial_1^- B$

$$(A *_1 B)(s,t) = \begin{cases} A(2s,t), & \text{for } 0 \leqslant s \leqslant 1/2, \\ B(2s-1,t), & \text{for } 1/2 \leqslant s \leqslant 1. \end{cases} \tag{7.13}$$

Symmetrically, there is a *vertical concatenation* $A *_2 C$, in the second variable, when $\partial_2^+ A = \partial_2^- C$.

We shall often use the following double homotopies, produced by a path $a\colon x \simeq y$

$$M(s,t) = a(\max(s,t)), \qquad L(s,t) = a(\min(s,t)), \tag{7.14}$$

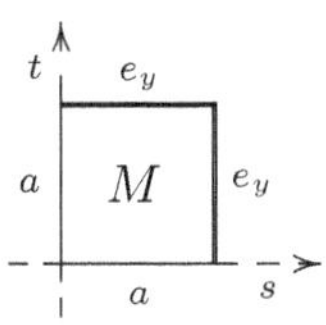


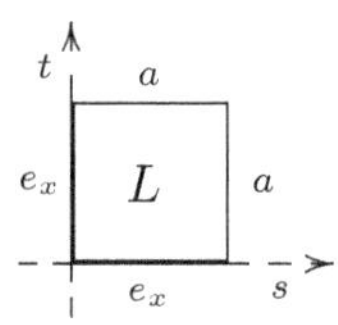


*These double homotopies come out of the higher structure of the interval mentioned in 1.4.3(d): the connections $g^\alpha\colon \mathbb{I}^2 \to \mathbb{I}$.*

### 7.1.8 Lemma

*The double homotopy $A\colon \mathbb{I}^2 \to X$ represented in* (7.12) *determines a 2-homotopy*

$$h\colon b_0 * a_1 \simeq_2 a_0 * b_1 \qquad (\partial_1^\alpha A = a_\alpha,\ \partial_2^\alpha A = b_\alpha). \tag{7.15}$$

*Proof* We let $L$ and $M$ be the double homotopies (7.14), associated to the paths $a_0$ and $a_1$.

We form the (regular) horizontal concatenation $L *_1 A *_1 M$, represented below, with different units on the axes

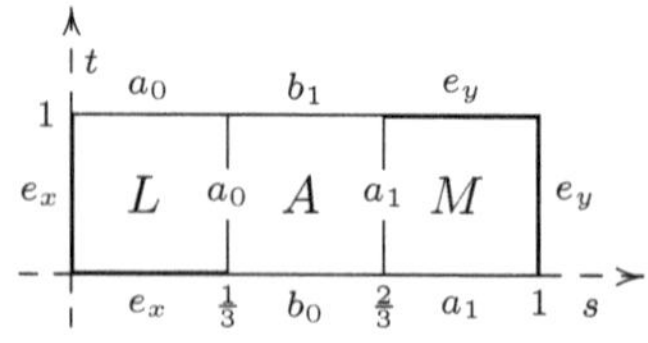


(7.16)

This is a 2-homotopy $e_x * b_0 * a_1 \simeq_2 a_0 * b_1 * e_y$. As we already know that

$$b_0 * a_1 \simeq_2 e_x * b_0 * a_1, \qquad a_0 * b_1 * e_y \simeq_2 a_0 * b_1,$$

the proof is done. □

### 7.1.9 Representative subsets

Let $A$ be a subset of the space $X$. We shall write as $\Pi_1 X|A$ the full subgroupoid of $\Pi_1 X$ with vertices in $A$. (R. Brown's notation is $\pi_1 XA$.)

We say that $A$ is a *representative subset* of $X$ if it meets all its path connected components. Then $\Pi_1 X|A$ is categorically equivalent to $\Pi_1 X$, by Exercise (b).

*Exercises and complements.* (a) If $A$ is a singleton $\{x_0\}$, $\Pi_1 X|A$ is the fundamental group $\pi_1(X, x_0)$.

(b) If $A$ is representative in $X$, the subgroupoid $\Pi_1 X|A$ is a *deformation retract* of $\Pi_1 X$: there exists a functor $P\colon \Pi_1 X \to \Pi_1 X|A$ such that, letting $J\colon \Pi_1 X|A \to \Pi_1 X$ be the inclusion, $PJ = \mathrm{id}\,(\Pi_1 X|A)$ and $JP \cong \mathrm{id}\,\Pi_1 X$. *Hint:* choose, for every $x \in X$, a convenient path $\lambda_x\colon x \simeq P(x)$ in $X$.

*This statement is a particular case of the characterisation 8.1.5(iii) of equivalence of categories: in fact, the subset $A$ is representative in $X$ if and only if the inclusion functor $J\colon \Pi_1 X|A \to \Pi_1 X$ is essentially surjective on objects. The general case is proved as this particular case.*

## 7.2 The fundamental group

At each basepoint $x \in X$, the fundamental group $\pi_1(X, x)$ is the group of automorphisms of the fundamental groupoid at the vertex $x$. Viewing these groups inside $\Pi_1 X$ makes evident how all of them are related.

We compute here the fundamental group of the circle and the $n$-torus $\mathbb{T}^n$.

Various exercises, in 7.2.6 and 7.2.7, show how one can get a concrete grasp of the fundamental group, making 'experiments' with elastic bands and strings in our 3-dimensional space, where we place obstructions (like rods or rings) that our material loops cannot trespass.

### 7.2.1 Definition

For every $x \in X$, the fundamental group $\pi_1(X, x)$ is the group of automorphisms $\Pi_1 X(x) = \Pi_1 X(x, x)$ of the groupoid $\Pi_1 X$, at the vertex $x$.

In other words, $\pi_1(X, x)$ is the set $\Omega(X, x)/\simeq_2$ of classes $[a]$ of loops of $X$ at the point $x$, modulo homotopy with fixed basepoint; this set is equipped with the operation induced by the concatenation of loops:

$$[a][b] = [a * b]. \tag{7.17}$$

By 7.1.2, we have the *fundamental-group functor*, defined on the category of pointed spaces, with values in the category of groups

$$\begin{aligned} &\pi_1\colon \mathsf{Top}_\bullet \to \mathsf{Gp}, && (X, x) \mapsto \pi_1(X, x), \\ &f_*\colon \pi_1(X, x) \to \pi_1(Y, y), && f_*([a]) = [fa]. \end{aligned} \tag{7.18}$$

Here a pointed map $f\colon (X, x) \to (Y, y)$ is sent to a group homomorphism $\pi_1(f) = f_*$; the latter takes the class of a loop $a \in \Omega(X, x)$ to the class of the loop $fa \in \Omega(Y, y)$.

### 7.2.2 Comments and complements

(a) Plainly, the group $\pi_1(X, x)$ only depends on the path component of $x$ in $X$. Basepoints in different path components of a space give unrelated fundamental groups.

More precisely, if $X'$ is the path component of $x$ in $X$, the inclusion $u\colon (X', x) \to (X, x)$ induces a canonical isomorphism $u_*\colon \pi_1(X', x) \to \pi_1(X, x)$. Identifying a loop $a\colon \mathbb{I} \to X'$ with $ua\colon \mathbb{I} \to X$ (as we generally did in the context of singular cubes), we also identify $\pi_1(X', x)$ with $\pi_1(X, x)$.

(b) Changing the basepoint within a path component of $X$ will give a non-canonical isomorphism, depending on the 2-homotopy class of the path 'along which we choose to move' (see 7.3.1).

(c) The functor $\pi_1$ can be identified with the homotopy functor associated to the pointed circle $(\mathbb{S}^1, e_1)$, as defined in (1.73)

$$\pi_1(X, x_0) = [(\mathbb{S}^1, e_1), (X, x_0)]_\bullet. \tag{7.19}$$

Indeed, a loop $a\colon \mathbb{I} \to X$ at $x_0$ amounts to a pointed map $\hat{a}\colon (\mathbb{S}^1, e_1) \to (X, x_0)$, and a homotopy $h\colon a \simeq_2 b$ with fixed basepoint amounts to a pointed homotopy $\hat{h}\colon \hat{a} \simeq \hat{b}$.

(d) The pointed singleton has a unique loop, and a trivial fundamental group.

(e) The pointed set $\pi_0(X, x_0)$, introduced in (1.72), will be dealt with in Section 7.5.

**7.2.3 *Exercises and complements*** (Homotopy invariance)

It is easy to prove the invariance of the fundamental group up to pointed homotopy. A more important result, about non-pointed homotopies, will be given in Theorem 7.3.2.

(a) For a pointed homotopy $\varphi\colon f \simeq g\colon (X, x_0) \to (Y, y_0)$ (see 1.4.5), we have $f_* = g_*\colon \pi_1(X, x_0) \to \pi_1(Y, y_0)$.

(b) (*Corollary*) A pointed homotopy equivalence $f\colon (X, x_0) \rightleftarrows (Y, y_0) : g$ induces two isomorphisms $\pi_1(X, x_0) \rightleftarrows \pi_1(Y, y_0)$, inverse to each other.

**7.2.4 *Proposition*** (Cartesian products)

*The functor $\pi_1\colon \mathsf{Top}_\bullet \to \mathsf{Gp}$ preserves cartesian products: for a product $(X, x) = \prod_{i\in I}(X_i, x_i)$ of pointed spaces (with projections $p_i$ and basepoint $x = (x_i)_{i\in I}$), there is a natural isomorphism of groups*

$$\varphi\colon \pi_1(X, x) \longrightarrow \textstyle\prod_{i\in I}\pi_1(X_i, x_i), \qquad \varphi([a]) = ([p_i a])_{i\in I}. \tag{7.20}$$

*Proof* It is a straightforward consequence of Proposition 7.1.6 on fundamental groupoids. □

**7.2.5 *Theorem*** (The fundamental group of the circle)

*There is an isomorphism*

$$\varphi\colon \mathbb{Z} \to \pi_1(\mathbb{S}^1, e_1), \qquad \varphi(k) = [a_k], \tag{7.21}$$

*where $a_k(t) = (\cos 2k\pi t,\ \sin 2k\pi t)$ is the usual loop winding $k$ times around the circle, with the usual convention for orientation (see (2.103)).*

**Note.* The isomorphism (7.21) agrees with the Hurewicz isomorphism $h\colon \pi_1(\mathbb{S}^1, e_1) \to H_1(\mathbb{S}^1)$ dealt with in 7.3.4.

*Proof* The argument is based on the exponential map

$$p\colon \mathbb{R} \to \mathbb{S}^1, \qquad p(t) = (\cos 2\pi t,\ \sin 2\pi t), \tag{7.22}$$

that can be represented as the projection of a helix on the plane: see figure (7.29).

We shall study a path $a\colon \mathbb{I} \to \mathbb{S}^1$ by a *lifting* $f\colon \mathbb{I} \to \mathbb{R}$, namely a path such that $pf = a$. In particular, the loop $a_k\colon e_1 \simeq e_1$ in $\mathbb{S}^1$ can be lifted to the path $f_k\colon 0 \simeq k$ in $\mathbb{R}$, where $f_k(t) = kt$ (for $k \in \mathbb{Z}$).

First, we prove that the mapping $\varphi$ defined in (7.21) is a homomorphism. For $k = h + h'$ in $\mathbb{Z}$, we lift the loop $a_h$ to the path $f = f_h$, the loop $a_k$ to the path $g = f_k$, and the loop $a_{h'}\colon e_1 \simeq e_1$ to the path $f'\colon h \simeq k$, where $f'(t) = h + h't$ (a path consecutive to $f$).

Then, in the group $\pi_1(\mathbb{S}^1, e_1)$

$$\begin{aligned}\varphi(h)\varphi(h') &= [a_h][a_{h'}] = [a_h * a_{h'}] = [pf * pf'] = [p(f * f')]\\ &= [pg] = [a_k] = \varphi(k),\end{aligned}$$

because $f * f' \simeq_2 g$: in the line, any two paths with the same endpoints are 2-homotopic, by means of the affine homotopy between them.

To prove that $\varphi$ is an isomorphism we use two crucial *lifting properties* of the exponential map $p$:

(i) every path $a\colon \mathbb{I} \to \mathbb{S}^1$ starting at $x_0 = a(0)$ has a unique lifting to a path $f\colon \mathbb{I} \to \mathbb{R}$ starting at a given point $y_0 \in p^{-1}\{x_0\}$,

(ii) every 2-homotopy $h\colon a \simeq_2 a'\colon \mathbb{I} \to \mathbb{S}^1$ of paths starting at $x_0$ has a unique lifting to a 2-homotopy $k\colon f \simeq_2 f'\colon \mathbb{I} \to \mathbb{R}$ of paths starting at a given point $y_0 \in p^{-1}\{x_0\}$.

Here we take these properties as granted. They will be proved in Theorem 7.2.9, as a consequence of a property of the map $p\colon \mathbb{R} \to \mathbb{S}^1$, namely being a 'covering map'.

Now, to show that $\varphi$ is surjective, we take a loop $a\colon e_1 \simeq e_1$ in $\mathbb{S}^1$; by (i) there exists a lifting $f\colon 0 \simeq k$ in $\mathbb{R}$, with $k \in p^{-1}\{e_1\} = \mathbb{Z}$, and the latter is 2-homotopic to $f_k\colon 0 \simeq k$ (by an affine homotopy). Then $k \in p^{-1}\{e_1\} = \mathbb{Z}$, and $\varphi(k) = [a_k] = [pf_k] = [pf] = [a]$.

Finally, to show that $\varphi$ is injective, suppose that $\varphi(k) = [a_k]$ is the identity of $\pi_1(\mathbb{S}^1, e_1)$: this means that $a_k$ is 2-homotopic to the trivial loop $a_0\colon e_1 \simeq e_1$; by (ii) there is a 2-homotopy between their liftings $f_k\colon 0 \simeq k$ and $f_0\colon 0 \simeq 0$, which means that $k = 0$. □

### *7.2.6 Exercises* (Playing with $\pi_1$)

The Poincaré group formalises aspects which – in some cases – can be made concrete, viewing loops as closed curves drawn in the plane, or elastic bands in our 3-dimensional space.

(a) We replace the circle with the (homotopy equivalent) pierced plane $X = \mathbb{R}^2 \setminus \{0\}$, where we can more clearly draw paths.

In the figure below the omitted point $(0,0)$ is marked with a cross. The loop $a$ of $(X, e_1)$

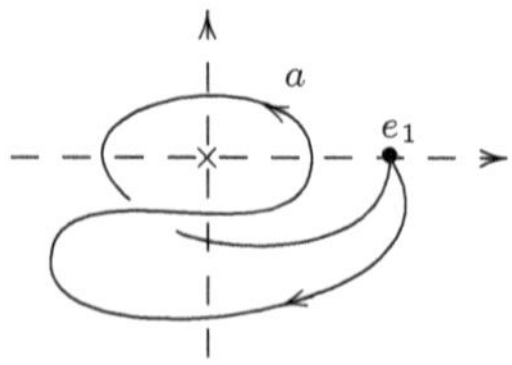


(7.23)

is 2-homotopic to the simple loop $a'$ turning once around the origin, counterclockwise. We can realise the loop by a stretched elastic band, and realise the deformation by letting the band come back to its natural form. On the other hand, the band cannot be contracted to the basepoint $e_1$ (staying in $X$): this 'shows' that $\pi_1(X, e_1)$ is not the trivial group.

The same 'experiment' can be more concretely realised in our three dimensional space, excluding a rod: in fact the complement $Y$ in $\mathbb{R}^3$ of the third axis contains $X$ and $\mathbb{S}^1$ as deformation retracts

$$Y = X \times \mathbb{R} = \{x \in \mathbb{R}^3 \mid (x_1, x_2) \neq (0,0)\} \supset X \supset \mathbb{S}^1.$$

(b) More interesting aspects appear piercing a plane in more points, or equivalently taking out of $\mathbb{R}^3$ some parallel lines. Consider the space $X = \mathbb{R}^2 \setminus \{e_1, -e_1\}$, pierced in two points and pointed at the origin (homotopy equivalent to the figure-eight space)

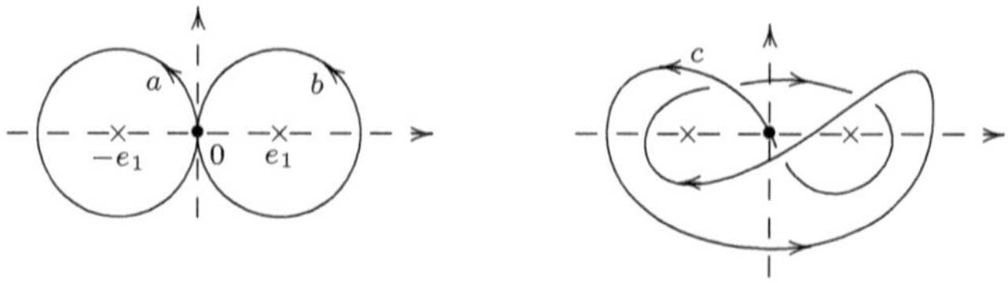


(7.24)

The left figure shows two loops $a, b$ at the origin, which will be shown in 7.4.4(b) to generate the fundamental group $\pi_1(X, 0)$. Concretely, we 'see' that they should determine different, non-trivial classes $[a], [b]$ in $\pi_1(X, 0)$. 'Experiment' also shows that these classes do not commute in $\pi_1(X, 0)$.

In fact, the right figure shows a path $c$ which is 2-homotopic to the concatenation $c' = a * b * a^\sharp * b^\sharp$. Realising this loop with a string winding around two parallel rods (in our 3-space), and tightening the loop, we see that the string cannot be taken 'outside' the rods: the loop $c$ is not trivial, and the class $[a][b][a]^{-1}[b]^{-1}$ is not the identity of $\pi_1(X, 0)$.

(c) The reader may note that, in the previous cases, deforming a loop by a 2-homotopy (keeping the basepoint fixed) or by a loop homotopy (allowing the latter to move) would not make a substantial difference.

This is not always the case: in the same space $X = \mathbb{R}^2 \setminus \{e_1, -e_1\}$, the loop $c$ represented below is 2-homotopic to $c' = b^\sharp * a * b$

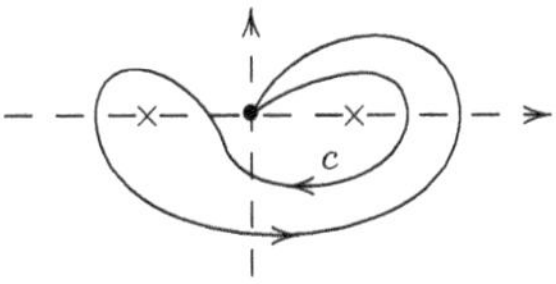


(7.25)

Now $c$ (and $c'$) is not 2-homotopic to $a$, but $c$ and $a$ are plainly loop homotopic: if we allow the basepoint of $c$ to leave the origin, the elastic band can be deformed to the simple loop $a$ around $-e_1$.

### *7.2.7 Exercises and complements* (The $n$-torus)

(a) For the $n$-torus $\mathbb{T}^n = \mathbb{S}^1 \times ... \times \mathbb{S}^1$, based at $x_0 = (e_1, ..., e_1)$, we have

$$\pi_1(\mathbb{T}^n, x_0) \cong \mathbb{Z}^n, \tag{7.26}$$

a straightforward consequence of Theorems 7.2.4 (on cartesian products) and 7.2.5 (on the fundamental group of the circle).

The fundamental group has a canonical basis $[b_i] = [u_i a_1]$, where $a_1$ is the standard loop of $\mathbb{S}^1$ based at $e_1$, and the mapping

$$u_i \colon \mathbb{S}^1 \to \mathbb{T}^n, \qquad u_i(x) = (e_1, ..., e_1, x, e_1, ... e_1), \tag{7.27}$$

is the $i$-th embedding (of pointed spaces), that inserts the variable $x$ in the $i$-th position, and sends $e_1 \in \mathbb{S}^1$ to $x_0 = (e_1, ..., e_1)$.

(b) The surface $\mathbb{T} = \mathbb{S}^1 \times \mathbb{S}^1$ is realised in $\mathbb{R}^3$ as in 3.2.2 by the rotation around the axis $z$ of a circle $C$, which lies in the plane $y = 0$, with centre $p = (2, 0, 0)$ and radius 1. The circle $C$ is a deformation retract of the pierced open half-plane $P = \{(x, z) \in \mathbb{R}^2 \mid x > 0\} \setminus \{(2, 0)\}$.

The torus $\mathbb{T}$ is thus a deformation retract of the space $X$ generated by $P$, by rotation around the axis $z$: in the figure below, this is the complement in $\mathbb{R}^3$ of a dashed line $L$ (the axis $z$) and a dotted circle $S$ (in the plane $xy$, with centre the origin and radius 2)

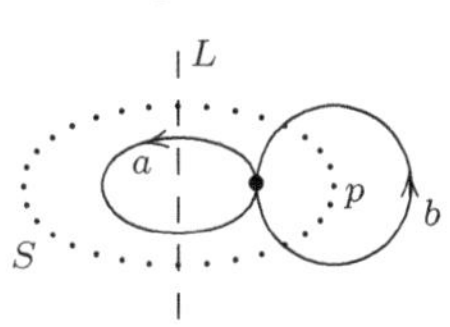


$$X = \mathbb{R}^3 \setminus (L \cup S). \tag{7.28}$$

Making these obstructions concrete, by a rod $L$ and a ring $S$ around the rod, we can realise the generators as the class $[a]$ of a loop around the rod (corresponding to the inner equator of the torus), and the class $[b]$ of a loop around the ring (corresponding to a meridian of the torus).

The loop $c = a * b * a^\sharp * b^\sharp$ can be materially realised by a string turning around the rod and the ring as prescribed (knotting its ends, of course); moving part of the string around the ring, one will find that the string can be taken 'outside' rod and ring: the loop $c$ is 2-homotopic to the trivial loop, and the classes $[a], [b]$ commute in the fundamental group.

### 7.2.8 Covering maps

We end this section with the basic definitions and examples of covering maps, proving (in 7.2.9) the lifting properties already used in 7.2.5 to determine the fundamental group of the circle.

*Their classification and relationship with the fundamental group are important facts, which can be studied in [Ha], Section 1.3, or [HiW], Chapter 6, or [Hu], Sections III.16 and III.17.*

Here we only give the basic definitions, and prove (in 7.2.9) the lifting properties of a covering map, already used in 7.2.5.

(a) A *covering map* $p\colon X \to B$ is a continuous surjective mapping between topological spaces such that every $b \in B$ has an open nbd $V$ whose inverse image $p^{-1}(V)$ is a union of disjoint open sets in $X$ (the *sheets* of $p$ over $V$), each of which is mapped homeomorphically onto $V$ by $p$

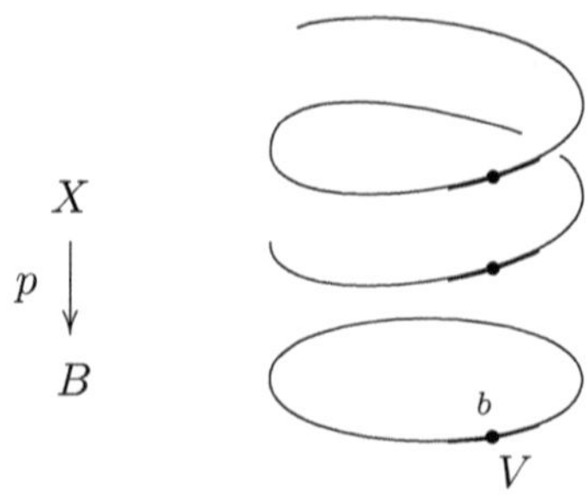


(7.29)

The nbd $V$ is said to be *evenly covered* by $p$. The mapping $p$ is a surjective local homeomorphism, and an open map.

(b) The prime examples, based on the standard circle $\mathbb{S}^1$, are the exponential map, represented in the figure above

$$p\colon \mathbb{R} \to \mathbb{S}^1, \qquad p(t) = e^{2\pi i t} = (\cos 2\pi t, \, \sin 2\pi t), \tag{7.30}$$

and the endomaps (for $n > 0$)

$$p_n\colon \mathbb{S}^1 \to \mathbb{S}^1, \qquad p_n(z) = z^n. \tag{7.31}$$

A *trivial covering map* is a cartesian projection $B \times F \to B$, where $B$ and $F$ are topological spaces and $F$ is discrete.

(c) Coming back to the general case, the spaces $X$ and $B$ are called the *total space* and the *base* of the covering. For every $b \in B$, the *fibre* $F_b = p^{-1}\{b\}$ is a discrete subspace of $X$.

Over an evenly covered nbd $V$ of $b$, the projection $p_V\colon p^{-1}(V) \to V$ is equivalent to a cartesian projection $p\colon V \times F_b \to V$, by a homeomorphism $h$ that agrees with these projections

$$\begin{array}{ccc} p^{-1}(V) & \xrightarrow{h} & V \times F_b \\ {\scriptstyle p_V}\downarrow & & \downarrow{\scriptstyle p} \\ V & =\!=\!= & V \end{array} \qquad (7.32)$$

The cardinal of the fibre $F_b$ is constant on every connected component of $B$. The exponential map (7.30) has a countable fibre, while the map (7.31) is an *n-fold covering*: the cardinal of 'the' fibre is $n$.

*(d) A covering map is the same as a fibre bundle with discrete fibres.

A *universal covering* $X \to B$ is a covering where the total space $X$ is simply connected; under suitable hypothesis, a space $B$ has a universal covering, which is a covering space of any other. The exponential map (7.30) is the universal covering of the circle.

### *7.2.9 Theorem* (Lifting properties)

*Let $p\colon X \to B$ be a covering map.*

*(i) Every path $b\colon \mathbb{I} \to B$ in the base, starting at $b_0 = b(0)$, has a unique lifting $a\colon \mathbb{I} \to X$ starting at a given point $x_0 \in F_{b_0} = p^{-1}\{b_0\}$.*

*(ii) Every 2-homotopy $h\colon b \simeq_2 b'\colon \mathbb{I} \to B$ of paths starting at $b_0$ has a unique lifting $k\colon a \simeq_2 a'\colon \mathbb{I} \to X$ to a 2-homotopy of paths starting at a given point $x_0 \in F_{b_0}$.*

*Proof* (i) For the given path $b\colon \mathbb{I} \to B$, we can find a decomposition $0 = t_0 < t_1 < \ldots < t_n = 1$ such that each interval $[t_{i-1}, t_i]$ is mapped by $b$ to some open set of $B$ evenly covered by $p$.

In fact, these sets form an open cover of $B$, and their preimages by $b$ form an open cover of $\mathbb{I}$. Choosing a Lebesgue number $\varepsilon$ of the latter, and $n \geqslant 1$ such that $1/n < \varepsilon$, we can take $t_i = i/n$.

Now, we lift $b_0 = b(0)$ at the given point $x_0 \in p^{-1}\{b_0\}$. We suppose we have uniquely lifted the restriction $b_i\colon [0, t_i] \to B$ of $b$ to a map $a_i\colon [0, t_i] \to X$ (with $pa_i = b_i$ and $a_i(0) = x_0$). This is certainly possible for $i = 0$; for every $i < n$ we can proceed one step further, because $b[t_i, t_{i+1}]$ is contained in an open set of $B$ evenly covered by $p$; our goal is thus reached in a finite number of steps.

Property (ii) is proved in a similar way, using a decomposition $0 = t_0 < t_1 < ... < t_n = 1$ such that each rectangle $[t_{i-1}, t_i] \times [t_{j-1}, t_j]$ is mapped by $h$ to some open set of $B$ evenly covered by $p$.

We begin by lifting $b_0 = h(0,0)$ at $x_0 = k(0,0)$ and complete the lifting in $n^2$ steps, following a lexicographic order on the set $\{1, ..., n\}^2$. □

## 7.3 Moving or dropping the basepoint

If we move the basepoint $x$ in a path component of the space $X$, the fundamental group $\pi_1(X, x)$ stays the same, up to an isomorphism that *depends* on the path we are following.

Because of this, if the space $X$ is 0-connected (non-empty and path connected), the symbol $\pi_1(X)$ is currently used for the 'isomorphism type' of the fundamental group of $X$ at any basepoint – a notation that should be used with prudence, because it is not related to canonical isomorphisms, nor denotes a functor: this is discussed in 7.3.1(b).

In the same hypotheses, each group $\pi_1(X, x)$ determines $H_1(X)$ in a canonical way, by the Hurewicz Theorem in degree 1.

### *7.3.1 The transition isomorphism*

Let $x_0$ and $x_1$ belong to the same path component of the space $X$: then $\pi_1(X, x_0)$ and $\pi_1(X, x_1)$ are isomorphic.

More precisely, every path $c\colon \mathbb{I} \to X$ from $x_0$ to $x_1$ determines, within the groupoid $\Pi_1 X$, a *transition isomorphism*

$$\hat{c}\colon \pi_1(X, x_0) \to \pi_1(X, x_1), \qquad [a] \mapsto [c]^{-1}[a][c] = [c^\sharp * b * c], \tag{7.33}$$

whose inverse is associated to the reversed path $c^\sharp$.

It is an *inner isomorphism* of the groupoid, see Exercise (c).

In the previous situation, a map $f\colon X \to Y$ gives a path $d = fc$ in $Y$, from $y_0 = f(x_0)$ to $y_1 = f(x_1)$, and two pointed maps (in $\mathsf{Top}_\bullet$) whose underlying map (in $\mathsf{Top}$) is $f$

$$f'\colon (X, x_0) \to (Y, y_0), \qquad f''\colon (X, x_1) \to (Y, y_1). \tag{7.34}$$

We have thus a square diagram of groups and homomorphisms

$$\begin{array}{ccc} \pi_1(X, x_0) & \xrightarrow{\hat{c}} & \pi_1(X, x_1) \\ {\scriptstyle f'_*}\downarrow & & \downarrow{\scriptstyle f''_*} \\ \pi_1(Y, y_0) & \xrightarrow[\hat{d}]{} & \pi_1(Y, y_1) \end{array} \tag{7.35}$$

This square is commutative: on a class $[a] \in \pi_1(X, x_0)$ we have:

$$\hat{d}f'_*[a] = \hat{d}[fa] = [fc]^{-1}[fa][fc] = f''_*([c]^{-1}[a][c]) = f''_*\hat{c}[a].$$

*Exercises and complements.* (a) Let $x_0, x_1$ belong to the same path component of $X$. The transition isomorphism $\hat{c}\colon \pi_1(X, x_0) \to \pi_1(X, x_1)$ only depends on the points $x_0$ and $x_1$ if and only if one of these groups is commutative (or both, equivalently).

(b) For a 0-connected space $X$, the notation $\pi_1(X)$ is used to denote the 'isomorphism type' of the fundamental group of $X$, which – as we have seen – does not depend on the choice of the basepoint. But let us stress once more that the isomorphism (7.33) depends on the choice of a path: the fundamental group does *not* give a functor $\mathsf{Top}' \to \mathsf{Gp}$ defined on the full subcategory of 0-connected spaces.

However, if $f\colon X \to Y$ is a map between *0-connected spaces with commutative fundamental group*, the symbol $f_*\colon \pi_1(X) \to \pi_1(Y)$ can be given a sense. First, choosing a point $x_0 \in X$, we have a precise homomorphism $f'_*\colon \pi_1(X, x_0) \to \pi_1(Y, f(x_0))$. Secondly, if we change the basepoint to $x_1 \in X$, any path $c\colon x_0 \simeq x_1$ in $X$ gives the commutative diagram (7.35), that links $f'_*$ with $f''_*\colon \pi_1(X, x_1) \to \pi_1(Y, f(x_1))$; this diagram is independent of $c$ and $d$, by (a).

All this will become clearer with the Hurewicz Theorem 7.3.4, which – in the previous hypotheses – gives a canonical isomorphism $\pi_1(X, x_0) \to H_1(X)$: in fact, *we are just looking at the functor* $H_1$.

*(c) An arbitrary groupoid $G$ presents facts similar to those described in (a): each arrow $c\colon x_0 \to x_1$ gives a group isomorphism

$$\hat{c}\colon G(x_0) \to G(x_1), \qquad a \mapsto c^{-1}ac, \tag{7.36}$$

which can be called an *inner automorphism* of the groupoid (with composition written in the 'straightforward' order). If $x_0 = x_1$ we are just considering an inner automorphism of the group $G(x_0)$.

## 7.3.2 Theorem (Non-pointed homotopies)

*(a) For a homotopy $\varphi\colon f \cong g\colon X \to Y$ and a point $x_0 \in X$, there is a commutative diagram*

$$\begin{array}{ccc} \pi_1(X, x_0) & \xrightarrow{f_*} & \pi_1(Y, y_1) \\ & \searrow^{g_*} & \downarrow{\scriptstyle \hat{c}} \\ & & \pi_1(Y, y_2) \end{array} \tag{7.37}$$

*where* $y_1 = f(x_0)$, $y_2 = g(x_0)$, *and the vertical arrow*

$$\hat{c}\colon \pi_1(Y, y_1) \to \pi_1(Y, y_2), \qquad [b] \mapsto [c]^{-1}[b][c] = [c^\sharp * b * c], \qquad (7.38)$$

*is the transition isomorphism of the path* $c = \varphi(x_0, -)$, *from* $y_1$ *to* $y_2$ *in* $Y$.

*(b) Suppose that the spaces* $X, Y$ *are 0-connected and homotopy equivalent. Then, for every* $x_0 \in X$ *and* $y_0 \in Y$

$$\pi_1(X, x_0) \cong \pi_1(Y, y_0). \qquad (7.39)$$

*(c) If the space* $X$ *is contractible, every group* $\pi_1(X, x_0)$ *is trivial. The same holds whenever the path component of* $x_0$ *in* $X$ *is contractible.*

*Proof* (a) It is an application of the naturality square (7.7) associated to the homotopy $\varphi\colon f \cong g$.

At the point $x_0 \in X$, the loop $a\colon x_0 \simeq x_0$ gives a commutative square in the groupoid $\Pi_1 Y$

$$\begin{array}{ccc} y_1 & \xrightarrow{[c]} & y_2 \\ {\scriptstyle [fa]}\downarrow & & \downarrow{\scriptstyle [ga]} \\ y_1 & \xrightarrow[{[c]}]{} & y_1 \end{array} \qquad [fa][c] = [c][ga], \qquad (7.40)$$

where $c = \varphi(x_0, -)$. Therefore

$$\hat{c}[fa] = [c]^{-1}[fa][c] = [ga].$$

(b) By 7.3.1, it is sufficient to prove the existence of an isomorphism $\pi_1(X, x_0) \cong \pi_1(Y, y_0)$, for a given pair of basepoints $x_0 \in X$ and $y_0 \in Y$.

We have a homotopy equivalence $f\colon X \rightleftarrows Y : g$. We choose a point $x_0$ in $X$; we let $y_0 = f(x_0)$ and $x_1 = g(y_0)$. By (a), the existence of a homotopy $\mathrm{id}\, X \simeq gf$ implies that the composed homomorphism

$$g_* f_*\colon \pi_1(X, x_0) \to \pi_1(Y, y_0) \to \pi_1(X, x_1) \qquad (7.41)$$

is a (transition) isomorphism (not the identity, in general), so that $f_*$ is mono and $g_*$ is epi.

Symmetrically, a homotopy $\mathrm{id}\, Y \simeq fg$ implies that $g_*$ is mono and $f_*$ is epi. Finally, they are both isomorphisms.

(c) A straightforward consequence of (b). □

### 7.3.3 The Hurewicz homomorphism

For a pointed space $(X, x_0)$, there is a natural relationship between the fundamental group $\pi_1(X, x_0)$ and the singular homology group $H_1(X)$: any loop $a \in \Omega(X, x_0)$ determines a homotopy class $[a] \in \pi_1(X, x_0)$ and a homology class $[a]_H \in H_1(X)$.

There is in fact a natural homomorphism, called the *Hurewicz homomorphism* (in degree 1)

$$h\colon \pi_1(X, x_0) \to H_1(X), \qquad h[a] = [a]_H \quad (a \in \Omega(X, x_0)), \tag{7.42}$$

examined in the following exercises and Theorem 7.3.4.

*Exercises and complements.* (a) The mapping (7.42) is a well-defined homomorphism of groups.

(b) This homomorphism is consistent with the transition isomorphisms (7.33): for any path $c\colon x_0 \simeq x_1$ in $X$ the following triangle commutes

$$\begin{array}{ccc} \pi_1(X, x_0) & \xrightarrow{\;h\;} & H_1(X) \\ {\scriptstyle \hat{c}}\downarrow & \nearrow{\scriptstyle h'} & \\ \pi_1(X, x_1) & & \end{array} \tag{7.43}$$

where $h'$ is the Hurewicz homomorphism at $x_1$. *Hint:* use Lemma 7.3.6.

(c) The Hurewicz homomorphism gives a natural transformation

$$h\colon \pi_1 \to H_1 U\colon \mathsf{Top}_\bullet \to \mathsf{Gp}, \tag{7.44}$$

where $U\colon \mathsf{Top}_\bullet \to \mathsf{Top}$ is the forgetful functor which drops the basepoint.

(d) To further investigate $h$ we have to recall some basic issues of group theory. In a group $G$, two elements $x, y$ commute (i.e. $xy = yx$) if and only if their *commutator*

$$[x, y] = x\, y\, x^{-1}\, y^{-1} \tag{7.45}$$

is the identity $1_G$. By definition, the subgroup $[G, G]$ *of commutators* of $G$ is the subgroup generated by all commutators $[x, y]$, for $x, y \in G$. Plainly, $G$ is commutative if and only if $[G, G]$ is the trivial subgroup.

Prove that the subgroup $[G, G]$ is normal in $G$.

(e) The quotient $G/[G, G]$ is a commutative group, determined by the following universal property:

(*) for every homomorphism of groups $f\colon G \to A$ with values in an abelian group there is a unique homomorphism $g\colon G/[G, G] \to A$ such that $gp = f$

$$\begin{array}{ccc} G & \xrightarrow{p} & G/[G,G] \\ & \searrow^{f} & \downarrow{\scriptstyle g} \\ & & A \end{array} \tag{7.46}$$

The quotient $G^{\mathrm{ab}} = G/[G,G]$ is called the *abelianised group* of $G$.

(f) The pair $(G^{\mathrm{ab}}, p \colon G \to G^{\mathrm{ab}})$ is thus a universal arrow from the object $G$ to the inclusion functor $U \colon \mathsf{Ab} \to \mathsf{Gp}$.

*The family of these universal arrows gives a functor $(-)^{\mathrm{ab}} \colon \mathsf{Gp} \to \mathsf{Ab}$, left adjoint to $U$ (see Section 8.1).*

### *7.3.4* ****Hurewicz Theorem*** (in dimension one)

*Let $(X, x_0)$ be a path connected pointed space.*

*Then the Hurewicz homomorphism $h \colon \pi_1(X, x_0) \to H_1(X)$ defined in (7.42) is surjective, and its kernel coincides with the subgroup of commutators of $\pi_1(X, x_0)$.*

*In other words, the Hurewicz homomorphism induces an isomorphism $h'$*

$$\begin{array}{ccc} \pi_1(X,x_0) & \xrightarrow{p} & \pi_1(X,x_0)^{\mathrm{ab}} \\ & \searrow^{h} & \downarrow{\scriptstyle h'} \\ & & H_1(X) \end{array} \tag{7.47}$$

*between the abelianised group $\pi_1(X, x_0)^{\mathrm{ab}}$ and $H_1(X)$.*

*Proof* See [HiW], Theorem 8.8.3, or [Hu], Theorem II.6.1. □

### *7.3.5* ***Remarks and complements***

(a) If a path connected pointed space $(X, x_0)$ has a commutative fundamental group $\pi_1(X, x_0)$, the Hurewicz homomorphism is an isomorphism. In this case we understand better the meaning of $\pi_1(X)$ discussed in 7.3.1(b): essentially, it refers to $H_1$, which for sure is a functor.

This is the case of the circle and of any torus $\mathbb{T}^n$.

(b) The homotopy groups $\pi_1(-)$ give a richer information than the singular homology groups $H_1(-)$: if $(X_i)$ is the partition of $X$ in path components, then

$$H_1(X) \cong \bigoplus_i H_1(X_i) \cong \bigoplus_i \pi_1(X_i, x_i)^{\mathrm{ab}}, \tag{7.48}$$

where $x_i$ is any point in the path component $X_i$. But we shall see that, in higher dimension, the homotopy groups of a space no longer determine its homology groups.

*(c) Every group $G$ can be realised as the fundamental group of a suitable *Eilenberg–Mac Lane space* $K(G,1)$: a 0-connected space whose fundamental group is isomorphic to $G$, while all higher homotopy groups are trivial (see [EiM2], [EiM3] and [Ha], Section 1.B).

If we take as $G$ a non-commutative *simple* group (i.e. a group whose only normal subgroups are the trivial and the total one), then $[G,G] = G$ and $G^{\mathrm{ab}} = 0$.

This shows that there are spaces $X$ with a quite complicated $\pi_1(X)$, and trivial $H_1(X)$.

### *7.3.6 Lemma*

*For a loop $a\colon x \simeq x$ and a path $c\colon x \simeq y$ in the space $X$, the loop $a$ and the loop $a' = c^\sharp * a * c\colon y \simeq y$ are loop homotopic.*

*Proof* (This extends an issue already considered in (7.25), for the plane pierced in two points.)

Using double homotopies, as in (7.14) and (7.16), the figure below shows how to construct a loop homotopy $H = L' *_1 A *_1 L$, from the loop $e_x * a * e_x$ (at $x$) to the loop $a' = c^\sharp * a * c$ (at $y$)

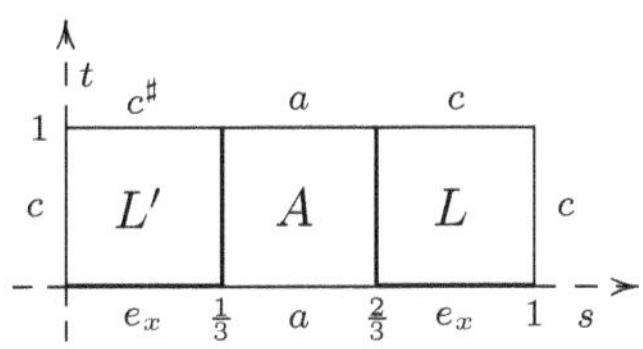


(7.49)

$$L'(s,t) = c(\min(1-s,t)), \quad A(s,t) = a(s), \quad L(s,t) = c(\min(s,t)).$$

Therefore the loop $a \simeq_2 e_x * a * e_x$ is loop homotopic to $c^\sharp * a * c$. □

### *7.3.7 Theorem* (Classification by degree)

*(a) Let $X$ be a 0-connected space with a commutative fundamental group $\pi_1(X, x_0)$ (at any point). Let $f, g\colon \mathbb{S}^1 \to X$ be two maps with the same induced homomorphism $u\colon H_1(\mathbb{S}^1) \to H_1(X)$. Then $f \simeq g$.*

*(b) Two maps $f, g\colon \mathbb{S}^1 \to \mathbb{S}^1$ are homotopic if (and only if)* $\deg f = \deg g$*: the topological degree* classifies *the homotopy class of these endomaps.*

*Proof* The argument uses the Hurewicz Theorem 7.3.4. It is sufficient to prove (a), as (b) is a straightforward consequence. For the standard basepoint $e_1 \in \mathbb{S}^1$ we let $x_1 = f(e_1)$ and $x_2 = g(e_1)$.

In the following diagram

$$\begin{array}{ccccc} \pi_1(\mathbb{S}^1, e_1) & \xrightarrow{h} & H_1(\mathbb{S}^1) & \xleftarrow{h} & \pi_1(\mathbb{S}^1, e_1) \\ {\scriptstyle f_*}\downarrow & & \downarrow{\scriptstyle u} & & \downarrow{\scriptstyle g_*} \\ \pi_1(X, x_1) & \xrightarrow{h'} & H_1(X) & \xleftarrow{h''} & \pi_1(X, x_2) \\ & & \xrightarrow[\hat{c}]{\qquad\qquad\qquad} & & \end{array} \tag{7.50}$$

- $h$, $h'$ and $h''$ are Hurewicz isomorphisms, by hypothesis,
- $\hat{c}\colon \pi_1(X, x_1) \to \pi_1(X, x_2)$ is the inner isomorphism associated to a chosen path $c\colon x_1 \simeq x_2$ in $X$,
- the squares commute, by the naturality of the Hurewicz homomorphism (in 7.3.3), and the triangle commutes, by (7.43): $h''\hat{c} = h'$.

We deduce that $\hat{c}f_* = g_* \colon \pi_1(\mathbb{S}^1, e_1) \to \pi_1(X, x_2)$.

For the exponential loop $p\colon \mathbb{I} \to \mathbb{S}^1$, the loops $a = pf \colon \mathbb{I} \to \mathbb{S}^1$ and $b = pg\colon \mathbb{I} \to \mathbb{S}^1$ correspond to the maps $f = \hat{a}$ and $g = \hat{b}$ (as in 1.4.2(a)).

We have thus, in $\pi_1(\mathbb{S}^1, e_1)$

$$[b] = g_*[p] = \hat{c}f_*[p] = [c^\sharp * a * c],$$

and $b \simeq_2 c^\sharp * a * c$; but the latter is loop homotopic to $a$, by Lemma 7.3.6. Therefore $a$ and $b$ are loop homotopic, and $f \simeq g$, by 1.4.8(b). □

## 7.4 The van Kampen Theorem

After some preliminaries on groups, we introduce one of the main tools to compute fundamental groups, the van Kampen Theorem. It plays a role similar to the Mayer–Vietoris sequence for homology groups: to determine $\pi_1(X)$ on the basis of a generalised open cover of the space $X$.

Yet, it is less effective: it needs stronger hypothesis (which do not apply to the circle), and cannot be easily extended to higher degree.

R. Brown's extension to groupoids (exposed in 7.4.5, 7.4.6) solves the first drawback: it is stated for any generalised open cover and can be used to determine the fundamental groupoid of the circle (and then the fundamental group). Its extension to higher dimensional groupoids has been developed in diverse forms, far beyond an elementary exposition.

Computing higher homotopy groups can be quite difficult or still unsolved (even for the spheres $\mathbb{S}^n$); the main tools, as the exact sequence of a fibration or the homotopy spectral sequences, are studied in [Hu].

### 7.4.1 Free products of groups

To state the van Kampen Theorem, we need some preliminary topics of group theory: free products and pushouts.

For a family of groups $(G_i)$, one can form a new group $G = *_i\, G_i$ called the *free product* of the family. It comes with a family of homomorphisms $u_i\colon G_i \to G$ $(i \in I)$, called *injections*, which satisfy the universal property we have considered in 1.6.2(i*)

$$\begin{array}{ccc} G_i & \xrightarrow{u_i} & G \\ & \searrow^{f_i} & \downarrow f \\ & & H \end{array} \qquad (7.51)$$

(*) for every group $H$ and every family of homomorphisms $f_i\colon G_i \to H$, there is a unique homomorphism $f\colon G \to H$ such that $fu_i = f_i$ (for $i \in I$).

In other words, we are considering the categorical sum in the category Gp (which had a name and a symbol in group theory, before the existence of categories).

The group $G$ is essentially built as follows (a more detailed construction is the subject of Exercise (a)). An element of $*_i\, G_i$ can be written as a finite word $(x_1, x_2, ..., x_n)$ on the *alphabet* $X = \sum_i |G_i|$, the disjoint union of the underlying sets, provided we replace every occurrence $x_h, x_{h+1}$ of two elements in the same group $G_i$ with their product in this group, and take out the unit element of any $G_i$.

Writing $x_1x_2 ... x_n$ the equivalence class of the word $(x_1, x_2, ..., x_n)$, two classes are multiplied by juxtaposition; the operation is associative, has a unit 1 (the class of the empty word) and inverses exist, with $(x_1x_2 ... x_n)^{-1} = x_n^{-1} ... x_2^{-1}x_1^{-1}$ (each inverse is computed in the corresponding $G_i$).

The homomorphism $u_i\colon G_i \to G$ obviously takes any $x \in G_i$ to the class of its unary word. The universal property (*) is plainly satisfied.

*Exercises and complements.* (a) Rewrite the construction of $*_i\, G_i$ in a more detailed form. *Hint:* start from the alphabet $\sum_{i\in I} |G_i|$, the disjoint union of the underlying sets of the groups.

(b) Verify the universal property of $*_i\, G_i$.

(c) The *free group* $G(X)$ on the set $X$ comes equipped with a mapping $\eta\colon X \to |G(X)|$ (the insertion of the base) with the usual universal property

(i) for every mapping $f\colon X \to |H|$ with values in a group $H$, there is a unique homomorphism $g\colon G(X) \to H$ such that $g\eta = f$.

The free group on the empty set is the trivial group, the free group on the singleton is the additive group $\mathbb{Z}$.

(d) In general, the free group $G(X)$ can be constructed as a free product of copies of the additive group $\mathbb{Z}$

$$G(X) = \ast_{x \in X} \mathbb{Z}. \tag{7.52}$$

The solution is obtained as a categorical sum of copies of the free object on the singleton, in the same way as the free $R$-module on a set, in (5.5). This is a general fact in category theory, as one can see in [G5], Exercise 5.1.3(e): a consequence of the fact that, in Set, each set is the sum of its singletons

(e) In the alphabet $\Sigma_{x \in X} |\mathbb{Z}|$, it is convenient to replace the $x$-indexed copy $\mathbb{Z} \times \{x\}$ with the free multiplicative group $\{x^k \mid k \in \mathbb{Z}\}$ generated by the singleton $\{x\}$, whose elements are formal powers of $x$. The elements of $G(X)$ can thus be described as equivalence classes $[w] \in W/{\sim}$

$$[w] = x_1^{k_1} x_2^{k_2} \ldots x_p^{k_p} \qquad (x_j \in X,\ k_j \in \mathbb{Z}), \tag{7.53}$$

of words in this alphabet, where we add up the exponents of consecutive terms with the same base, and cancel all occurrences of the form $x^0$.

The free group $\mathbb{Z} * \mathbb{Z}$ on two elements is already non commutative.

### *7.4.2 Pushouts of groups*

There are given two homomorphisms $f_1, f_2$ with the same domain $G_0$. Their pushout, defined in 1.7.1(b), is a group $G$ equipped with two maps $u_i \colon G_i \to G$ $(i = 1, 2)$ which form a commutative square $u_1 f_1 = u_2 f_2$, in a universal way:

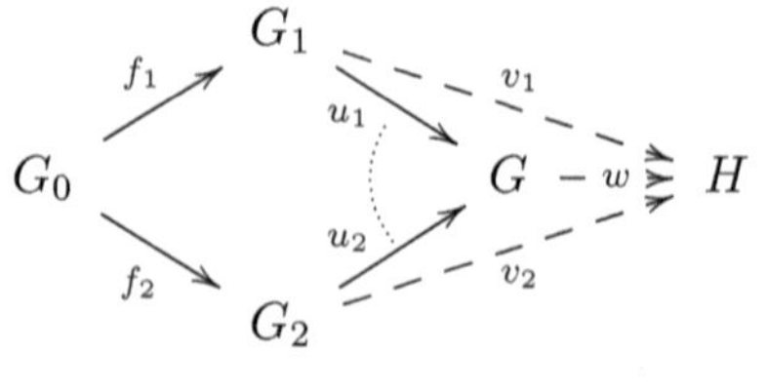


(7.54)

(*) for every triple $(H, v_1, v_2)$ such that $v_1 f_1 = v_2 f_2$ there exists a unique homomorphism $w \colon G \to H$ such that $wu_1 = v_1$, $wu_2 = v_2$.

The solution always exists, as shown below.

*Exercises and complements.* (a) The elementary case where $G_0$ is the trivial group leads to the free product $G = G_1 * G_2$, with its canonical injections.

(b) In the general case, one builds a solution as a quotient $G = (G_1 * G_2)/K$ of the free product: $K$ is the normal subgroup of $G_1 * G_2$ generated by all the elements of the form $u_1 f_1(x)\, u_2 f_2(x^{-1})$, for $x \in G_0$.

### 7.4.3 *The van Kampen Theorem*

*Let $X$ be a space; let $X_1, X_2$ be two subspaces, and $X_0 = X_1 \cap X_2$. Suppose that $X = \operatorname{int} X_1 \cup \operatorname{int} X_2$ and $x_0 \in X_0$.*

*Suppose also that $X_0$, $X_1$ and $X_2$ are path connected (whence $X$ is also). The following diagram of groups and inclusions is a pushout in* Gp

$$\begin{array}{ccc} \pi_1(X_0, x_0) & \xrightarrow{u_1} & \pi_1(X_1, x_0) \\ {\scriptstyle u_2}\downarrow & & \downarrow{\scriptstyle v_1} \\ \pi_1(X_2, x_0) & \xrightarrow[v_2]{} & \pi_1(X, x_0) \end{array} \qquad (7.55)$$

*Proof* This theorem, also named after Seifert and van Kampen, was proved around 1930. Here it will be obtained as a corollary of its version for groupoids, in 7.4.6. □

### 7.4.4 *Exercises and complements* (Computations)

(a) Prove that $\pi_1(\mathbb{S}^n, e_1) = 0$, for $n > 1$.

(b) (*The figure-eight space*) Calculate the fundamental group of the figure-eight space (in 3.1.1), using the same open cover used for its homology groups. This exercise makes precise what we have already 'experimented' in 7.2.6(b).

The result is easily extended to the fundamental group of a bouquet of circles $\mathbb{S}^1 \vee ... \vee \mathbb{S}^1$, or a bouquet of spheres.

(c) In a space $X$ with a non-commutative fundamental group one can always find two loops, at the same basepoint, which are loop homotopic and not 2-homotopic. *Hint:* use Lemma 7.3.6.

(d) Compute the fundamental group of the space $P_n = \mathbb{D}^2/R_n$ defined in 3.1.8 (including the projective plane, for $n = 2$).

*(e) The fundamental group of the Klein bottle $\mathbb{K}$ is generated by two elements $\alpha, \beta$ under the relation $\alpha = \beta\alpha\beta$. The generators can be realised as $\alpha = [a]$, $\beta = [b]$, where $a$ and $b$ are the loops used in 3.2.4.

We recall that their homology classes $[a]_H$ and $[b]_H$ generate $H_2(\mathbb{K}) \cong \mathbb{Z} \oplus \mathbb{Z}_2$, under the relation $2[b]_H = 0$. The fundamental group of $\mathbb{K}$ is not commutative.

The reader should not find it difficult to complete the missing points, using the open cover $(U, V)$ of Exercise 3.2.4(a).

### 7.4.5 The van Kampen Theorem for groupoids

*Let $X$ be a space; let $X_1, X_2$ be two subspaces, and $X_0 = X_1 \cap X_2$. Suppose that $X = \operatorname{int} X_1 \cup \operatorname{int} X_2$.*

*The following diagram of small groupoids and functors (induced by inclusions) is a pushout of groupoids (as defined in 1.7.1(b))*

$$\begin{array}{ccc} \Pi_1 X_0 & \xrightarrow{u_1} & \Pi_1 X_1 \\ {\scriptstyle u_2}\downarrow & & \downarrow{\scriptstyle v_1} \\ \Pi_1 X_2 & \xrightarrow[v_2]{} & \Pi_1 X \end{array} \qquad (7.56)$$

Note. All limits and colimits exist in the category of small groupoids, but this is not needed here.

*Proof* This is the groupoid version of the van Kampen Theorem, by Ronald Brown [Br1] (also exposed in [Br2], [Br3]). It does not need the hypothesis of path-connectedness of the spaces $X_0$, $X_1$ and $X_2$.

The proof is written in 7.4.9. □

### 7.4.6 Corollary

*We assume the same hypotheses on $X$, $X_1$, $X_2$ and $X_0$, as in the previous theorem.*

*(a) We also assume that $A \subset X$ meets all the path components of $X_0$, $X_1$, $X_2$ and therefore of $X$; in other words, each $A_i = X_i \cap A$ is representative in $X_i$ (as defined in 7.1.9). Then, restricting all the groupoids of (7.56) to the vertices in $A$, we still have a pushout of groupoids*

$$\begin{array}{ccc} \Pi_1 X_0|A_0 & \xrightarrow{u_1} & \Pi_1 X_1|A_1 \\ {\scriptstyle u_2}\downarrow & & \downarrow{\scriptstyle v_1} \\ \Pi_1 X_2|A_2 & \xrightarrow[v_2]{} & \Pi_1 X|A \end{array} \qquad (7.57)$$

*where $\Pi_1 X|A$ is the full subgroupoid of $\Pi_1 X$ with vertices in $A$, a deformation retract of $\Pi_1 X$ (see 7.1.9); and so on.*

*(b) If $X_0$, $X_1$ and $X_2$ are path connected (as well as $X$), and $x_0 \in X_0$, we can take $A = \{x_0\}$ as the representative set. Then the four groupoids fall down to groups, and our statement gives the ordinary van Kampen's Theorem 7.4.3.*

*Proof* (a) We write $X_3 = X$ and $A_3 = A$.

In the given hypotheses, we have seen in 7.1.9 that there are deformation retracts (for $i = 0, ..., 3$)

$$J_i \colon \Pi_1 X_i|A_i \rightleftarrows \Pi_1 X_i : P_i, \qquad P_i J_i = \mathrm{id}\,, \qquad \lambda_i \colon \mathrm{id} \cong J_i P_i, \tag{7.58}$$

determined by choosing a path $\lambda_x \colon x \simeq P(x)$ in $X$ which, for $x \in X_i$, stays in $X_i$. (We begin with a choice for $x \in X_0 = X_1 \cap X_2$, and extend it to $X_1 \setminus X_0$ and $X_2 \setminus X_0$, covering also $X$.) If $x \in A$, we take the trivial path $\lambda_x = e_x \colon x \simeq x = P(x)$.

Forgetting the natural isomorphisms $\lambda_i$, it is now sufficient to prove that 'a retract of a pushout is a pushout', a general fact of category theory proved in Lemma 7.4.8. (The argument is straightforward; an interested reader can state and prove independently this result.)

(b) A straightforward consequence of (a): a commutative square of groups is a pushout in Gp if (and only if) it is in Gpd. □

### *7.4.7 The fundamental groupoid of the circle*

The groupoid version of van Kampen's Theorem can be applied to the usual open arcs $X_1$ and $X_2$ of the circle, whose intersection has two connected components (note that the group version does not apply).

It is sufficient (and convenient) to take a representative set $A = \{p, q\}$ formed of two points, one in each connected component of $X_1 \cap X_2$

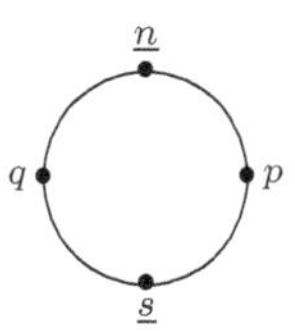


$$\begin{aligned} X_1 &= \mathbb{S}^1 \setminus \{\underline{s}\} \simeq \{*\}, \\ X_2 &= \mathbb{S}^1 \setminus \{\underline{n}\} \simeq \{*\}, \\ X_0 &= X_1 \cap X_2 \simeq A = \{p, q\}. \end{aligned} \tag{7.59}$$

Then $G_1 = \Pi_1 X_1|A$ and $G_2 = \Pi_1 X_2|A$ are two copies of the indiscrete groupoid (see 1.3.1) on two vertices $p, q$, while $G_0 = \Pi_1 X_0|A$ is the discrete groupoid on the same vertices.

The pushout $G = \Pi_1 \mathbb{S}^1|A$ of $G_1$ and $G_2$ over $G_0$ is the groupoid on the objects $p, q$ freely generated by two arrows $\lambda \colon p \to q$ and $\mu \colon q \to p$.

In particular, $G(p)$ is the free group generated by the endomorphism $\alpha = \lambda\mu \colon p \to p$, and is isomorphic to $\mathbb{Z}$; similarly $G(q)$ is the free group generated by the endomorphism $\beta = \mu\lambda \colon q \to q$.

We find again that the fundamental group of the circle is the infinite cyclic group. This proof does not use the theory of covering maps.

### *7.4.8* **Lemma** (Pushouts and retracts)

*In a category* $\mathsf{C}$ *we have two commutative squares*

$$\begin{array}{ccc} X_0 & \xrightarrow{u_1} & X_1 \\ {\scriptstyle u_2}\downarrow & & \downarrow{\scriptstyle v_1} \\ X_2 & \xrightarrow[v_2]{} & X_3 \end{array} \qquad\qquad \begin{array}{ccc} A_0 & \xrightarrow{u'_1} & A_1 \\ {\scriptstyle u'_2}\downarrow & & \downarrow{\scriptstyle v'_1} \\ A_2 & \xrightarrow[v'_2]{} & A_3 \end{array} \qquad (7.60)$$

*We also have four retractions* $j_i\colon A_i \rightleftarrows X_i : p_i$ $(0 \leqslant i \leqslant 3)$ *consistent with the previous squares (with* $i = 1, 2$*)*

$$u_i j_0 = j_i u'_i, \qquad v_i j_i = j_3 v'_i, \qquad p_i u_i = u'_i p_0, \qquad p_3 v_i = v'_i p_i \qquad (7.61)$$

*which means that the sections* $j_i\colon A_i \to X_i$ *form a commutative cube, from the right square to the left, and the retractions* $p_i\colon X_i \to A_i$ *form a commutative cube, from the left square to the right.*

*Then, if the left square is a pushout, the right one is also.*

**Note.* Similar facts hold for all colimits.

*Proof* There are given two morphisms $h_i\colon A_i \to Y$ $(i = 1, 2)$ such that $h_1 u'_1 = h_2 u'_2$. We define $f_i = h_i p_i\colon A_i \to Y$ and get a commutative square: $f_i u_i = h_i p_i u_i = h_i u'_i p_0$ does not depend on $i = 1, 2$.

There is thus a unique $f\colon X_3 \to Y$ such that $f v_i = f_i$ (for $i = 1, 2$), and a morphism $h = f j_3\colon A_3 \to Y$ satisfying the required conditions

$$h v'_i = f j_3 v'_i = f v_i j_i = f_i j_i = h_i p_i j_i = h_i,$$

for $i = 1, 2$. Moreover, if also $h'\colon A_3 \to Y$ gives $h' v'_i = h_i$ (for $i = 1, 2$), then $h' p_3 v_i = h' v'_i p_i = h_i p_i = f_i$, whence $h' p_3 = f$ and $h' = f j_3 = h$. □

### *7.4.9* **Proof of Theorem** *7.4.5*

(a) We shall use the regular concatenation of consecutive paths (in (1.76)), written as $a_1 * ... * a_n$.

Let $G$ be a small groupoid (with composition written in the same order as in fundamental groupoids) and take two functors $F_i\colon \Pi_1 X_i \to G$ which coincide on $\Pi_1 X_0$ $(F_1 u_1 = F_2 u_2)$; we have to prove that they have a unique 'extension' $F\colon \Pi_1 X \to G$. On the objects this is obvious, since $X = X_1 \cup X_2$ and $X_0 = X_1 \cap X_2$.

(b) Let then $a\colon x \simeq x'$ be a path in $X$. By the Lebesgue Number Lemma 1.4.9, there is a finite decomposition $0 < 1/n < 2/n < ... < 1$ of the standard interval such that each subinterval $[(i-1)/n, i/n]$ is mapped by

$a$ to $X_1$ or $X_2$ (possibly both); we call it a *suitable* decomposition for our data.

Thus, $a$ is a regular concatenation $a_1 * ... * a_n$ where each path $a_i\colon [0,1] \to X$ is contained in $X_{k_i}$ (with $k_i = 1, 2$).

We define

$$F[a] = F_{k_1}[a_1] * ... * F_{k_n}[a_n] \in G(F(x), F(x')). \tag{7.62}$$

First, this morphism $F[a]$ does not depend on the choice of $k_i$: if $\mathrm{Im}\,(a_i) \subset X_1 \cap X_2 = X_0$, then $F_1 u_1 = F_2 u_2$ shows that $F_1[a_i] = F_2[a_i]$.

Second, $F[a]$ does not depend on the choice of $n$: if also $m$ gives a suitable decomposition, one can use the regular decomposition in $mn$ subintervals to prove that $m$ and $n$ give the same result.

Third, $F[a]$ does not depend on the representative path $a$. It will be sufficient to show this for a second path $a'\colon x \to x'$, linked to the first by a 2-homotopy $H\colon a \simeq_2 a'$; in other words, $H\colon \mathbb{I}^2 \to X$ has degenerate 1-indexed faces, and 2-indexed faces $a, a'$.

(c) Applying the Lebesgue Number Lemma to the compact metric square $[0,1]^2$, there is some integer $n > 0$ such that each elementary square

$$[(i-1)/n, i/n] \times [(j-1)/n, j/n] \qquad\qquad (i, j = 1, ..., n),$$

is mapped by $H$ to $X_1$ or $X_2$.

In this way, $H$ can be rewritten as a pasting of its $n^2$ restrictions to these squares, reparametrised as $H_{ij}\colon \mathbb{I}^2 \to X_{k(i,j)} \subset X$ (with $k(i,j) = 1, 2$)

$$H = (H_{11} *_1 H_{21} *_1 ... *_1 H_{n1}) *_2 ... *_2 (H_{1n} *_1 H_{2n} *_1 ... *_1 H_{nn}), \tag{7.63}$$

where $*_1$ denotes a horizontal concatenation (in the first coordinate) and $*_2$ a vertical one.

Every 2-cube $K = H_{ij}$ yields, by Lemma 7.1.8, a 2-homotopy relation in $X_{k(i,j)}$

$$\partial_2^- K * \partial_1^+ K \simeq_2 \partial_1^- K * \partial_2^+ K. \tag{7.64}$$

Using the fact that the 1-indexed faces on the boundary, $\partial_1^- H_{1i}$ and $\partial_1^+ H_{ni}$, are constant paths, and the coincidence of faces between contiguous little squares, we can gradually move from the path $a$ to $a'$, following a lexicographic order.

We have thus:

$$\begin{aligned}F[a] &= F_{k(1,1)}[\partial_2^- H_{11}] * \dots * F_{k(n,1)}[\partial_2^- H_{n1}] \\ &= F_{k(1,1)}[\partial_2^- H_{11}] * \dots * F_{k(n,1)}[\partial_2^- H_{n1}] * F_{k(n,1)}[\partial_1^+ H_{n1}] \\ &= F_{k(1,1)}[\partial_2^- H_{11}] * \dots * F_{k(n,1)}[\partial_1^- H_{n1}] * F_{k(n,1)}[\partial_2^+ H_{n1}] \\ &= F_{k(1,1)}[\partial_2^- H_{11}] * \dots * F_{k(n,1)}[\partial_1^+ H_{n-1,1}] * F_{k(n,2)}[\partial_2^- H_{n2}] \\ &= \dots = F_{k(1,2)}[\partial_2^- H_{12}] * \dots * F_{k(n,2)}[\partial_2^- H_{n2}] \\ &= \dots = F_{k(1,n)}[\partial_2^- H_{1n}] * \dots * F_{k(n,n)}[\partial_2^- H_{nn}] \\ &= \dots = F_{k(1,n)}[\partial_2^+ H_{1n}] * \dots * F_{k(n,n)}[\partial_2^+ H_{nn}] = F[a'].\end{aligned} \quad (7.65)$$

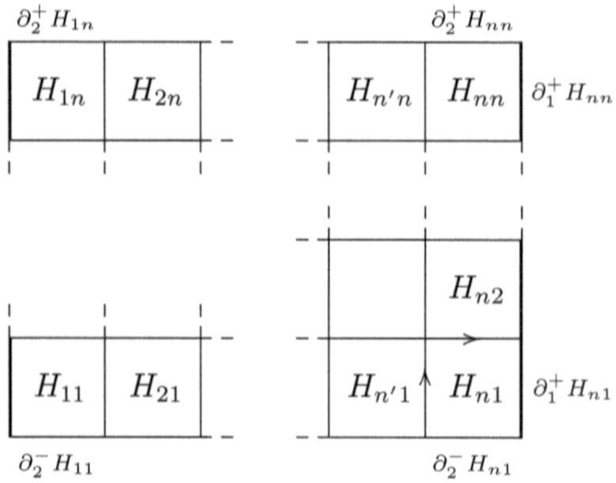


The first line of formula (7.65) is represented in the bottom of the figure (with $n' = n - 1$). The double homotopy $H_{n1}$ allows the passage from this line to the fourth line of (7.65), where the path $\partial_2^- H_{n1} \simeq_2 \partial_2^- H_{n1} * \partial_1^+ H_{n1}$ has been replaced with the path

$$\partial_1^- H_{n1} * \partial_2^+ H_{n1} = \partial_1^+ H_{n'1} * \partial_2^- H_{n2},$$

(marked with arrows), applying (7.64) and contiguity with the near squares.

At the end of the process the first line of formula (7.65) is replaced with the last, represented in the top row of the figure.

(d) Now $F\colon \Pi_1 X \to G$ is also well defined on arrows. To show that it preserves composition just note that, if two consecutive paths $a, b$ have a suitable decomposition on $n$ subintervals, then $a * b$ inherits a suitable decomposition $a * b = a_1 * \dots * a_n * b_1 * \dots * b_n$ which separates the original paths.

Finally, the uniqueness of the functor $F$ is obvious. □

## 7.5 Hints at the higher homotopy groups

We sketch now the definition and basic properties of the higher homotopy groups $\pi_n(X, x_0)$, and the relative homotopy groups $\pi_n(X, A, x_0)$, stating the higher dimensional Hurewicz Theorem. All constructions are based on the geometry of the cube, which allows an easy definition of the group operations.

The reader is referred to the literature cited in the introduction of this chapter: [Sp2, Ha, Hu].

### *7.5.1 Homotopy groups*

Let $(X, x_0)$ be a pointed space. For every $n \geqslant 0$ we consider the set of relative maps

$$\Omega_n(X, x_0) = \mathsf{Top}_2((\mathbb{I}^n, \underline{\partial}\mathbb{I}^n), (X, x_0)), \tag{7.66}$$

formed of all $n$-cubes $a\colon \mathbb{I}^n \to X$ that take the border $\underline{\partial}\mathbb{I}^n$ (in $\mathbb{R}^n$) to the basepoint $x_0$, called *cubes at* $x_0$. This set will be pointed at the map $u_n\colon \mathbb{I}^n \to X$, constant at $x_0$.

We define $\pi_n(X, x_0)$ as the quotient (pointed) set

$$\pi_n(X, x_0) = \Omega_n(X, x_0)/\simeq, \tag{7.67}$$

modulo relative homotopy: two cubes $a, a'\colon (\mathbb{I}^n, \underline{\partial}\mathbb{I}^n) \to (X, x_0)$ give the same class $[a] = [a']$ if there is a relative homotopy $h\colon a \simeq a'$ (see (4.1)). This is an ordinary homotopy $a \simeq a'$ with *fixed border*

$$\begin{gathered} h\colon \mathbb{I}^n \times \mathbb{I} \to X, \qquad h(s,0) = a(s), \quad h(s,1) = a'(s) \quad (s \in \mathbb{I}^n), \\ h(s,t) = x_0, \quad \text{for } s \in \underline{\partial}\mathbb{I}^n \text{ and } t \in \mathbb{I}. \end{gathered} \tag{7.68}$$

The case $n = 0$ gives a mere pointed set, examined in Exercise 7.5.2(a).

For $n \geqslant 1$, the set $\Omega_n(X, x_0)$ is equipped with $n$ operations, the concatenations $a *_i b$ in each direction $i = 1, ..., n$ (already considered for $i$-consecutive 2-cubes, in (7.13))

$$(a *_i b)(t_1, ..., t_n) = \begin{cases} a(t_1, ..., 2t_i ..., t_n), & \text{for } 0 \leqslant t_i \leqslant 1/2, \\ b(t_1, ..., 2t_i - 1, ..., t_n), & \text{for } 1/2 \leqslant t_i \leqslant 1. \end{cases} \tag{7.69}$$

These operations are easily seen to be well defined: cf. Exercise 7.5.2(b).

For $n = 1$ we have the single operation $a * b$ of loop concatenation, which induces the group structure $[a][b] = [a * b]$ of $\pi_1(X, x_0)$.

For $n \geqslant 2$, all these operations induce the same operation $[a]+[b] = [a *_i b]$, making $\pi_n(X, x_0)$ into an abelian group: this is proved in Exercise 7.5.2(d).

### 7.5.2 Exercises and complements

(a) For $n = 0$ the previous definition gives again the pointed set $\pi_0(X, x_0) = (|X|/\simeq, [x_0])$ of path components of $X$, introduced in (1.72).

(b) For $n \geqslant 1$, each operation $*_i$ is well defined in the set $\Omega_n(X, x_0)$ and induces an operation $[a *_i b]$ on $\pi_n(X, x_0)$, making it into a group. In each case, the identity is the class $[u_n]$ of the constant map $\mathbb{I}^n \to X$ at $x_0$.

(c) For $n \geqslant 2$, two distinct operations $*_i$, $*_j$ of $\Omega_n(X, x_0)$ satisfy the *middle-four interchange* property

$$(a *_i b) *_j (c *_i d) = (a *_j c) *_i (b *_j d). \tag{7.70}$$

and the induced operations in $\pi_n(X, x_0)$ also satisfy it.

(d) For $n \geqslant 2$, all the induced operations $[a] +_i [b] = [a *_i b]$ in $\pi_n(X, x_0)$ coincide, giving a unique operation $[a] + [b]$ which makes $\pi_n(X, x_0)$ into an abelian group. *Hint:* apply the exercise below.

(e) If a set $A$ is equipped with two operations, say $+$ and $\bullet$, which have the same identity $e$ and satisfy the middle-four interchange property

$$(x + y) \bullet (z + u) = (x \bullet z) + (y \bullet u), \qquad \text{for all } x, y, z, u \in A, \tag{7.71}$$

then these operations are commutative and coincide.

*Note.* The associativity of the operations is not assumed. Nevertheless, this result does not apply to the operations of $\Omega_n(X, x_0)$, which do not have an identity.

*(f) Formally, the sequence of structures of all $\pi_n(X, x_0)$ is made clearer by these remarks:

(i) $\pi_0(X, x_0)$ is a pointed set,

(ii) $\pi_1(X, x_0)$ is a group, which is a group *in* the category $\mathsf{Set}_\bullet$,

(iii) $\pi_2(X, x_0)$ is an abelian group, which is a group *in* the category $\mathsf{Gp}$,

(iv) $\pi_n(X, x_0)$ is an abelian group, which is a group *in* the category $\mathsf{Ab}$ (for $n \geqslant 3$).

All this makes sense defining a *group in* a category $\mathsf{C}$ with finite products, as an object $G$ equipped with mappings $m \colon G \times G \to G$ (*multiplication*), $i \colon G \to G$ (*inversion*), $e \colon G^0 \to G$ (*identity*) satisfying some axioms.

An interested reader can develop these issues (which are not difficult), or see [M2], p. 3.

### 7.5.3 Homotopy functors

(a) Extending the case $n = 1$ (in 7.2.1), we have a sequence of functors defined on the category of pointed spaces

$$\begin{aligned} \pi_n\colon \mathsf{Top}_\bullet \to \mathsf{C}_n, \qquad & (X, x) \mapsto \pi_n(X, x), \\ f_*\colon \pi_n(X, x) \to \pi_n(Y, y), \qquad & f_*([a]) = [fa], \end{aligned} \tag{7.72}$$

with values in the category $\mathsf{C}_n$ of pointed sets (for $n = 0$), or groups (for $n = 1$), or abelian groups (for $n \geqslant 2$). A pointed map $f\colon (X, x) \to (Y, y)$ is sent to the morphism $\pi_n(f) = f_*$ which takes the class of a cube $a \in \Omega_n(X, x)$ to the class of $fa \in \Omega_n(Y, y)$.

For $n = 0$, this mapping preserves the basepoint. For $n \geqslant 1$ it preserves the structural operation: at the level of cubes, we already have that $f(a *_i b) = fa *_i fb$, for $i = 1, ..., n$.

(b) We can now extend an issue already noted for $n = 1$ (in 7.2.2(c)): the functor $\pi_n$ can be identified with the homotopy functor associated to the pointed space $(\mathbb{S}^n, e_1)$, defined in (1.73)

$$\pi_n(X, x_0) = [(\mathbb{S}^n, e_1), (X, x_0)]_\bullet, \tag{7.73}$$

so that $\pi_n$ explores pointed spaces by probes defined on the pointed spheres (see 0.4).

In fact, a cube $a\colon \mathbb{I}^n \to X$ at $x_0$ amounts to a pointed map $\hat{a}\colon (\mathbb{S}^n, e_1) \to (X, x_0)$, and a homotopy $h\colon a \simeq b$ with fixed border amounts to a pointed homotopy $\hat{h}\colon \hat{a} \simeq \hat{b}$.

This also works for $n = 0$: a point $a\colon \{0\} \to X$ is the same as a pointed map $\hat{a}\colon (\mathbb{S}^0, 1) \to (X, x_0)$ with $\hat{a}(-1) = a(0)$, and the path component of $a(0)$ corresponds to the class of $\hat{a}$ up to pointed homotopy.

### 7.5.4 Complements

The rest of this section is meant to give some further information on the higher homotopy groups and the relative ones. Complex proofs will be referred to the books [Ha, Hu].

(a) (*The transition isomorphism*) Plainly, the pointed set $\pi_0(X, x_0) = (|X|/\simeq, [x_0])$ stays the same when $x_0$ varies in a path component of $X$.

For $n \geqslant 1$, each path $c\colon x_0 \simeq x_1$ in the space $X$ gives a transition isomorphism

$$\hat{c}_n\colon \pi_n(X, x_0) \to \pi_n(X, x_1), \tag{7.74}$$

which extends the 1-dimensional case (in 7.3.1).

For a 0-connected space $X$, the symbol $\pi_n(X)$ can thus denote the isomorphism type of each $\pi_n(X, x_0)$, with due care (see the introduction of Section 7.3).

The case $n = 2$ is sufficient to show the general pattern: a cube $A\colon \mathbb{I}^2 \to X$ at $x_0$ is transformed into a cube at $x_1$, pasting around $A$ various double homotopies built with the path $c$, as in the following picture

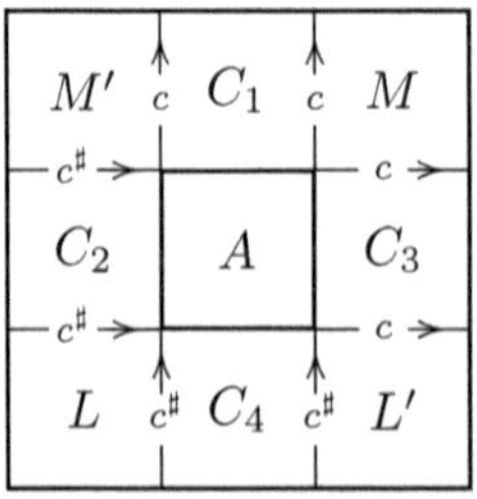


$$\begin{aligned} &M'(s,t) = c(\max(1-s,t)), \\ &M(s,t) = c(\max(s,t)), \\ &C_1(s,t) = c(t), \\ &\dots \\ &L(s,t) = c^\sharp(\min(s,t)), \\ &L'(s,t) = c^\sharp(\min(1-s,t)). \end{aligned} \qquad (7.75)$$

(b) (*Cartesian products*) The functor $\pi_n\colon \mathsf{Top}_\bullet \to \mathsf{C}_n$ preserves products, for all $n \geqslant 0$.

The case $n = 0$ is an easy consequence of the fact that a product $\Pi_i X_i$ of non-empty spaces is path connected if and only if all the factors $X_i$ are. The case $n = 1$ was proved in 7.2.4 (and deduced from the fundamental groupoid case in 7.1.6); for higher $n$, the proof is as in 7.1.6.

*(c) The higher homotopy groups $\pi_k(\mathbb{S}^1)$ are trivial, for $k \geqslant 2$.

This is proved in [Ha], Proposition 4.1, as a consequence of the lifting properties of the exponential map $p\colon \mathbb{R} \to \mathbb{S}^1$.

*(d) It follows that $\pi_k(\mathbb{T}^n)$ is trivial, for $n \geqslant 1$ and $k \geqslant 2$. On the other hand, we have seen that the homology groups $H_k(\mathbb{T}^n)$ are only trivial for $k > n$ (in Exercise 6.7.7(e)).

In particular, any map $\mathbb{S}^2 \to \mathbb{T}$ is homotopic to a constant: the homotopy groups do not detect the cavity of dimension 2 in the torus.

### 7.5.5 **Hurewicz Theorem* (in higher dimension)

*Suppose that the 0-connected space $X$ has*

$$\pi_1(X) = \ldots = \pi_{n-1}(X) = 0, \qquad (7.76)$$

*for $n \geqslant 2$. Then, for every $x_0 \in X$, there is a Hurewicz isomorphism*

$$h_n\colon \pi_n(X, x_0) \to H_n(X), \qquad h_n[a] = [a]_H, \qquad (7.77)$$

*that takes the class of a cube $a\colon (\mathbb{I}^n, \underline{\partial}\mathbb{I}^n) \to (X, x_0)$ to its homology class.*

*Proof* We only remark that the $n$-cube $a$ is a cycle, because its faces are constant cubes of degree $\geqslant 1$. The proof of the theorem can be found in [Ha], Theorem 4.32, or worked out as in [Hu], Chapter V, Exercise C. □

### 7.5.6 Higher connectedness

(a) Let $X$ be a 1-connected space, which means that it is non-empty, path connected, and has $\pi_1(X) = 0$. By Theorem 7.5.5, $\pi_2(X) \cong H_2(X)$.

We say that $X$ is *2-connected* if $\pi_2(X) = 0$, or equivalently $H_2(X) = 0$.

(b) More generally we say that a space $X$ is *n-connected*, for $n \geqslant 1$, if it is non-empty, path connected, and satisfies the following equivalent conditions:

(i) $\pi_1(X) = \pi_2(X) = \ ... \ = \pi_n(X) = 0$,
(ii) $\pi_1(X) = 0, \qquad H_2(X) = \ ... \ = H_n(X) = 0$.

Then $\pi_{n+1}(X) \cong H_{n+1}(X)$.

The equivalence readily follows from the Hurewicz Theorem, by induction. Let us note that, in (ii), we cannot replace $\pi_1$ with $H_1$: we have seen in 7.3.5(c) that $H_1(X)$ can be trivial when $\pi_1(X)$ is not.

(c) Let $n \geqslant 2$. We have seen in 7.4.4(a) that $\pi_1(\mathbb{S}^n) = 0$. Since $H_k(\mathbb{S}^n) = 0$ for $0 < k < n$, we deduce that

$$\pi_k(\mathbb{S}^n) = 0 \text{ for } 0 < k < n, \qquad \pi_n(\mathbb{S}^n) \cong \mathbb{Z}. \tag{7.78}$$

For $n \geqslant 1$, the sphere $\mathbb{S}^n$ is $(n-1)$-connected and we have an isomorphism

$$\begin{aligned} &h_n \colon \pi_n(\mathbb{S}^n, e_1) \to H_n(X), \\ h_n[a] = [a]_H \qquad &(\text{for } a \colon (\mathbb{I}^n, \underline{\partial}\mathbb{I}^n) \to (X, x_0)). \end{aligned} \tag{7.79}$$

*(d) The higher homotopy groups of $\mathbb{S}^n$ (far from vanishing!) are still the subject of research: there are infinitely many pairs $(n, k)$ for which $\pi_k(\mathbb{S}^n)$ is known, and infinitely many for which it is not.

*(e) The group $\pi_3(\mathbb{S}^2, e_1)$ is infinite cyclic, generated by the *Hopf map* $p \colon \mathbb{S}^3 \to \mathbb{S}^2$: see [Hu], Section III.5 and Proposition V.6.5.

### 7.5.7 Pointed relative pairs

We introduce now the relative homotopy groups $\pi_n(X, A, x_0)$, for $n \geqslant 2$, and the relative pointed set $\pi_1(X, A, x_0)$. The classical reference is [Hu], Section 4.3.

We work in the obvious category $\mathsf{Top}_{2\bullet}$ *of pointed relative pairs*. An object $(X, A, x_0)$ is a relative pair $(X, A)$ pointed at some $x_0 \in A \subset X$.

A (pointed relative) *map* $f\colon (X, A, x_0) \to (Y, B, y_0)$ is a relative map $(X, A) \to (Y, B)$ that takes $x_0$ to $y_0$.

A (pointed relative) *homotopy* $\varphi\colon f \simeq g\colon (X, A, x_0) \to (Y, B, y_0)$ is a relative homotopy $(X, A) \to (Y, B)$ such that $\varphi(x_0, t) = y_0$, for $t \in \mathbb{I}$.

$\mathsf{Top}_\bullet$ is fully embedded in $\mathsf{Top}_{2\bullet}$, sending the pointed space $(X, x_0)$ to $(X, \{x_0\}, x_0)$. The new constructions will extend the absolute ones, with respect to this embedding. Moreover, $\mathsf{Top}_{2\bullet}$ can be embedded in the category $\mathsf{Top}_3$ *of relative triples of spaces* $(X, A, B)$, where $B$ is a subspace of $A$, which is a subspace of $X$ (already used in Theorem 4.1.8).

The construction will again be based on the geometry of the standard cubes:

$$\partial\mathbb{I}^n = \mathbb{I}^{n-1} \cup J^{n-1}, \qquad \mathbb{I}^{n-1} \cap J^{n-1} = \partial\mathbb{I}^{n-1} \qquad (n \geqslant 1), \tag{7.80}$$

where, as usual, $\mathbb{I}^{n-1}$ is nested in $\mathbb{I}^n$ as the face $\delta_n^-(\mathbb{I}^{n-1})$, in the hyperplane $t_n = 0$, while $J^{n-1}$ is the union of the other $2n - 1$ faces.

### 7.5.8 Relative homotopy groups

(a) For every $n \geqslant 1$ we consider the set of relative maps

$$\Gamma_n(X, A, x_0) = \mathsf{Top}_3((\mathbb{I}^n, \mathbb{I}^{n-1}, J^{n-1}), (X, A, \{x_0\})), \tag{7.81}$$

formed of all the *relative* $n$-cubes $a\colon \mathbb{I}^n \to X$, that send $\mathbb{I}^{n-1}$ into $A$, and $J^{n-1}$ to the basepoint $x_0$. This set is pointed at the map $u_n\colon \mathbb{I}^n \to X$, constant at $x_0$. (In $\mathsf{Top}_\bullet$, $\Gamma_n(X, \{x_0\}, x_0) = \Omega_n(X, x_0)$, see (7.66).)

We define $\pi_n(X, A, x_0)$ as the quotient pointed set

$$\pi_n(X, A, x_0) = \Gamma_n(X, A, x_0)/\simeq, \tag{7.82}$$

modulo relative homotopy: two relative $n$-cubes $a, a'$ of $(X, A, x_0)$ give the same class $[a] = [a']$ when there exists a pointed relative homotopy $h\colon a \simeq a'$. In other words:

$$\begin{aligned} &h\colon \mathbb{I}^n \times \mathbb{I} \to X, \quad h(s, 0) = a(s), \quad h(s, 1) = a'(s) \qquad (s \in \mathbb{I}^n), \\ &h(s, t) \in A, \qquad h(s', t) = x_0 \quad (s \in \mathbb{I}^{n-1}, s' \in J^{n-1}, t \in \mathbb{I}). \end{aligned} \tag{7.83}$$

(b) The case $n = 1$ gives a mere pointed set. An element $[a]$ is the class of a path $a\colon \mathbb{I} \to X$ with $a(0) \in A$ and $a(1) = x_0$, modulo homotopies which keep the starting point in $A$ and the ending point at $x_0$. (These paths cannot be concatenated, generally.)

For $n \geqslant 2$, the set $\Gamma_n(X, A, x_0)$ is equipped with $n - 1$ operations, the concatenation $a *_i b$ in each direction $i = 2, ..., n$, defined as in (7.69). (Again, direction $i = 1$ cannot be used, because a relative cube sends the face $\mathbb{I}^{n-1}$ to $A$, not to the basepoint, generally.)

Applying Exercise 7.5.2(b), these operations are well defined and each of them produces a group structure. For $n = 2$ we have *one* operation $a *_2 b$, which induces the group structure $[a][b]$ of $\pi_2(X, A, x_0)$. For $n \geqslant 3$, all these operations induce the same operation $[a] + [b] = [a *_i b]$, making $\pi_n(X, A, x_0)$ into an abelian group, as in 7.5.2(d).

(c) We have thus a sequence of functors, with values in $\mathsf{Ab}$ for $n \geqslant 3$

$$\pi_1 \colon \mathsf{Top}_{2\bullet} \to \mathsf{Set}_\bullet, \quad \pi_2 \colon \mathsf{Top}_{2\bullet} \to \mathsf{Gp}, \quad \pi_n \colon \mathsf{Top}_{2\bullet} \to \mathsf{Ab}, \tag{7.84}$$

where the map $f \colon (X, A, x_0) \to (Y, B, y_0)$ is sent to

$$f_* \colon \pi_n(X, A, x_0) \to \pi_n(Y, B, y_0), \quad f_*[a] = [fa] \quad (n \geqslant 1). \tag{7.85}$$

These functors are invariant up to pointed relative homotopy. They extend the absolute functors $\pi_n$, although $\pi_1$ only takes values in $\mathsf{Set}_\bullet$.

*(d) The Relative Hurewicz Theorem can be found – in different forms – in [Hu], Exercise C of Chapter V, and in [Sp2], Theorem 7.5.4.

### 7.5.9 The exact sequence of a pointed relative pair

(a) A pointed relative pair $(X, A, x_0)$ has a *sequence of homotopy groups* (or pointed sets, in low degree), natural for pointed relative maps

$$\begin{array}{ccccc}
 & & \cdots & \longrightarrow & \pi_2(X, A, x_0) \\
 & & & D_2 & \\
\pi_1(A, x_0) & \xrightarrow{u_{*1}} & \pi_1(X, x_0) & \xrightarrow{v_{*1}} & \pi_1(X, A, x_0) \\
 & & & D_1 & \\
\pi_0(A, x_0) & \xrightarrow{u_{*0}} & \pi_0(X, x_0). & &
\end{array} \tag{7.86}$$

Here the map $u \colon (A, x_0) \to (X, x_0)$ is the inclusion, while $v \colon (X, \{x_0\}, x_0) \to (X, A)$ is defined by the identity of $X$. The connecting homomorphism (or pointed mapping) $D_n$, for $n \geqslant 1$, is computed as:

$$D_n \colon \pi_n(X, A, x_0) \to \pi_{n-1}(A, x_0), \quad D_n[a]_{(X,A)} = [a|\mathbb{I}^{n-1}]_A. \tag{7.87}$$

where $a \colon (\mathbb{I}^n, \mathbb{I}^{n-1}, J^{n-1}) \to (X, A, x_0)$ is a relative $n$-cube, and $a|\mathbb{I}^{n-1}$ is its restriction $(\mathbb{I}^n, \partial\mathbb{I}^{n-1}) \to (A, x_0)$.

It is easy to see that $D_n$ is a well-defined pointed mapping, and a homomorphism for $n \geqslant 2$ ([Hu], IV.5.4).

*(b) Exactness Theorem. *The homotopy sequence of a pointed relative pair $(X, A, x_0)$ is exact (that is, an exact sequence of pointed sets).*

The proof can be seen in [Hu], IV.5.7.

We already remarked, in 2.1.2(f), that the exactness of a sequence in $\mathsf{Gp}$ is assessed in the same way in $\mathsf{Set}_\bullet$. We also note that one could complete the sequence (7.86) defining $\pi_0(X, A, x_0) = \mathrm{Cok}\, u_{*0}$.

## 7.6 Weak homotopy equivalence and Whitehead theorems

We end this chapter dealing with weak homotopy equivalence, and providing a synopsis of its interplay with singular homology and CW-complexes. Most proofs will be referred to Spanier's book [Sp2].

The homotopy groups are invariant up to weak homotopy, by definition. Singular homology is also, by a well-known Whitehead Theorem (see 7.6.3), but Alexander–Spanier and Čech cohomology theories are not (as shown by elementary examples, in 7.6.2). Loosely speaking, invariance up to weak homotopy is related to exploring a space $X$ by *probes* $M^n \to X$, but can fail when we explore $X$ by *functions*.

Another celebrated Whitehead Theorem says that, for CW-complexes, weak homotopy equivalences and homotopy equivalences coincide (in 7.6.4). Eilenberg–Mac Lane spaces will be briefly introduced at the end.

### *7.6.1 Definition*

(a) As already mentioned in 1.4.4(b), a map $f\colon X \to Y$ of topological spaces is said to be a *weak homotopy equivalence* if it induces isomorphisms in all homotopy sets and groups

$$\pi_n(f)\colon \pi_n(X, x) \to \pi_n(Y, f(x)), \tag{7.88}$$

for all $x \in X$ and $n \geqslant 0$.

Plainly, every homotopy equivalence is a weak homotopy equivalence; the converse fails, as shown below.

By definition, the homotopy functors $\pi_n$ are *invariant up to weak homotopy*, that is, they turn every weak homotopy equivalence into an isomorphism. The same holds for singular homology, as we shall see in 7.6.3.

*We already said in 1.4.4(b) that the *weak homotopy category* $\mathsf{HoTop}$ is defined by formally inverting, in $\mathsf{Top}$, the weak homotopy equivalences.*

### *7.6.2 Exercises and complements*

(a) For the discrete space $D = \mathbb{N} \cup \{\infty\}$, the map $f\colon D \to \mathbb{N}^\bullet$ that lifts the identity is a weak homotopy equivalence and not a homotopy equivalence. Another example will be given in (c).

(b) On this map, Alexander–Spanier cohomology gives a non-surjective homomorphism $\overline{H}^0(f)\colon \overline{H}^0(\mathbb{N}^\bullet) \to \overline{H}^0(D)$ (see 4.5.8(e)), showing again that $f\colon D \to \mathbb{N}^\bullet$ is not a homotopy equivalence. More interestingly, it shows that Alexander–Spanier cohomology is not invariant up to weak homotopy equivalence.

Čech cohomology is neither, as it coincides with Alexander–Spanier's on all paracompact Hausdorff spaces: see 4.7.4.

*(c) The *Warsaw circle* $S_W \subset \mathbb{R}^2$, studied by Borsuk and his school, is obtained inserting infinite oscillations in a circle ([Sp2], Example 6.9.17). Taking out a point (far from the oscillations, for simplicity), $S_W$ is still connected (behaving as a circle) but no longer path connected (behaving as a line).

Alexander–Spanier and Čech cohomologies detect the first behaviour: $\overline{H}^n(S_W) \cong \check{H}^n(S_W)$ is isomorphic to the singular cohomology of the circle, as proved in [Sp2], *loc. cit.*; in particular $\overline{H}^1(S_W) \cong \mathbb{Z}$.

On the other hand, weak homotopy equivalence (and therefore singular homology) distinguishes the second: it is easy to construct a bijective map $f\colon [0,1[\, \to S_W$ which is a weak homotopy equivalence. In particular, the induced homomorphism $f^*: \overline{H}^1(S_W) \to \overline{H}^1([0,1[\,) = 0$ shows that $f$ is not a homotopy equivalence, and that $S_W$ is not a CW-complex.

### 7.6.3 *Weak homotopy equivalence and singular homology

(a) The relationship between homotopy groups and singular homology is complex. Loosely speaking, Hurewicz theorem says that the former determine the latter *for an initial segment of degrees*, but this can drastically change in higher degree: for instance, the $n$-dimensional torus $\mathbb{T}^n$ has trivial homotopy groups $\pi_k(\mathbb{T}^n)$ for all $k > 1$, and rich homology groups up to degree $n$, that accurately describe its geometry.

Working again on an initial segment of degrees, there is an important result by J.H.C. Whitehead (not to be confused with another celebrated Whitehead Theorem (see 7.6.4(a)).

(b) Whitehead Theorem. *Let $f\colon (X, x_0) \to (Y, y_0)$ be a pointed map between path-connected pointed spaces, and $n \geqslant 1$.*

*If $f_{*k}\colon \pi_k(X, x_0) \to \pi_k(Y, y_0)$ is an isomorphism for $1 \leqslant k < n$ and is surjective for $k = n$, then the same holds for $f_{*k}\colon H_k(X) \to H_k(Y)$, in singular homology. The converse holds when $X$ and $Y$ are 1-connected.*

One can find a proof in [Sp2], Theorem 7.5.9 (using the relative Hurewicz theorem and the mapping cylinder of $f$, see 8.2.1). Or also in [Hu], Exercise D of Chapter 5 (using the same theorem and the exact sequences of a pointed pair).

(c) Corollary. *Singular homology is invariant up to weak homotopy equivalence: a weak homotopy equivalence $f\colon X \to Y$ induces isomorphism in singular homology, in every degree.*

It is an easy consequence of the theorem. Plainly, we only need to consider a weak homotopy equivalence $f\colon X \to Y$ between path connected spaces. Choosing a point $x_0 \in X$, the map $f\colon (X, x_0) \to (Y, f(x_0))$ induces an isomorphism $\pi_n(f)$ for all $n \geqslant 1$. By the previous theorem, the same holds for $H_n(f)$, in singular homology with integral coefficients; or any coefficient group, by 6.2.7(a).

### 7.6.4 *Weak homotopy equivalence and CW-complexes

We state, without proof, some important results.

(a) Whitehead Theorem on CW-complexes. *A weak homotopy equivalence between CW-complexes is a homotopy equivalence.*

These properties are thus equivalent for CW-complexes; we have already seen that this does not hold for general spaces. The theorem was proved by J.H.C. Whitehead, in 1949 [W1, W2]. The proof can be seen in [Sp2], Theorem 7.6.22 and Corollary 7.6.24.

(b) Cellular Approximation Theorem. *Every topological space is weakly homotopy equivalent to a CW-complex. The latter is determined up to homotopy equivalence, by (a).*

See [Sp2], Theorem 7.8.1 (in a more general, relative form).

### 7.6.5 *Homotopy groups and inductive limits

(a) Theorem 3.4.7, about singular homology and inductive limits, can be easily adapted to homotopy groups (or pointed sets).

We start from be the inductive limit $(X, x_0)$ of a nested sequence $(X_n, x_0)$ of pointed $T_1$-spaces. Then the homotopy group $\pi_k(X, x_0)$ is the inductive limit of a system in $\mathsf{Gp}$ (or $\mathsf{Set}_\bullet$)

$$\pi_k(X_0, x_0) \to \pi_k(X_1, x_0) \to \dots \to \pi_k(X_n, x_0) \to \dots \tag{7.89}$$

The proof is similar to that of Theorem 3.4.7: every cube of $(X, x_0)$ lies in some $(X_n, x_0)$, and every relative homotopy of cubes also does.

(b) It is now easy to show that *the infinite dimensional sphere $\mathbb{S}^\infty$ has trivial homotopy groups, and is contractible.*

In fact, $\pi_k(\mathbb{S}^n, e_1) = 0$, for $0 \leqslant k < n$ (see 7.5.6(c)), and $\pi_k(\mathbb{S}^\infty, e_1) = 0$ for all $k \geqslant 0$. The map $p\colon \mathbb{S}^\infty \to \{*\}$ is thus a weak homotopy equivalence. Since $\mathbb{S}^\infty$ is a CW-complex (see 4.9.4(c)), $p$ is also a homotopy equivalence, applying Whitehead Theorem 7.6.4(a).

### 7.6.6 *Eilenberg–Mac Lane spaces

The *Eilenberg–Mac Lane spaces* $K(G, n)$ extend the spaces $K(G, 1)$ mentioned in 7.3.5(c). For this celebrated topic we refer to [Sp2], Section 8.1, [Ha], Section 4.2, and the original article [EiM4].

(a) For every $n > 0$ and any non-trivial group $G$ (assumed to be commutative when $n \geqslant 2$), one can construct a 0-connected space $K(G, n)$ with a single non-trivial homotopy group in degree $n$

$$\pi_n(K(G, n)) \cong G, \qquad \pi_k(K(G, n)) = 0, \text{ for } k \neq 0, n. \tag{7.90}$$

$K(G, n)$ can always be constructed as a CW-complex, so that the solution is determined up to homotopy equivalence: cf. [Ha], Proposition 4.30.

For instance, one can construct a space $K(\mathbb{Z}, n)$ as the 'infinite symmetric product' $\mathrm{SP}(\mathbb{S}^n)$: see [Ha], p. 365.

(b) Every sequence $(G_n)_{n \geqslant 1}$ of groups, commutative for $n \geqslant 2$, can be realised as the sequence $\pi_n(X)$ of homotopy groups of a 0-connected space, taking the cartesian product of all $K(G_n, n)$, for $n \geqslant 1$.

(c) The homotopy construction of *additive* generalised homology and cohomology theories for CW-complexes is an interesting subject, linked to the Eilenberg–Mac Lane spaces. Some presentations take this approach from the start, only relying on homological algebra in a few points [AGP]. Others arrive to this construction after the usual ones, like [Ha], Section 4.3, [Sw], Chapter 8, and [May2], Chapter 22.

# 8
# Interactions of homotopy and homology

This chapter draws together various lines of the previous ones, studying the links of homotopy and homology, for spaces and pointed spaces (after those discussed in Chapter 7). Reduced homology and cohomology are closely related to homotopy constructions like the suspension and the cofibre sequence of a map, built here in Section 8.2; in a dually similar way, the homotopy groups are related to loop spaces and the fibre sequence of a pointed map, as shown in 8.6.8.

We begin with a preliminary Section 8.1 on categorical topics that will be used throughout this chapter: adjoint functors and general exponential laws, linking cartesian or tensor products to function spaces.

Homotopy pushouts of spaces, mapping cones and suspension are studied in Section 8.2, building the cofibre sequence of a map, related to reduced homology and cohomology.

Path spaces and homotopy pullbacks are dealt with in Section 8.3, after examining a basic problem: the exponential law only holds for topological spaces under suitable conditions (sufficient for our purposes) of 'generalised local compactness'; formally, the category $\mathsf{Top}$ is not cartesian closed.

The last three sections are about pointed spaces. The exponential law is now based on the *smash product* $X \wedge Y$, a sort of tensor product studied in Sections 8.4 and 8.5, together with the spaces of pointed maps. Finally Section 8.6 studies homotopy pushouts of pointed spaces, pointed cones and pointed suspension, highlighting their close relationship with the smash product. Homotopy pullbacks, cocones and loop spaces are also presented, together with the fibre sequence of a pointed map and its relation to the homotopy groups.

This chapter is one of the main addition with respect to [G6]. References will be given in the appropriate section.

## 8.1 Adjoint functors and exponential laws

Adjunctions, a global way of presenting universal properties, form the underlying layout of various aspects of Algebra and Topology, and have already appeared in several points of the previous chapters. They also allow us to formalise the exponential laws we encountered in 5.1.7, and will study here. The cylinder and path functors of spaces and pointed spaces (Sections 8.2 and 8.6) are also adjoint functors, as many pairs of homotopy constructions, in Section 8.6.

*Literature.* Adjoint functors were introduced in 1958 by D. Kan [Ka], with various applications in Topology and Algebraic Topology – another instance of the dialectic relationship between these subjects and Category Theory.

All books on categories, like [M2, Bo1, AHS, G5], have chapters on adjunctions. Applications in Topology, Algebraic Topology and Homological Algebra can be found in all of them, and form a large part of [G5], in Chapters 5 and 6. For monoidal categories and exponential laws we refer to [M2], Chapter VII and [G5], Section 3.4; their theory is specifically dealt with in [EiK, Ky2].

### *8.1.1 Introducing adjunctions*

We begin with two well known situations, which help us to introduce adjoint functors.

(a) The forgetful functor $U\colon \mathsf{Top} \to \mathsf{Set}$ has two well-known 'extreme approximations' to an inverse which does not exist, the functors $D, C\colon \mathsf{Set} \to \mathsf{Top}$, where $DX$ (resp. $CX$) is the set $X$ with the discrete (resp. indiscrete) topology. In fact they provide the set $X$ with the finest (resp. the coarsest) topology.

For every set $X$ and every space $Y$ we can identify the following hom-sets

$$\mathsf{Top}(DX, Y) = \mathsf{Set}(X, UY), \qquad \mathsf{Set}(UY, X) = \mathsf{Top}(Y, CX), \qquad (8.1)$$

since every mapping $X \to Y$ becomes continuous if we put on $X$ the discrete topology, and every mapping $Y \to X$ becomes continuous if we put on $X$ the indiscrete one.

The first fact will say that $D$ is left adjoint to $U$ (and $U$ right adjoint to $D$), which is written as $D \dashv U$. The second says that $U \dashv C$. The name comes from the analogy with adjoint operators between Hilbert spaces.

(b) In other cases it may be convenient to follow another approach, based on universal arrows.

To wit, let us start from the forgetful functor $U\colon R\,\mathsf{Mod} \to \mathsf{Set}$. We have seen, in (5.5), that, for every set $X$, there is a universal arrow from $X$ to the functor $U$

$$F(X) = \bigoplus_{x \in X} R, \qquad \eta X \colon X \to UF(X), \tag{8.2}$$

consisting of the free $R$-module $F(X)$ on the set $X$, with the insertion of the canonical base $\eta X$.

The fact that this universal arrow exists *for every* $X$ *in* Set allows us to construct a 'backward' functor $F \colon \mathsf{Set} \to R\,\mathsf{Mod}$, left adjoint to $U$.

$F$ is already defined on the objects. For a mapping $f \colon X \to Y$ in Set

$$\begin{array}{ccccc} X & \xrightarrow{\eta X} & UF(X) & \qquad & F(X) \\ {\scriptstyle f}\downarrow & & \downarrow{\scriptstyle U(Ff)} & & \downarrow{\scriptstyle Ff} \\ Y & \xrightarrow[\eta Y]{} & UF(Y) & & F(Y) \end{array} \tag{8.3}$$

we define $F(f)$ by the universal property of $\eta X$, as the unique morphism $F(X) \to F(Y)$ in $R\,\mathsf{Mod}$ such that $U(Ff)$ makes the previous square commute. (In this example $F(f)$ is just the linear extension of $f$; but the general pattern is more evident if we proceed in a formal way.)

Further applications of the universal property readily show that $F$ preserves composition and identities (as is obvious in this particular case). Let us note that we have extended the given object-function $F(X)$ in the only way that makes the family of mappings $(\eta_X)$ into a natural transformation

$$\eta \colon 1 \to UF \colon \mathsf{Set} \to \mathsf{Set},$$

called the *unit* of the adjunction.

(c) The relationship between these functors $U$ and $F$ (assuming now that we have defined both) can be equivalently expressed as in (a), by a family of bijective mappings, (for $X$ in Set and $A$ in $R\,\mathsf{Mod}$)

$$\begin{gathered} \varphi_{XA} \colon R\,\mathsf{Mod}(FX, A) \to \mathsf{Set}(X, UA), \\ \varphi_{XA}(g \colon FX \to A) = Ug.\eta X \colon X \to UFX \to UA, \end{gathered} \tag{8.4}$$

which is natural in the variables $X$ and $A$ (on the category $\mathsf{Set}^{\mathrm{op}} \times R\,\mathsf{Mod}$).

Concretely, $\varphi_{XA}(g \colon FX \to A)$ is the restriction of the $R$-homomorphism $g$ to the base $X$ of $F(X)$; its bijectivity means that every mapping $f \colon X \to UA$ has a unique extension to a homomorphism $g \colon FX \to A$.

On the other hand, the adjunction $D \dashv U$ in (a) can also be expressed by the universal arrow $(DX, \eta X \colon X \to UDX)$ from the set $X$ to the functor $U \colon \mathsf{Top} \to \mathsf{Set}$, letting $\eta X = \mathrm{id}\, X$. Similarly, the adjunction $U \dashv C$ can also be expressed by the universal arrow $(UY, \eta' Y \colon Y \to CUY)$ from the space $Y$ to the functor $C\colon \mathsf{Set} \to \mathsf{Top}$, where $\eta' Y$ lifts the identity of the underlying sets.

### 8.1.2 Main definitions

An *adjunction* $F \dashv G$, with a functor $F\colon \mathsf{C} \to \mathsf{D}$ *left adjoint* to a functor $G\colon \mathsf{D} \to \mathsf{C}$, can be equivalently presented in four forms.

(i) We assign two functors $F\colon \mathsf{C} \to \mathsf{D}$ and $G\colon \mathsf{D} \to \mathsf{C}$ together with a family of bijections

$$\varphi_{XY}\colon \mathsf{D}(FX, Y) \to \mathsf{C}(X, GY) \qquad (X \text{ in } \mathsf{C},\ Y \text{ in } \mathsf{D}),$$

which is natural in $X, Y$. More formally, the family $(\varphi_{XY})$ is a functorial isomorphism

$$\varphi\colon \mathsf{D}(F(-), -) \longrightarrow \mathsf{C}(-, G(-))\colon \mathsf{C}^{\mathrm{op}} \times \mathsf{D} \longrightarrow \mathsf{Set}.$$

(ii) We assign a functor $G\colon \mathsf{D} \to \mathsf{C}$ and, for every object $X$ in $\mathsf{C}$, a universal arrow (see 1.6.6)

$$(F_0X, \eta X\colon X \to GF_0X), \qquad \text{from the object } X \text{ to the functor } G.$$

(ii*) We assign a functor $F\colon \mathsf{C} \to \mathsf{D}$ and, for every object $Y$ in $\mathsf{D}$, a universal arrow

$$(G_0Y, \varepsilon Y\colon FG_0Y \to Y), \qquad \text{from the functor } F \text{ to the object } Y.$$

(iii) We assign two functors $F\colon \mathsf{C} \to \mathsf{D}$ and $G\colon \mathsf{D} \to \mathsf{C}$, together with two natural transformations

$$\eta\colon \mathrm{id}\,\mathsf{C} \to GF \ \ (\text{the } \textit{unit}), \qquad \varepsilon\colon FG \to \mathrm{id}\,\mathsf{D} \ \ (\text{the } \textit{counit}),$$

which satisfy the *triangular identities*: $\varepsilon F.F\eta = \mathrm{id}\,F$ and $G\varepsilon.\eta G = \mathrm{id}\,G$

$$F \xrightarrow{F\eta} FGF \xrightarrow{\varepsilon F} F \ \ (\text{composite } \mathrm{id}\,F) \qquad G \xrightarrow{\eta G} GFG \xrightarrow{G\varepsilon} G \ \ (\text{composite } \mathrm{id}\,F) \tag{8.5}$$

A proof of the equivalence can be found in [M2], Section IV.1, Theorem 2, or in [G5], Theorem 3.1.5. Essentially:

- given (i) one defines $\eta X = \varphi_{X,FX}(\mathrm{id}\,FX)\colon X \to GFX$ and $\varepsilon Y = (\varphi_{GY,Y})^{-1}(\mathrm{id}\,GY)\colon FGY \to Y$,
- given (ii) one defines $F(X) = F_0X$, the morphism $F(f\colon X \to X')$ by the universal property of $\eta X$, and the morphism $\varphi_{XY}(g\colon FX \to Y)$ as $Gg.\eta X\colon X \to GY$,
- given (iii) one defines the mapping $\varphi_{XY}\colon \mathsf{D}(F(X), Y) \to \mathsf{C}(X, G(Y))$ as

above, and also a backward mapping $\psi_{XY}(f\colon X \to GY) = \varepsilon Y.Ff$; then one proves that these mappings are inverse to each other, by the triangular identities.

These forms have different features.

Form (i) is the classical definition of an adjunction, and is at the origin of the name, by analogy with adjoint operators of Hilbert spaces.

Form (ii) is used when we start from a given functor $G$ and want to construct its left adjoint (possibly less easy to define). Form (ii*) is used in a dual way.

The 'algebraic' form (iii) leads to the formal theory of adjunctions. *In fact, it makes sense in an abstract 2-category.*

To better understand the global picture of an adjunction one can now review the three examples of 8.1.1, completing their expression in each of the previous forms.

### *8.1.3 Examples and comments*

In the previous chapters there are various cases where the general existence of universal arrows ensures the existence of an adjoint.

(a) (*Free objects*) As we have seen in 1.6.7(b), given a functor $U\colon \mathsf{A} \to \mathsf{Set}$, the free $\mathsf{A}$-object over a set $X$ is an object $F(X)$ equipped with a universal arrow $\eta\colon X \to UF(X)$ from the object $X$ to the functor $U$. The existence of all free objects in $\mathsf{A}$ amounts to the existence of the left adjoint $F \dashv U$.

In particular, we have constructed the free abelian group in 1.2.5, the free monoid in 1.6.7(c), the free $R$-module in 5.1.1, and the free group in 7.4.1. In fact, all algebraic structures defined by 'equational axioms' have free objects; the following point is an extension of this fact.

(b) Many interesting constructions in algebra arise as a left adjoint to a forgetful functor $U\colon \mathsf{A} \to \mathsf{B}$, between categories of algebraic structures of which the second is 'weaker' than the first – in the sense of having less structure, or less properties, or both. This includes point (a), letting $\mathsf{B} = \mathsf{Set}$, the category of trivial algebraic structures.

Here we have also considered the inclusion functor $U\colon \mathsf{Ab} \to \mathsf{Gp}$, whose left adjoint is the *abelianisation functor* $(-)^{\mathrm{ab}}\colon \mathsf{Gp} \to \mathsf{Ab}$, with unit the canonical projection $pG\colon G \to G^{\mathrm{ab}} = G/[G,G]$ (see 7.3.3(e), (f)).

A general result in this sense can be found in [G5], Theorem 4.4.5; several concrete instances are computed in Section 4.1, *loc. cit.*

(c) (*Limit functor as a right adjoint*) The description of the limit of a functor $X\colon \mathsf{S} \to \mathsf{C}$ as a universal arrow $(LX, \varepsilon_X\colon D(LX) \to X)$ from the diagonal functor $D\colon \mathsf{C} \to \mathsf{C}^{\mathsf{S}}$ to the object $X$ of $\mathsf{C}^{\mathsf{S}}$ (in 1.7.4(c)) shows that the existence of all S-based limits in C amounts to the existence of a right adjoint to $D$, the *limit functor* $L\colon \mathsf{C}^{\mathsf{S}} \to \mathsf{C}$, with counit $\varepsilon\colon DL \to \mathrm{id}\,\mathsf{C}$.

In particular, if the category C has all products indexed by the set $I$, the diagonal functor $D\colon \mathsf{C} \to \mathsf{C}^I$ has a right adjoint, the functor $\prod\colon \mathsf{C}^I \to \mathsf{C}$ *of $I$-indexed products.*

(d) Dually, the existence of all S-based colimits in C amounts to the existence of a left adjoint to $D$, the colimit functor $L'\colon \mathsf{C}^I \to C$, with unit $\eta\colon \mathrm{id}\, C \to DL'$. In particular, if the category C has all sums indexed by the set $I$, there is a functor $\sum\colon \mathsf{C}^I \to \mathsf{C}$ *of $I$-indexed sums*, left adjoint to the diagonal functor $D\colon \mathsf{C} \to \mathsf{C}^I$.

Comparing (c) and (d), we see that the existence of a left or a right adjoint to the diagonal functor $D$ are independent facts.

(e) The theory of adjunctions is selfdual, as shown below; but the concrete examples, based on categories of structured sets, are not.

Thus, the forgetful functor $U\colon \mathsf{A} \to \mathsf{Set}$ of an equational algebraic structure has always a left adjoint, but generally lacks a right one.

### *8.1.4 Main properties of adjunctions*

(a) (*Uniqueness and existence*) Given a functor, its left adjoint (or its right adjoint) is uniquely determined up to isomorphism: this follows from the uniqueness property of universal arrows, in 1.6.6.

A standard way of proving the existence (under suitable hypothesis) is the Adjoint Functor Theorem of P. Freyd: see [M2], Section V.6, Theorem 2, or [G5], Section 3.5.

(b) (*Composition*) As a consequence of the composition of universal arrows (in 1.6.6(d)), two consecutive adjunctions

$$\begin{aligned} &F\colon \mathsf{C} \rightleftarrows \mathsf{D} : G, \qquad && \eta\colon 1 \to GF, \quad && \varepsilon\colon FG \to 1, \\ &H\colon \mathsf{D} \rightleftarrows \mathsf{E} : K, \qquad && \rho\colon 1 \to KH, \quad && \sigma\colon HK \to 1, \end{aligned} \tag{8.6}$$

give a composed adjunction from the first to the third category, with the following unit and counit

$$\begin{gathered} HF\colon \mathsf{C} \rightleftarrows \mathsf{E} : GK, \\ (G\rho F).\eta\colon 1 \to GF \to GK\,HF, \\ \sigma.(H\varepsilon K)\colon HF\,GK \to HK \to 1. \end{gathered} \tag{8.7}$$

An adjunction can thus be written as an arrow

$$(F, G, \eta, \varepsilon)\colon \mathsf{C} \twoheadrightarrow \mathsf{D}, \tag{8.8}$$

here *directed as the left adjoint*, and distinguished by a dot. (According to the context, the other direction can be preferable.)

*There is thus a category AdjCat of small categories and adjunctions, with the previous composition and obvious identities.*

(c) (*Duality*) Applying duality on categories, functors and transformations, an adjunction $(F, G, \eta, \varepsilon)\colon \mathsf{C} \dot{\to} \mathsf{D}$ is transformed into an adjunction

$$(G^{\mathrm{op}}, F^{\mathrm{op}}, \varepsilon^{\mathrm{op}}, \eta^{\mathrm{op}})\colon \mathsf{D}^{\mathrm{op}} \dot{\to} \mathsf{C}^{\mathrm{op}}, \tag{8.9}$$

with left adjoint $G^{\mathrm{op}}\colon \mathsf{D}^{\mathrm{op}} \to \mathsf{C}^{\mathrm{op}}$ and unit $\varepsilon^{\mathrm{op}}\colon \mathrm{id}\,\mathsf{D}^{\mathrm{op}} \to F^{\mathrm{op}}G^{\mathrm{op}}$.

(d) (*Preservation properties*) A left adjoint preserves all the existing colimits, while a right adjoint preserves the existing limits. The proof is straightforward, cf. [G5], Exercise 6.3.4(g).

(e) These properties are often used as necessary conditions for the existence of an adjoint.

We have seen in 8.1.1(a) that the forgetful functor $U\colon \mathsf{Top} \to \mathsf{Set}$ has both adjoints: $D \dashv U \dashv C$, where $DX$ (resp. $CX$) is the set $X$ with the discrete (resp. indiscrete) topology. This chain *cannot be extended*: $D$ does not preserve arbitrary products and cannot have a left adjoint, while $C$ does not preserve (even binary) sums and cannot have a right adjoint.

*Non-trivial unbounded chains of adjunctions can be seen in [G5], 1.7.7 and 3.2.8.*

### *8.1.5 Equivalence and adjunctions*

An *adjoint equivalence* is an adjunction $(F, G, \eta, \varepsilon)\colon \mathsf{C} \dot{\to} \mathsf{D}$ where the unit $\eta\colon 1 \to GF$ and the counit $\varepsilon\colon FG \to 1$ are functorial isomorphisms. Equivalently, it is an equivalence of categories where the isomorphisms $GF \cong \mathrm{id}\,\mathsf{C}$ and $FG \cong \mathrm{id}\,\mathsf{D}$ are bound by the triangular identities. (Note that $F \dashv G \dashv F$.)

An important theorem states that the following conditions on a functor $F\colon \mathsf{C} \to \mathsf{D}$ are equivalent, forming a simple *characterisation of the equivalence of categories*:

(i) the functor $F$ is an equivalence of categories,

(ii) the functor $F$ can be completed to an adjoint equivalence $(F, G, \eta, \varepsilon)$,

(iii) the functor $F$ is faithful, full and *essentially surjective on objects*.

The last property means that: for every object $Y$ of $\mathsf{D}$ there exists some object $X$ in $\mathsf{C}$ such that $F(X)$ is isomorphic to $Y$ in $\mathsf{D}$. The proof of the equivalence of these conditions requires the axiom of choice: see [M2], Section V.4, or [G5], Theorem 1.5.8.

The reader can set up the proof, extending the solution of Exercise 7.1.9(b).

*Examples and comments.* (a) The (large) category of finite sets is equivalent to its full subcategory of finite cardinals, the sets $n = \{0, 1, ..., n-1\}$, which is small (and cannot be isomorphic to the former).

In fact, it is sufficient to apply characterisation (iii) to the embedding of the full subcategory of finite cardinals.

(b) Similarly, the category of finite dimensional vector fields on the field $K$ is equivalent to its full subcategory on the objects $K^n$, for $n \geqslant 0$.

*The latter can be realised, up to categorical isomorphism, as a category $\mathrm{Mat}(K)$ whose objects are the natural numbers, and whose arrows $n \to m$ are the matrices of type $m \times n$, with entries in $K$ (by the usual matrix representation of linear mappings); the composition is matrix product.

The only tricky, or amusing, point is that we have to introduce a zero matrix $0_{m,n}$ of type $m \times n$, to represent the zero morphism $K^n \to K^m$, even when $m = 0$ or $n = 0$. This 'formal matrix' can hardly be written as a table, which should have no rows and some columns, or some rows and no columns, or nothing – all these cases being distinguished!*

### 8.1.6 Cartesian exponentials

The exponential laws considered in 5.1.7 appear now to be families of adjunctions, based on the cartesian product or – more generally – a 'tensor product'. We begin by recalling the cartesian case.

(a) In $\mathsf{Set}$, the paradigmatic case, every set $A$ determines a natural bijection in the variables $X, Y$

$$\begin{gathered} \varphi^A_{XY} \colon \mathsf{Set}(X \times A, Y) \to \mathsf{Set}(X, \mathsf{Set}(A, Y)), \\ (f \colon X \times A \to Y) \mapsto (g \colon X \to \mathsf{Set}(A, Y)), \\ g(x) = f(x, -) \colon A \to Y. \end{gathered} \tag{8.10}$$

This is a family of adjunctions, depending on the 'parameter' $A$: the functor $- \times A \colon \mathsf{Set} \to \mathsf{Set}$ is left adjoint to the functor $\mathsf{Set}(A, -) \colon \mathsf{Set} \to \mathsf{Set}$. The hom-set $\mathsf{Set}(A, Y)$, also written as $Y^A$, is called an *exponential set.*

The counit of the adjunction is the *evaluation mapping*

$$\varepsilon Y \colon Y^A \times A \to Y, \qquad \varepsilon Y(g, a) = g(a), \tag{8.11}$$

which evaluates the mapping $g \colon A \to Y$ at $a \in A$.

(b) More generally, let $\mathsf{C}$ be a category with finite products. An object $A$ is said to be *exponentiable* in $\mathsf{C}$, for the cartesian product, if the functor $- \times A \colon \mathsf{C} \to \mathsf{C}$ has a right adjoint, called the *exponential functor* $(-)^A \colon \mathsf{C} \to \mathsf{C}$. This means that we have a natural bijection in the variables $X, Y$

$$\varphi_{XY} \colon \mathsf{C}(X \times A, Y) \to \mathsf{C}(X, Y^A). \tag{8.12}$$

Equivalently, for every $Y$ there is a universal arrow $(Y^A, \varepsilon Y)$ from the functor $- \times A$ to $Y$

$$\varepsilon Y \colon Y^A \times A \to Y \qquad (evaluation). \tag{8.13}$$

A necessary condition is that the functor $- \times A$ preserve the existing colimits of $\mathsf{C}$.

(c) The category $\mathsf{C}$ is said to be *cartesian closed* if all its objects are exponentiable in this sense. Set is our first example, others can be found below. Top is not cartesian closed, but the exponentiable spaces (for the cartesian product) play an important role, analysed in Section 8.3.

Pointed categories, as $R\,\mathsf{Mod}$ in 5.1.7(b), will need a different approach, discussed below.

### *8.1.7 Exercises and complements*

(a) Let $\mathsf{C}$ be a category with finite products. The terminal object $\top$ gives a forgetful functor (possibly non-faithful)

$$U \colon \mathsf{C} \to \mathsf{Set}, \qquad U = \mathsf{C}(\top, -), \tag{8.14}$$

which plays a role in the exponential law: if $A$ is an exponentiable object, there is a canonical bijection between $U(Y^A)$ and $\mathsf{C}(A, Y)$, making $Y^A$ into an *enriched version of the hom-set* $\mathsf{C}(A, Y)$.

(b) For instance, in the category pOrd of preordered sets and monotone mappings, the exponential $Y^A$ is the set $\mathsf{pOrd}(A, Y)$ of monotone mappings $A \to Y$, *equipped* with the pointwise preorder: $h \prec h'$ if and only if $h(a) \prec_Y h'(a)$, for every $a \in A$. All this can be restricted to the full subcategory Ord of ordered sets. Both categories are cartesian closed.

(c) Let $\mathsf{C}$ be a category with finite products. If $A, B$ are exponentiable objects, also $A \times B$ is exponentiable and there is a canonical isomorphism of functors

$$(-)^{A \times B} = ((-)^B)^A, \tag{8.15}$$

so that the functors $(-)^A$ and $(-)^B$ commute. *Hint*: use the composition of right adjoints, in 8.1.4(b).

(d) In a *pointed* category $\mathsf{C}$, exponentials can only have an interest with respect to a non-cartesian product. In fact in this case the only exponentiable object for the cartesian product is the zero object, and $\mathsf{C}$ is cartesian closed if and only if it is equivalent to $\mathbf{1}$. The forgetful functor $U = \mathsf{C}(\top, -)$ of point (a) is also trivial: it is constant (up to isomorphism) at the singleton set.

*(e) The category $\mathsf{Cat}$ is cartesian closed: $Y^A$ is the category of functors $A \to Y$ and their natural transformations, defined in 1.5.4. The exponential law is an extension of those of $\mathsf{Set}$ and pOrd, that one can easily write down, or find in [G5], Exercise 3.4.7(a). We also note that the forgetful functor $\mathsf{Cat}(\mathbf{1}, X)$ represented by the terminal category gives the set of objects of $X$, and is not faithful.

### 8.1.8 *Tensor product and exponentials*

As we have seen in 5.1.7(b), exponentials in $R\,\mathsf{Mod}$ make sense with respect to the tensor product of $R$-modules.

The general theory (including the cartesian case) deals with a category $\mathsf{C}$ equipped with a functor in two variables, usually called and denoted as a *tensor product*

$$- \otimes -\colon \mathsf{C} \times \mathsf{C} \to \mathsf{C}, \qquad (X, Y) \mapsto X \otimes Y, \tag{8.16}$$

whose regularity property, considered below, *are not yet assumed.*

An object $A$ is said to be *exponentiable* in $\mathsf{C}$, for the given product, if the functor $- \otimes A\colon \mathsf{C} \to \mathsf{C}$ has a right adjoint, the *exponential functor* $(-)^A\colon \mathsf{C} \to \mathsf{C}$. This means that we have a natural bijection in the variables $X, Y$

$$\varphi_{XY}\colon \mathsf{C}(X \otimes A, Y) \to \mathsf{C}(X, Y^A). \tag{8.17}$$

Equivalently, for every $Y$ there is a universal arrow $(Y^A, \varepsilon Y)$ from the functor $- \otimes A$ to $Y$

$$\varepsilon Y\colon Y^A \otimes A \to Y \qquad (\textit{evaluation}). \tag{8.18}$$

A necessary condition is that the functor $- \otimes A$ preserve the existing colimits of $\mathsf{C}$. The object $Y^A$ is called an *exponential*; it is also written as $\mathrm{Hom}(A, Y)$ and called an *internal hom* – typically in categories of algebraic structures.

### 8.1.9 *Monoidal categories*

Tensor products often satisfy regularity conditions, which will not be really important in this chapter and will only partially hold for pointed spaces. We only sketch the main definitions, referring to the literature previously quoted [M2, G5, EiK, Ky2].

(i) A *monoidal category* is a category $\mathsf{C}$ equipped with a *tensor product* $- \otimes -$, as in (8.16), and an object $E$, called the *unit*, satisfying several properties.

It is assumed that the tensor product is associative up to a natural isomorphism

$$\kappa(X,Y,Z)\colon X\otimes(Y\otimes Z)\to(X\otimes Y)\otimes Z \qquad (\mathit{associator}), \tag{8.19}$$

and the object $E$ acts as an identity, up to natural isomorphisms

$$\lambda(X)\colon E\otimes X\to X, \qquad \rho(X)\colon X\otimes E\to X \qquad (\mathit{unitors}). \tag{8.20}$$

These isomorphisms are further required to form a 'coherent' system, which allows one to 'forget' them and write

$$(X\otimes Y)\otimes Z = X\otimes Y\otimes Z = X\otimes(Y\otimes Z), \quad E\otimes X = X = X\otimes E.$$

(ii) A *symmetric* monoidal category is also equipped with a symmetry isomorphism, coherent with the other ones:

$$s(X,Y)\colon X\otimes Y\to Y\otimes X. \tag{8.21}$$

The latter cannot be omitted: the isomorphism $s(X,X)\colon X\otimes X\to X\otimes X$ is not the identity, generally.

(iii) A *symmetric monoidal closed* category is a symmetric monoidal category where all the objects are exponentiable.

*Examples and complements.* (a) $R\,\mathsf{Mod}$ is a symmetric monoidal closed category with respect to the usual tensor product; the unit is $R$, as a module on itself. The structural isomorphisms have been constructed in 5.1.5(a)–(c). The exponential $Y^A = \mathrm{Hom}_R(A,Y)$ is a module of homomorphisms.

(b) We shall see in Sections 8.4–8.6 that $\mathsf{Top}_\bullet$ has a partially associative 'smash product', that determines its exponentiable objects. Forgetting topologies, $\mathsf{Set}_\bullet$ is a symmetric monoidal closed category for its smash product.

(c) A category with finite products has a symmetric monoidal structure given by the categorical product, with unit the terminal object $E = \top$. The structural isomorphisms are determined by the canonical projections, and are automatically coherent. This structure is called *cartesian.*

(d) For instance, the cartesian (closed) structure of $\mathsf{Set}$ has terminal object $E = \{*\}$. The structural isomorphisms are obvious:

$$\begin{aligned} &\kappa_{XYZ}\colon X\times(Y\times Z)\to(X\times Y)\times Z, && (x,(y,z))\mapsto((x,y),z),\\ &\lambda X\colon (*,x)\mapsto x, && \rho X\colon (x,*)\mapsto x,\\ &s(X,Y)\colon X\times Y\to Y\times X, && (x,y)\mapsto(y,x). \end{aligned} \tag{8.22}$$

The cartesian structures of $\mathsf{pOrd}$ and $\mathsf{Ord}$, in 8.1.7(b), work in the same way.

(e) Extending Exercise 8.1.7(a), in a monoidal category $\mathsf{C}$ the unit $E$ gives a forgetful functor

$$U\colon \mathsf{C} \to \mathsf{Set}, \qquad U = \mathsf{C}(E, -), \tag{8.23}$$

and, for every exponentiable object $A$, there is a canonical bijection $U(Y^A) \to \mathsf{C}(A, Y)$, making $Y^A$ into an enriched hom-set.

## 8.2 Homotopy pushouts, cones and suspension

In homotopy theory, pullbacks and pushouts of spaces are often replaced by 'homotopy stable' constructions, called homotopy pullbacks and homotopy pushouts; they were introduced by M. Mather [Mat], in 1976.

We deal here with homotopy pushouts, an extension of the cylinder functor $IX = X \times \mathbb{I}$ which gives a structural definition of cones and suspension. Their homotopical stability is made precise in 8.2.2(e).

The derived cofibre sequence of a map, in 8.2.7, is closely related to reduced homology.

### *8.2.1 Homotopy pushouts*

Let $f\colon X \to Y$ and $g\colon X \to Z$ be two maps with the same domain. The *standard homotopy pushout*, or *h-pushout*, *from $f$ to $g$* is a four-tuple $(W; u, v; \lambda)$, formed by a topological space, two maps and a homotopy as below, that satisfies the following universal property:

$$\begin{array}{ccc} X & \xrightarrow{\;g\;} & Z \\ {\scriptstyle f}\downarrow & \overset{\lambda}{\nearrow} & \downarrow{\scriptstyle v} \\ Y & \xrightarrow[\;u\;]{} & W \end{array} \qquad \lambda\colon uf \simeq vg\colon X \to W, \tag{8.24}$$

- for every similar four-tuple $(W'; u', v'; \lambda')$ there is precisely one map $h\colon W \to W'$ such that $u' = hu$, $v' = hv$, $\lambda' = h\lambda$.

The h-pushout object $W$ is determined up to homeomorphism; it will be denoted as $I(f, g)$ and also called the *double mapping cylinder* of $f$ and $g$. When $g$ or $f$ is an identity we have a *mapping cylinder*, $I(f, X)$ or $I(X, g)$; in particular, $IX = I(X, X)$: see Exercise 8.2.2(b).

We shall see that $I(f, g)$ can be equivalently defined by an ordinary colimit, in Exercise 8.2.2(a). Concretely, it can be constructed pasting the cylinder $IX = X \times \mathbb{I}$ with the spaces $Y$ and $Z$, under the equivalence relation $\sim$ generated by identifying each point $(x, 0)$ of the lower base of

$IX$ with $f(x) \in Y$, and each point $(x, 1)$ of the upper base with $g(x) \in Z$

$$\begin{aligned} I(f,g) = (Y + IX + Z)/\!\sim, \quad [x,0] = [f(x)], \quad [x,1] = [g(x)], \\ u(y) = [y], \quad v(z) = [z], \quad \lambda(x,t) = [x,t], \end{aligned} \tag{8.25}$$

(for $x \in X$, $y \in Y$, $z \in Z$, $t \in \mathbb{I}$). $I(f, g)$ has the final topology for the maps

$$uf\colon X \to I(f,g), \qquad vg\colon X \to I(f,g), \qquad \lambda\colon IX \to I(f,g). \tag{8.26}$$

In fact, given the four-tuple $(W'; u', v'; \lambda')$, the map $h\colon W \to W'$ is (and must be) defined as:

$$h[y] = u'(y), \qquad h[z] = v'(z), \qquad h[x,t] = \lambda'(x,t).$$

The construction is made clear by the following pictures. In the first $f$ and $g$ are injective, and we have drawn a path from $y \in Y$ to $z \in Z$, through the cylinder $IX$

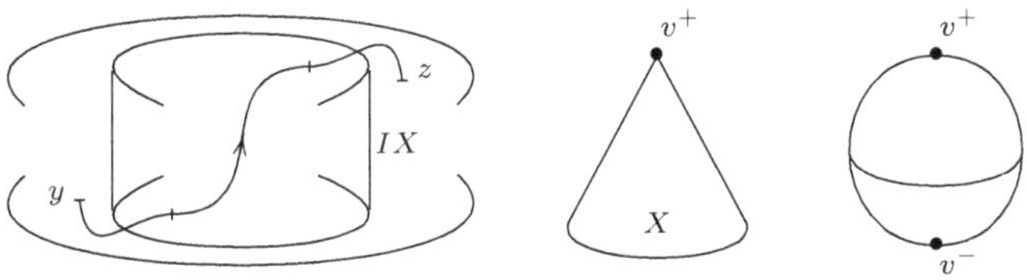


(8.27)

In the second picture, $f = \mathrm{id}\,X$, $g$ is the terminal map $p_X\colon X \to \{*\}$, and $I(X, p_X)$ is the upper cone $C^+(X)$. In the third, $I(p_X, p_X) = \Sigma X$ is the suspension of $X$; these constructions are studied below.

We shall see that the structural maps $u\colon Y \to I(f,g)$ and $v\colon Z \to I(f,g)$ are topological embeddings (Exercise 8.2.2(c)). The map $\lambda\colon IX \to I(f,g)$ is always injective *outside* of the bases $\partial^\alpha X$ of $IX$; it is injective 'everywhere' if and only if both $f$ and $g$ are.

### 8.2.2 Exercises and complements

(a) The homotopy pushout $I(f,g)$ is also given by the colimit of the solid left diagram below, where we have introduced the cylinder $IX$ to represent homotopies

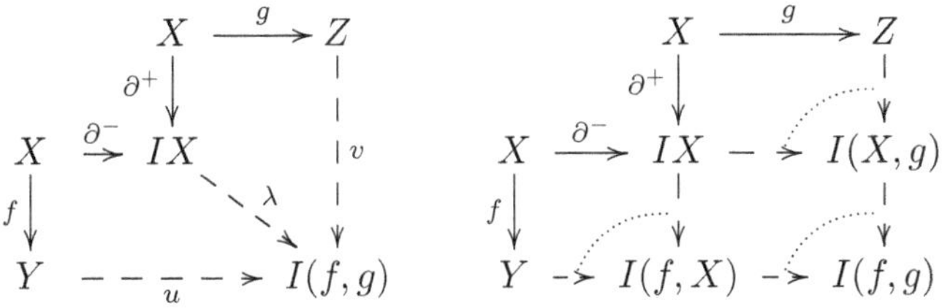


(8.28)

This colimit – a formalisation of the construction of (8.25) – amounts to three ordinary pushouts, shown in the right diagram. (One can also 'paste' the two lower squares in one pushout, or alternatively the two right-hand squares, according to Exercise 1.7.2(b).)

(b) The cylinder $IX = X \times \mathbb{I}$ is the h-pushout of the pair $(\mathrm{id}\, X, \mathrm{id}\, X)$, by means of the structural homotopy $\partial$ represented by the identity of the cylinder

$$\begin{array}{ccc} X & \xrightarrow{\mathrm{id}} & X \\ {\scriptstyle \mathrm{id}}\downarrow & \overset{\partial}{\nearrow} & \downarrow{\scriptstyle \partial^+} \\ X & \xrightarrow[\partial^-]{} & IX \end{array} \qquad \begin{array}{c} \partial\colon \partial^- \simeq \partial^+ \colon X \to IX, \\ \\ \hat{\partial} = \mathrm{id}\, IX. \end{array} \tag{8.29}$$

(c) In the homotopy pushout (8.24) the structural maps $u\colon Y \to I(f, g)$ and $v\colon Z \to I(f, g)$ are topological embeddings.

(d) Homotopy pushouts give a functor

$$I(-, -)\colon \mathsf{Top}^{\vee} \to \mathsf{Top}, \tag{8.30}$$

where $\vee$ is the category $1 \leftarrow 0 \rightarrow 2$, already used in 1.7.2(a) for ordinary pushouts.

*(e) A more complex argument proves that a coherent triple of homotopies 'generates' a homotopy, as one can see in [G1], Theorem 4.1.6, in a much more general context. This stability property explains while homotopy pushouts are more important than ordinary pushouts in homotopy theory.

*(f) Find an example of an ordinary pushout which is not homotopically stable. *Hint*: use two homotopic maps $\mathbb{S}^1 \to \mathbb{D}^2$.

### 8.2.3 Mapping cones

(a) A map $f\colon X \to Y$ has an *upper mapping cone* $C^+f$, also called an *upper homotopy cokernel*, or *upper h-cokernel*, defined as the h-pushout $I(f, p_X)$

$$\begin{array}{ccc} X & \xrightarrow{p} & \{*\} \\ {\scriptstyle f}\downarrow & \overset{\gamma}{\nearrow} & \downarrow{\scriptstyle v^+} \\ Y & \xrightarrow[u]{} & C^+f \end{array} \qquad C^+f = I(f, p_X).$$

Its structural maps are the *lower base* $u = \mathrm{hcok}^+(f)\colon Y \to C^+f$ and the *upper vertex* $v^+\colon \{*\} \to C^+f$; we also have a *structural homotopy* $\gamma\colon uf \to v^+p_X \colon X \to C^+f$, which links $uf$ to a constant map, in a universal way (explaining the name of 'homotopy cokernel'). The upper h-cokernel is a functor $C^+ \colon \mathsf{Top}^{\mathbf{2}} \to \mathsf{Top}$, defined on the category of morphisms of Top (see 1.5.4).

Symmetrically, $f$ has a *lower mapping cone* $C^-f = I(p_X, f)$, with a *lower vertex* $v^-$ and an *upper base* $u = \mathrm{hcok}^-(f) \colon Y \to C^-f$

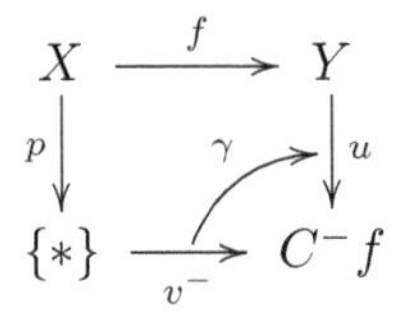

$$C^-f = I(p_X, f). \tag{8.31}$$

The terms *upper* and *lower* agree with the vertices. This is consistent with *future* and *past* contractibility in Directed Algebraic Topology, where the two constructions are not isomorphic, generally [G1]; in the present reversible setting they are, but it is still convenient to distinguish them.

*Exercises.* (a) Define, for a pointed space $W$, a functor $[-, W] \colon \mathsf{Top}^{\mathrm{op}} \to \mathsf{Set}_\bullet$, using (1.64).

(b) For every pointed space $W$, the previous functor takes the sequence $(f, u)$ to an exact sequence of pointed sets (see 2.1.2(f))

$$X \xrightarrow{f} Y \xrightarrow{u} C^+f \qquad\qquad [C^+f, W] \xrightarrow{u^*} [Y, W] \xrightarrow{f^*} [X, W]. \tag{8.32}$$

Because of this property, the sequence $(f, u)$ can be said to be (homotopically) *coexact*: see [Sp2], Section 7.1 and [Sw], Section 2.30 (where the definition is formulated in $\mathsf{Top}_2$ and $\mathsf{Top}_\bullet$, respectively).

### *8.2.4 Cones*

In particular, the *upper cone* $C^+X = C^+(\mathrm{id}\, X) = I(X, p_X)$ of a space $X$ (already represented in figure (8.27)) is given by the h-pushout in the left diagram below, or equivalently by the ordinary pushout at the right

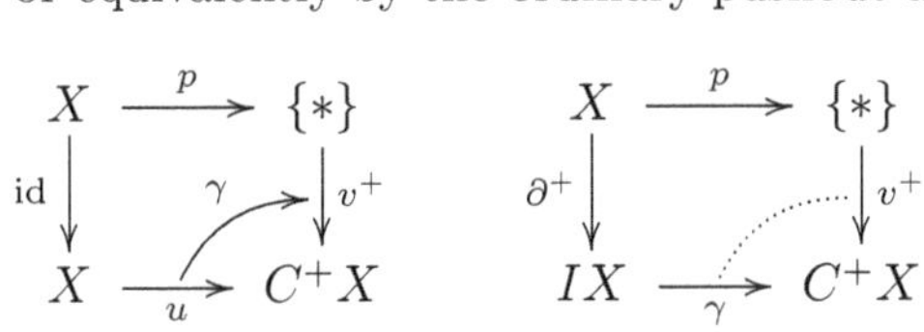

It can be expressed as a topological quotient, where the upper base of the cylinder $IX$ is collapsed to an upper vertex $v^+ = v^+(*)$, but the lower base $u$ is still an embedding

$$\begin{aligned} C^+X &= (IX + \{*\})/(\partial^+X + \{*\}), \qquad && \gamma(x, t) = [x, t], \\ u &= \gamma\partial^- \colon X \to IX \to C^+X, && u(x) = [x, 0]. \end{aligned} \tag{8.33}$$

The homotopy $\gamma\colon IX \to C^+X$ allows one to deform – in the cone – the lower base $u\colon X \to C^+X$ to the map $\gamma\partial^+ = v^+p_X\colon X \to C^+X$, constant at the upper vertex. In particular, $C^+\emptyset = \{v^+\}$ is a singleton. If $X$ is not empty, $C^+X$ is a quotient of $IX$

$$C^+X = IX/\partial^+X, \qquad \gamma(x,t) = [x,t], \qquad v^+(*) = [x,1]. \tag{8.34}$$

We have thus defined a functor $C^+\colon \mathsf{Top} \to \mathsf{Top}$. (It can take values in $\mathsf{Top}_\bullet$, with basepoint $v^+$.)

Dually, the *lower cone* $C^-X = I(p_X, X)$ collapses the lower base of $IX$ to a lower vertex $v^- = v^-(*)$. Again, the upper and lower cones are homeomorphic, but it is often convenient to distinguish them; in Directed Algebraic Topology this is necessary [G1].

*Exercises and complements.* (a) The cone $C^+X$ is always contractible.

(b) A space $X$ is contractible if and only if the base $u\colon X \to C^+X$ of its upper cone has a retraction $h\colon C^+X \to X$.

*(c) Given a homotopy $\varphi\colon f \simeq g\colon X \to Y$, the functor $C^+$ 'generates' a homotopy $C^+f \simeq C^+g$, as one can see in [G1], Theorem 4.1.7.

### 8.2.5 Suspension

(a) The *suspension* $\Sigma X$ of the space $X$ can be structurally defined as the h-pushout $I(p_X, p_X)$

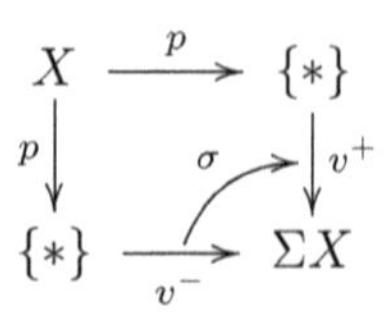


$$\sigma\colon v^-p \simeq v^+p\colon X \to \Sigma X,$$

$$\Sigma X = (\{*\} + IX + \{*\})/R,$$

and amounts to the quotient displayed above, modulo the equivalence relation $R$ that collapses the lower base of $IX$ to a lower vertex $v^- = v^-(*)$ and the upper base to an upper vertex $v^+ = v^+(*)$. It was already represented in figure (8.27).

This gives a functor $\Sigma\colon \mathsf{Top} \to \mathsf{Top}$.

(b) This construction agrees with the previous 'elementary' definition in 2.4.5:

- if $X$ is not empty, $\Sigma X$ is the quotient of $IX$ that collapses each base to a (distinct) point, $v^-$ or $v^+$,
- otherwise, $\Sigma\emptyset$ is a discrete two-point space, usually identified to $\mathbb{S}^0$ (as in 2.4.5(c)): $\Sigma\emptyset = \{v^-, v^+\} = \mathbb{S}^0$.

(c) As a particular case of 8.2.2(a), the suspension $\Sigma X = I(p_X, p_X)$ can be obtained as the colimit of the solid left diagram below

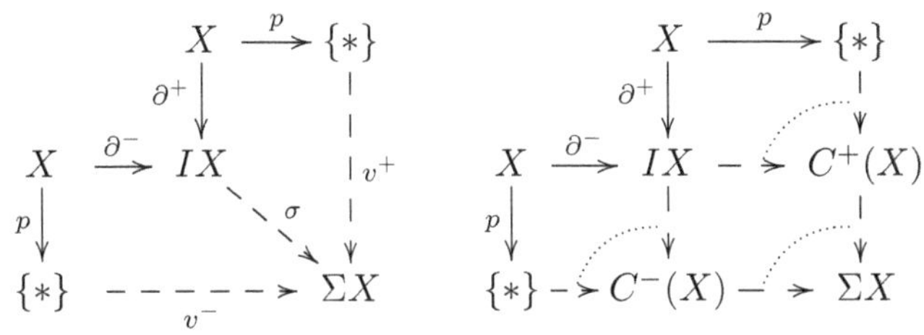


(8.35)

or equivalently by three (ordinary) pushouts, in the right diagram above.

### 8.2.6 Suspension, cones and homology

(a) (*Suspension and cones*) Pasting the lower pushouts *or* the right-hand pushouts in the last diagram above (as in Exercise 1.7.2(b)), $\Sigma X$ is linked to the cones $C^+(X)$ and $C^-(X)$ by the following pushouts

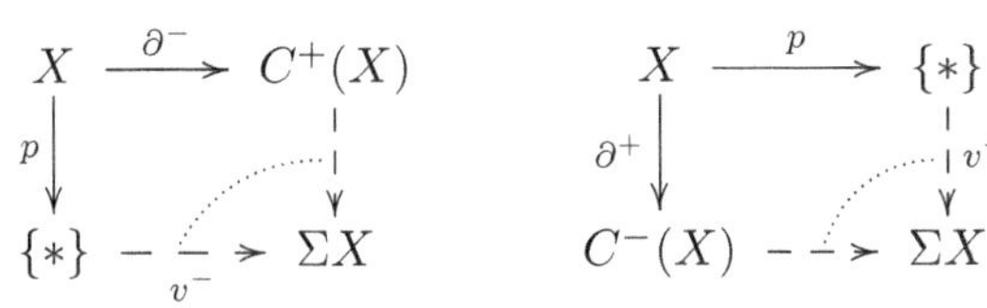


In other words, $\Sigma X$ is the quotient of each cone that collapses its base to a point (also for $X = \emptyset$)

$$\Sigma X = C^+(X)/\partial^- X = C^-(X)/\partial^+ X.$$

(b) (*Cones in a suspension*) The space $X$ embeds in its suspension

$$i\colon X \to \Sigma X, \qquad i(x) = [x, 1/2], \tag{8.36}$$

separating $\Sigma X$ in two cones, canonically homeomorphic to the standard ones:

$$\begin{aligned} C_-X &= \{[x,t] \in \Sigma X \mid 0 \leqslant t \leqslant 1/2\} \;\cong\; C^-(X), \\ C_+X &= \{[x,t] \in \Sigma X \mid 1/2 \leqslant t \leqslant 1\} \;\cong\; C^+(X). \end{aligned} \tag{8.37}$$

$\Sigma X$ is the pasting of these cones on their base (another pushout).

(c) (*Suspension and reduced homology*) We recall the canonical isomorphism in reduced singular homology (see 3.6.3)

$$\delta_n X\colon \tilde{H}_{n+1}(\Sigma X) \to \tilde{H}_n(X) \qquad (n \geqslant -1). \tag{8.38}$$

### *8.2.7 The cofibre sequence of a map*

A map $f\colon X \to Y$ has a natural (upper) *cofibre sequence*, or Puppe sequence

$$X \xrightarrow{f} Y \xrightarrow{u} C^+f \xrightarrow{d} \Sigma X \xrightarrow{\Sigma f} \Sigma Y \xrightarrow{\Sigma u} \Sigma(C^+f) \xrightarrow{\Sigma d} \Sigma^2 X ... \tag{8.39}$$

formed by its h-cokernel $u = \mathrm{hcok}^+(f)$ and the *differential* $d = d^+(f)\colon C^+f \to \Sigma X$.

The latter is defined by the universal property of the upper cone $C^+f$

$$\begin{gathered} du = v^- p_Y \colon Y \to \Sigma X, \qquad dv^+ = v^+ \colon \{*\} \to \Sigma X, \\ d\gamma = \sigma \colon v^- p_X \simeq v^+ p_X \colon X \to \Sigma X. \end{gathered} \tag{8.40}$$

**Complements.* (a) In this sequence, every map is the h-cokernel of the previous one, up to homotopy equivalence: see [G1], Theorem 4.7.5(d) (using the fact that, here, upper and lower h-cokernels are homeomorphic). Applying Exercise 8.2.3(b), this sequence is taken to an exact sequence of pointed sets by the functor $[-, W]$, for any pointed space $W$.

(b) The relationship of reduced homology with suspension and the cofibre sequence is the base for 'single-space' homology theories, introduced by G.M. Kelly [Ky1] (in a form much reviewed later) and proved to be equivalent to Eilenberg–Steenrod relative theories.

The crucial distinction between these two presentations of homology theories is the fact of being based:

- on short exact sequences of $\mathsf{Top}_2$ (see 4.1.9), in the relative case,
- on suspension and cofibre sequences, in the absolute reduced case,

pertaining thus to homological *or* homotopical algebra.

## 8.3 Exponentiable spaces and function spaces

This section addresses a critical problem we were mentioning: the exponential law of sets cannot be totally transported to topological spaces. However, every locally compact space $A$ is exponentiable, in the sense that the functor $- \times A\colon \mathsf{Top} \to \mathsf{Top}$ has a right adjoint $(-)^A$, formed by function spaces with the compact-open topology: this is proved in Theorem 8.3.3.

In particular, the cylinder functor $I(X) = X \times \mathbb{I}$ has a right adjoint, the path functor $P(Y) = Y^{\mathbb{I}}$; homotopies of maps $X \to Y$ are thus equivalently represented by maps $I(X) \to Y$ or $X \to P(Y)$. Homotopy pullbacks, an extension of the path functor, are briefly considered in 8.3.5, and will be more important for pointed spaces, in Section 8.6.

The main results in this domain, including the characterisation of exponentiable spaces, are reviewed in 8.3.6.

*Literature.* The exponential law in topology is frequently approached in an alternative way, aiming to replace the category Top with some full subcategory which is cartesian closed. Several 'convenient categories' of topological spaces have been introduced in this sense, at the price of modifying spaces, their cartesian products and subspaces. The category $\mathsf{Top}_\bullet$ is similarly replaced.

An interested reader can begin with an article by Escardó, Lawson and Simpson [EsLS], which studies and compares various candidates, like compactly generated Hausdorff spaces (a.k.a. Kelley spaces), compactly generated space, locally compactly generated spaces, core compactly generated spaces, and sequential spaces – following their terminology. $\Delta$-generated spaces, also called D-spaces in the domain of diffeology [IZ], are used in [ShYH].

### *8.3.1 Exponentiable spaces*

Let us come back to the cartesian structure of the category Top, defined by the cartesian product $X \times Y$ and the terminal object $\{*\}$ (see 8.1.6 and 8.1.7).

(a) A space $A$ is said to be *exponentiable* if the functor $- \times A \colon \mathsf{Top} \to \mathsf{Top}$ has a right adjoint $(-)^A \colon \mathsf{Top} \to \mathsf{Top}$.

We may assume, by 8.1.7(a), that the *exponential space* $Y^A$ is the function set $\mathsf{Top}(A, Y)$ equipped with the *exponential topology*, strictly determined by the adjunction: it is the finest topology of $\mathsf{Top}(A, Y)$ that makes all transpose mappings $g \colon X \to \mathsf{Top}(A, Y)$ continuous, for every topological space $X$ and every map $f \colon X \times A \to Y$.

(b) The adjunction is expressed by a natural bijection in the variables $X, Y$

$$\begin{gathered} \varphi_{XY} \colon \mathsf{Top}(X \times A, Y) \to \mathsf{Top}(X, Y^A), \\ (f \colon X \times A \to Y) \mapsto (g \colon X \to Y^A), \qquad g(x) = f(x, -) \colon A \to Y, \end{gathered} \tag{8.41}$$

a restriction of the exponential law in Set, in 8.1.6(a). The evaluation map

$$\varepsilon Y \colon Y^A \times A \to Y, \qquad \varepsilon Y(g, a) = g(a), \tag{8.42}$$

is also a restriction of the evaluation mapping of sets.

(c) If $A, B$ are exponentiable spaces, we have already seen in Exercise 8.1.7(c) that also $A \times B$ is exponentiable and there is a canonical isomorphism of functors

$$(-)^{A \times B} = ((-)^B)^A, \tag{8.43}$$

where the endofunctors $(-)^A$ and $(-)^B$ commute.

### *8.3.2 The compact-open topology*

The set of maps $\mathsf{Top}(X, Y)$ between two topological spaces can be given diverse topologies: we are mainly interested in the *compact-open topology.*

Its open subsets are generated by the sets

$$W(K, V) = \{h \in \mathsf{Top}(X, Y) \mid h(K) \subset V\}, \tag{8.44}$$

where $K$ is any compact subspace of $X$ and $V$ is any open subset of $Y$. A general open subset is a union of finite intersections of open sets of the previous form. The latter form thus a canonical subbase of the topology, and are called the *subbasic* open sets.

Now, suppose that the space $A$ is *locally compact*, in the sense that every point has a fundamental system of compact neighbourhoods. (If $A$ is Hausdorff, this is well known to be equivalent to the fact that every point has a compact neighbourhood.) We prove that $A$ is exponentiable and that the function space has the compact-open topology.

### *8.3.3 Theorem*

*Every locally compact space $A$ is exponentiable. Then the functor $F = - \times A\colon \mathsf{Top} \to \mathsf{Top}$ has a right adjoint*

$$G\colon \mathsf{Top} \to \mathsf{Top}, \qquad G(Y) = Y^A, \qquad G(g) = g_\sharp\colon h \mapsto gh, \tag{8.45}$$

*where $Y^A$ is the set $\mathsf{Top}(A, Y)$, with the compact-open topology.*

*The counit of the adjunction is the evaluation map*

$$\varepsilon_Y\colon FG(Y) = Y^A \times A \longrightarrow Y, \qquad \varepsilon_Y(h, a) = h(a). \tag{8.46}$$

*The cartesian product by $A$ preserves quotient maps and all colimits.*

*Proof* (a) We already know that the last point is a consequence of the adjunction. We have defined the space $G(Y) = Y^A = \mathsf{Top}(A, Y)$, and the mapping $\varepsilon_Y$. We have to prove that the latter is continuous, and the pair $(Y^A, \varepsilon_Y)$ is a universal arrow from the functor $F$ to the object $Y$. In other words, for every map $f\colon X \times A \to Y$, there is a unique map $g\colon X \to Y^A$ such that $g(x)(a) = f(x, a)$, for all $x \in X$ and $a \in A$.

(b) First we prove that, for every $Y$, the mapping $\varepsilon_Y$ is continuous at any point $(h_0, a_0) \in Y^A \times A$.

Let $V$ be an open nbd of $h_0(a_0)$ in $Y$. As $A$ is locally compact, there is a compact nbd $K$ of $a_0$ in $A$ such that $h_0(K) \subset V$. The subbasic open set $W(K, V) \times K$ of $Y^A \times A$ is a nbd of $(h_0, a_0)$, which $\varepsilon_Y$ takes to $V$.

(c) Now, we prove that, for every map $f\colon X \times A \to Y$, the mapping

$$g\colon X \to Y^A, \qquad g(x) = f(x, -)\colon A \to Y, \tag{8.47}$$

is well defined, i.e. $f(x, -)$ is continuous on $A$, for every $x \in X$.

We fix $x_0 \in X$ and prove the continuity of $h = g(x_0)$ at $a_0 \in A$. Let $V$ be an open nbd of $h(a_0) = f(x_0, a_0)$ in $Y$. As $f\colon X \times A \to Y$ is continuous, there is a nbd $U \times K$ of $(x_0, a_0)$ such that $f(U \times K) \subset V$, and $h(K) = f(\{x_0\} \times K) \subset V$.

(d) Finally, the mapping $g$ is continuous. (Its uniqueness under the condition $f = \varepsilon Y.(g \times A)$ is already obvious at the level of sets.)

It is sufficient to prove that the preimage

$$g^{-1}(W(K, V)) = \{x \in X \mid f(\{x\} \times K) \subset V\},$$

of any subbasic open set $W(K, V)$ is open in $X$.

Let $x \in g^{-1}(W(K, V))$; then $f^{-1}(V)$ is an open nbd of the compact subset $\{x\} \times K$ of $X \times A$, and there exists an open subset $U$ in $X$ such that $\{x\} \times K \subset U \times K \subset f^{-1}(V)$. Thus $x \in U \subset g^{-1}(W(K, V))$. □

### *8.3.4 The path functor*

(a) In particular, the cylinder functor $I = - \times \mathbb{I}\colon \mathsf{Top} \to \mathsf{Top}$ has a right adjoint

$$\begin{gathered} P\colon \mathsf{Top} \to \mathsf{Top}, \\ P(Y) = Y^{\mathbb{I}}, \qquad P(g)(a) = ga \quad (\text{for } g\colon Y \to Y',\ a\colon \mathbb{I} \to Y), \end{gathered} \tag{8.48}$$

called the *cocylinder*, or *path functor*. $P(Y)$ is the space of all paths $\mathbb{I} \to Y$, with the compact-open topology. The counit is the evaluation of paths

$$\varepsilon Y\colon P(Y) \times \mathbb{I} \to Y, \qquad (a, t) \mapsto a(t). \tag{8.49}$$

The universal property states that, for every map $f\colon X \times \mathbb{I} \to Y$ there is a unique map $g\colon X \to PY$ such that $f(x, t) = g(x)(t)$ (for all $x \in X$ and $t \in \mathbb{I}$).

In other words, it says that a homotopy $\varphi$ between maps $X \to Y$ can be equivalently represented by a map $\hat{\varphi}\colon IX \to Y$ or a map $\check{\varphi}\colon X \to PY$.

(b) The functor $P$ inherits from the interval $\mathbb{I}$, contravariantly, a dual structure, which we write with the same symbols used for the cylinder functor, in (1.58) and (1.87):

$$\begin{aligned} &\partial^\alpha\colon P(Y) \to Y, && \partial^\alpha(a) = a(\alpha) && (\textit{faces}), \\ &e\colon Y \to P(Y), && e(y)(t) = y && (\textit{degeneracy}), \\ &r\colon P(Y) \to P(Y), && r(a)(t) = a(1 - t) && (\textit{reversion}). \end{aligned} \tag{8.50}$$

In this description a homotopy $\check{\varphi}\colon X \to PY$ has faces $\partial^\alpha\check{\varphi}\colon X \to Y$.

(c) *Another example.* Replacing $\mathbb{I}$ with $\mathbb{S}^1$ we get a new adjunction $\mathrm{An} \dashv \Lambda$

$$\begin{aligned} &\mathrm{An}\colon \mathsf{Top} \to \mathsf{Top}, \quad \mathrm{An}(X) = X \times \mathbb{S}^1 \quad \text{(the \textit{annulus} functor)},\\ &\Lambda\colon \mathsf{Top} \to \mathsf{Top}, \quad \Lambda(Y) = Y^{\mathbb{S}^1} \quad \text{(the \textit{free-loop} functor)}, \end{aligned} \tag{8.51}$$

which represents endo-homotopies $f \simeq f\colon X \to Y$.

### 8.3.5 Homotopy pullbacks

Dualising 8.2.1 and 8.2.2, homotopy pullbacks extend the path space and can be constructed with it.

(a) Let $f\colon Y \to X$ and $g\colon Z \to X$ be two maps with the same codomain.

The *standard homotopy pullback*, or *h-pullback, from f to g* is a four-tuple $(W; u, v; \lambda)$ as in the left diagram below, where $\lambda\colon fu \to gv\colon W \to X$ is a homotopy satisfying the following universal property

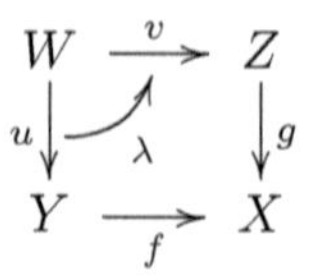


(8.52)

- for every similar four-tuple $(W'; u', v'; \lambda')$ there is precisely one map $h\colon W' \to W$ such that $u' = uh$, $v' = vh$, $\lambda' = \lambda h$.

(b) The solution $W$ is determined up to homeomorphism, and will be denoted as $P(f, g)$. It can be constructed with the path space $PX$ and the limit of the solid left diagram below, which amounts to three ordinary pullbacks, as in the right diagram (containing the mapping cocylinders $P(f, X) = P(f, \mathrm{id}\,X)$ and $P(X, g) = P(\mathrm{id}\,X, g)$)

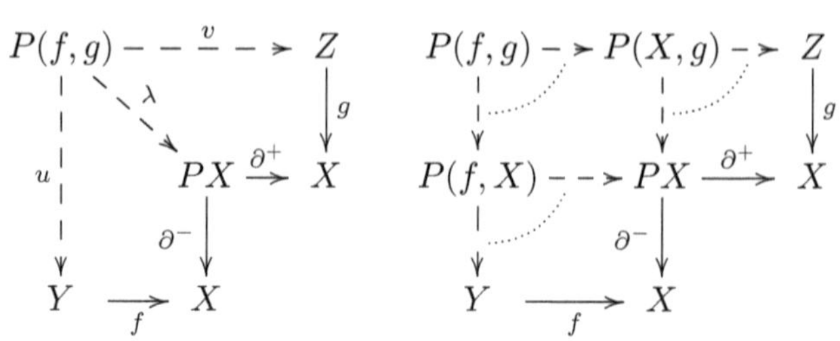


(8.53)

$P(f, g)$ is a subspace of the cartesian product $Y \times PX \times Z$

$$P(f, g) = \{(y, a, z) \in Y \times PX \times Z \mid a(0) = f(y),\ a(1) = g(z)\}. \tag{8.54}$$

Standard homotopy pullbacks form a functor $\mathsf{Top}^\wedge \to \mathsf{Top}$, where $\wedge$ is the category $1 \to 0 \leftarrow 2$, already used in 1.7.2(a) for ordinary pullbacks.

(c) In particular, the path space $PX$ is the homotopy pullback $P(\mathrm{id}\,X, \mathrm{id}\,X)$, with the structural homotopy represented by the identity of $PX$

$$\begin{array}{ccc} PX & \xrightarrow{\partial^+} & X \\ {\scriptstyle\partial^-}\downarrow & \nearrow_{\partial} & \downarrow{\scriptstyle \mathrm{id}} \\ X & \xrightarrow[\mathrm{id}]{} & X \end{array} \qquad \begin{array}{l} \partial\colon \partial^- \to \partial^+\colon PX \to X, \\ \\ \check{\partial} = \mathrm{id}\,PX. \end{array} \tag{8.55}$$

On the other hand, the construction of cones and suspension cannot be dualised in Top: if $Y$ or $Z$ (or both) are the initial object $\emptyset$, we always get an empty space $P(f, g)$. The cone and suspension endofunctors of Top do not preserve sums and do not have a right adjoint.

*(d) *Remarks.* The homotopy pullback functor $\mathsf{Top}^\wedge \to \mathsf{Top}$ is homotopically stable, by [G1], Theorem 4.1.6 applied to the opposite category $\mathsf{Top}^{\mathrm{op}}$.

To show that this fails for the ordinary pullback functor $\mathsf{Top}^\wedge \to \mathsf{Top}$ it is sufficient to consider two points $x' \simeq x''\colon \{*\} \to X$ connected by a path, in a space $X$. The pullback of the pair of maps $(x, x')$ is empty if $x' \neq x''$, and the singleton otherwise. On the other hand, the homotopy pullback $P(x', x'')$ can be realised as a subspace of the path space $PX$

$$P(x', x'') = \{a \in PX \mid a(0) = x',\, a(1) = x''\},$$

which is homotopically equivalent to the loop space $P(x, x)$, for any $x \simeq x' \simeq x''$.

### 8.3.6 *Complements and literature

(a) Exponentiable spaces $A$ are characterised as *core compact* spaces, which means that every open nbd $V$ of a point $x \in A$ contains an open nbd $U$ of $x$ such that every open cover of $V$ has a finite subcover of $U$.

The characterisation theorem is essentially due to Day and Kelly [DaK]. One can find a recent exposition in [EsH].

(b) Less explicitly, the space $A$ is exponentiable if and only if the functor $- \times A$ preserves quotient maps. The necessity of this condition is already known (see 8.1.6(a)). Conversely, if $- \times A$ preserves quotient maps it also preserves coequalisers (by 1.7.4(d), (e)) and all colimits (the preservation of sums being obvious). The existence of the right adjoint can now be deduced from the Adjoint Functor Theorem (cited in 8.1.4(a)), as the 'solution set condition' is easily verified.

(c) As a partial converse to Theorem 8.3.3, every exponentiable space which is Hausdorff must be locally compact, as proved in [He], 7.3.18. In particular, the rational line $\mathbb{Q}$ is not exponentiable; this will also be proved in

8.5.7. We also recall that a compact non-Hausdorff space need not be locally compact; for instance, the one-point compactification of $\mathbb{Q}$ is not.

(d) A *bug-eyed interval.* On the other hand, a compact, locally compact space (is exponentiable and) need not be Hausdorff. One can easily form an example as a compact subspace $A$ of the *bug-eyed line*, the usual instance of a locally euclidean space which is not Hausdorff. The latter is the quotient $L = (\mathbb{R} \times \{0,1\})/R$ modulo the equivalence relation that identifies all pairs $(x,0)$ and $(x,1)$ for $x \neq 0$, leaving two points $0_\alpha = [(0,\alpha)]$ with no disjoint pair of nbds (the *bug-eyes*, for $\alpha = 0,1$). $L$ is locally compact, being locally euclidean. The image $A$ of $\mathbb{I} \times \{0,1\}$ in $L$ gives a compact, locally compact subspace which is still non-Hausdorff.

(e) For a locally compact $A$ and all spaces $X, Y$, the canonical bijection $\mathsf{Top}(X \times A, Y) \to \mathsf{Top}(X, Y^A)$ is a homeomorphism between the compact-open topologies of these function sets ([Du], Theorem XII.5.3).

## 8.4 Smash product of pointed spaces

In the last three sections we work in the category $\mathsf{Top}_\bullet$ of pointed spaces and pointed maps. The basepoint of a pointed space $X$ is written as $0_X$, or $0$, and the underlying topological space is written as $|X|$. The notation as a pair $(|X|, 0_X)$ is rarely used. Any map and homotopy between pointed space is understood to be pointed, by default. Unpointed spaces *can* be denoted by different letters, like $S, T$, etc.

The smash product $X \wedge Y$ of pointed spaces is introduced in 8.4.1; it is the universal solution for 'bipointed' maps – corresponding to the bilinear mappings which define the tensor product of $R$-modules, in Section 5.1. But topology makes things complicated: the smash product is only associative under suitable assumptions, studied in the next section.

For the sake of concreteness, we study the product $X \wedge Y$ of two non-degenerate euclidean intervals, pointed at any point. If $X$ and $Y$ are compact, the space $X \wedge Y$ is compact Hausdorff, hence locally compact and exponentiable; it is also first countable (see 8.4.3). Otherwise, if one of them at least is not compact, $X \wedge Y$ is still Hausdorff but it has none of the other previous properties and cannot be realised in the euclidean plane; nevertheless, it gives interesting spaces (letting $X$ and $Y$ vary as specified).

*Literature.* The smash product and the exponential law in $\mathsf{Top}_\bullet$ are studied in [Mau], Section 6.2, and [G7]. Other expositions (e.g. [Wh2, May2, ShYH]) are developed in a 'convenient category' of pointed topological spaces, as specified in the Introduction of Section 8.3. The smash product of pointed objects over a general ground category is studied in [CaJ].

### 8.4.1 Definition

(a) We recall that $\mathsf{Top}_\bullet$ is a pointed category, with zero object the pointed singleton $\{0\}$ (also written 0), and zero maps $X \to \{*\} \to Y$; kernels and cokernels are defined in 2.1.2(f).

(b) We are interested in a defective monoidal structure, based on the *smash product* $X \wedge Y$, also called a *reduced product*

$$
\begin{gathered}
- \wedge -\colon \mathsf{Top}_\bullet \times \mathsf{Top}_\bullet \to \mathsf{Top}_\bullet, \\
X \wedge Y = (X \times Y)/(X \vee Y), \qquad X \vee Y = X \times \{0\} \cup \{0\} \times Y,
\end{gathered}
\tag{8.56}
$$

which is symmetric but not associative, unless particular assumptions are made.

$X \wedge Y$ is the quotient of $X \times Y$ that collapses the wedge $X \vee Y$ (see 1.6.2(d)) to the new basepoint, written as 0. In other words, $X \wedge Y$ is the cokernel of the canonical embedding $X \vee Y \to X \times Y$. The class of the pair $(x, y)$ is written $x \wedge y$, and $x \wedge 0 = 0 = 0 \wedge y$.

The unit of this operation is the discrete space $\mathbb{S}^0 = \{-1, 1\}$, pointed (as usual) at $e_1 = 1$.

A well known counterexample to associativity comes from the triple of spaces $(\mathbb{Q}, \mathbb{Q}, \mathbb{N})$, proposed by D. Puppe; a proof of the associativity failure, in this case, is reported in [MayS], Section 1.7. A variety of examples, including the previous one, will be obtained in 8.5.6 and 8.5.7.

(c) This failure is only concerned with topology. The smash product in $\mathsf{Set}_\bullet$, defined as above, is easily seen to be associative, up to a canonical bijection – the set-theoretical *associator*

$$
\kappa\colon X \wedge (Y \wedge Z) \to (X \wedge Y) \wedge Z, \qquad \kappa(x \wedge (y \wedge z)) = (x \wedge y) \wedge z, \tag{8.57}
$$

so that $\mathsf{Set}_\bullet$ is a symmetric monoidal category, and the forgetful functor $\mathsf{Top}_\bullet \to \mathsf{Set}_\bullet$ preserves the smash product.

(d) We say that a triple $(X, Y, Z)$ of pointed spaces is *associative* (for the smash product) if its set-theoretical associator is a (pointed) homeomorphism, called a *structural homeomorphism*.

More generally, the terms *structural map* and *structural homeomorphism* of pointed spaces will refer to a bijection of the symmetric monoidal closed structure of $\mathsf{Set}_\bullet$ which happens to be continuous or a homeomorphism. A structural homeomorphism will be denoted by the equality sign.

Let us note that $\mathsf{Top}_\bullet$ inherits for free the coherence of the comparisons of $\mathsf{Set}_\bullet$ which happen to be continuous.

We begin to study the problem, but the main results will be given in the next section.

### 8.4.2 Exercises and complements

(a) In a similar way to the tensor product of $R$-modules, the smash product $X \wedge Y$ is determined by a universal *bipointed* map (i.e. pointed in each variable)

$$\eta \colon X \times Y \twoheadrightarrow X \wedge Y, \qquad \eta(x, y) = x \wedge y. \tag{8.58}$$

The universal property says that, for every bipointed map $\varphi \colon X \times Y \twoheadrightarrow Z$ of pointed spaces there is a unique pointed map $h \colon X \wedge Y \to Z$ such that $\varphi = h\eta$. (But let us note that a bipointed map is also pointed, while a bilinear mapping is not a homomorphism, generally.)

(b) The forgetful functor $U \colon \mathsf{Top}_{\bullet} \to \mathsf{Top}$, which forgets the basepoint, has a left adjoint

$$(-)_{\bullet} \colon \mathsf{Top} \to \mathsf{Top}_{\bullet}, \qquad S_{\bullet} = S + \{*\}, \tag{8.59}$$

which adds to a space a basepoint $*$, open and closed. It embeds $\mathsf{Top}$ in $\mathsf{Top}_{\bullet}$ as a coreflective subcategory.

(c) The functor $(-)_{\bullet}$ transforms the cartesian product of spaces into the smash product of pointed spaces (preserving the unit object)

$$(S \times T)_{\bullet} = S_{\bullet} \wedge T_{\bullet}, \qquad (x, y) \mapsto x \wedge y. \tag{8.60}$$

(d) For arbitrary topological spaces $X, Y$ with subspaces $A, B$, the spaces $X/A$ and $Y/B$ are pointed (as made clear in 4.2.2) and

$$X/A \wedge Y/B \;=\; (X \times Y)/(X \times B \cup A \times Y). \tag{8.61}$$

This formula extends (8.56) (if $A$ and $B$ are singletons) and the previous (8.60) (with $A = B = \emptyset$).

(e) If the pointed spaces $X, Y$ are compact or Hausdorff, $X \wedge Y$ is also, as is obvious or easily proved.

We shall see that the smash product of compact Hausdorff pointed spaces is associative (Theorem 8.5.4(b)): their full subcategory has a symmetric monoidal structure. On the other hand, locally compact and exponentiable pointed spaces are not closed under smash product: we shall see that $\mathbb{R} \wedge \mathbb{I}$ and $\mathbb{R} \wedge \mathbb{R}$ are not locally compact, in 8.5.8.

(f) Characterise the smash product $X \wedge Y$, where $Y$ is a discrete pointed space; in particular, $\mathbb{S}^n \wedge Y$.

(g) For (pointed) homotopies $\varphi \colon f \to f' \colon X \to X'$ and $\psi \colon g \to g' \colon Y \to Y'$ one can define a homotopy

$$\begin{gathered} \varphi \wedge \psi \colon f \wedge g \to f' \wedge g' \colon X \wedge Y \to X' \wedge Y', \\ (\varphi \wedge \psi)(x \wedge y, t) = \varphi(x, t) \wedge \psi(y, t). \end{gathered} \tag{8.62}$$

### 8.4.3 Smash product of intervals

To get a more concrete idea of this operation, it is interesting to examine the smash product $X \wedge Y$ of two non-degenerate euclidean intervals, pointed at any point.

*Exercises and complements.* (a) (*The easy case*) If $X$ and $Y$ are both compact, $X \wedge Y$ can be realised in the euclidean plane. It is a compact Hausdorff space, exponentiable and first countable.

In fact, up to pointed homeomorphism, $X$ and $Y$ are either the standard interval $\mathbb{I}$ or the standard 1-disc $\mathbb{D} = \mathbb{D}^1 = [-1, 1]$, both pointed at 0. This gives three spaces $X \wedge Y$

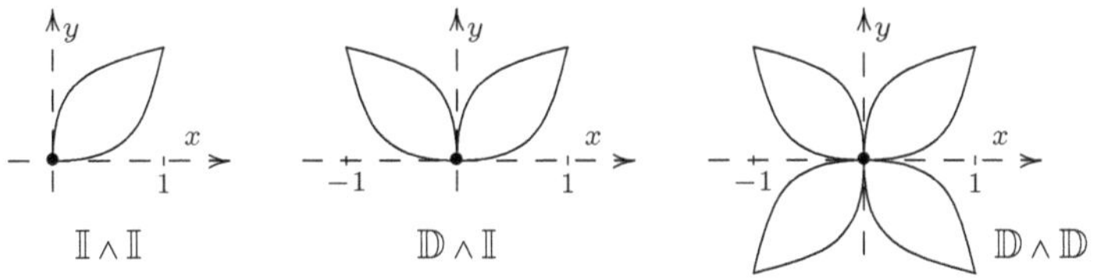


(8.63)

They can be realised in the euclidean plane as above, and are respectively homeomorphic to the wedge of one, two or four compact discs $\mathbb{D}^2$, pointed at a point of the boundary.

*Hints.* The proof, an exercise of topology, can be based on the following map

$$f \colon X \times Y \to \mathbb{R}^2, \qquad f(x, y) = (|x| \wedge |y|)(x, y), \tag{8.64}$$

where $X$ and $Y$ are either $\mathbb{I}$ or $\mathbb{D}$; $|x| \wedge |y|$ is the distance of $(x, y)$ from the subspace $X \vee Y$.

(b) If $X$ or $Y$ is not compact, $X \wedge Y$ is still Hausdorff but we shall see that it has none of the other previous properties. By 8.4.4(b) it is not first countable, and cannot be embedded in the plane. By Theorem 8.5.8 it is not locally compact; being Hausdorff, it is neither compact nor exponentiable. Nevertheless, the spaces $X \wedge Y$ obtained in this way, and explored below, are quite interesting.

*(c) Extending point (a), each $n$-ary smash product of non-degenerate compact intervals is a wedge of compact $n$-discs pointed as above – at most $2^n$ of them, for the smash power $\mathbb{D}^{\wedge n}$ (see 8.5.1(c)).

### 8.4.4 Neighbourhoods of the basepoint

We are still considering the product $X \wedge Y$ of two non-degenerate euclidean intervals, pointed at 0. The property of being first countable depends on compactness.

(a) Let $X$ and $Y$ be compact. We already know that $X \wedge Y$ is first countable. As a more direct argument, it is sufficient to prove that the following subsets form a local bases of nbds of $X \vee Y$ in $X \times Y$, for $n > 0$

$$L_n = \{(x, y) \in X \times Y \mid |x| \wedge |y| < 1/n\}. \tag{8.65}$$

Indeed, if $W$ is any open nbd of $X \vee Y$ in $X \times Y$, the disjoint compact subsets $X \vee Y$ and $(X \times Y) \setminus W$ have a positive distance $\varepsilon$, and $W \subset L_n$ for $n > 1/\varepsilon$.

(b) Let $X$ be non-compact. We prove that $X \wedge Y$ has no countable local basis at the basepoint (and cannot be realised in the euclidean plane, nor in any metric space).

We can assume that $X$ and $Y$ contain the interval $[0, 1[$. For the sake of simplicity we take $X = [0, 1[$ (the argument below can be easily modified if $0_X$ is an interior point). Supposing we have a sequence of open nbds $(W_n)$ of $X \vee Y$ in $X \times Y$, we shall construct an open nbd $W$ that does not contain any of them.

We use the strictly increasing sequence $x_n = (n+1)/(n+2)$ in $X$, which begins at $x_0 = 1/2$ and converges to 1 in $\mathbb{I}$. We begin by forming an open nbd $W'_\varepsilon$ of $X \times \{0\}$ in $X \times Y$, as in the following picture

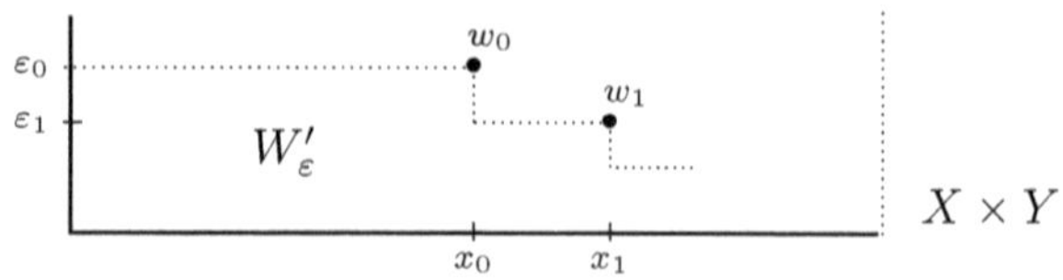


(8.66)

choosing a decreasing sequence $\varepsilon = (\varepsilon_n)_{n \in \mathbb{N}}$ in $]0, 1/2[$ such that $w_n = (x_n, \varepsilon_n) \in W_n$. Letting

$$W'_n = [0, x_n[ \times [0, \varepsilon_n[, \qquad W'_\varepsilon = \cup W'_n,$$

the subset $W'_\varepsilon$ is an open nbd of $X \times \{0\}$ in $X \times Y$, and every $W_n$ has a point $w_n \notin W'_\varepsilon$. Finally $W'_\varepsilon \cup ([0, 1/2[ \times Y)$ is an open nbd of $X \vee Y$ in $X \times Y$ and does not contain any $W_n$.

*(c) Taking $X = Y = [0, 1[$, one can similarly construct a local base at the basepoint of $X \wedge X$, indexed by the continuum $\mathcal{E}$ of all (weakly) decreasing sequences $\varepsilon = (\varepsilon_n)$ of the interval $]0, 1/2[$.

In fact, using again the sequence $x_n = (n+1)/(n+2)$, the following family of open sets of $X \times X$

$$W_\varepsilon = \cup_n ([0, x_n[ \times [0, \varepsilon_n[) \cup ([0, \varepsilon_n[ \times [0, x_n[) \qquad (\varepsilon \in \mathcal{E}), \tag{8.67}$$

forms a base of open nbds of $X \vee X$ in $X^2$. Their projections in $X \wedge X$ form thus a local base at the basepoint, indexed by $\mathcal{E}$.

The proof can be found in [G7], Section 4.4.

## 8.5 Partial associativity and function spaces

We deal now with the partial associativity properties of the smash product, introducing a property of 'regular associativity' (in 8.5.2) related to the *colax* monoidal structure of the product (in 8.5.5).

Theorem 8.5.4 shows that this property is strongly connected to exponentiability and compactness of pointed spaces, making clearer what we have already seen in the previous section; it will play a relevant role for the homotopy constructions of pointed spaces, in Section 8.6.

We also determine a large class of non-associative triples, including the well known case $(\mathbb{Q}, \mathbb{Q}, \mathbb{N})$.

### *8.5.1 Symmetry, associativity and smash powers*

The symmetry of the smash product strongly interacts with associativity, simplifying this issue.

(a) It is easy to see that the triple $(X, Y, Z)$ is associative if and only if $(Z, Y, X)$ is.

In fact, if we assume that the following set-theoretical associator is a homeomorphism

$$\kappa\colon X \wedge (Y \wedge Z) \to (X \wedge Y) \wedge Z, \qquad \kappa(x \wedge (y \wedge z)) = (x \wedge y) \wedge z,$$

we get the structural homeomorphism of $(Z, Y, X)$ as $\psi\kappa^{-1}\varphi$, using the composed symmetries:

$$\varphi\colon Z \wedge (Y \wedge X) \to Z \wedge (X \wedge Y) \to (X \wedge Y) \wedge Z, \quad z \wedge (y \wedge x) \to (x \wedge y) \wedge z,$$
$$\psi\colon X \wedge (Y \wedge Z) \to (Y \wedge Z) \wedge X \to (Z \wedge Y) \wedge X, \quad x \wedge (y \wedge z) \to (z \wedge y) \wedge x.$$

(b) Every triple $(X, Y, X)$ of pointed spaces is associative. Indeed, in this case the set-theoretical associator is a composed symmetry:

$$X \wedge (Y \wedge X) \to (Y \wedge X) \wedge X \to (X \wedge Y) \wedge X.$$

(c) The *smash power* $X^{\wedge n}$ is inductively defined as

$$X^{\wedge 0} = \mathbb{S}^0, \qquad X^{\wedge n+1} = X^{\wedge n} \wedge X \qquad (n \geqslant 0). \tag{8.68}$$

The symmetric formula $X^{\wedge n+1} = X \wedge X^{\wedge n}$ can also be used, up to structural homeomorphism.

In fact, assuming that $X^{\wedge n} \wedge X = X \wedge X^{\wedge n}$ (true for $n = 0, 1$) and applying (b):

$$\begin{aligned} X^{\wedge n+1} \wedge X &= (X^{\wedge n} \wedge X) \wedge X = (X \wedge X^{\wedge n}) \wedge X \\ &= X \wedge (X^{\wedge n} \wedge X) = X \wedge X^{\wedge n+1}. \end{aligned}$$

(d) On the other hand, a triple $(X, X, Y)$ need not be associative, as it happens for $(\mathbb{Q}, \mathbb{Q}, \mathbb{N})$.

### 8.5.2 Regular associativity

Let $(X, Y, Z)$ be a triple of pointed spaces. The ternary smash products $X \wedge (Y \wedge Z)$ and $(X \wedge Y) \wedge Z$ can be viewed as two topologies on the quotient *set*

$$(X \times Y \times Z)/(X \vee Y \vee Z),$$

each of them (weakly) coarser than the quotient topology, and we shall give two separate regularity conditions, forming a stronger pair.

This property is also considered in an article by Carboni and Janelidze [CaJ], in a general context of pointed categories, with a different goal: studying conditions under which the property universally holds.

(a) The ternary smash products of the given triple are covered by two surjective maps $\rho'$ and $\rho''$, defined on the cartesian product $X \times Y \times Z$

$$\begin{array}{lll} X \times Y \times Z \xrightarrow{\rho'} X \wedge (Y \wedge Z) & & \rho'(x, y, z) = x \wedge (y \wedge z), \\ \quad\quad \searrow^{\rho''} & & \\ \quad\quad\quad (X \wedge Y) \wedge Z & & \rho''(x, y, z) = (x \wedge y) \wedge z. \end{array} \tag{8.69}$$

For instance, $\rho'$ is the composite of the canonical surjective maps

$$X \times Y \times Z \xrightarrow{1 \times p} X \times (Y \wedge Z) \xrightarrow{q} X \wedge (Y \wedge Z), \tag{8.70}$$

and is a quotient map if and only if $1 \times p$ sends every open (or closed) subset of $X \times Y \times Z$ *saturated for* $\rho'$ to an open (or closed) subset of $X \times (Y \wedge Z)$.

(b) With some abuse of terminology, we say that a particular instance $X \wedge (Y \wedge Z)$ or $(X \wedge Y) \wedge Z$ is a *regular smash product* if the corresponding map $\rho'$ or $\rho''$, in (8.69), is a quotient map.

One could say that the triple $(X, Y, Z)$ is *left regular* or *right regular*, which would be formally correct but liable to confusion.

In the first case the set-theoretical associator $\kappa$ is a (structural) map, making the following diagram commutative in $\mathsf{Top}_\bullet$

$$\begin{array}{lll} X \times Y \times Z \xrightarrow{\rho'} X \wedge (Y \wedge Z) & & \\ \quad\quad \searrow^{\rho''} \quad\quad \downarrow \kappa & & \kappa(x \wedge (y \wedge z)) = (x \wedge y) \wedge z. \\ \quad\quad\quad (X \wedge Y) \wedge Z & & \end{array} \tag{8.71}$$

In the second case there is a structural map $\kappa'$ the other way round.

(c) If both occurrences are regular, $\kappa$ is a structural homeomorphism and we say that the triple $(X, Y, Z)$ is *regularly associative.*

This property is strictly stronger than associativity: for instance, the triple $(\mathbb{Q}, \mathbb{N}, \mathbb{Q})$ is associative, by symmetry (in 8.5.1(b)), but is not regularly associative: we shall see $(\mathbb{Q} \wedge \mathbb{N}) \wedge \mathbb{Q} = (\mathbb{N} \wedge \mathbb{Q}) \wedge \mathbb{Q}$ is not a regular smash product, in Theorem 8.5.7(a).

It is also useful to note that, if one of $X \wedge (Y \wedge Z)$ and $(X \wedge Y) \wedge Z$ is a regular smash product but the other is not, the triple $(X, Y, Z)$ is not associative.

### 8.5.3 *Exponentiable pointed spaces*

(a) The pointed space $A$ is said to be *exponentiable* if the functor $- \wedge A \colon \mathsf{Top}_\bullet \to \mathsf{Top}_\bullet$ has a right adjoint $(-)^A$

$$- \wedge A \dashv (-)^A, \qquad \varphi_{XY} \colon \mathsf{Top}_\bullet(X \wedge A, Y) \to \mathsf{Top}_\bullet(X, Y^A), \tag{8.72}$$

characterised by a natural bijection $\varphi_{XY}$, the exponential law of pointed spaces.

(b) If the underlying space $|A|$ is exponentiable (for the cartesian product), it is easy to see that $A$ is exponentiable in $\mathsf{Top}_\bullet$ (for the smash product): the exponential $Y^A$ is the set of pointed maps $\mathsf{Top}_\bullet(A, Y)$ with the topology induced by the exponential topology of the whole space of (unpointed) maps $|Y|^{|A|}$. The natural bijection $\varphi_{XY}$ (corresponding to the structure of $\mathsf{Set}_\bullet$) is a restriction of the unpointed one, in $\mathsf{Top}$ (recalled in (8.41)).

Less obviously, it is also true that a pointed space which is exponentiable in $\mathsf{Top}_\bullet$ for the smash product has an underlying space exponentiable in $\mathsf{Top}$ [Ca]. The terms *exponentiable pointed space* and *pointed exponentiable space* are thus equivalent and non-ambiguous.

(c) If $A$ and $B$ are exponentiable pointed spaces, $A \times B$ is also, by 8.3.1(c), but $A \wedge B$ need not be, even in the case of a smash power, like $\mathbb{R} \wedge \mathbb{R}$ (see 8.5.8).

However, it is still true that the endofunctors $(-)^A$ and $(-)^B$ commute, by canonical homeomorphisms induced by the unpointed exponentials. Their left adjoints $- \wedge A$ and $- \wedge B$ also commute. (Playing with symmetries, this corresponds to part of 8.5.4(a).)

(d) We now study the relationship of exponentiable pointed space with associative triples. The main results are gathered in the following theorem, which depends on Theorem 8.5.9, a property of compactness deferred to the end of this section.

***8.5.4 Theorem*** (Exponentiability and regular associativity)
*The letters $X, Y, Z$ always denote pointed spaces.*

*(a) If $X$ and $Z$ are exponentiable (e.g. locally compact), each triple $(X, Y, Z)$ is regularly associative (in the sense of 8.5.2), and the set-theoretical associator $\kappa\colon X \wedge (Y \wedge Z) \to (X \wedge Y) \wedge Z$ is a structural homeomorphism.*

*(b) If $Y$ is compact and $X$ or $Z$ is compact and exponentiable (e.g. compact Hausdorff), the triple $(X, Y, Z)$ is regularly associative. (We recall that a compact exponentiable space need not be Hausdorff: see 8.3.6(d).)*

*(c) If $Y$ and $Z$ are exponentiable and all triples $(X, Y, Z)$ are associative, the pointed space $Y \wedge Z$ is exponentiable.*

*(d) If $Y$ and $Z$ are compact and exponentiable, $Y \wedge Z$ is also and there are structural homeomorphisms, for all pointed spaces $X$ (given by canonical bijections of* $\mathsf{Set}_\bullet$*.)*

$$(X \wedge Y) \wedge Z = X \wedge (Y \wedge Z) = X \wedge (Z \wedge Y) = (X \wedge Z) \wedge Y, \qquad (8.73)$$

$$(X^Z)^Y = X^{Y \wedge Z} = X^{Z \wedge Y} = (X^Y)^Z, \qquad (8.74)$$

*One can replace $Y$ and $Z$ with $Y^{\wedge m}$ and $Z^{\wedge n}$, for $m, n \geqslant 0$.*

*(e) If $Y$ and $Z$ are compact and exponentiable, the functors $F = - \wedge Y$ and $G = - \wedge Z$ commute, and any finite composite of them is structurally isomorphic to $H = - \wedge (Y^{\wedge m} \wedge Z^{\wedge n})$, for suitable $m, n \geqslant 0$. The right adjoints $(-)^Y$ and $(-)^Z$ interact similarly.*

*(f) The full subcategory of compact exponentiable pointed spaces is a symmetric monoidal category, for the smash product.*

*Proof* (a) If $X$ is exponentiable, the functor $X \times -$ preserves quotients in (8.70) and $X \wedge (Y \wedge Z)$ is a regular smash product. If also $Z$ is exponentiable, $(X \wedge Y) \wedge Z$ is regular as well.

(b) First we suppose that $Y$ and $Z$ are compact. Applying Theorem 8.5.9 to the pointed space $H = Y \times Z$ and its compact subspace $K = Y \vee Z$, the map $1 \times p\colon X \times Y \times Z \to X \times (Y \wedge Z)$ is a quotient map, and $X \wedge (Y \wedge Z)$ is a regular smash product. Adding the hypothesis that $Z$ is exponentiable, $(X \wedge Y) \wedge Z$ is also regular, as in the previous point. The other case is symmetrical.

(c) The composed endofunctor $F = (- \wedge Y) \wedge Z$ has a composed right adjoint $G = ((-)^Z)^Y$. But we are assuming that $F \cong - \wedge (Y \wedge Z)$, so that $Y \wedge Z$ is exponentiable.

(d) Now $Y$ and $Z$ are supposed to be compact and exponentiable. From (b) and symmetry we get (8.73). As to (8.74), the composed endofunctor

$F = (- \wedge Y) \wedge Z$ has a composed right adjoint $G = ((-)^Z)^Y$. But $F = - \wedge (Y \wedge Z)$, by (8.73), which implies that $Y \wedge Z$ is exponentiable and $G = (-)^{Y \wedge Z}$. The last statement follows from the fact that the pointed spaces $Y^{\wedge m}$ and $Z^{\wedge n}$ are also compact and exponentiable.

(e), (f) Follow from (d). Letting $H_{mn} = - \wedge (Y^{\wedge m} \wedge Z^{\wedge n})$, the functor $FH$ (resp. $GH$) is structurally isomorphic to $H_{m+1,n}$ (resp. $H_{m,n+1}$). □

### 8.5.5 A colax monoidal structure

Our approach to regular associativity, in 8.5.2, can be rewritten in a more formal way, extending the smash product to make $\mathsf{Top}_\bullet$ into a symmetric *colax monoidal* category. (This is also explored in [G7], and more briefly in [G5], Section 5.6. For the general theory of lax monoidal categories one can see Leinster's book [Ln], Section 3.1.)

The (regular) *n-ary smash product* $X_1 \wedge ... \wedge X_n$ is defined by the universal multipointed mapping (i.e. pointed in each variable)

$$\eta\colon X_1 \times ... \times X_n \twoheadrightarrow X_1 \wedge ... \wedge X_n \qquad \eta(x_1, ..., x_n) = x_1 \wedge ... \wedge x_n. \tag{8.75}$$

The solution is a quotient map, the cokernel of the embedding of the subspace $H$ of coordinate hyperplanes

$$H \subset X_1 \times ... \times X_n, \qquad H = \{(x_i) \in \Pi X_i \mid \exists i\colon x_i = 0\}. \tag{8.76}$$

The term $x_1 \wedge ... \wedge x_n$ annihilates if and only if some coordinate $x_i$ is zero; all the other equivalence classes are singletons.

As described in [Ln] (for the dual lax case), we have a coherent system of associativity comparisons which need not be invertible. In particular, a triple of spaces has two comparisons

$$\begin{aligned} &\gamma'\colon X \wedge Y \wedge Z \to X \wedge (Y \wedge Z), \quad \gamma'(x \wedge y \wedge z) = x \wedge (y \wedge z), \\ &\gamma''\colon X \wedge Y \wedge Z \to (X \wedge Y) \wedge Z, \quad \gamma''(x \wedge y \wedge z) = (x \wedge y) \wedge z, \end{aligned} \tag{8.77}$$

produced by the universal property of the tripointed mapping $\eta\colon X \times Y \times Z \twoheadrightarrow X \wedge Y \wedge Z$. The tripointed mapping $\rho'$ of (8.69) is thus the composite

$$X \times Y \times Z \xrightarrow{\eta} X \wedge Y \wedge Z \xrightarrow{\gamma'} X \wedge (Y \wedge Z) \qquad \rho' = \gamma'\eta, \tag{8.78}$$

so that $X \wedge (Y \wedge Z)$ is a regular smash product (meaning that $\rho'$ is a quotient map) if and only if $\gamma'$ is invertible.

Similarly, $(X \wedge Y) \wedge Z$ is a regular smash product if and only if $\gamma''$ is invertible. Plainly, these properties make sense in the general theory of (co)lax monoidal categories.

### *8.5.6 Counterexamples*

The rest of this section is of a technical interest, although the problems highlighted here cannot be ignored.

First, we want to form a class of *non-associative* triples of pointed spaces of the form $(X, Y, \mathbb{Q})$ (or $(\mathbb{Q}, Y, X)$, by symmetry).

We assume that $X$ and $Y$ satisfy the following conditions (*others will be added*):

(i) $X$ has an infinite closed discrete subspace $D$ (and is not compact),

(ii) $Y$ has an infinite discrete subspace $E$ and $\overline{E} = E \cup \{0\}$ is a first-countable subspace with $0 \notin E$.

For instance these conditions are satisfied when:

- $X$ is $\mathbb{N}$, $\mathbb{Z}$, $\mathbb{Q}$, $\mathbb{R}$ or any non-compact real interval pointed at any point,
- $Y$ is $\mathbb{Q}$, $\mathbb{I}$, $\mathbb{R}$, $\mathbb{S}^n$ $(n > 0)$ or any non-degenerate real interval pointed at any point.

Each of these lists can be closed under finite non-empty products.

On the other hand, we have seen in Theorem 8.5.4(b) that any triple $(X', Y', Z)$ of pointed spaces where $X', Y'$ are compact and exponentiable (e.g. compact Hausdorff) is regularly associative.

**8.5.7 Theorem** (Non-associative triples)

*Let $X$ and $Y$ be pointed spaces verifying the previous conditions (i), (ii), respectively.*

*(a) The functor $- \times \mathbb{Q}$ does not preserve the quotient $p \colon X \times Y \to X \wedge Y$, and the smash product $(X \wedge Y) \wedge \mathbb{Q}$ is not regular.*

*(b) If moreover $X$ is exponentiable, the triple $(X, Y, \mathbb{Q})$ is not associative. By symmetry, $(\mathbb{Q}, Y, X)$ is neither, including the case $(\mathbb{Q}, \mathbb{Q}, \mathbb{N})$.*

*Proof* See [G7], Theorem 4.8. □

**8.5.8 Theorem** (Non-exponentiable smash products)

*Let $X$ and $Y$ be pointed spaces verifying, respectively, the conditions (i), (ii) of 8.5.6. We also suppose that all the points of $E$ are closed in $\overline{E}$.*

*Then $X \wedge Y$ is not locally compact: the basepoint has no compact nbd.*

*If moreover $X, Y$ are Hausdorff, $X \wedge Y$ is also, and is not exponentiable.*

*Proof* Again $p \colon X \times Y \to X \wedge Y$ is the canonical projection. An arbitrary nbd of $[0]$ in $X \wedge Y$ is the image $p(W)$ of a nbd $W$ of $X \vee Y$ in $X \times Y$, and we prove that $p(W)$ cannot be compact, constructing a closed subset $p(C)$ which is not.

We fix an injective sequence $(x_n)$ in $D \setminus \{0\}$, and we assume that $D$ is the support of this sequence, which is still an infinite closed discrete subspace of $X$; we choose a family of open subsets $(U_n)$ of $X$ such that $U_n \cap D = \{x_n\}$.

For every $n \geqslant 0$, $W \cap (\{x_n\} \times Y)$ is a nbd of $(x_n, 0)$ in $\{x_n\} \times Y$ and meets $\{x_n\} \times E$ in some point $w_n = (x_n, y_n)$. We form the subset $C$

$$C = \{w_n \mid n \geqslant 0\} \subset D \times E, \qquad w_n = (x_n, y_n) \in W \setminus (X \vee Y). \qquad (8.79)$$

The space $C$ is infinite and discrete, because every point $(x_n, y) \in D \times Y$ has a nbd $U_n \times Y$ in $X \times Y$ that only meets $C$ at $(x_n, y_n)$. It is also closed in $D \times \overline{E}$ (and $X \times Y$): using the fact that all $y_n$ are closed in $\overline{E}$, every point $(x_n, y)$ with $y \neq y_n$ has a nbd $U_n \times (\overline{E} \setminus \{y_n\})$ in $X \times \overline{E}$ that does not meet $C$.

Now the space $p(C)$ is closed in $X \wedge Y$ (because $C$ does not meet $X \vee Y$) and in $p(W)$. The compactness of the latter would imply the compactness of $p(C)$, which is discrete and infinite (because the projection $p$ is injective on $C$).

Finally, if $X, Y$ are Hausdorff, we know that $X \wedge Y$ is also and cannot be exponentiable because it is not locally compact. □

### 8.5.9 Theorem

*Let $K$ be a compact subspace of the space $H$. For every space $X$, the surjective map $f = 1 \times p \colon X \times H \to X \times (H/K)$ is a quotient map.*

*Proof* This result has been used in Theorem 8.5.4. It can be found as Theorem 6.2.4(b) of Maunder's book [Mau], not often available.

Let $W \subset X \times H$ be an open $f$-saturated subset. We have to prove that $f(W)$ is open in $X \times (H/K)$. Equivalently, every $(x, y) \in W$ has a basic nbd $U \times V \subset W$ (a product of open sets) such that $f(U \times V) = U \times p(V)$ is open in $X \times (H/K)$; this will be certainly true if $V$ is $p$-saturated.

There are two cases.

(a) If $y \in K$, then $\{x\} \times K \subset W$ (because $W$ is $f$-saturated). The subspace $\{x\} \times K$ is covered by a family of basic nbds $U_z \times V_z \subset W$ of the points $(x, z) \in \{x\} \times K$, and there is a finite subfamily $U_i \times V_i \subset W$, where $(V_i)$ still covers $K$. We have thus an open nbd $U = \bigcap_i U_i$ of $x$ and an open subset $V = \bigcup_i V_i$ containing $K$, which is $p$-saturated.

Finally $U \times V$ is an open nbd of $(x, y)$ contained in $W$

$$U \times V = \bigcup_i U \times V_i \subset \bigcup_i U_i \times V_i \subset W.$$

(b) If $y \notin K$, there is an open nbd $U \times V$ of $(x, y)$ in $W$. If $V$ does not meet $K$, it is $p$-saturated. Otherwise, let $y' \in V \cap K$. We apply the previous point to $(x, y') \in U \times V \subset W$, and we find an open nbd $U' \times V' \subset W$ with $V' \supset K$. Then $(U \cap U') \times (V \cup V')$ is an open nbd of $(x, y)$ contained in $W$, with $V \cup V' \supset K$. □

## 8.6 From pointed cylinder to loop spaces

The cylinder functor, path functor, cones and suspension for pointed spaces can be defined adapting the unpointed case. But cylinder, cones and suspension can be equivalently obtained in an internal way, using the smash product. The Associativity Theorem 8.5.4 steps in, to deal with the powers (or composites) of these functors, including the formula $\Sigma^n X = X \wedge \mathbb{S}^n$ for the pointed suspension (in 8.6.3 and 8.6.6).

All these functors have *here* a right adjoint: path space, cocones and loop spaces, all of them obtained as pointed exponentials. The cofibre and fibre sequence of a pointed map are exposed in 8.6.4 and 8.6.8.

*Literature.* Most of these homotopy constructions are rarely studied for general pointed spaces; one can see [Mau], Section 6.4 and [Di], Chapter 4.

### *8.6.1 Pointed cylinder*

Homotopies of pointed spaces were defined in 1.4.5, as ordinary homotopies of pointed maps that keep the basepoint fixed. Working inside $\mathsf{Top}_\bullet$, we can represent them by the *cylinder functor of pointed spaces*, or *pointed cylinder*.

On the pointed space $X$ this is defined as the quotient of the unpointed cylinder $|X| \times \mathbb{I}$ that collapses the fibre at $0_X$, forming the new basepoint

$$IX = (|X| \times \mathbb{I})/(\{0_X\} \times \mathbb{I}), \quad 0_{IX} = [0_X, t] \qquad (t \in \mathbb{I}). \tag{8.80}$$

The pointed cylinder is an endofunctor of $\mathsf{Top}_\bullet$, with an obvious action on each map $f \colon X \to Y$

$$\begin{aligned} &I \colon \mathsf{Top}_\bullet \to \mathsf{Top}_\bullet, \\ &If \colon IX \to IY, \quad (If)[x, t] = [f(x), t] \qquad (x \in X,\, t \in \mathbb{I}). \end{aligned} \tag{8.81}$$

The faces, degeneracy and reversion of the unpointed cylinder (see 1.4.3, 1.5.3) induce similar natural transformations for the pointed cylinder. We use the same notation (for $\alpha = 0, 1$)

$$\begin{aligned} &\partial^\alpha\colon \mathrm{id} \to I\colon \mathsf{Top}_\bullet \to \mathsf{Top}_\bullet, && \partial^\alpha(x) = [x, \alpha] && (\textit{faces}), \\ &e\colon I \to \mathrm{id} : \mathsf{Top}_\bullet \to \mathsf{Top}_\bullet, && e[x,t] = x && (\textit{degeneracy}), \\ &r\colon I \to I\colon \mathsf{Top}_\bullet \to \mathsf{Top}_\bullet, && r[x,t] = [x, 1-t] && (\textit{reversion}). \end{aligned} \tag{8.82}$$

Now, a (pointed) map $\varphi\colon IX \to Y$ represents a homotopy $\varphi\colon f_0 \simeq f_1$ between the maps $f_\alpha = \varphi\partial^\alpha$ (and amounts to a continuous mapping $|X| \times \mathbb{I} \to |Y|$ taking all pairs $(0_X, t)$ to $0_Y$). The reversed homotopy $\varphi^\sharp\colon f_1 \simeq f_0$ is represented by $\varphi r$. The trivial homotopy $e_f\colon f \simeq f$ of the map $f$ is represented by $fe$.

Also here, when useful, we distinguish a homotopy $\varphi\colon f_0 \simeq f_1$ from the map $\hat{\varphi}\colon IX \to Y$ that represents it.

*An exercise.* (a) One can easily prove that

$$I^n X = (|X| \times \mathbb{I}^n)/(\{0_X\} \times \mathbb{I}^n), \tag{8.83}$$

using the fact that $- \times \mathbb{I}$ preserves quotient maps. Another proof will be given in 8.6.3.

### *8.6.2 Pointed path functor*

More simply, the *path space* of a pointed space

$$P(Y) = (|Y|^{\mathbb{I}}, \omega_0), \qquad \omega_0(t) = 0_Y \qquad (t \in \mathbb{I}), \tag{8.84}$$

is the path space $|Y|^{\mathbb{I}}$ of the underlying topological space (with compact-open topology, see 8.3.4), pointed at the trivial loop of the basepoint.

We have thus the *path functor*, or *cocylinder functor*, of pointed spaces, right adjoint to the pointed cylinder ($I \dashv P$)

$$P\colon \mathsf{Top}_\bullet \to \mathsf{Top}_\bullet, \quad P(g)(a) = ga \quad (\text{for } g\colon Y \to Y',\ a \in |Y|^{\mathbb{I}}). \tag{8.85}$$

In fact, for every (pointed) map $\hat{\varphi}\colon IX \to Y$ there is a unique map $\check{\varphi}\colon X \to PY$ such that $\hat{\varphi}[x,t] = \check{\varphi}(x)(t)$ (for all $x \in X$ and $t \in \mathbb{I}$). In other words, a pointed homotopy $\varphi$ can be equivalently described by a map $\hat{\varphi}\colon IX \to Y$ or a map $\check{\varphi}\colon X \to PY$, in $\mathsf{Top}_\bullet$.

The counit of the adjunction is again the evaluation of paths (induced by the unpointed evaluation, in (8.49))

$$\varepsilon Y\colon IPY \to Y, \qquad [a,t] \mapsto a(t), \tag{8.86}$$

which is well defined, since the fibre $\{\omega_0\} \times \mathbb{I} \subset PY \times \mathbb{I}$ is sent to $0_Y$.

As in (8.50), the basic structure of the functor $P\colon \mathsf{Top}_\bullet \to \mathsf{Top}_\bullet$ consists of the following natural transformations

$$\begin{aligned} &\partial^\alpha\colon P(Y) \to Y, && \partial^\alpha(a) = a(\alpha) && (faces), \\ &e\colon Y \to P(Y), && e(y)(t) = y && (degeneracy), \\ &r\colon P(Y) \to P(Y), && r(a)(t) = a(1-t) && (reversion). \end{aligned} \tag{8.87}$$

In this description, the faces of a homotopy $\check{\varphi}\colon X \to PY$ are defined as $\partial^\alpha\check{\varphi}\colon X \to Y$.

### 8.6.3 The pointed interval

We now introduce the *pointed interval*, adding a basepoint as in 8.4.2(b)

$$\mathbb{I}_\bullet = \mathbb{I} + \{*\}. \tag{8.88}$$

The cylinder and path functor of $\mathsf{Top}_\bullet$ can thus be expressed as the adjoint functors

$$\begin{aligned} &I\colon \mathsf{Top}_\bullet \to \mathsf{Top}_\bullet, && IX = X \wedge \mathbb{I}_\bullet, \\ &P\colon \mathsf{Top}_\bullet \to \mathsf{Top}_\bullet, && PY = Y^{\mathbb{I}_\bullet} && (I \dashv P), \end{aligned} \tag{8.89}$$

using the structure of $\mathsf{Top}_\bullet$, smash product and pointed exponentials, instead of the cartesian structure of $\mathsf{Top}$, as previously.

Applying 8.4.2(c), the smash powers of $\mathbb{I}_\bullet$ correspond to the standard cubes $\mathbb{I}^n$

$$(\mathbb{I}_\bullet)^{\wedge n} = (\mathbb{I}^n)_\bullet, \tag{8.90}$$

and Theorem 8.5.4 gives again the formula (8.83):

$$I^nX = X \wedge (\mathbb{I}_\bullet)^{\wedge n} = X \wedge (\mathbb{I}^n)_\bullet = (|X| \times \mathbb{I}^n)/(\{0_X\} \times \mathbb{I}^n). \tag{8.91}$$

Moreover, also by 8.5.4, the iterated path space

$$P^nY = Y^{(\mathbb{I}_\bullet)^{\wedge n}} = Y^{(\mathbb{I}^n)_\bullet}, \tag{8.92}$$

is the set $\mathsf{Top}_\bullet((\mathbb{I}^n)_\bullet, Y) = \mathsf{Top}(\mathbb{I}^n, U(Y))$ with the compact-open topology, pointed at the constant map at $0_Y$.

### 8.6.4 Pointed cones

All the constructions of Section 8.2 for topological spaces can be adapted to pointed spaces: replacing spaces, maps, homotopies and cylinders with the pointed analogues, we obtain the homotopy pushouts, mapping cones, suspension and cofibre sequences of $\mathsf{Top}_\bullet$. For the sake of simplicity we shall focus on pointed cones and suspension; the mapping cones and the cofibre sequence are briefly dealt with in Exercise (f).

Without using h-pushouts, the *upper (pointed) cone* $C^+X$, or *reduced cone*, of a pointed space can be constructed as an ordinary pushout (or a cokernel) in $\mathsf{Top}_\bullet$, based on the pointed cylinder $IX$

$$\begin{array}{ccc} X & \xrightarrow{p} & \{*\} \\ {\scriptstyle\partial^+}\downarrow & & \downarrow{\scriptstyle v^+} \\ IX & \xrightarrow[\gamma]{} & C^+X \end{array} \qquad \begin{array}{c} C^+X = \mathrm{Cok}\,(\partial^+\colon X \to IX) = IX/\partial^+X, \\ \gamma[x,t]_I = [x,t]_C, \\ v^+(*) = v^+ = [x,1]_C. \end{array}$$

In other words, $C^+X$ is the quotient of the pointed cylinder $IX$ which collapses the upper base $\partial^+X$ (certainly non-empty) to an upper vertex $v^+$. The map $\gamma\colon IX \to C^+X$ is the canonical projection (writing as $[x,t]_I$ and $[x,t]_C$ the equivalence classes in $IX$ and $C^+X$).

We have thus a functor $C^+$ and a natural transformation:

$$C^+\colon \mathsf{Top}_\bullet \to \mathsf{Top}_\bullet, \qquad \gamma\colon I \to C^+, \tag{8.93}$$

characterised by the following universal property:

(*) for every (pointed) homotopy $\varphi\colon f \simeq 0$ reaching the zero-map $X \to Y$, there is precisely one map $h\colon C^+X \to Y$ such that $\varphi = h\gamma\colon IX \to C^+X \to Y$.

Dually, the *lower cone* $C^-X = \mathrm{Cok}\,(\partial^-\colon X \to IX)$ is obtained by collapsing the lower base of $IX$ to a lower vertex $v^- = v^-(*)$.

*Remarks and complements.* (a) The cones of the pointed space $X$ can be obtained as smash products

$$C^\alpha X = X \wedge \mathbb{I}_\alpha = (X \times \mathbb{I})/(X \times \{\alpha\} \cup \{0\} \times \mathbb{I}) \quad (\alpha = 0, 1), \tag{8.94}$$

where $\mathbb{I}_\alpha$ is the standard interval, *pointed at* $\alpha$.

(b) $C^+X$ can also be viewed as the quotient of the unpointed cone $C^+|X|$ that collapses the fibre of the basepoint, namely the image of $\{0_X\} \times \mathbb{I}$ in $C^+|X|$, forming the new basepoint.

(c) The map $\gamma\colon IX \to C^+X$ is the cokernel of $\partial^+\colon X \to IX$ in the pointed category $\mathsf{Top}_\bullet$, as defined in 2.1.2(f).

(d) The lower base of $C^+X$ gives a natural transformation $\partial^-\colon 1 \to C^+$

$$\partial^- = \gamma\partial^-\colon X \to IX \to C^+X, \qquad \partial^-(x) = [x,0]_C, \tag{8.95}$$

whose components are topological embeddings (as in 8.2.2(c)).

(e) The upper vertex gives a natural transformation

$$v^+ \colon 0 \to C^+ \colon \mathsf{Top}_\bullet \to \mathsf{Top}_\bullet, \tag{8.96}$$

where 0 denotes the constant endofunctor at the zero object $\{*\}$; but this transformation is determined by the pointed structure and carries no further information.

(f) The mapping cones of a pointed map $f \colon X \to Y$ (defined as in 8.2.3) can also be derived from the mapping cones of $X$, by the following pushouts

$$\begin{array}{ccc} X & \xrightarrow{\partial^-} & C^+(X) \\ {\scriptstyle f}\downarrow & & \downarrow \\ Y & \xrightarrow[u]{} & C^+(f) \end{array} \qquad\qquad \begin{array}{ccc} X & \xrightarrow{f} & Y \\ {\scriptstyle \partial^+}\downarrow & & \downarrow{\scriptstyle u} \\ C^-(X) & \longrightarrow & C^-(f) \end{array}$$

*As in 8.2.7, the pointed map $f$ has a natural (upper) *cofibre sequence*, or Puppe sequence

$$X \xrightarrow{f} Y \xrightarrow{u} C^+f \xrightarrow{d} \Sigma X \xrightarrow{\Sigma f} \Sigma Y \xrightarrow{\Sigma u} \Sigma(C^+f) \xrightarrow{\Sigma d} \Sigma^2 X ... \tag{8.97}$$

with the same exactness properties.*

### 8.6.5 Pointed suspension

The (pointed) *suspension* $\Sigma X$ of a pointed space (also called *reduced suspension*) can be obtained by ordinary colimits in $\mathsf{Top}_\bullet$, as in (8.35) (based on the pointed cylinder $IX$), and in (8.2.6) (based on the pointed cones $C^\alpha(X)$), noting that the vertices $v^-, v^+$ of the unpointed case are here identified: there is *one* vertex $v \colon \{*\} \to \Sigma X$, the zero map.

Again, $\Sigma X$ is the quotient of each (pointed) cone of $X$ that collapses its base to a point

$$\begin{aligned} \Sigma X &= \mathrm{Cok}\,(\partial^- \colon X \to C^+(X)) = C^+(X)/\partial^- X \\ &= \mathrm{Cok}\,(\partial^+ \colon X \to C^-(X)) = C^-(X)/\partial^+ X. \end{aligned} \tag{8.98}$$

We have thus the (pointed) *suspension functor* $\Sigma$ and a natural transformation $\sigma$

$$\Sigma \colon \mathsf{Top}_\bullet \to \mathsf{Top}_\bullet, \qquad \sigma \colon I \to \Sigma, \tag{8.99}$$

characterised by the following universal property:

(*) for every (pointed) endo-homotopy $\varphi \colon 0 \simeq 0$ of the zero-map $X \to Y$ there is precisely one map $h \colon \Sigma X \to Y$ such that $\varphi = h\sigma \colon IX \to Y$.

*Remarks.* (a) The suspension of pointed spaces can be expressed as a smash product

$$\Sigma X = X \wedge \mathbb{S}^1 = (X \times \mathbb{I})/(X \times \{0, 1\} \cup \{0\} \times \mathbb{I}). \qquad (8.100)$$

(b) $\Sigma X$ can thus be viewed as the quotient of the unpointed cylinder $I|X|$ that collapses the subspace $\partial^- X \cup (\{0_X\} \times \mathbb{I}) \cup \partial^+ X$ to the basepoint $v$, or the quotient of the unpointed suspension $\Sigma|X|$ that collapses the fibre of the basepoint (the image of $\{0_X\} \times \mathbb{I}$ in $\Sigma|X|$).

(c) The following endofunctors of $\mathsf{Top}_\bullet$ (and their powers)

$$I = - \wedge \mathbb{I}_\bullet, \quad C^+ = - \wedge \mathbb{I}_+, \quad C^- = - \wedge \mathbb{I}_-, \quad \Sigma = - \wedge \mathbb{S}^1,$$

commute with each other and with each functor $- \wedge A$ produced by an exponentiable pointed space, by 8.5.3(c).

(d) The map $\sigma \colon IX \to \Sigma X$ is the colimit in $\mathsf{Top}_\bullet$ of the diagram

$$\partial^-, 0, \partial^+ \colon X \rightrightarrows IX,$$

a generalised coequaliser.

### 8.6.6 The pointed spheres

Let us recall that the pointed $n$-sphere can be defined as the topological quotient of the euclidean $n$-cube $\mathbb{I}^n$

$$\mathbb{S}^n = \mathbb{I}^n / \underline{\partial}\mathbb{I}^n \qquad (n \geqslant 0), \qquad (8.101)$$

that collapses the border of $\mathbb{I}^n$ in $\mathbb{R}^n$ to the basepoint $e_1$. (For the case $\mathbb{S}^0 = \mathbb{I}^0/\emptyset$, see 4.2.2.)

The pointed $n$-sphere is a smash power of the pointed circle

$$\mathbb{S}^n = \mathbb{S}^1 \wedge ... \wedge \mathbb{S}^1 = (\mathbb{S}^1)^{\wedge n}, \qquad (8.102)$$

as a consequence of Exercise 8.4.2(d).

Applying 8.5.4, the iterated pointed suspension $\Sigma^n X$ is produced by the smash product with $\mathbb{S}^n$

$$\Sigma^n X = (...(X \wedge \mathbb{S}^1)...) \wedge \mathbb{S}^1 = X \wedge (\mathbb{S}^1)^{\wedge n} = X \wedge \mathbb{S}^n, \qquad (8.103)$$

and all spheres are pointed suspensions of $\mathbb{S}^0$

$$\Sigma^n \mathbb{S}^0 = \mathbb{S}^0 \wedge \mathbb{S}^n = \mathbb{S}^n. \qquad (8.104)$$

### *8.6.7 Homotopy pullbacks, pointed cocones and loop spaces*

Let $f\colon Y \to X$ and $g\colon Z \to X$ be two pointed maps with the same codomain. The *homotopy pullback* $P(f,g)$ in $\mathsf{Top}_\bullet$ is defined as in (8.52) and constructed as in $\mathsf{Top}$: a subspace of the cartesian product $Y \times PX \times Z$ of pointed spaces

$$P(f,g) = \{(y,a,z) \in Y \times PX \times Z \mid a(0) = f(y),\ a(1) = g(z)\}, \qquad (8.105)$$

pointed at $(0, \omega_0, 0)$. Here, in the pointed case, we can also dualise mapping cones, cones and suspension.

The *mapping cocones* $E^+ f = P(f, i_X)$ and $E^- f = P(i_X, f)$ are particular homotopy pullbacks, where $i_X\colon \{*\} \to X$ is the zero map

$$\begin{aligned} E^+ f &= P(f, i_X) = \{(y,a) \in Y \times PX \mid a(0) = f(y),\ a(1) = 0\}, \\ E^- f &= P(i_X, f) = \{(a,y) \in PX \times Y \mid a(0) = 0,\ a(1) = f(y)\}. \end{aligned} \qquad (8.106)$$

The *upper* and *lower cocone* $E^\alpha X$ and the *loop space* $\Omega X = P(i_X, i_X)$ of the pointed space $X$ are also particular homotopy pullbacks, and kernels in $\mathsf{Top}_\bullet$

$$\begin{array}{ccc} E^+X & \xrightarrow{p} & \{*\} \\ {\scriptstyle u}\downarrow & \nearrow_{\gamma} & \downarrow{\scriptstyle i} \\ X & \xrightarrow[\mathrm{id}]{} & X \end{array} \qquad \begin{array}{ccc} E^-X & \xrightarrow{u} & X \\ {\scriptstyle p}\downarrow & \nearrow_{\gamma} & \downarrow{\scriptstyle \mathrm{id}} \\ \{*\} & \xrightarrow[i]{} & X \end{array} \qquad \begin{array}{ccc} \Omega X & \xrightarrow{p} & \{*\} \\ {\scriptstyle p}\downarrow & \nearrow_{\omega} & \downarrow{\scriptstyle g} \\ \{*\} & \xrightarrow[i]{} & X \end{array} \qquad (8.107)$$

$$\begin{aligned} E^+X &= P(\mathrm{id}\,X, i_X) = \{a \in PX \mid a(1) = 0\} = \mathrm{Ker}\,(\partial^+\colon PX \to X), \\ E^-X &= P(i_X, \mathrm{id}\,X) = \{a \in PX \mid a(0) = 0\} = \mathrm{Ker}\,(\partial^-\colon PX \to X), \\ \Omega X &= P(i_X, i_X) = \{a \in PX \times X \mid a(0) = 0 = a(1)\} \\ &= \mathrm{Ker}\,(\partial^-\colon E^+X \to X) = \mathrm{Ker}\,(\partial^+\colon E^-X \to X). \end{aligned}$$

*Remarks.* (a) The following endofunctors of $\mathsf{Top}_\bullet$

$$P = (-)^{\mathbb{I}_\bullet}, \quad E^+ = (-)^{\mathbb{I}_+}, \quad E^- = (-)^{\mathbb{I}_-}, \quad \Omega = (-)^{\mathbb{S}^1}, \qquad (8.108)$$

are right adjoint to the endofunctors of 8.6.5(c). They commute with each other, and with each functor $(-)^A$ produced by an exponentiable pointed space $A$.

(b) For any pointed space $X$ the iterated loop space

$$\Omega^n(X) = X^{\mathbb{S}^n}, \qquad (8.109)$$

can be identified with the pointed set $\Omega_n(X)$ of (7.66), equipped with the compact-open topology

$$\Omega_n(X) = \mathsf{Top}_2((\mathbb{I}^n, \underline{\partial}\mathbb{I}^n), (X, 0_X)) = \mathsf{Top}_\bullet(\mathbb{S}^n, X). \qquad (8.110)$$

### 8.6.8 *Complements

(a) Mapping cocones and loop spaces form another Puppe sequence of a pointed map $f\colon X \to Y$, its (upper) *fibre sequence*

$$... \Omega^2 Y \xrightarrow{\Omega d} \Omega(E^+f) \xrightarrow{\Omega u} \Omega X \xrightarrow{\Omega f} \Omega Y \xrightarrow{d} E^+f \xrightarrow{u} X \xrightarrow{f} Y. \qquad (8.111)$$

Now, every map is the h-kernel of the following one, up to homotopy equivalence: see [G1], 1.8.5 and the dual to Theorem 4.7.5(d).

(b) The adjunction $\Sigma \dashv \Omega$ is consistent with (pointed) homotopies. In fact, applying Theorem 8.5.4, a homotopy $\varphi\colon f \to g\colon \Sigma X \to Y$ of (pointed) maps can be equivalently represented by maps

- $I(\Sigma X) = \Sigma X \wedge \mathbb{I}_\bullet = X \wedge \mathbb{I}_\bullet \wedge \mathbb{S}^1 \longrightarrow Y$,
- $X \longrightarrow (Y^{\mathbb{S}^1})^{\mathbb{I}_\bullet} = P(\Omega Y)$,

which means a homotopy $\varphi'\colon f' \to g'\colon X \to \Omega Y$ between the transpose maps.

(c) Taking the previous point into account, the homotopy groups of a pointed space $X$ are linked to the loop spaces $\Omega^n X$

$$\begin{gathered} \pi_{n+1}(X) = [\mathbb{S}^{n+1}, X]_\bullet = [\Sigma\mathbb{S}^n, X]_\bullet = [\mathbb{S}^n, \Omega X]_\bullet = \pi_n(\Omega X), \\ \pi_n(X) = \pi_0(\Omega^n X), \end{gathered} \qquad (8.112)$$

using the notation $[X, Y]_\bullet$ for the sets of homotopy classes of pointed maps, as in 1.4.6. This relation should be made precise as to the algebraic structure (for $n > 0$), working with the H-structure of any loop space under concatenation: see [Hu], Section III.11, or [Du], Chapter XIX.

The first relation above is dually similar to that of reduced homology groups and suspension, in 8.2.6(c). But we note that the pair $(H_n, \Sigma)$ determines the homology functors of *lower* degree, while $(\pi_0, \Omega)$ determines all the homotopy groups.

# 9

# Solution of the exercises

Easy exercises and exercises marked with * may be left to the reader.

## 9.1 Exercises of Chapter 1

### *9.1.1 Solutions of 1.1.1*

(a) Let $a < b$ in $\mathbb{R}$. We want to prove that there are homeomorphisms, for the following forms of non-degenerate intervals:

$$\begin{aligned} &\text{(i)} \quad \mathbb{R} \cong \,]a, +\infty[\, \cong \,]-\infty, b[\, \cong \,]a, b[\, \cong \,]0, 1[ && (\textit{open}), \\ &\text{(ii)} \quad [a, +\infty[\, \cong \,]-\infty, b] \cong [a, b[\, \cong \,]a, b] \cong [0, 1[ && (\textit{semiopen}), \\ &\text{(iii)} \quad [a, b] \cong [0, 1] && (\textit{compact}). \end{aligned}$$

First let us note that, for every $a \in \mathbb{R}$ and $\lambda \neq 0$, the affine mapping

$$f \colon \mathbb{R} \to \mathbb{R}, \qquad f(x) = \lambda x + a, \tag{9.1}$$

is a homeomorphism.

Point (iii) is already proved: the compact interval $[0, 1]$ is homeomorphic to its $f$-image $[a, a + \lambda]$ (with $\lambda > 0$), and therefore to any non-degenerate compact interval.

Similarly, the open interval $]0, 1[$ is homeomorphic to any bounded open interval $]a, b[$. But $\mathbb{R}$ is homeomorphic to the open interval $J = ]-\pi/2, \pi/2[$, as shown by the invertible function $\tan \colon J \to \mathbb{R}$, and to the open interval $J' = ]0, +\infty[$, as shown by the invertible function $\exp \colon \mathbb{R} \to J'$. To conclude point (i), $J'$ is homeomorphic to any interval $]a, +\infty[$ by a translation, and the latter is homeomorphic to $]-\infty, -a[$ by $f(x) = -x$.

Point (ii) is proved in the same way. The affine mapping (9.1) proves that all the intervals $[a, b[$ are homeomorphic to each other, and all the intervals $]a, b]$ are also. But the former are homeomorphic to the latter, since $[0,1[\, \cong\, ]0,1]$, via $f(x) = 1 - x$. Taking into account that $[0, \pi/2[\, \cong [0, +\infty[$, via the tangent function, the last verifications are obvious.

(b) The idea is to distinguish the presence of the end-points by properties of connectedness.

Let $X$ be a connected space. We say that a point $x \in X$ *disconnects* the space if $X \setminus \{x\}$ is no longer connected. Plainly, given a homeomorphism $f\colon X \to Y$, the point $x$ disconnects $X$ if and only if $f(x)$ disconnects $Y$.

We can say something more: $f$ restricts to a bijection $f'\colon A \to B$ between the subsets of the points which disconnect their space; and of course it also restricts to a bijection between their complements, the subsets of points which do not disconnect the space. (If useful, we can also remark that we have two homeomorphisms between the corresponding subspaces.)

Now, in the interval $]0,1[$ every point disconnects the space. The interval $[0,1[$ has precisely one point which does not disconnect it, namely 0. The interval $[0,1]$ has two such points.

(c) Trivially, a singleton space is only homeomorphic to a singleton. To prove that $\mathbb{R}$ is not homeomorphic to any higher $\mathbb{R}^n$ we proceed as in (b): any point $x \in \mathbb{R}$ disconnects the line, while each pierced space $\mathbb{R}^n \setminus \{x\}$ is path connected, for $n > 1$.

(d) A degenerate interval is not equipotent to the finite sphere $\mathbb{S}^0 = \{-1, 1\}$. Any non-degenerate interval $J$ is disconnected by any interior point, while any sphere $\mathbb{S}^n$ stays path connected if we take out a point.

### *9.1.2 Solutions of 1.1.2*

(a) A map $x\colon \{*\} \to X$ amounts to a point of $X$, and a homotopy $\varphi\colon x \simeq y\colon \{*\} \to X$ amounts to a path in $X$, from $x$ to $y$. All this will be better analysed in Section 1.4.

(b) $\mathbb{R}^n$ is a topological vector space on the topological field $\mathbb{R}$, which means that the linear combination

$$c\colon \mathbb{R}^n \times \mathbb{R}^n \times \mathbb{R} \times \mathbb{R} \longrightarrow \mathbb{R}^n, \qquad c(v, w, \lambda, \mu) = \lambda v + \mu w,$$

is a continuous mapping.

The mapping $\varphi$ can be written as a composite of continuous mappings

$$X \times \mathbb{I} \longrightarrow \mathbb{R}^n \times \mathbb{R}^n \times \mathbb{I} \times \mathbb{I} \longrightarrow \mathbb{R}^n,$$
$$(x, t) \mapsto (f(x), g(x), 1 - t, t) \mapsto (1 - t)f(x) + tg(x),$$

whose image, in the given hypotheses, is contained in $Y$.

### 9.1.3 Solutions of 1.1.3

(a) Using the $\mathbb{R}$-linear structure of $\mathbb{R}^n$, the origin is a strong deformation retract of $\mathbb{R}^n$

$$p\colon \mathbb{R}^n \to \{0\}, \qquad \varphi\colon ip \simeq \mathrm{id}\,\mathbb{R}^n, \quad \varphi(x,t) = tx. \tag{9.2}$$

This point is a particular case of the next.

(b) If $X \subset \mathbb{R}^n$ is starred with respect to $x_0 \in X$, the singleton $\{x_0\}$ is a deformation retract of $X$, by the affine homotopy $\varphi\colon mp \simeq \mathrm{id}\,X$ (see (1.3)) that travels on the line segments of $\mathbb{R}^n$, from $x_0$ to each $x \in X$

$$\begin{aligned} &p\colon X \rightleftarrows \{x_0\} : m, && pm_0 = \mathrm{id}\,, && \varphi\colon mp \simeq \mathrm{id}\,X, \\ &\varphi\colon X \times \mathbb{I} \to X, && && \varphi(x,t) = (1-t)x_0 + tx. \end{aligned} \tag{9.3}$$

(c) Plainly, (i) is equivalent to (ii):

- if $X$ is contractible, there is a homotopy $\varphi\colon mp \simeq \mathrm{id}\,X$, as in (1.4),
- conversely, if $\mathrm{id}\,X$ is homotopic to a map $X \to X$ constant at $x_0 \in X$, then $\{x_0\}$ is a deformation retract of $X$.

The rest follows easily, by whisker composition of homotopies and maps (see (1.61)).

(d) The normalisation mapping $N$ is obviously the identity on $\mathbb{S}^n$. Moreover we have an affine homotopy $\varphi\colon mN \simeq \mathrm{id}\,X$

$$\varphi(x,t) = (1-t)N(x) + tx \qquad ((x,t) \in X \times \mathbb{I}), \tag{9.4}$$

that moves on each line segment from $N(x)$ to $x$, staying in the pierced space $X$.

(e) If $A$ is a deformation retract of $X$, any homotopy $\varphi\colon mp \simeq \mathrm{id}\,X$ satisfies the required conditions. Conversely, if these are satisfied, we have a map $p = \varphi(-,0)\colon X \to A$ that satisfies $p(a) = \varphi(a,0) = a$ (for $a \in A$); moreover $\varphi(x,0) = p(x) = mp(x)$ and $\varphi(x,1) = x$ (for $x \in X$).

An accurate reader will note a light abuse of notation in this argument: a more formal writing would distinguish the endomap $e = \varphi(-,0)\colon X \to X$ from its codomain-restriction $p\colon X \to A$, satisfying $e = mp$. According to the context, an abuse of notation can be harmless or a source of errors; yet, an excessive formalisation would make mathematics boring and unreadable.

### 9.1.4 Solutions of 1.1.6

We begin by proving (b). First, $\mathbb{S}^n \subset \mathbb{R}^{n+1}$, but the group $H_n(\mathbb{S}^n) \cong \mathbb{Z}$ cannot be embedded in $H_n(\mathbb{R}^{n+1}) = 0$ (for $n > 0$).

Second, let us suppose that the inclusion $m\colon A \to X$ has a (continuous!) retraction $p\colon X \to A$. Then $pm = \text{id}\,A$, and the functor $H_n$ gives

$$H_n(A) \xrightarrow{m_*} H_n(X) \xrightarrow{p_*} H_n(A) \qquad p_*m_* = \text{id}\,H_n(A), \tag{9.5}$$

showing that the homomorphism $m_*$ has a left inverse, and is injective.

Putting together these two facts, we have verified (a): $\mathbb{S}^n$ is not a retract of a contractible space, for $n > 0$.

For $n = 0$ the argument can be adapted: the group $H_0(\mathbb{S}^0) \cong \mathbb{Z}^2$ cannot be embedded in $H_0(\mathbb{R}) \cong \mathbb{Z}$.

However, the computation of the group $H_0(X)$ (in Exercise 2.2.5(d)) will depend on the partition of the space $X$ in path components, which usually comes after the study of connected spaces and the Intermediate Value Theorem; we already remarked that the latter says that $\mathbb{S}^0$ is not a retract of $\mathbb{R}$.

### 9.1.5 Solutions of 1.1.8

We begin with point (c). By hypothesis, $U$ is open in $X$ if and only if every trace $u_i^{-1}(U)$ is open in $X_i$. If all the restrictions $f_i = fu_i\colon X_i \to Y$ are continuous and $V$ is open in $Y$, then every $u_i^{-1}f^{-1}(V) = f_i^{-1}(V)$ is open in $X_i$, and $f^{-1}(V)$ is open in $X$. The converse is trivial: all restrictions of continuous mappings are continuous.

Points (a) and (b) are a consequence. If $(X_i)_{i\in I}$ is a generalised open cover, a subset $A$ of $X$ is open if and only if all its traces $A \cap \text{int}X_i$ are open in $X_i$. Similarly, if $(X_i)_{i\in I}$ is a finite closed cover, a subset $A$ of $X$ is closed if and only if all its traces are closed in $X_i$.

### 9.1.6 Solutions of 1.2.2

(a) The euclidean line is homotopy equivalent to the singleton space, but its underlying set is infinite.

(b) Let $\varphi\colon X \times \mathbb{I} \to Y$ be a homotopy $f \simeq g\colon X \to Y$. For every $x \in X$, the mapping $\varphi(x, -)\colon \mathbb{I} \to Y$ is continuous and its image is a path connected subspace of $Y$ containing $f(x)$ and $g(x)$.

Therefore, if $A$ is the connected (resp. path connected) component of $x$ in $X$, $f(A)$ and $g(A)$ are contained in the same connected (resp. path connected) component of $Y$; this means that $\text{Cc}(f) = \text{Cc}(g)$ and $\Pi_0(f) = \Pi_0(g)$.

The same holds for the functor $\mathcal{A} = \mathsf{Top}(-, \mathbb{Z})$, because a map $a\colon Y \to \mathbb{Z}$ is sent to the maps

$$\mathcal{A}(f)(a) = af\colon X \to \mathbb{Z}, \qquad \mathcal{A}(g)(a) = ag\colon X \to \mathbb{Z},$$

which coincide: for every $x \in X$, the path $a\varphi(x, -)\colon \mathbb{I} \to \mathbb{Z}$ is constant, and its endpoints $af(x)$, $ag(x)$ coincide.

### 9.1.7 Solutions of 1.2.3

(b) A map $a\colon X \to \mathbb{Z}$ gives such a decomposition, letting $X_k = a^{-1}\{k\}$. This correspondence is plainly bijective.

(c) If $X$ is empty we only have the empty mapping $\emptyset \to \mathbb{Z}$. Otherwise, every constant mapping $X \to \mathbb{Z}$ is continuous, and $\mathcal{A}(X)$ is infinite.

(d) We can assume that $X$ is non-empty, because this is a consequence of all the conditions we are considering.

(i) $\Rightarrow$ (ii). If $X$ is connected, each map $a\colon X \to \mathbb{Z}$ is constant, and any point $x \in X$ gives a canonical isomorphism of rings (independent of $x$)

$$\varphi\colon \mathcal{A}(X) \to \mathbb{Z}, \qquad \varphi(a) = a(x). \tag{9.6}$$

Obviously, (ii) $\Rightarrow$ (iii) and (iv). Conversely, if $X$ is not connected, we have partition $X = U \cup V$ in disjoint non-empty open sets. This gives two non-zero maps $a, b\colon X \to \mathbb{Z}$

$$a(U) = \{1\}, \quad a(V) = \{0\}, \qquad\qquad b(U) = \{0\}, \quad a(V) = \{1\},$$

with $ab = 0$. Therefore $\mathcal{A}(X)$ is not an integral domain.

Choosing two elements $x \in U$, $x' \in V$, we have a surjective homomorphism $\mathcal{A}(X) \to \mathbb{Z} \times \mathbb{Z}$, $a \mapsto (a(x), a(x'))$ of rings (and of the underlying additive groups), which shows that $\mathcal{A}(X)$ is not a cyclic group.

(e) It is an extension of the previous points. We choose an element $x_i \in X_i$ in each connected component. This gives a homomorphism with values in the product ring $\mathbb{Z}^I$

$$\varphi\colon \mathcal{A}(X) \to \mathbb{Z}^I, \qquad \varphi(a) = (a(x_i))_{i \in I}. \tag{9.7}$$

This homomorphism is canonical and injective, because $a$ is constant on each connected component. It is also surjective, in our hypothesis: any family $(\lambda_i)_{i \in I}$ of integers defines a mapping $a\colon X \to \mathbb{Z}$, letting $a(x) = \lambda_i$ when $x \in X_i$; this mapping is continuous, because the preimage of any $\lambda \in \mathbb{Z}$ is a union of (open) connected components of $X$.

(f) A function $a\colon \mathbb{N}^{\bullet} \to \mathbb{Z}$ is continuous, i.e. locally constant, if and only if it is eventually constant. The restriction to $\mathbb{N}$ gives thus an injective ring-homomorphism

$$\mathcal{A}(X) \longrightarrow \mathbb{Z}^{\mathbb{N}} = \mathsf{Set}(\mathbb{N}, \mathbb{Z}), \qquad a \mapsto (a(n))_{n \in \mathbb{N}}, \tag{9.8}$$

whose image is the subring $R \subset \mathbb{Z}^{\mathbb{N}}$ formed by the eventually constant sequences of integers.

$R$ is countable: its underlying additive group $|R|$ is the internal direct sum of two countable subgroups of $\mathbb{Z}^{\mathbb{N}}$

$$|R| = (\bigoplus_{n\in\mathbb{N}} \mathbb{Z}) \oplus \mathbb{Z}e,$$

namely the ideal of eventually null sequences and the subring generated by the unit element $e = (1, 1, ..., 1, ...)$ of $\mathbb{Z}^{\mathbb{N}}$.

$R$ is thus the unital ring canonically associated to the non-unital ring $\bigoplus_{n\in\mathbb{N}} \mathbb{Z}$. We also note that the additive group $|R|$ is isomorphic to $\bigoplus_{n\in\mathbb{N}} \mathbb{Z}$. The polynomial ring $\mathbb{Z}[X]$ in one indeterminate has the 'same' underlying additive group, but is an integral ring, quite different from $R$.

*(g) Every irrational number $\xi$ determines a subset

$$A_\xi = ]-\infty, \xi[ \cap \mathbb{Q} = ]-\infty, \xi] \cap \mathbb{Q},$$

which is open and closed in $\mathbb{Q}$. Its characteristic function $\mathbb{Q} \to \mathbb{Z}$ is continuous.

This proves that $\mathcal{A}(\mathbb{Q})$ is at least a continuum. On the other hand, $\mathcal{A}(\mathbb{Q}) \subset \mathsf{Set}(\mathbb{Q}, \mathbb{Z})$ cannot have a greater cardinal.

### *9.1.8 Solutions of 1.2.4*

Points (a) and (b) are obvious.

(c) We have seen in Exercise 1.2.3(f) that the ring $\mathcal{A}(\mathbb{N}^\bullet)$ is countable, while $\mathcal{A}(\mathbb{N}) \cong \mathbb{Z}^{\mathbb{N}}$ is a continuum.

### *9.1.9 Solutions of 1.3.3*

(a) In $\mathsf{Set}$, if $x_0 \in A \subset X$, we can define a retraction $p\colon X \to A$ letting $p(x) = x$ for $x \in A$, and $p(x) = x_0$ otherwise. In $\mathsf{Set}_\bullet$ we proceed in the same way, using the basepoint.

(b) If the subgroup $A \subset X$ has a retraction $p\colon X \to A$ in $\mathsf{Ab}$, the subgroup $B = \mathrm{Ker}\, p$ meets our conditions. In fact, if $x \in A \cap B$, then $x = p(x) = 0$; if $x \in X$, then $x = p(x) + (x - p(x)) \in A + B$.

Conversely, if there exists a subgroup $B \subset X$ such that $A \cap B = \{0\}$ and $A + B = X$, each element $x \in X$ can be uniquely written as $x = a + b$, with $a \in A$ and $b \in B$. This gives a canonical isomorphism $X \cong A \oplus B$, and a retraction $p\colon X \to A$.

(c) A cyclic group of prime order has no other subgroups than the necessary ones.

As to the group $\mathbb{Z}$, any two non-trivial subgroups $h\mathbb{Z}$ and $k\mathbb{Z}$ meet in a non-trivial subgroup. Yet $\mathbb{Z} \cong h\mathbb{Z} \cong h\mathbb{Z} \oplus 0$, which is not an internal direct sum of subgroups (for $h > 0$).

(d) We have to check that $U$ is open in $A$ (if and) only if it is the $m$-preimage of an open subset of $X$. In fact, $U = (pm)^{-1}(U) = m^{-1}(p^{-1}(U))$ is the $m$-preimage of $p^{-1}(U)$.

(e) We have to check that $U$ is open in $A$ if (and only if) its $p$-preimage is open in $X$. In fact, if $p^{-1}(U)$ is open in $X$, $U = m^{-1}p^{-1}(U)$ is open in $A$.

### *9.1.10 Solutions of 1.4.2*

(a) The mappings $a(2t)$ and $b(2t-1)$ have the same outcome at $t = 1/2$, namely $a(1) = b(0)$; the mapping $a * b \colon \mathbb{I} \to X$ is thus well defined. Its continuity follows from the Finite Closed Cover Lemma, applied to the cover of $\mathbb{I}$ formed by the closed intervals $[0, 1/2]$ and $[1/2, 1]$.

One can also use the left and right continuity of $a * b$, at $t = 1/2$.

(d) A consequence of the canonical homeomorphism $p' \colon \mathbb{I}/\underline{\partial}\mathbb{I} \to \mathbb{S}^1$, $p'[t] = e^{2\pi i t}$, in (1.51).

### *9.1.11 Solutions of 1.4.3*

(a) The mappings $\varphi(x, 2t)$ and $\psi(x, 2t-1)$ take the same value when $t = 1/2$, namely $g(x)$. Moreover they are continuous on the subspaces $X \times [0, 1/2]$ and $X \times [1/2, 1]$, respectively, and the latter form a finite closed cover of $IX$.

Here, continuity cannot be reduced to left and right continuity, as in Exercise 1.4.2(a): the Finite Closed Cover Lemma is the appropriate tool to construct maps 'by cases', in homotopy theory.

(d) We have a homotopy equivalence between $X$ and the singleton space

$$\begin{aligned} &p \colon X \rightleftarrows \{*\} : m, \qquad pm = \mathrm{id}\,, \qquad \varphi \colon mp \simeq \mathrm{id}\, X, \\ &\varphi \colon X \times \mathbb{I} \to X, \qquad \varphi(x, 0) = m(*), \quad \varphi(x, 1) = x \quad (x \in X). \end{aligned} \tag{9.9}$$

Therefore, for each $x \in X$, the map $\varphi(x, -) \colon \mathbb{I} \to X$ is a path from $x_0 = m(*)$ to $x$, and $X$ is path connected.

We have seen that the singleton $\{x_0\}$ is a deformation retract of $X$. But in fact any point $x_1 \in X$ would do: if $m' \colon \{*\} \to X$ corresponds to the point $x_1$ and $a \colon \mathbb{I} \to X$ is a path from $x_1$ to $x_0$, the map $\psi(x, t) = a(t)$ is a homotopy $\psi \colon m'p \simeq mp \colon X \to X$ (between constant maps), and the concatenation $\psi * \varphi \colon m'p \simeq \mathrm{id}\, X$ shows that $\{x_1\}$ is also a deformation retract of $X$.

### *9.1.12 Solutions of 1.4.6*

(a) We already know from 1.2.2(c) that $\Pi_0(X) = \mathrm{hoTop}(\{*\}, X)$.

(b) A point $x$ of the pointed space $X$ is identified with the pointed map $\hat{x}\colon \mathbb{S}^0 \to X$ which reaches $x$ (and takes the basepoint 1 to $0_X$); the path component $[x]_P$ is identified with the pointed homotopy class of $\hat{x}$.

In this way we identify the functor $\pi_0$ with the (canonically isomorphic) hom-functor $[\mathbb{S}^0, -]_\bullet\colon \mathrm{hoTop}_\bullet \to \mathsf{Set}_\bullet$.

### *9.1.13 Solutions of 1.4.7*

(a) The (convex) space $\mathbb{I}$ is contractible (by 1.1.3(b)), whence a path $a\colon \mathbb{I} \to X$ is homotopic to a constant path $e_x\colon \mathbb{I} \to X$ in the same path component (by 1.1.3(c)). Plainly, two constant paths $e_x, e_y\colon \mathbb{I} \to X$ are homotopic if and only if the points $x, y$ are, which means that they belong to the same path component of $X$.

(b) The first point is proved as for ordinary homotopies of general maps: trivial homotopies, reversed homotopies and homotopy concatenation (defined in 1.4.3) can be restricted to path homotopies with fixed endpoints.

Moreover, if $h\colon a \simeq_2 b\colon x \simeq y$, then $k(s,t) = h(1-s,t)$ is a 2-homotopy $a^\sharp \simeq_2 b^\sharp\colon y \simeq x$.

(c) The functions $\mathrm{id}\,\mathbb{I}$ and $f$ are paths in $\mathbb{I}$, with the same endpoints: 0 and 1. As a natural choice, we can use the affine homotopy

$$h\colon \mathrm{id}\,\mathbb{I} \simeq_2 f\colon \mathbb{I} \to \mathbb{I}, \quad h(s,t) = (1-t)s + tf(s) \qquad (s,t \in \mathbb{I}), \tag{9.10}$$

which is a 2-homotopy. By whisker composition, we have a 2-homotopy $ah\colon a \simeq_2 af\colon \mathbb{I} \to X$.

(d) Using the previous point, it is sufficient to use a strictly increasing reparametrisation $f\colon \mathbb{I} \to \mathbb{I}$ that transforms the regular decomposition $0 < 1/n < ... < 1$ into an arbitrary decomposition $0 < t_1 < ... < 1$; for instance the piecewise affine function $f(t) = (nt-k)(t_k - t_{k-1}) + t_k$, where $k = 1, ..., n$ and $t \in [(k-1)/n, k/n]$.

### *9.1.14 Solutions of 1.5.1*

(a) For a scalar $\lambda \in R$, every $R$-homomorphism $f\colon A \to B$ gives a commutative square

$$\begin{array}{ccc} A & \xrightarrow{\lambda} & A \\ {\scriptstyle f}\downarrow & & \downarrow{\scriptstyle f} \\ B & \xrightarrow[\lambda]{} & B \end{array} \qquad \lambda f(x) = f(\lambda x) \quad (\text{for } x \in A). \tag{9.11}$$

(b) The $R$-homomorphism $\varphi_R\colon R \to R$ is the multiplication by a scalar $\lambda = \varphi(1)$. For every $R$-module $A$, the naturality of $\varphi$ on the homomorphism $f\colon R \to A$ that sends 1 to an element $x \in A$ gives the relation $\varphi_A(x) = \varphi_A(f(1)) = f(\varphi_R(1)) = f(\lambda) = \lambda x$.

### 9.1.15 Solutions of 1.5.2

(d) The claim follows from the naturality of $\psi\colon H \to K$ on the general component $\varphi X\colon FX \to GX$

$$\begin{array}{ccc} HFX & \xrightarrow{\psi FX} & KFX \\ {\scriptstyle H(\varphi X)}\downarrow & & \downarrow{\scriptstyle K(\varphi X)} \\ HGX & \xrightarrow[\psi GX]{} & KGX \end{array} \tag{9.12}$$

### 9.1.16 Solutions of 1.5.3

(a) A map $f\colon X \to Y$ gives squares of maps

$$\begin{array}{ccc} X & \xrightarrow{\partial^\alpha} & IX \\ {\scriptstyle f}\downarrow & & \downarrow{\scriptstyle If} \\ Y & \xrightarrow[\partial^\alpha]{} & IY \end{array} \qquad \begin{array}{ccc} IX & \xrightarrow{e} & X \\ {\scriptstyle If}\downarrow & & \downarrow{\scriptstyle f} \\ IY & \xrightarrow[e]{} & Y \end{array} \qquad \begin{array}{ccc} IX & \xrightarrow{r} & IX \\ {\scriptstyle If}\downarrow & & \downarrow{\scriptstyle If} \\ IY & \xrightarrow[r]{} & IY \end{array} \tag{9.13}$$

which are easily seen to be commutative.

(b) We have a natural isomorphism $\varphi\colon F \to G$. If $f \simeq g\colon X \to Y$ in $\mathsf{Top}$, the naturality condition of $\varphi$ on $f$ and $g$ gives two commutative squares

$$\begin{array}{ccc} FX & \xrightarrow{\varphi X} & GX \\ {\scriptstyle Ff}\downarrow\downarrow{\scriptstyle Fg} & & {\scriptstyle Gf}\downarrow\downarrow{\scriptstyle Gg} \\ FY & \xrightarrow[\varphi Y]{} & GY \end{array} \tag{9.14}$$

By hypothesis, $Ff = Fg$ and the components of $\varphi$ are invertible in $\mathsf{D}$. It follows that $Gf = Gg$.

### 9.1.17 Solutions of 1.5.4

(c) A natural transformation $\varphi\colon F \to G\colon \mathsf{C} \to \mathsf{D}$ gives a functor

$$\begin{gathered} \Phi\colon \mathsf{C} \times \mathbf{2} \to \mathsf{D}, \qquad \Phi(-,0) = F, \qquad \Phi(-,1) = G, \\ \Phi(f\colon X \to X', \iota\colon 0 \to 1) = \varphi(f)\colon F(X) \to G(X'), \end{gathered} \tag{9.15}$$

where $\iota\colon 0 \to 1$ is the non-identity arrow of **2** and $\varphi(f)$ is defined in (1.82). Similarly, we have a functor

$$\Phi'\colon \mathsf{C} \to \mathsf{D}^{\mathbf{2}}, \qquad \Phi'(X) = \varphi_X\colon F(X) \to G(X),$$
$$\Phi'(f\colon X \to X') = (Ff, Gf)\colon \varphi_X \to \varphi_{X'}. \tag{9.16}$$

Both procedures $\varphi \mapsto \Phi$ and $\varphi \mapsto \Phi'$ are invertible.

### *9.1.18 Solutions of 1.5.5*

(a) If $\mathsf{C}$ is non-empty and indiscrete, any object $X_0$ of $\mathsf{C}$ gives a constant functor $G\colon \mathbf{1} \to \mathsf{C}$ which is easily seen to be weakly inverse to $F$: we have $FG = \mathrm{id}\,\mathbf{1}$ and $GF \simeq \mathrm{id}\,\mathsf{C}$. Conversely, such a weak inverse gives a bijection between each hom-set $\mathsf{C}(X, Y)$ and a singleton.

### *9.1.19 Solutions of 1.6.4*

(a) First (i) implies (ii): the relation $up + vq = \mathrm{id}\,C$ is ensured by $p(up + vq) = p$ and $q(up + vq) = q$. Conversely, to show that (ii) implies (i), take two $R$-homomorphisms $f\colon X \to A$ and $g\colon X \to B$. If $h\colon X \to C$ satisfies $ph = f$ and $qh = g$, then $h = (up+vq)h = uf + vg$; on the other hand, the morphism $h = uf + vg\colon X \to C$ does have $f, g$ as components.

One proves in the same way that (i*) is equivalent to (ii).

### *9.1.20 Solutions of 1.6.5*

(a) The equivalence of these conditions on the object $Z$ is obvious. If there is such an object, any morphism $A \to B$ that factorises through $Z$ is the additive identity of the group $\mathsf{C}(A, B)$, by the distributivity of composition.

(b) Any full subcategory of $\mathsf{Ab}$ inherits a preadditive structure; taking all non-trivial groups, any object $A$ has at least two distinct endomorphisms: the trivial one and the identity.

(c) In $\mathsf{Rng}$ the polynomial ring $\mathbb{Z}[X]$ on one indeterminate is the initial object, while the trivial ring $\{0\}$ is the terminal object. In $\mathsf{Rng}'$ the trivial ring $\{0\}$ is initial and terminal.

(d) Using the description by components and co-components, the morphisms

$$f_1 = (1, 0)\colon X \to X \times Y, \qquad f_2 = (0, 1)\colon Y \to X \times Y,$$

give a morphism $f = [f_1, f_2]\colon X + Y \to X \times Y$ as required. It is not invertible in $\mathsf{Set}_\bullet$ and $\mathsf{Top}_\bullet$, except trivial cases, as we have seen in 1.6.2(d).

(e) In the same way as a category on a single object amounts to a monoid (cf. 1.3.2(c)).

*(f) A sum of sets $(X, u_i \colon X_i \to X)$ in Set is (easily proved to be) still a sum in the category of relations of sets. In the latter, by selfduality, $X$ is also the product of the family $(X_i)$, with projections given by the reverse relations $u_i{}^\sharp \colon X \to X_i$.

### *9.1.21 Solutions of 1.6.7*

(c) For a set $X$, we view an $n$-tuple $w = (x_1, x_2, ..., x_n) \in X^n$ as an $n$-ary *word* in the alphabet $X$. Two words are multiplied by *juxtaposition* (or *concatenation*): $(x_1, ..., x_p)(y_1, ..., y_q) = (x_1, ..., x_p, y_1, ..., y_q)$.

We are also considering the empty word $e = (\ ) \in X^0$, of length 0, which acts as a unit for this operation: $ew = w = we$, for every word $w$.

The disjoint union $M(X) = \sum_{n \in \mathbb{N}} X^n$ of all sets $X^n$ is thus a monoid. It is actually the *free monoid* on the set $X$, by the mapping 'of unary words'

$$\eta \colon X \to |M(X)|, \qquad \eta(x) = (x). \tag{9.17}$$

It is easy to verify that $(M(X), \eta)$ is a universal arrow from the set $X$ to the forgetful functor $U \colon \mathsf{Mon} \to \mathsf{Set}$ of the category of monoids.

Also here we can identify the element $x \in X$ with the unary word $\eta(x) = (x)$, so that each element of $M(X)$ can be written in a unique way as a finite product $x_1 x_2 ... x_n$ of elements of the base (including the empty product $e = \prod_{i \in \emptyset} x_i$).

### *9.1.22 Solutions of 1.7.2*

(a) These diagonal functors take an object $X$ of $\mathsf{C}$ to the pair $DX = (\mathrm{id}\, X, \mathrm{id}\, X)$, viewed in $\mathsf{C}^\wedge$ or $\mathsf{C}^\vee$. The rest is obvious.

(b) Suppose, for instance, that both squares are pushouts. Given two maps $w \colon Z \to T$ and $w' \colon X' \to T$ such that $whg = w'f$, the left pushout gives a morphism $k \colon Y' \to T$ such that $ku = wh$ and $ku' = w'$; then the right pushout gives a morphism $k' \colon Z' \to T$ such that $k'v = w$ and $k'v' = k$.

This is the morphism we are looking for: $k'v = w$ and $k'v'u' = ku' = u'$. Its uniqueness is verified in the same way, backwards.

### *9.1.23 Solutions of 1.7.5*

(a) In the notation of 1.7.1(a), the pullback-object $X$ of the pair of maps $f_i \colon X_i \to X_0$ is constructed as the equaliser of the obvious maps $f_i p_i \colon X_1 \times X_2 \to X_0$ (for $i = 1, 2$).

(b) The sum of two ordered lines is an ordered set having all pullbacks and pushouts, but lacks finite products and coproducts.

### *9.1.24 Solutions of 1.7.8*

(a) The filtered set $\mathcal{P}_{\mathrm{f}}(X)$ can be viewed as an inductive system of sets, indexed by itself. The set $X$, equipped with the inclusions $K \to X$ of its finite subsets, is the inductive limit of this system, because every $x \in X$ belongs to its singleton.

(b) This works in the same way, taking into account that the join $A' + A''$ of two finitely generated subgroups is finitely generated, and every $x \in A$ belongs to the subgroup generated by itself.

(c) Plainly, the join $A' + A''$ of two finite subgroups of the abelian group $A$ is finite. But the only finite subgroup of $\mathbb{Z}$ is the trivial one, and the inductive limit of this system is the trivial group.

In fact, the inductive limit of the system of finite subgroups of $A$ is its torsion subgroup.

(e) Every finite sum $X_J = \Sigma_{i\in J} X_i$ comes with a universal cocone $u_{iJ}\colon X_i \to X_J$ (for $i \in J$). If $J \subset K$ in $\mathcal{P}_{\mathrm{f}}(I)$, there is a unique consistent morphism $u_{JK}\colon X_J \to X_K$ (such that $u_{JK}u_{iJ} = u_{iK}$, when $i \in J \subset K$). Plainly, their family forms an inductive system over the filtered set $\mathcal{P}_{\mathrm{f}}(I)$, and we take its limit $X = \varinjlim_J X_J$, with universal cocone $u_J\colon X_J \to X$.

By straightforward verifications, the map $u_i = u_J u_{iJ}\colon X_i \to X$ does not depend on $J$ (for $i \in J$), and their family forms a universal cocone for the sum of all $X_i$.

## 9.2 Exercises of Chapter 2

### *9.2.1 Solutions of 2.1.2*

(a) First we prove that property (i) determines its solution, up to a coherent isomorphism (as any universal property). In fact, if we have two of them, $k_i\colon K_i \to A$ $(i = 1, 2)$, there are unique morphisms $u, v$ such that

$$u\colon K_1 \to K_2, \qquad v\colon K_2 \to K_1, \qquad k_2 u = k_1, \qquad k_1 v = k_2,$$

and $vu\colon K_1 \to K_1$ is the identity, because $k_1(vu) = k_2 u = k_1 = k_1(\mathrm{id}\, K_1)$. Similarly $uv = \mathrm{id}\, K_2$, so that $u, v$ are isomorphisms, inverse to each other.

Now, it is sufficient to remark that the embedding $\ker f\colon \mathrm{Ker}\, f \rightarrowtail A$ does satisfy (i): if $fh = 0$ the image of $h$ is contained in $\mathrm{Ker}\, f$, and we can (and must) define $u$ as the codomain-restriction of $h$.

(b) The uniqueness of the solution to (i*), up to isomorphism, is proved as above: we reverse all arrows, including $f$, and start from two solutions $c_i\colon B \to C_i$.

Then we remark that the projection $\mathrm{cok}\, f\colon B \twoheadrightarrow B/f(A)$ satisfies (i*): if $hf = 0$ the kernel of $h$ contains $f(A)$, and we can (and must) define $u$ as the homomorphism induced by $h$.

(c) The existence and uniqueness of the morphism $u'$ comes from the universal property of $\ker g$: in fact the composite $u(\ker f)$ annihilates $g$, because $gu(\ker f) = vf(\ker f) = 0$.

Similarly, the existence and uniqueness of the morphism $v'$ comes from the universal property of $\mathrm{cok}\, f$.

(d) We apply (c), taking into account that $\mathrm{Coim}\, f = \mathrm{Cok}\,(\ker f)$ and $\mathrm{Im}\, f = \mathrm{Ker}\,(\mathrm{cok}\, f)$.

(f) In $\mathsf{Set}_\bullet$ and $\mathsf{Top}_\bullet$ the kernel of a morphism is the preimage of the base-point, while the cokernel is the quotient of the codomain which collapses the image.

### 9.2.2 Solutions of 2.1.3

(a)–(f) Obvious.

(g) The canonical factorisations $f = mp$ and $g = nq$ of two consecutive homomorphisms give a commutative diagram (2.12), where $\mathrm{Ker}\, q = \mathrm{Ker}\, g$.

In this commutative diagram, the pair $(f, g)$ is exact if and only if $(m, q)$ is exact, $p$ is epi and $n$ is mono; adding $\mathrm{Ker}\, p$ and $\mathrm{Cok}\, n$, these three conditions are expressed by short exact sequences.

(h) In this exact sequence, $f$ is epi if and only if $\mathrm{Im}\, f = \mathrm{Ker}\, g$ is $B$, if and only if $g = 0$. Similarly $h$ is mono if and only if $\mathrm{Im}\, g = \mathrm{Ker}\, h$ is 0, if and only if $g = 0$. The rest is a consequence

### 9.2.3 Solutions of 2.1.6

(a) A straightforward consequence of Exercise 2.1.2(d). More concretely, one can apply condition (i′) of the Induction Lemma 2.1.4: the formula $\partial_n f_n(x) = f_{n-1}\partial_n(x)$ says that:

- a cycle $x \in Z_n(A)$ is sent to $f_n(x) \in Z_n(B)$,
- a boundary $\partial_n(x) \in B_{n-1}(A)$ is sent to the boundary $\partial_n f_n(x)$ in $B$.

(b) It is a straightforward application of the Induction Lemma 2.1.4: $B_n(f)$ is a restriction of $Z_n(f)$, and the latter induces $H_n(f)$ on the homology quotients.

### 9.2.4 Solutions of 2.1.7

(a) The reflexivity property $f \simeq f$ comes from the sequence of zero homomorphisms $0\colon A_n \to B_{n+1}$. For the symmetry property, we take the opposite components $-\varphi_n\colon A_n \to B_{n+1}$. For transitivity, we add the components of two homotopies $\varphi\colon f \to g$ and $\psi\colon g \to h$

$$\varphi + \psi\colon f \to h, \qquad (\varphi + \psi)_n = \varphi_n + \psi_n. \tag{9.18}$$

Consistence with composition can be readily verified on the whisker composition of $\varphi\colon f \to g\colon A \to B$ with two chain morphisms $h\colon A' \to A$, $k\colon B \to B'$. Then the sequence of homomorphisms

$$\psi_n = k_{n+1}\varphi_n h_n\colon A'_n \to B'_{n+1} \tag{9.19}$$

gives a homotopy $\psi\colon kfh \to kgh$

$$\begin{aligned}
\partial_{n+1}\psi_n + \psi_{n-1}\partial_n &= \partial_{n+1}(k_{n+1}\varphi_n h_n) + (k_n\varphi_{n-1}h_{n-1})\partial_n \\
&= (k_n\partial_{n+1}\varphi_n h_n) + (k_n\varphi_{n-1}\partial_n h_n) \\
&= k_n(\partial_{n+1}\varphi_n + \varphi_{n-1}\partial_n)h_n = k_n(g_n - f_n)h_n.
\end{aligned}$$

(b) We have a homotopy $\varphi\colon f \to g\colon A \to B$. For every cycle $z \in Z_n(A)$, the homology class $[z] \in H_n(A)$ gives

$$H_n(g)[z] - H_n(f)[z] = [g_n(z) - f_n(z)] = [\partial_{n+1}\varphi_n(z) + \varphi_{n-1}\partial_n(z)] = 0,$$

because $z$ is an $n$-cycle and $\partial_{n+1}\varphi_n(z)$ is an $n$-boundary.

### 9.2.5 Solutions of 2.1.9

(a)–(c). Omitting degrees, we have:

$$\begin{aligned}
\partial\partial(a, x, b) &= \partial(\partial a - x,\ -\partial x,\ \partial b + x) \\
&= (\partial\partial a - \partial x + \partial x,\ \partial\partial x,\ \partial\partial b + \partial x - \partial x) = 0,
\end{aligned}$$

$$\begin{aligned}
\partial(If)(a, x, b) &= (\partial f(a) - f(x),\ -\partial f(x),\ \partial f(b) + f(x)) \\
&= (f\partial a - f(x),\ -f\partial(x),\ f\partial b + f(x)) = (If)\partial(a, x, b),
\end{aligned}$$

$$\partial^-\partial(a) = (\partial a, 0, 0) = \partial\partial^-(a), \quad \partial^+\partial(a) = (0, 0, \partial a) = \partial\partial^+(a),$$

$$e\partial(a, x, b) = e(\partial a - x,\ -\partial x,\ \partial b + x) = \partial a + \partial b = \partial e(a, x, b),$$

$$\begin{aligned}
r\partial(a, x, b) &= r(\partial a - x,\ -\partial x,\ \partial b + x) = (\partial b + x,\ \partial x,\ \partial a - x) \\
&= \partial(b, -x, a) = \partial r(a, x, b).
\end{aligned}$$

(d) We take a general chain morphism $\Phi = [f, \varphi, g] \colon IA \to B$, where the component of degree $n$

$$\begin{aligned}\Phi_n = [f_n, \varphi_{n-1}, g_n] \colon (IA)_n &\to B_n, \\ (a, x, b) &\mapsto f_n(a) + \varphi_{n-1}(x) + g_n(b),\end{aligned} \tag{9.20}$$

is written as in 1.6.2, by its co-components in Ab

$$f_n \colon A_n \to B_n, \quad \varphi_{n-1} \colon A_{n-1} \to B_n, \quad g_n \colon A_n \to B_n.$$

We now analyse the condition $\partial_{n+1}\Phi_{n+1} = \Phi_n\partial_{n+1}$ on a general element $(a, x, b) \in (IA)_{n+1}$

$$\begin{aligned}\partial_{n+1}\Phi_{n+1}(a, x, b) &= \partial_{n+1}(f_{n+1}a + \varphi_n x + g_{n+1}b) \\ &= \partial_{n+1}f_{n+1}a + \partial_{n+1}\varphi_n x + \partial_{n+1}g_{n+1}b,\end{aligned}$$

$$\begin{aligned}\Phi_n\partial_{n+1}(a, x, b) &= \Phi_n(\partial_{n+1}a - x, -\partial_n x, \partial_{n+1}b + x) \\ &= f_n\partial_{n+1}a - f_n x - \varphi_{n-1}\partial_n x + g_n\partial_{n+1}b + g_n x.\end{aligned}$$

Their equality is equivalent to saying that:

- $\partial_{n+1}f_{n+1}(a) = f_n\partial_{n+1}(a)$,
- $\partial_{n+1}\varphi_n(x) + \varphi_{n-1}\partial_n(x) = g_n(x) - f_n(x)$,
- $\partial_{n+1}g_{n+1}(b) = g_n\partial_{n+1}(b)$,

for all $a \in A_{n+1}$, $x \in A_n$ and $b \in A_{n+1}$.

Finally, giving the chain morphism $\Phi$ is equivalent to giving two chain morphisms $f, g \colon A \to B$ and a chain homotopy $\varphi \colon f \simeq g$.

### *9.2.6 Solutions of 2.2.1*

(a) Letting $j \leqslant i$ and applying $\delta_j^\beta\delta_i^\alpha$ to $(t_1, ..., t_{n-1})$ we get:

$$\begin{aligned}\delta_j^\beta\delta_i^\alpha(t_1, ..., t_{n-1}) &= \delta_j^\beta(t_1, ..., t_{i-1}, \alpha, ..., t_{n-1}) \\ &= (t_1, ..., t_{j-1}, \beta, ..., t_{i-1}, \alpha, ..., t_{n-1}),\end{aligned}$$

where, in the last step, the inserted $\alpha$ has been shifted to position $i + 1$. This proves that $\delta_j^\beta\delta_i^\alpha = \delta_{i+1}^\alpha\delta_j^\beta$.

The other verifications are similar.

### *9.2.7 Solutions of 2.2.3*

(a) We form a diagram as in (2.14)

$$\begin{array}{ccccc} \mathbb{Z}\mathrm{Deg}_n K & \rightarrowtail & \mathbb{Z}K_n & \twoheadrightarrow & C_n(K) \\ {\scriptstyle f'}\downarrow & & \downarrow{\scriptstyle f} & & \downarrow{\scriptstyle f''} \\ \mathbb{Z}\mathrm{Deg}_{n-1} K & \rightarrowtail & \mathbb{Z}K_{n-1} & \twoheadrightarrow & C_{n-1}(K) \end{array} \qquad (9.21)$$

where $f = \partial_n$ is defined on the canonical basis as

$$\partial_n(a) = \Sigma_{i,\alpha}\,(-1)^{i+\alpha}\,\partial_i^\alpha a.$$

This homomorphism has a restriction $f'$, because of (2.45). By the Induction Lemma 2.1.4, there is a unique homomorphism $f''\colon C_n(K) \to C_{n-1}(K)$ that makes diagram (9.21) commutative.

(b) For a generator $a \in K_{n+1}$ we have

$$\partial_n\partial_{n+1}(a) = \Sigma_{i,j}\,(-1)^{i+j+\alpha+\beta}\,\partial_i^\alpha\partial_j^\beta(a), \qquad (9.22)$$

where $\alpha, \beta$ take the values $0, 1$, and $i = 1, ..., n$, $j = 1, ..., n+1$.

This sum can be split in two parts, for $j \leqslant i$ and $j > i$, respectively

$$\begin{aligned} \partial_n\partial_{n+1}(a) = {} & \Sigma_{j \leqslant i}\,(-1)^{i+j+\alpha+\beta}\,\partial_i^\alpha\partial_j^\beta(a) \\ & + \Sigma_{j>i}\,(-1)^{i+j+\alpha+\beta}\,\partial_i^\alpha\partial_j^\beta(a). \end{aligned}$$

Working on the left addendum, we apply the cubical relations of the faces ($\partial_i^\alpha\partial_j^\beta = \partial_j^\beta\partial_{i+1}^\alpha$ for $j \leqslant i$), and then a change of variables on the indices (replacing $j$ with $i$, and $i$ with $j-1$)

$$\begin{aligned} \Sigma_{j \leqslant i}\,(-1)^{i+j+\alpha+\beta}\,\partial_i^\alpha\partial_j^\beta(a) &= \Sigma_{j \leqslant i}\,(-1)^{i+j+\alpha+\beta}\,\partial_j^\beta\partial_{i+1}^\alpha(a) \\ &= -\,\Sigma_{i<j}\,(-1)^{i+j+\alpha+\beta}\,\partial_i^\alpha\partial_j^\beta(a). \end{aligned}$$

The left addendum is thus the additive inverse of the right one, and $\partial_n\partial_{n+1}(a) = 0$.

(c) If $A = B \cup C$ is a disjoint union of sets, $\mathbb{Z}A$ is the internal direct sum of two subgroups, which can be identified with $\mathbb{Z}B$ and $\mathbb{Z}C$, so that $\mathbb{Z}A/\mathbb{Z}B \cong \mathbb{Z}C$. This is applied to the set $A = K_n$, its subset $B = \mathrm{Deg}_n K$ and the complement $C = A \setminus B$.

### *9.2.8 Solutions of 2.2.5*

(a) There are no mappings $\mathbb{I}^n \to \emptyset$. Therefore $C_+(\emptyset)$ is the trivial chain complex: all its components are 0, and all homology groups are 0.

For the singleton, there is a unique mapping $a_n\colon \mathbb{I}^n \to \{*\}$, for $n \geqslant 0$.

The differential of the *non-normalised* chain complex always involves an even number of summands (the cube $\mathbb{I}^n$ has $2n$ faces)

$$\partial_n(a_n) = \Sigma_{i,\alpha}(-1)^{i+\alpha}a_{n-1} = 0,$$

and all its homology groups are isomorphic to $\mathbb{Z}$. But we are interested in the normalised chain complex, and every generator $a_n$ is degenerate in positive degree: $a_n = e_i(a_{n-1})$, for $n > 0$ and any $i$.

Thus the cubical chain complex $C_+\{*\}$ of the singleton is isomorphic to

$$\ldots\ 0 \to 0 \to 0 \to \mathbb{Z} \tag{9.23}$$

and its homology is $\mathbb{Z}$ in degree 0, and trivial elsewhere.

(b) Here it is convenient to use the 'augmented' chain complex

$$\ldots\ C_2(X) \xrightarrow{\partial_2} C_1(X) \xrightarrow{\partial_1} C_0(X) \xrightarrow{\tilde{\partial}_0} \mathbb{Z} \tag{9.24}$$

where we have added the homomorphism $\tilde{\partial}_0(\Sigma\,\lambda_i x_i) = \Sigma\,\lambda_i$. Plainly $\tilde{\partial}_0\partial_1 = 0$, because the boundary of a 1-cube is the formal difference of two points, with coefficients 1 and $-1$.

The homomorphism $\tilde{\partial}_0$ is surjective because $X \neq \emptyset$. Moreover $\operatorname{Ker}\tilde{\partial}_0 = \operatorname{Im}\partial_1$ because $X$ is path-connected: if $c = \Sigma\,\lambda_i x_i$ $(i = 1, ..., n)$ and $\Sigma\,\lambda_i = 0$, we fix a point $x_0 \in X$ and a path $a_i$ from $x_0$ to $x_i$; then the chain $c' = \Sigma\,\lambda_i a_i$ gives $\partial_1(c') = \Sigma\,\lambda_i x_i - \Sigma\,\lambda_i x_0 = c$, proving that $c \in \operatorname{Im}\partial_1$.

Therefore $\tilde{\partial}_0$ induces a Noether isomorphism on $H_0(X) = C_0(X)/\operatorname{Im}\partial_1$

$$H_0(X) \to \mathbb{Z}, \qquad [x] \mapsto 1_{\mathbb{Z}}. \tag{9.25}$$

(c) A cube $a\colon \mathbb{I}^n \to X$ has a path-connected image, contained in one path component $X_i$; its faces 'belong' to the same component. We also note that $a$ is degenerate in $X$ if and only if it is as a cube of $X_i$.

All this proves the canonical isomorphism of chain complexes in (2.58). Applying the functor $H_n$ we get an isomorphism

$$H_n(\textstyle\bigoplus_{i\in I} C_+(X_i)) \ \to\ H_n(X), \qquad [(z_i)_{i\in I}] \mapsto \Sigma_{i\in I}\,[z_i]. \tag{9.26}$$

We also have the isomorphism (2.29)

$$H_n(\textstyle\bigoplus_{i\in I} C_+(X_i)) \to \bigoplus_{i\in I} H_n(C_+(X_i)), \qquad [(z_i)_{i\in I}] \mapsto ([z_i])_{i\in I}, \tag{9.27}$$

and composing the inverse of the latter with (9.26) we obtain the isomorphism we are looking for: it sends the essentially finite family $([z_i])_{i\in I}$ to the homology class $\Sigma_{i\in I}\,[z_i]$ of $H_n(X)$.

(d) A consequence of (b) and (c). Without using (c), one can extend the argument of (b) using the homomorphism

$$\tilde{\partial}_0\colon C_0(X) \to \mathbb{Z}(\Pi_0(X)), \qquad x \mapsto [x]_P,$$

(whose kernel is the image of $\partial_1 \colon C_1(X) \to C_0(X)$), and its naturality in the variable $X$.

(e) A consequence of (a), (b) and (c).

### 9.2.9 Solutions of 2.2.8

(b) Let $h \colon \mathbb{I}^2 \to X$ be a homotopy with fixed endpoints, from $a$ to $b$, as in (1.75). Then the boundary of the 2-cube $h$ is the 1-chain $a - b$

$$\partial_2(h) = -e_x + e_y + a - b = a - b, \tag{9.28}$$

as the degenerate paths $e_x = e_1(x)$ and $e_y = e_1(y)$ are zero in $C_1(X)$.

(c) Similarly, we take a loop homotopy $h \colon \mathbb{I}^2 \to X$ from $a$ to $b$, as in (1.80). The boundary of the 2-cube $h$ is again the 1-chain $a - b$

$$\partial_2(h) = -c + c + a - b = a - b, \tag{9.29}$$

but here $a$ and $b$ are loops, and we have proved that $[a] = [b]$.

(d) A consequence of (c), as a constant loop is a degenerate 1-cube.

(e) Starting from the concatenated path $c = a * b$ of $X$, we construct a 2-cube $h \colon \mathbb{I}^2 \to X$ whose boundary $\partial_2 h$ is precisely the chain $a + b - c$

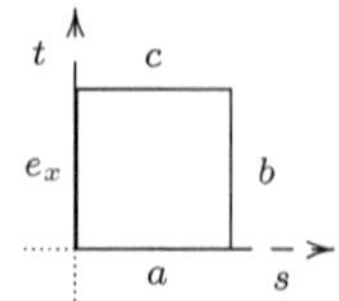


$$h(s, t) = c(s(1 + t)/2).$$

(f) The following 2-cube $h \colon \mathbb{I}^2 \to X$ has $\partial_2 h = a + a^\sharp$

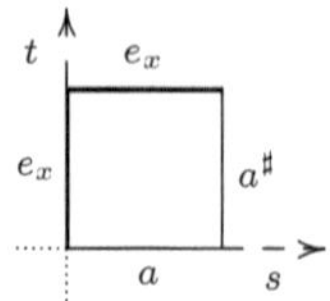


$$h(s, t) = a(\min(s, 1 - t)).$$

### 9.2.10 Solution of 2.3.6(a)

The pullback property is plainly satisfied, because $U, V$ are subspaces of $X$ and $A = U \cap V$. The pushout property is obviously satisfied in Set, and also in Top by the Cover Lemma 1.1.8(a).

### 9.2.11 Solutions of 2.3.8

(a) The composite $k'_{*n}h'_{*n}\colon H_n(A) \to H_n(C)$ is obviously zero, because $k'_n h'_n\colon A_n \to C_n$ already is.

Suppose now that $b \in Z_n(B)$ and $k'_{*n}[b] = [k'_n(b)] = 0$. Thus $k'_n(b) = \partial_{n+1}(\hat{c})$, for some $\hat{c} \in C_{n+1}$, which can be lifted to $\hat{b} \in B_{n+1}$. We have formed the right part of the following diagram

$$\begin{array}{ccccccccc}
 & & & & & \hat{b} & \mapsto & \hat{c} & (n+1) \\
 & & & & & & & \downarrow & \\
\hat{a} & \mapsto & b - \partial\hat{b} & \mapsto & 0 & b & \mapsto & c & (n) \\
\downarrow & & \downarrow & & & \downarrow & & & \\
\partial\hat{a} & \mapsto & \partial b = 0 & & & 0 & & & (n-1)
\end{array}$$

The cycle $b - \partial_{n+1}\hat{b}$ belongs to $\operatorname{Ker} k'_n = \operatorname{Im} h'_n$, and can be lifted to an element $\hat{a} \in A_n$; the commutative left square shows that $k'_{n-1}(\partial_n\hat{a}) = \partial_n(b - \partial_{n+1}\hat{b}) = 0$; therefore $\partial_n\hat{a} = 0$ (as $k'_{n-1}$ is mono), and the cycle $\hat{a}$ works as expected

$$h'_{*n}[\hat{a}] = [b - \partial_{n+1}\hat{b}] = [b].$$

(b) The composite $D_n k'_{*n}\colon H_n(B) \to H_{n-1}(A)$ is zero, as shown by the following diagram, where we are assuming that $b \in B_n$ is a cycle, so that $D_n k'_{*n}[b] = D_n[c] = [0]$

$$\begin{array}{cccccc}
 & & b & \mapsto & c & (n) \\
 & & \downarrow & & \downarrow & \\
0 & \mapsto & 0 & \mapsto & 0 & (n-1)
\end{array}$$

Suppose now that $c \in Z_n(C)$ and $D_n[c] = [a] = 0$. Thus $a = \partial_{n+1}(\hat{a})$, for some $\hat{a} \in A_{n+1}$

$$\begin{array}{cccccc}
\hat{a} & & b & \mapsto & c & (n) \\
\downarrow & & \downarrow & & \downarrow & \\
a & \mapsto & \partial b & \mapsto & 0 & (n-1)
\end{array}$$

It follows that $b - h'_n(\hat{a})$ is a cycle, and

$$k'_{*n}[b - h'_n(\hat{a})] = [k'_n(b) - k'_n h'_n(\hat{a})] = [k'_n(b)] = [c].$$

### 9.2.12 Solutions of 2.4.5

(a) Apply the MV-sequence, with $H_n(U) = H_n(V) = 0$ for $n \geqslant 1$.

(b) For $H_1(\Sigma X)$ we use the following portion of the MV-sequence, where $A \simeq X$ is path connected and $h_0$ is mono

$$0 \to H_1(\Sigma X) \xrightarrow{D_1} H_0(A) \xrightarrow{h_0} H_0(U) \oplus H_0(V).$$

### 9.2.13 Solutions of 2.5.2

(a) Obvious, as $f_{*n} = g_{*n}$.

(b) In fact:

$$(fg)_{*n}(\zeta) = f_{*n}g_{*n}(\zeta) = (\deg f)\,(\deg g)\,\zeta, \qquad (\mathrm{id}\,\mathbb{S}^n)_{*n} = \mathrm{id}\,H_n(\mathbb{S}^n).$$

(c) Let $g\colon \mathbb{S}^n \to \mathbb{S}^n$ be a weak inverse of $f$. By the previous point $(\deg f)\,(\deg g) = 1$, and these numbers are both 1, or both $-1$.

(d) Same proof as for 2.4.4(b).

(e) We use the generator $\zeta = [a] \in H_1(\mathbb{S}^1)$, given by the loop $a(t) = e^{2\pi it}$ (as in 2.4.1(d)).

Thus $fk_*[a] = [f_k a] = [a_k]$ is the homology class of the loop $a_k(t) = e^{2k\pi it}$, computed as $k\zeta$ in Exercise 2.4.1(e).

(f) It is an obvious consequence of the naturality of the isomorphism $\delta_n X\colon H_{n+1}(\Sigma X) \to H_n(X)$, in 2.4.5

$$\begin{array}{ccc} H_{n+1}(\Sigma X) & \longrightarrow & H_n(X) \\ {\scriptstyle (\Sigma f)_*}\downarrow & & \downarrow{\scriptstyle f_*} \\ H_{n+1}(\Sigma X) & \longrightarrow & H_n(X) \end{array} \qquad (9.30)$$

### 9.2.14 Solutions of 2.5.5

(a) Applying Exercises 2.2.8(f) and 2.2.8(e)

$$[z_1] = [a' + b'^{\sharp} + a^{\sharp} + b] = [a' * b'^{\sharp} * a^{\sharp} * b],$$

we find a loop in $S^1$, that corresponds to the loop $t \mapsto e^{2\pi it}$ $(t \in \mathbb{I})$ of the circle, used in Exercise 2.4.1(d) as a generator of $H_1(\mathbb{S}^1)$.

(b) For $n \geqslant 2$ we remark that the lower face $\partial^-_{n+1}$ of $f_{n+1} = \mathrm{id}\,\mathbb{I}^{n+1}$

$$\partial^-_{n+1} f_{n+1}\colon \mathbb{I}^n \to \mathbb{I}^{n+1}, \qquad (t_1, ..., t_n) \mapsto (t_1, ..., t_n, 0), \qquad (9.31)$$

coincides with $f_n = \mathrm{id}\,\mathbb{I}^n$, as an $n$-cube of $\mathbb{I}^{n+1}$ under the usual nesting of all $\mathbb{R}^n$.

We suppose that $[\partial_n f_n]$ is a generator of $H_{n-1}(S^{n-1})$ and prove that the isomorphism $D_n\colon H_n(\mathbb{S}^n) \to H_{n-1}(A)$ of the MV-sequence used in Theorem 2.4.2 takes $[\partial_{n+1} f_{n+1}]$ to a generator of $H_{n-1}(A) \cong H_{n-1}(S^{n-1})$.

In fact the cycle $z_n = \partial_{n+1} f_{n+1} \in C_n(S^n)$, a linear combination of $n$-cubes, can be split leaving $\partial^-_{n+1} f_{n+1} = f_n$ in $C_n(V)$ and all the others in $C_n(U)$

$$\partial_{n+1} f_{n+1} = c - c',$$
$$c = \Sigma_{i \leqslant n} (-1)^{i+\alpha} \, \partial_i^\alpha f_{n+1} + (-1)^n \, \partial^+_{n+1} f_{n+1} \in C_n(U),$$
$$c' = (-1)^{n+1} f_n \in C_n(V).$$

Finally $D_n[\partial_{n+1} f_{n+1}] = [\partial_n c'] = (-1)^{n+1} [\partial_n f_n] = (-1)^{n+1} [z_{n-1}]$.

(c) We form the composite of two homeomorphisms:

$$h_n = h''h' \colon \mathbb{I}^n \to [-1,1]^n \to \mathbb{D}^n,$$

$$h'(x) = 2x - (1,1,...,1), \qquad h''(x) = \begin{cases} (||x||_\infty/||x||)x, & \text{for } x \neq 0, \\ 0 & \text{for } x = 0. \end{cases}$$

Here $||x||_\infty = \max(|x_1|, ..., |x_n|)$ is the $l_\infty$-norm of $\mathbb{R}^n$, which determines the cube

$$[-1,1]^n = \{x \in \mathbb{R}^n \mid ||x||_\infty \leqslant 1\}.$$

The map $h'' \colon [-1,1]^n \to \mathbb{D}^n$ takes every point $x$ of the border of $[-1,1]^n$ to its normalised point $x/||x|| \in \mathbb{S}^{n-1}$, and is linear on every line through the origin.

(d) We use now, for $\mathbb{S}^n$, the MV-sequence of Theorem 2.4.2. As in (b), we suppose that $[z'_{n-1}]$ is a generator of $H_{n-1}(\mathbb{S}^{n-1})$ and prove that the isomorphism $D_n \colon H_n(\mathbb{S}^n) \to H_{n-1}(A)$ takes $[z'_n]$ to a generator of $H_{n-1}(A) \cong H_{n-1}(\mathbb{S}^{n-1})$.

Here $z'_n = a_n - i_{n+1} a_n$ already has $a_n \in C_n(U)$ and $i_{n+1} a_n \in C_n(V)$. We already remarked that $a_n = u_n h_n \colon \mathbb{I}^n \to \mathbb{D}^n \to \mathbb{S}^n$ coincides with $h_n$ on the border of $\mathbb{I}^n$, so that $D_n[z'_n] = [\partial_n a_n] = [\partial_n h_n]$ is the generator of $H_{n-1}(\mathbb{S}^{n-1})$ constructed in the previous point.

(e) The map $\sigma_n \colon \mathbb{S}^n \to \mathbb{S}^n$ can be defined as a composite $\sigma''_n \sigma'_n \colon \mathbb{S}^n \to S'^n \to \mathbb{S}^n$, through the sphere $S'^n$ of centre $x_0 = (0, ..., 0, 1/2)$ and radius 1/2.

On the upper hemisphere, $\sigma'_n(x)$ is the intersection of the half-line from the origin to $x$, with the inner sphere $S'^n$

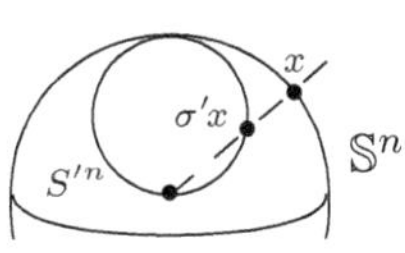

while the lower hemisphere is collapsed to the origin, $\sigma'_n(\mathbb{S}^{n-1}) = 0$; then we send $S'^n$ to $\mathbb{S}^n$, by the obvious homeomorphism $\sigma''_n(x) = 2(x - x_0)$.

This map is homotopic to the identity, and induces the identity of $H_n(\mathbb{S}^n)$, sending $a_n$ to $\sigma_n a_n = b_n$ and collapsing $i_{n+1}a_n$ to a constant cube. The generator $[z'_n]$ is thus sent to $[z'_n] = [b_n]$, the same generator.

## 9.3 Exercises of Chapter 3

### 9.3.1 Solutions of 3.1.2

(a) If $X$ is empty, $H_0(X) = 0$ and there is nothing to prove.

Otherwise $H_0(X)$ is the infinite cyclic group generated by the homology class $[x]_X$ of a point $x \in X$; this is sent by $f_{*0}\colon H_0(X) \to H_0(Y)$ to the homology class $[f(x)]_Y$, an element of the canonical basis of $H_0(Y)$ (by Exercise 2.2.5(d)). This proves that $f_{*0}$ is a monomorphism.

(b) For any $x \in X$, the homology class $f_{*0}([x]_X) = [f(x)]_Y$ generates $H_0(Y)$.

(c) The first claim follows from (a): the components of $h_0 = (i_{*0}, j_{*0})$ are injective. The rest is a consequence (as in 2.1.3(h)) of the exact sequence

$$H_1(U) \oplus H_1(V) \xrightarrow{k_1} H_1(X) \xrightarrow{D_1} H_0(A) \xrightarrow{h_0} H_0(U) \oplus H_0(V).$$

### 9.3.2 Solutions of 3.1.3

(a) After the homology groups of the circle and the figure-eight space, we can easily guess that $H_1(C_n)$ is isomorphic to $\mathbb{Z}^n$, with a basis formed by the homology classes of $n$ simple loops, one in each circle, while the higher homology groups of $C_n$ are trivial.

This can be proved by induction on $n \geqslant 1$. The initial step being granted, we assume that our claim holds for $n-1$ and prove it for $n \geqslant 2$, using the MV-sequence for the following open cover $(U, V)$ (drawn for $n = 3$)

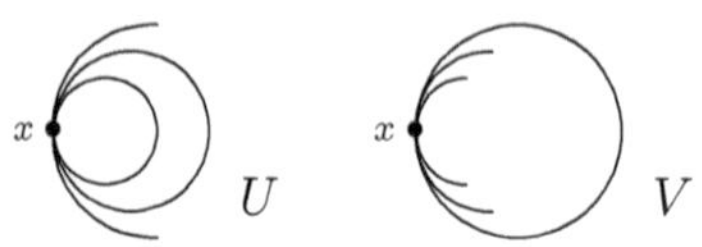


$$U \simeq C_{n-1}, \quad V \simeq \mathbb{S}^1, \quad A = U \cap V \simeq \{*\}, \tag{9.32}$$

where $U$ is the union of $C_{n-1}$ with a contractible open nbd $A$ of $x$ in $C_n$, and $V$ is the union of the last circle of $C_n$ with $A$.

All homology groups $H_k(C_n)$ are trivial for $k > 1$, as in 3.1.1.

For $H_1(C_n)$ we proceed as in 3.1.1, with the exact sequence

$$H_1(A) \xrightarrow{h_1} H_1(U) \oplus H_1(V) \xrightarrow{k_1} H_1(C_n).$$

Again, $H_1(A) = 0$ and $k_1$ is mono. But $k_1$ is also epi, because $A$ is path connected (we apply Exercise 3.1.2(c)). Finally, using the inductive hypothesis, we have an isomorphism

$$H_1(C_n) \cong H_1(C_{n-1}) \oplus H_1(\mathbb{S}^1) \cong \mathbb{Z}^{n-1} \oplus \mathbb{Z},$$

that also proves our claim on the basis of $H_1(C_n)$.

(b) One can easily find an open cover $(U, V)$ of $X$ with:

$$U \simeq C_2, \qquad V \simeq \mathbb{S}^1, \qquad A = U \cap V \simeq \{*\},$$

which gives the same results that we found for the space $C_3$, namely:

$$H_1(X) \cong \mathbb{Z}^3, \qquad H_k(X) = 0 \quad \text{for } k > 1.$$

To prove that $C_3$ and $X$ are not homeomorphic it is sufficient to note that $C_3$ has a single point that disconnects it, while $X$ has two of them.

### *9.3.3 Solutions of 3.1.4*

(a) Let us note that, for $j \neq i$, the subset $u_j^{-1}(u_i(U_i))$ is $\{0_{X_j}\}$ in case (i) and $\emptyset$ in case (ii).

A subset $V$ of form (i) or (ii) is always open in $X$, because in both cases we have, for all $j$: $u_j^{-1}(V) = \bigcup_i u_j^{-1}(u_i(U_i)) = U_j$.

Conversely, every open subset $V$ of $X$ can be written as:

$$V = \bigcup_i (V \cap \operatorname{Im} u_i) = \bigcup_i u_i(U_i),$$

where each $U_i = u_i^{-1}(V)$ is open in $X_i$. Thus $V$ is either of form (i) (when it contains the basepoint), or of form (ii) (when it does not).

(b) Let $U$ be open in $X_i$ and $V = u_i(U)$. If $0_{X_i} \notin U$, we have seen in (a) that the subset $V$ is open in $X$ (of form (ii)), and also open in $u_i(X_i)$.

Otherwise $0_{X_i} \in U$ and $0_X \in V$. The subset $V' = V \cup (\bigcup_{j \neq i} u_j(X_j))$ is an open subset of $X$ of form (i). Now $V = V' \cap u_i(X_i)$ is just open in $u_i(X_i)$.

(c) A consequence of (b): the pointed sum $X$ is the union of its subspaces $u_i(X_i)$, which meet at the basepoint $0_X$; each of them is homeomorphic to $X_i$, and path connected (resp. connected).

(d) The space $\bigvee_{n>0} \mathbb{S}^0$ is homeomorphic to the discrete space $\mathbb{N}$. Its canonical image in the countable product $\prod_{n>0} \mathbb{S}^0$ has the cofinite topology, strictly coarser.

### 9.3.4 Solutions of 3.1.7

(a) Applying Theorem 3.1.6, the pointed sum $\mathbb{S}^1 \vee \mathbb{S}^2 \vee ... \vee \mathbb{S}^n$ satisfies our condition.

(b) One can take the wedge of $\beta_1$ copies of $\mathbb{S}^1$, $\beta_2$ copies of $\mathbb{S}^2$, ..., and $\beta_n$ copies of $\mathbb{S}^n$. Finally, one can add $\beta_0 - 1$ copies of $\mathbb{S}^0$, in the pointed sum.

### 9.3.5 Solutions of 3.1.8

(a) We use the following open cover $(U, V)$ of $P_n$

$$\begin{aligned} U = P_n \setminus \{0\} \simeq \mathbb{S}^1, \qquad V = \text{int}\,\mathbb{D}^2 \simeq \{*\}, \\ A = U \cap V = \text{int}\,\mathbb{D}^2 \setminus \{0\} \simeq \mathbb{S}^1. \end{aligned} \tag{9.33}$$

The only non-obvious claim is the relation $P_n \setminus \{0\} \simeq \mathbb{S}^1$. We begin by observing that the homotopy equivalence $\mathbb{D}^2 \setminus \{0\} \simeq \mathbb{S}^1$ we used in 1.1.3(d) does not move the points of $\mathbb{S}^1$; therefore it induces a homotopy equivalence $P_n \setminus \{0\} \simeq \mathbb{S}^1/\mathbb{Z}_n$, with the orbit space of the action of $\mathbb{Z}_n$ on $\mathbb{S}^1$.

Now, this space is homeomorphic to $\mathbb{S}^1$: the surjective endomap $f(z) = z^n$ (a power in $\mathbb{C}$) induces a bijective map $\mathbb{S}^1/\mathbb{Z}_n \to \mathbb{S}^1$, which is a homeomorphism (from a compact space to a Hausdorff one).

Loosely speaking, winding the circle on itself, $n$ times, we still get the circle.

To determine $H_1(P_n)$ we use the following exact sequence (omitting the trivial group $H_1(V)$)

$$H_1(A) \xrightarrow{h_1} H_1(U) \xrightarrow{k_1} H_1(P_n)$$

where $k_1$ is epi (by Exercise 3.1.2(c)), and therefore a cokernel of $h_1$.

We have thus to compute the homomorphism $h_1$. For $H_1(A) \cong H_1(\mathbb{S}^1)$ we choose the generator $\zeta = [a]_A$, where $a$ is the simple loop drawn below, at the left

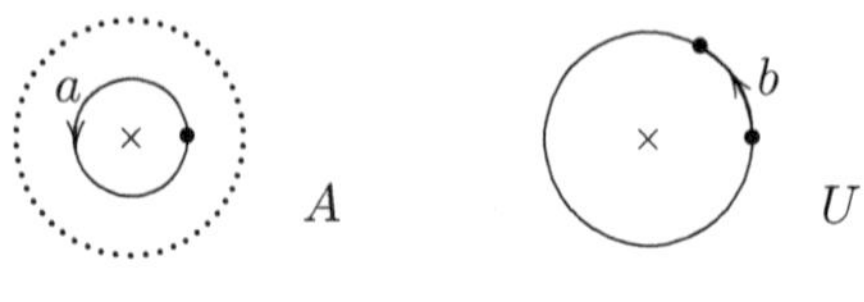


(9.34)

For $H_1(U)$ we choose the generator $\zeta' = [b]_U$, where $b$ is a simple loop of $U$, the image of a path in the border of $\mathbb{D}^2$, from 1 to $e^{2\pi i/n}$. (The right figure is drawn for $n = 6$.)

Now $h_1[a]_A = [a]_U = n[b]_U$, because in $U$ the loop $a$ is loop homotopic to the image in $P_n$ of the simple loop around the border of $\mathbb{D}^2$, which is the concatenation of $n$ loops equal to $b$.

The homomorphism $h_1$ is thus, essentially, the multiplication by $n$ in an infinite cyclic group. Its cokernel can be realised as $\mathbb{Z}_n$

$$H_1(A) \xrightarrow{h_1} H_1(U) \xrightarrow{p} \mathbb{Z}_n$$

by the projection $p$ taking $\zeta' = [b]_U$ to the class $\overline{1}$: an epimorphism whose kernel is generated by $n[b]_U = h_1[a]_A$.

This proves that $H_1(P_n) \cong \mathbb{Z}_n$, with generator $[b]_{P_n}$.

(b) For $i > 1$ we have the exact sequence

$$H_i(U) \xrightarrow{k_i} H_i(P_n) \xrightarrow{D_i} H_{i-1}(A) \xrightarrow{h_{i-1}} H_{i-1}(U) \oplus H_{i-1}(V)$$

where $H_iU = 0$. We distinguish two cases, each of them giving $H_i(P_n) = 0$

- if $i > 2$: $H_{i-1}(A) = 0$, and $H_i(P_n) = 0$,

- if $i = 2$: $h_{i-1} = h_1$ has been computed above, and is a monomorphism (the multiplication by $n$ in an infinite cyclic group); therefore $H_2(P_n) = \operatorname{Im} D_2 = \operatorname{Ker} h_1 = 0$.

### 9.3.6 Solutions of 3.1.9

(a) We can use the pointed sum

$$X = P_{\tau_1} \vee P_{\tau_2} \vee ... \vee P_{\tau_k}. \tag{9.35}$$

In fact, the origin has a contractible nbd in $P_n$, the open disc $\operatorname{int} \mathbb{D}^2$. Applying Theorem 3.1.6 and Exercise 3.1.8(a) we get, for $n > 1$

$$H_1(X) \cong \mathbb{Z}_{\tau_1} \oplus \mathbb{Z}_{\tau_2} \oplus ... \oplus \mathbb{Z}_{\tau_k}, \qquad H_n(X) = 0. \tag{9.36}$$

(b) One can use the suspension $Y = \Sigma^{n-1}X$ of the previous space. Applying (2.121) and (2.123):

$$H_k(Y) \cong H_{k-n}(X), \quad \text{for } k \geqslant n, \qquad H_k(Y) = 0, \quad \text{for } 0 < k < n.$$

(c) Applying again Theorem 3.1.6, the spaces $P_4$ and $P_2 \vee P_2$ have

$$H_1(P_4) \cong \mathbb{Z}_4, \qquad H_1(P_2 \vee P_2) \cong \mathbb{Z}_2 \oplus \mathbb{Z}_2,$$

with torsion coefficients 4 and 2, 2 in degree 1 (and no others).

(d) Always applying Theorem 3.1.6, the spaces $P_6$ and $P_2 \vee P_3$ have indeed torsion coefficient 6 in degree 1

$$H_1(P_6) \cong \mathbb{Z}_6, \qquad H_1(P_2 \vee P_3) \cong \mathbb{Z}_2 \oplus \mathbb{Z}_3 \cong \mathbb{Z}_6,$$

and no other.

These two spaces are not homeomorphic: the basepoint disconnects $P_2 \vee P_3$, but no point of $P_6$ disconnects it.

One can form diverse path-connected spaces $X$ having $H_1(X) \cong \mathbb{Z}_{30}$ (with torsion coefficient 30) and trivial homology in higher degree, like:

$$P_{30}, \qquad P_2 \vee P_{15}, \qquad P_3 \vee P_{10}, \qquad P_5 \vee P_6, \qquad P_2 \vee P_3 \vee P_5.$$

(e) To apply Theorem 3.1.6, it is sufficient to note that, if the basepoint of $(X, x_0)$ has a contractible open nbd $U$, then the suspension $\Sigma X$ (of the unpointed space) can be pointed at $(x_0, 1/2)$, with a contractible open nbd $U \times \,]0,1[$. (This is simpler than using the reduced suspension $\Sigma(X, x_0)$, that will be introduced in 8.6.5.)

### *9.3.7 Solutions of 3.2.3*

(a) We use an open cover $(U, V)$ of the space $\mathbb{T}$, where $y$ is the image of the centre of $\mathbb{I}^2$

$$U = \mathbb{T} \setminus \{y\}, \qquad V = \operatorname{int} \mathbb{I}^2 \simeq \{*\}, \qquad A = U \cap V \simeq \mathbb{S}^1, \tag{9.37}$$

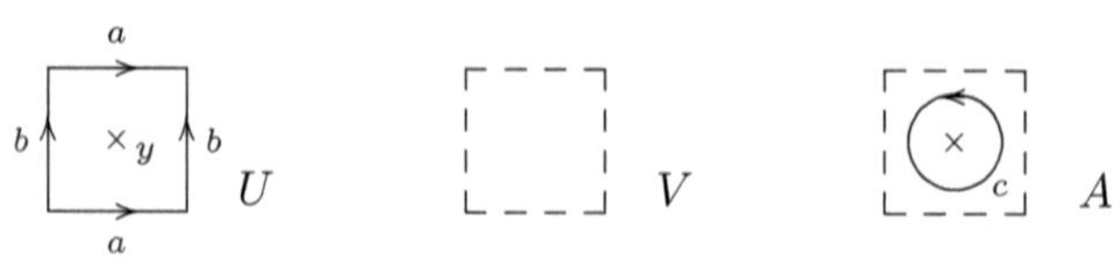


The space $U = \mathbb{T} \setminus \{y\}$ is homotopy equivalent to its retract formed by $\operatorname{Im} a \cup \operatorname{Im} b$ (the union of the equator and the meridian subspaces), which is the figure-eight space $\mathbb{S}^1 \vee \mathbb{S}^1$ examined in 3.1.1, with $H_1(\mathbb{S}^1 \vee \mathbb{S}^1) \cong \mathbb{Z}^2$.

The following portion of the MV-sequence shows that $H_i(\mathbb{T}) = 0$ for $i > 2$

$$H_i(U) \xrightarrow{h_i} H_i(\mathbb{T}) \xrightarrow{D_i} H_{i-1}(A).$$

To compute $H_1(\mathbb{T})$ and $H_2(\mathbb{T})$ we use the exact sequence

$$0 \to H_2(\mathbb{T}) \xrightarrow{D_2} H_1(A) \xrightarrow{h_1} H_1(U) \xrightarrow{k_1} H_1(\mathbb{T}) \tag{9.38}$$

where $H_1(V) = 0$, and we know that $k_1$ is epi (by Exercise 3.1.2(c)).

The homomorphism $h_1$ is trivial, as shown by its value on a generator

$$\begin{aligned} &h_1 = i_{*1} \colon H_1(A) \to H_1(U), \\ &h_1([c]) = [c]_U = [a] + [b] - [a] - [b] = 0. \end{aligned} \tag{9.39}$$

Here $c$ is realised as a simple loop in $A$ around the missing point $y$, as in the figure above; in $U$, $c$ is loop homotopic to the loop $a * b * a^\sharp * b^\sharp$, whose homology class is trivial.

Coming back to the exact sequence (9.38), $D_2$ and $k_1$ are isomorphisms, and our claim is proved.

(b) The generators of $H_1(\mathbb{T})$ are the images of the two obvious generators of $H_1(\mathbb{S}^1 \vee \mathbb{S}^1)$, namely the homology classes $[a], [b]$ of the equator and meridian loops.

As a generator of $H_2(\mathbb{T})$, our intuition suggests to use the canonical projection $p\colon \mathbb{I}^2 \to \mathbb{T}$, which enfolds (and detects) the 2-dimensional cavity of the torus. This 2-cube is readily seen to be a cycle

$$\partial_2 p = -b + b + a - a = 0.$$

To prove that $[p]$ is a generator we are to prove that $D_2[p]$ is a generator of $H_1(A)$, rewriting $[p] = [c + c']$ in a form on which we can compute $D_2$

$$c = -u + u' + v - v' \in C_2(U), \qquad c' \in C_2(V), \qquad (9.40)$$

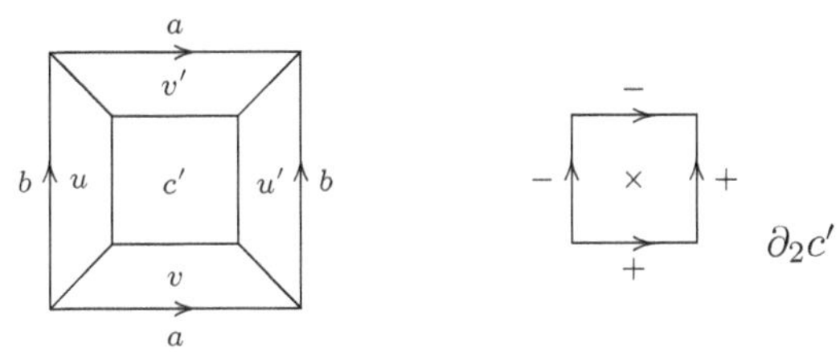


To make this precise, we start from a 3-cube $q\colon \mathbb{I}^3 \to \mathbb{T}$, whose image is represented in the figure above: first we squeeze the standard 3-cube on $\mathbb{I}^2$, as if we were orthogonally projecting a truncated pyramid on its basis, then we apply the projection $p\colon \mathbb{I}^2 \to \mathbb{T}$.

In this way:

- the lateral faces $u = \partial_1^- q$, $u' = \partial_1^+ q$, $v = \partial_2^- q$, $v' = \partial_2^+ q$ are homeomorphisms onto the quadrilaterals represented above, and form the chain $c \in C_2(U)$,

- the upper basis $c' = \partial_3^+ q$ is a homeomorphism onto the square $[1/4, 3/4]^2$, contained in $V = \operatorname{int} \mathbb{D}^2$,

- the lower basis $\partial_3^- q$ is the projection $p\colon \mathbb{I}^2 \to \mathbb{T}$ we are interested in (its image is covered by the other faces).

The 2-chain $\partial_3 q = c + c' - p$ is a boundary of the torus. The 2-cycle $p$ is thus homologous to the cycle $c + c' = p + \partial_3 q$, on which we know how to compute $D_2$

$$D_2[p] = D_2[c + c'] = [\partial_2 c] = -[\partial_2 c'].$$

The proof is now complete: $\partial_2 c'$ is a cycle of $A = V \setminus \{*\}$, and $[\partial_2 c']$ is indeed a generator of $H_1(A) \cong H_1(\mathbb{S}^1)$, as shown in 2.5.5(a).

### 9.3.8 Solution of 3.2.4(a)

We use the MV-sequence of the Klein bottle $\mathbb{K}$, with an open cover $(U, V)$ similar to that used for the torus

$$U = \mathbb{T} \setminus \{y\}, \qquad V = \operatorname{int} \mathbb{I}^2 \simeq \{*\}, \qquad A = U \cap V \simeq \mathbb{S}^1, \tag{9.41}$$

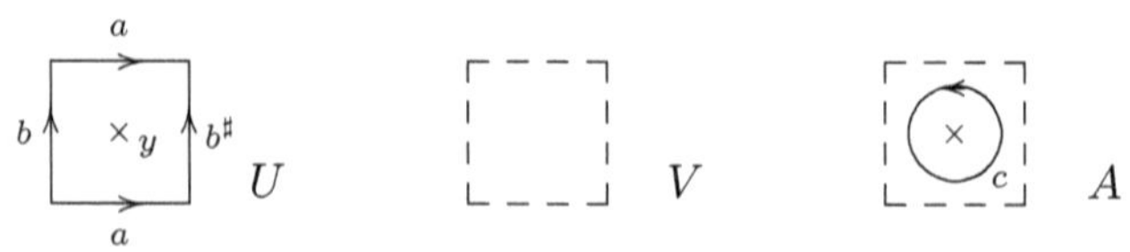


Again, $U \simeq \mathbb{S}^1 \vee \mathbb{S}^1$: the homotopy types of $U, V$ and $A$ are characterised as in 9.3.7. We find in the same way that $H_i(\mathbb{K}) = 0$ for $i > 2$.

Then we use the exact sequence

$$0 \to H_2(\mathbb{K}) \xrightarrow{D_2} H_1(A) \xrightarrow{h_1} H_1(U) \xrightarrow{k_1} H_1(\mathbb{K}) \to 0 \tag{9.42}$$

where again $H_1(V) = 0$ is omitted, and $k_1$ is epi.

Here the homomorphism $h_1 \colon H_1(A) \to H_1(U)$ is not trivial: for a simple loop $c$ in $A$ around the missing point $y$, we have

$$h_1([c]) = [c]_U = [a] + [b^\sharp] - [a] - [b] = -2[b]. \tag{9.43}$$

Therefore $h_1$ is a monomorphism and $H_2(\mathbb{K}) = 0$. Moreover $\operatorname{Im} h_1$ is the subgroup of $\mathbb{Z}[a] \oplus \mathbb{Z}[b]$ generated by $2[b]$, and

$$H_1(\mathbb{K}) \cong \operatorname{cok} h_1 \cong \mathbb{Z} \oplus \mathbb{Z}_2,$$

with generators $[a]_\mathbb{K}$ (of infinite order) and $[b]_\mathbb{K}$ (of order 2).

### 9.3.9 Solution of 3.2.6(a)

The spaces $U, V$ and $A$ can be described, up to homotopy equivalence, as

$$U \simeq \mathbb{T} \setminus \{*\}, \qquad V \simeq \mathbb{T} \setminus \{*\}, \qquad A \simeq \mathbb{S}^1, \tag{9.44}$$

and we have already seen, in 9.3.7(a), that the pierced torus $\mathbb{T} \setminus \{*\}$ is homotopy equivalent to $\mathbb{S}^1 \vee \mathbb{S}^1$, the figure-eight space.

Moreover, as in (9.39):

$$i_{*1} = 0 \colon H_1(A) \to H_1(U), \qquad j_{*1} = 0 \colon H_1(A) \to H_1(V).$$

To compute $H_1(X)$ and $H_2(X)$ we use the exact sequence

$$0 \to H_2(X) \xrightarrow{D_2} H_1(A) \xrightarrow{0} H_1(U) \oplus H_1(V) \xrightarrow{k_1} H_1(X)$$

where we have proved that $h_1 = (i_{*1}, j_{*1}) = 0$, and we know that $k_1$ is epi, by Exercise 3.1.2(c).

Thus $D_2 \colon H_2(X) \cong H_1(A) \cong \mathbb{Z}$, while $k_1$ gives $H_1(X) \cong \mathbb{Z}^4$.

As generators of $H_1(X)$ we can take loops corresponding to two generators for each torus.

### 9.3.10 Solution of 3.3.1(e)

First, let $X = \mathbb{R}^{n+1} \setminus \{0\}$. The embedding $\mathbb{S}^n \subset X$ induces a bijective map $f \colon \mathbb{S}^n/T \to X/\mathbb{R}^*$. Proving that the codomain is Hausdorff would conclude the argument, but it is even simpler to show directly that $f$ is an open map: if $U$ is an open subset of $\mathbb{S}^n$ saturated for $T$, the pierced cone $V = \mathbb{R}^*U$ is plainly open in $X$ and saturated for $\mathbb{R}^*$, whence $f(p_nU)$ is open in $X/\mathbb{R}^*$.

Second, the embedding $u_n \colon \mathbb{D}^n \to \mathbb{S}^n$ obviously induces a bijective map $\mathbb{D}^n/R \to \mathbb{S}^n/T$ from a compact to a Hausdorff space.

### 9.3.11 Solutions of 3.4.1

(a) For the first point, it is sufficient to prove that $X_0$ is a subspace of $X$ (we can always forget an initial segment of the sequence). If $U_0$ is open in $X_0$, we can choose some $U_1$ open in $X_1$ such that $U_1 \cap X_0 = U_0$; then we choose some (larger) $U_2$ open in $X_2$ such that $U_2 \cap X_1 = U_1$ (and therefore $U_2 \cap X_0 = U_0$); and so on. Finally $U = \bigcup U_n$ is open in $X$, and $U \cap X_0 = U_0$. The second claim is obvious.

(b) The second point is obvious: any path in $X_n$ gives a path in $X$. As to the first, suppose that $X$ has an open and closed subset $A \neq \emptyset$; each trace $A_n = A \cap X_n$ is open and closed in $X_n$. Eventually, $A_n$ is non-empty and coincides with $X_n$, so that $A = X$.

### 9.3.12 Solutions of 3.5.3

(b) We start from an element $c \in C$ with $w(c) = 0$, and we want to prove that $c = 0$. As $v'$ is mono, $h(c) = 0$ in $D$: see the left diagram below

$$\begin{array}{ccc} c & \longmapsto & h(c) \\ \downarrow{\scriptstyle w} & & \downarrow{\scriptstyle v'} \\ 0 & \longmapsto & 0 \end{array} \qquad\qquad \begin{array}{ccccccc} a & & b & \longmapsto & c & \longmapsto & 0 \\ {\scriptstyle u}\downarrow & & {\scriptstyle v}\downarrow & & \downarrow{\scriptstyle w} & & \downarrow{\scriptstyle v'} \\ a' & \longmapsto & b' & \longmapsto & 0 & \longmapsto & 0 \end{array} \tag{9.45}$$

We move to the right diagram above; by the exactness of the upper row, there is some $b \in B$ such that $g(b) = c$; then $b' = v(b) \in B'$ goes to $w(c) = 0$ in $C'$.

There is now some $a' \in A'$ such that $f'(a') = b'$; but $u$ is epi, and there is some $a \in A$ such that $u(a) = a'$.

By the commutativity of the left square (in the right diagram), $vf(a) = f'u(a) = b' = v(b)$; but $v$ is mono, and $f(a) = b$. Finally $c = gf(a) = 0$.

(c) We start from an element $c' \in C'$ and we want to prove that $c' \in \operatorname{Im} w$.

We form the left diagram below, where $d' = h'(c') \in D' = \operatorname{Im} v'$ and there is some $d \in D$ such that $d' = v'(d)$; moreover, $k(d) = 0$ in $E$, because $u'$ is mono. By the exactness of the upper row, there is some $c \in C$ such that $h(c) = d$

$$\begin{array}{ccccc} c & \longmapsto & d & \longmapsto & 0 \\ & & \big\downarrow{\scriptstyle v'} & & \big\downarrow{\scriptstyle u'} \\ c' & \longmapsto & d' & \longmapsto & 0 \end{array} \qquad \begin{array}{ccccc} b & \longmapsto & g(b) & & \\ {\scriptstyle v}\big\downarrow & & \big\downarrow{\scriptstyle w} & & \\ b' & \longmapsto & c' - w(c) & \longmapsto & 0 \end{array} \tag{9.46}$$

Moving to the right diagram above, the element $c' - w(c)$ goes to 0 in $D'$ (because $h'(c') = d' = v'h(c) = h'w(c)$); there is thus some $b' \in B'$ such that $g'(b') = c' - w(c)$. As $v$ is epi, there is some $b \in B$ such that $v(b) = b'$, and we have an element $g(b) \in C$ such that $wg(b) = c' - w(c)$. Finally the element $c + g(b) \in C$ gives $w(c + g(b)) = w(c) + c' - w(c) = c'$.

### *9.3.13 Solutions of 3.5.4*

(a) Obviously, (iii) implies (i) and (ii).

(i) $\Rightarrow$ (iii). For $b \in B$, the element $b - nq(b)$ belongs to $\operatorname{Ker} q = \operatorname{Im} m$, and we define $p(b) \in A$ as the unique element such that $mp(b) = b - nq(b)$.

This mapping $p$ satisfies the relation $mp + nq = \operatorname{id} B$; it is a homomorphism, because $mp$ is, and $m$ is injective. The relation $pm = \operatorname{id} A$ holds, because $mpm = m - nqm = m$ and $m$ is injective.

(ii) $\Rightarrow$ (iii). There exists a mapping $n\colon C \to B$ such that $mp + nq = \operatorname{id} B$: for every $c \in C$ we can choose $b \in B$ such that $q(b) = c$, and we let $n(c) = b - mp(b)$. This mapping is a homomorphism, because $nq$ is and $q$ is surjective. (By the same reason, $n$ is determined by $p$). Moreover $qn = \operatorname{id} C$, cancelling $q$ from the relation $qnq = q - qmp = q$.

(b) From (iii) it follows easily that $\operatorname{Im} m + \operatorname{Im} n = B$ and $\operatorname{Im} m \cap \operatorname{Im} n = 0$.

(c) Let $(c_i)_{i \in I}$ be a basis of $C$. For every $i \in I$ we choose some $b_i \in B$ such that $q(b_i) = c_i$, and define $n\colon C \to B$ as the unique homomorphism such that $n(c_i) = b_i$. This gives $qn = \operatorname{id} C$.

In general there are different possible choices. For instance, if $C$ is infinite cyclic, with generator $c$, and $q(b) = c$, any element $b'$ of the lateral $m(A) + b$ can be chosen as $n(c)$. This lateral is equipotent to $A$.

### 9.3.14 Solutions of 3.5.5

(a) Assuming 3.5.5(*), the morphism $f = up + vq - \mathrm{id}\, C$ gives $pf = 0$ and factorises through $v = \ker p$, as $f = vh$. It follows that $h = qvh = qf = 0$ and $f = 0$.

Conversely, if (3.69) is satisfied, let us prove that $v = \ker p$. We take $f\colon X \to C$ such that $pf = 0$; then $f = (up + vq)f = vqf$, and we have proved that $f$ factorises (uniquely) through the monomorphism $v$.

(b) The composite (3.70) is computed as follows (where $(f, g)\colon A \to B \times B$)

$$\nabla(f \oplus g)\Delta = \nabla(f \times g)\Delta = \nabla(f, g) = f + g.$$

### 9.3.15 Solutions of 3.6.2

(b) The epimorphism $\rho X\colon \tilde{C}(X) \twoheadrightarrow C_+(X)$ in (3.78) gives a (natural) short exact sequence of unbounded chain complexes

$$\mathrm{Ker}\,\rho \overset{\iota X}{\rightarrowtail} \tilde{C}(X) \overset{\rho X}{\twoheadrightarrow} C_+(X) \qquad (\mathrm{Ker}\,\rho)_{-1} = \tilde{C}_{-1}(X) = \mathbb{Z}, \qquad (9.47)$$

where $\mathrm{Ker}\,\rho \subset \tilde{C}(X)$ has one non-trivial component, in degree $-1$, and $H_{-1}(\mathrm{Ker}\,\rho) = (\mathrm{Ker}\,\rho)_{-1} = \mathbb{Z}$.

The derived exact homology sequence

$$0 \to \tilde{H}_0(X) \xrightarrow{\rho X} H_0(X) \xrightarrow{DX} \mathbb{Z} \xrightarrow{pX} \tilde{H}_{-1}(X) \to 0$$

is natural on $\mathsf{Top}$.

Its connecting morphism $DX$ (in degree 0) and $pX = \iota_{*-1}X$ are computed as:

$$DX\colon H_0(X) \to \mathbb{Z}, \qquad (DX)[\Sigma_i\, \lambda_i x_i] = [\tilde{\partial}_0(\Sigma_i\, \lambda_i x_i)] = \Sigma_i\, \lambda_i,$$
$$pX\colon H_{-1}(\mathrm{Ker}\,\rho) \to \tilde{H}_{-1}(X), \qquad (pX)(k) = [\iota X(k)] = [k].$$

(d), (e) Obvious

(g) One simply replaces, in the proof of 2.3.7, the ordinary chain complexes with the reduced ones.

(h) For $n > 0$: $f_{*n} = 0$ and $\tilde{H}_n(X) = H_n(X)$. Otherwise $f_{*0}$ is computed as:

$$f_{*0}[\Sigma\, \lambda_i x_i] = \Sigma\, \lambda_i[*] = \Sigma\, \lambda_i \in H_0(\{*\}) = \mathbb{Z}, \qquad (9.48)$$

and coincides with the homomorphism $DX\colon H_0(X) \to \mathbb{Z}$ of the exact sequence (3.81). The kernel and cokernel of the latter are indeed $\tilde{H}_0(X)$ and $\tilde{H}_{-1}(X)$.

### 9.3.16 Solutions of 3.6.7

(a) For the singleton, there is a unique mapping $a_n\colon \Delta^n \to \{*\}$, for every $n \geqslant 0$. The differential of the chain complex involves $n+1$ summands

$$\partial_n(a_n) = \Sigma_i\,(-1)^i\,\partial_i a = \Sigma_i\,(-1)^i\,a_{n-1},$$

and is 0 for $n$ odd, and an isomorphism for $n$ even.

The simplicial chain complex $S_+(\{*\})$ of the singleton is isomorphic to

$$\ldots\ \mathbb{Z} \xrightarrow{1} \mathbb{Z} \xrightarrow{0} \mathbb{Z} \xrightarrow{1} \mathbb{Z} \xrightarrow{0} \mathbb{Z}. \tag{9.49}$$

Its homology is $\mathbb{Z}$ in degree 0, and trivial elsewhere.

## 9.4 Exercises of Chapter 4

### 9.4.1 Solutions of 4.1.2

(a) The embedding $C_n(A) \rightarrowtail C_n(X)$ comes from an inclusion of canonical bases $c_n(A) \subset c_n(X)$ where

- $c_n(X)$ is the set of all non-degenerate $n$-cubes $a\colon \mathbb{I}^n \to X$,
- $c_n(A) = \{a \in c_n(X) \mid \operatorname{Im} a \subset A\}$.

Therefore $C_n(X, A)$ is the free abelian group generated by the complement $c_n(X, A) = c_n(X) \setminus c_n(A)$.

(b) The short exact sequence $C_n(A) \rightarrowtail C_n(X) \twoheadrightarrow C_n(X, A)$ certainly splits, because $C_n(X, A)$ is a free abelian group. Here we can say something more: we have a canonical section $C_n(X, A) \rightarrowtail C_n(X)$ given by the inclusion $c_n(X, A) \subset c_n(X)$ of canonical bases, as in (a).

### 9.4.2 Solutions of 4.1.6

(a) In the exact sequence of the pair $(X, A)$

$$H_nA \xrightarrow{u_{*n}} H_nX \xrightarrow{v_{*n}} H_n(X, A) \xrightarrow{D_n} H_{n-1}A \xrightarrow{u_{*n-1}} H_{n-1}X$$

the morphisms $u_{*n}$ and $u_{*n-1}$ are isomorphisms, and $H_n(X, A) = 0$.

(b) The exact homology sequence of the pair shows that $H_0(X, A)$ is the cokernel of the homomorphism $H_0(A) \to H_0(X)$ induced by the inclusion.

(d) The exact sequence

$$H_n(\{x\}) \to H_n(X) \to H_n(X, x) \to H_{n-1}(\{x\})$$

gives a canonical isomorphism $H_n(X) \to H_n(X, x)$ for $n > 1$. In lower degree we have:

$$0 \to H_1X \xrightarrow{k_1} H_1(X, x) \xrightarrow{D_1} H_0\{x\} \xrightarrow{h_n} H_0X \xrightarrow{h_0} H_0(X, x) \to 0$$

where $h_0$ is injective, $D_1 = 0$ and $k_1$ is an isomorphism. On the right hand, the same sequence shows that $H_0(X, x) \cong H_0(X)/\mathrm{Im}\, h_0$, where $\mathrm{Im}\, h_0$ is generated by $[x] \in H_0(X)$.

Therefore $H_0(X, x)$ is freely generated by the path components of $X$, excluding that of $x$, which makes it isomorphic to $\tilde{H}_0(X)$, by Exercise 3.6.2(d).

### 9.4.3 Solutions of 4.1.7

(b) The commutative diagram of (unbounded) chain complexes

$$\begin{array}{ccccc} \tilde{C}(A) & \rightarrowtail & \tilde{C}(X) & \twoheadrightarrow & C_+(X, A) \\ \downarrow{\scriptstyle \rho A} & & \downarrow{\scriptstyle \rho X} & & \| \\ C_+(A) & \rightarrowtail & C_+(X) & \twoheadrightarrow & C_+(X, A) \end{array} \tag{9.50}$$

has short exact rows, and induces in homology the commutative diagram (4.18), with exact rows.

### 9.4.4 Solutions of 4.4.3

(a) As for 4.1.6(a).

(c) A space $X$ has a unique map $f\colon X \to \{*\}$ with values in the singleton. We define:

$$\begin{aligned} \tilde{H}_n(X) &= \mathrm{Ker}\,(f_{*n}\colon H_n(X) \to H_n(\{*\})) \quad (n \geqslant 0), \\ \tilde{H}_{-1}(X) &= \mathrm{Cok}\,(f_{*0}\colon H_0(X) \to H_0(\{*\})). \end{aligned} \tag{9.51}$$

Because of the dimension axiom, we readily have that:

$$\tilde{H}_n(X) = H_n(X), \quad \text{for } n > 0, \tag{9.52}$$

so that we are only interested in $\tilde{H}_0(X)$ and $\tilde{H}_{-1}(X)$. Moreover, by (b)

$$\begin{aligned} \tilde{H}_0(\emptyset) = 0, \qquad \tilde{H}_{-1}(\emptyset) &= H_0(\{*\}) = G, \\ \tilde{H}_{-1}(X) = 0, \quad &\text{for } X \neq \emptyset. \end{aligned} \tag{9.53}$$

For the last point, it is sufficient to show that $f_{*0}\colon H_0(X) \to H_0(\{*\})$ is an epimorphism, for $X \neq \emptyset$. In fact, we have a retraction $\{*\} \to X \to \{*\}$, with induced homomorphisms $(G \to H_0(X) \to G) = \mathrm{id}\, G$.

Many other properties of reduced homology can now be proved, following the outline of 3.6.2 and 3.6.3.

### 9.4.5 Solutions of 4.5.3

(a) We use, as suggested, the following projections

$$\varepsilon_i \colon X^n \to X^{n-1}, \qquad \varepsilon_i(x_1, ..., x_n) = (x_1, ..., \hat{x}_i, ..., x_n),$$
$$\varepsilon_i \varepsilon_j = \varepsilon_j \varepsilon_{i+1}, \quad \text{for } j \leqslant i.$$

In this way:

$$d^{n-1}\lambda = \Sigma_{1 \leqslant i \leqslant n+1} (-1)^i \lambda \varepsilon_i,$$
$$d^n d^{n-1}\lambda = \Sigma_{i \leqslant n+1, j \leqslant n+2} (-1)^{i+j} \lambda \varepsilon_i \varepsilon_j$$
$$= \Sigma_{i<j} (-1)^{i+j} \lambda \varepsilon_i \varepsilon_j + \Sigma_{j \leqslant i} (-1)^{i+j} \lambda \varepsilon_i \varepsilon_j = 0.$$

In fact, the second addendum is the opposite of the first, by the usual change of variables:

$$\Sigma_{j \leqslant i} (-1)^{i+j} \lambda \varepsilon_i \varepsilon_j = \Sigma_{j \leqslant i} (-1)^{i+j} \lambda \varepsilon_j \varepsilon_{i+1} = -\Sigma_{i<j} (-1)^{i+j} \lambda \varepsilon_i \varepsilon_j.$$

### 9.4.6 Solutions of 4.5.5

(a) For the singleton $X$, a function $g \colon X^{n+1} \to G$ amounts to its value, and is locally zero if and only if it is zero. We identify all components $\overline{C}^n(X; G) = C^n(X; G) = G$.

The differential of the cochain complex $\overline{C}^+(X; G) = C^+(X; G)$

$$d^n \colon C^n(X; G) \to C^{n+1}(X; G), \qquad d^n g = \Sigma_{1 \leqslant i \leqslant n+2} (-1)^i g, \tag{9.54}$$

is zero for $n$ even $\geqslant 0$ and $-\mathrm{id}\, G$ for $n$ odd.

Therefore our cochain complex is exact in positive degree, and $\overline{H}^n(X; G) = 0$, for $n > 0$. Moreover we have got a canonical isomorphism

$$\overline{H}^0(X; G) = \mathrm{Ker}\,(d^0 \colon C^0(X; G) \to C^1(X; G)) = C^0(X; G) \cong G,$$

that sends the function $\lambda \colon X \to G$ to $\lambda(*) \in G$.

### 9.4.7 Solutions of 4.5.8

(a) The relative cochain complex $\overline{C}^+(X, A; G)$ has an augmentation

$$0 \to \mathcal{A}(X, A; G) \xrightarrow{\eta} \overline{C}^0(X, A; G) \xrightarrow{d^0} \overline{C}^1(X, A; G) \tag{9.55}$$

where a locally constant functions $\lambda \colon X \to G$ vanishing on $A$ is sent to the class $\overline{\lambda} \in \overline{C}^0(X, A; G)$.

It is sufficient to prove that this sequence is exact, so that $\eta$ embeds $\mathcal{A}(X, A; G)$ as $\mathrm{Ker}\, d^0 = \overline{H}^0(X, A; G)$.

First, $\eta$ is injective: if $\eta(\lambda) = 0$, the (locally constant) function $\lambda$ is locally zero, and everywhere 0. Secondly, $\mathrm{Ker}\, d^0 = \mathrm{Im}\, \eta$ because, for any function $\lambda \colon X \to G$ vanishing on $A$, the function

$$d^0(\lambda) \colon X^2 \to G, \qquad d^0(\lambda)(x_1, x_2) = \lambda(x_2) - \lambda(x_1),$$

is locally zero (at the diagonal) if and only if $\lambda$ is locally constant.

(d) If the connected components of the space $X$ are open, a function $\lambda \colon X \to G$ is locally constant if and only if it is constant on each connected component; this amounts to a family $(g_i) \in G^I$.

### *9.4.8 Solutions of 4.6.3*

(a) A consequence of formula (4.96), which ensures that:

- the product of two cocycles is a cocycle,

- the product of a coboundary and a cocycle, in any order, is a coboundary:

$$dy = 0 \;\Rightarrow\; (dx)y = d(xy), \qquad dx = 0 \;\Rightarrow\; x(dy) = \pm\, d(xy).$$

### *9.4.9 Solutions of 4.6.4*

(a) We have to verify that, for $\lambda \in C^p(X;G)$ and $\mu \in C^q(X;G)$

$$d^n(\lambda{\bullet}\mu) = d^p(\lambda){\bullet}\mu + (-1)^p \lambda{\bullet}d^q(\mu). \tag{9.56}$$

The left term can be computed as a sum of two summands:

$$\begin{aligned}
d^n(\lambda{\bullet}\mu)(x_1, ..., x_{n+1}) &= \textstyle\sum_{1 \leqslant i \leqslant n+2} (-1)^i\, (\lambda{\bullet}\mu)(x_1, ..., \hat{x}_i, ..., x_{n+2}) \\
&= \textstyle\sum_{1 \leqslant i \leqslant p+1} (-1)^i\, (\lambda{\bullet}\mu)(x_1, ..., \hat{x}_i, ..., x_{n+2}) \\
&\quad + \textstyle\sum_{p+2 \leqslant i \leqslant n+2} (-1)^i\, (\lambda{\bullet}\mu)(x_1, ..., \hat{x}_i, ..., x_{n+2}) \\
&= \textstyle\sum_{1 \leqslant i \leqslant p+1} (-1)^i\, \lambda(x_1, ..., \hat{x}_i, ..., x_{p+2})\, \mu(x_{p+2}, ..., x_{n+2}) \\
&\quad + \textstyle\sum_{p+2 \leqslant i \leqslant n+2} (-1)^i\, \lambda(x_1, ..., x_{p+1})\, \mu(x_{p+1}, ..., \hat{x}_i, ..., x_{n+2}).
\end{aligned}$$

Now, computing the two summands on the right, we single out a term in each of them:

$$\begin{aligned}
&(d^p \lambda{\bullet}\mu)(x_1, ..., x_{n+2}) \\
&\quad = \textstyle\sum_{1 \leqslant i \leqslant p+2} (-1)^i\, \lambda(x_1, ..., \hat{x}_i, ..., x_{p+2})\, \mu(x_{p+2}, ..., x_{n+2}) \\
&\quad = \textstyle\sum_{1 \leqslant i \leqslant p+1} (-1)^i\, \lambda(x_1, ..., \hat{x}_i, ..., x_{p+2})\, \mu(x_{p+2}, ..., x_{n+2}) \\
&\qquad + (-1)^p\, \lambda(x_1, ..., x_{p+1})\, \mu(x_{p+2}, ..., x_{n+2}),
\end{aligned}$$

$$\begin{aligned}
&(-1)^p(\lambda\bullet d^q\mu)(x_1, ..., x_{n+2})\\
&\quad = \textstyle\sum_{1\leqslant i\leqslant q+2} (-1)^{p+i}\, \lambda(x_1, ..., x_{p+1})\, \mu(x_{p+1}, ..., \hat{x}_{p+i}, ..., x_{n+2})\\
&\quad = (-1)^{p+1}\lambda(x_1, ..., x_{p+1})\, \mu(x_{p+2}, ..., x_{n+2})\\
&\qquad + \textstyle\sum_{2\leqslant i\leqslant q+2} (-1)^{p+i}\, \lambda(x_1, ..., x_{p+1})\, \mu(x_{p+1}, ..., \hat{x}_{p+i}, ..., x_{n+2}).
\end{aligned}$$

Adding these results, the two terms we have put apart cancel, and we find the same result as above.

(b) The graded homomorphism $f^\sharp\colon C^*(X;R) \to C^*(Y;R)$ already preserves the product:

$$\begin{aligned}
&(f^{\sharp n}(\lambda\bullet\mu))(y_1, ..., y_{n+1}) = \lambda(fy_1, ..., fy_{p+1})\, \mu(fy_{p+1}, ..., fy_{n+1})\\
&\quad = ((f^{\sharp p}\lambda)\bullet(f^{\sharp q}\mu))(y_1, ..., y_{n+1}).
\end{aligned}$$

### *9.4.10 Solutions of 4.7.2*

(a) The group $H^0(\mathcal{U};G) = \mathrm{Ker}\,(d^0\colon C^0(\mathcal{U};G) \to C^1(\mathcal{U};G))$ coincides with $G(\mathcal{U})$, because $(d^0\varphi)(U,U') = \varphi(U') - \varphi(U)$, for $U \cap U' \neq \emptyset$.

(b) For every open cover $\mathcal{U}$ we have a homomorphism

$$(\hat{\ })_{\mathcal{U}}\colon G(\mathcal{U}) \to \mathcal{A}(X;G), \quad \hat{\varphi}(x) = \varphi(U), \quad \text{for } x \in U \in \mathcal{U},$$

where the continuity of the function $\hat{\varphi}\colon X \to G$ is ensured by the fact that it is constant on every $U \in \mathcal{U}$.

This family of homomorphisms is obviously consistent with the inductive system $H^0(\mathcal{U};G)$ indexed by $\mathrm{Cov}(X)$, and determines a homomorphism

$$\check{H}^0(X;G) \to \mathcal{A}(X;G), \qquad [\varphi] \mapsto \hat{\varphi}.$$

This homomorphism is surjective: a continuous function $\xi\colon X \to G$ determines an open cover $\mathcal{U}_\xi$ and a function $\varphi$

$$\mathcal{U}_\xi = \{U_g = \xi^{-1}\{g\} \mid g \in G\}, \quad \varphi\colon \mathcal{U}_\xi \to G, \quad \varphi(U_g) = g,$$

which belongs to $G(\mathcal{U}_\xi)$ and gives back $\hat{\varphi} = \xi$.

It is also injective: if $\hat{\varphi}(x) = 0$ for all $x \in X$, the original $\varphi\colon \mathcal{U} \to G$ is also constant at 0.

### *9.4.11 Solutions of 4.8.3*

(a) Cancelling the zero terms (given by repeated variables) and reordering the exterior products, we get:

$$\begin{aligned}
\omega \wedge \omega' &= \lambda_1\mu_1\, dx \wedge dydz + \lambda_2\mu_2\, dy \wedge dxdz + \lambda_3\mu_3\, dz \wedge dxdy\\
&= (\lambda_1\mu_1 - \lambda_2\mu_2 + \lambda_3\mu_3)\, dxdydz.
\end{aligned} \tag{9.57}$$

(b) The properties (i)–(iii) of 4.6.2 are easily verified: consistency and distributivity hold by definition. Associativity can be verified on the canonical bases, and is obvious using formula (4.120), based on the extended symbols $dx^{i_1}dx^{i_2}...dx^{i_p}$.

With these symbols, it is also obvious that the products

$$\begin{aligned}(dx^{i_1}...dx^{i_p})\wedge(dx^{j_1}...dx^{j_q}) &= dx^{i_1}...dx^{i_p}dx^{j_1}...dx^{j_q},\\ (dx^{j_1}...dx^{j_q})\wedge(dx^{i_1}...dx^{i_p}) &= dx^{j_1}...dx^{j_q}dx^{i_1}...dx^{i_p},\end{aligned} \tag{9.58}$$

are related by a factor $(-1)^{pq}$.

### 9.4.12 Solutions of 4.8.4

(b) Beginning with the product $\rho = \lambda\mu$ of two $C^\infty$-functions, we get the expected 1-form

$$\begin{aligned} d(\lambda\mu) &= d(\Sigma_{1\leqslant k\leqslant n}(D_k\rho)\,dx^k)\\ &= \Sigma_{1\leqslant k\leqslant n}((D_k\lambda)\mu + \lambda(D_k\mu))\,dx^k = (d\lambda)\mu + \lambda(d\mu).\end{aligned}$$

Now we apply definition (4.123) to a product $\omega\wedge\omega' = \Sigma_{\mathbf{i},\mathbf{j}}\lambda_{\mathbf{i}}\mu_{\mathbf{j}}(dx^{\mathbf{i}}\wedge dx^{\mathbf{j}})$, taking into account that each term $dx^{\mathbf{i}}\wedge dx^{\mathbf{j}}$ is either 0 or $\sigma(\mathbf{i},\mathbf{j})dx^{\mathbf{k}}$, which is a basic differential, up to sign

$$\begin{aligned} d^{p+q}(\omega\wedge\omega') &= \Sigma_k\, d^0(\lambda_{\mathbf{i}}\mu_{\mathbf{j}})\wedge(\Sigma_{\mathbf{i},\mathbf{j}}\,dx^{\mathbf{i}}\wedge dx^{\mathbf{j}})\\ &= ((d\lambda_{\mathbf{i}})\mu_{\mathbf{j}} + \lambda_{\mathbf{i}}d(\mu_{\mathbf{j}}))\wedge(\Sigma_{\mathbf{i},\mathbf{j}}\,dx^{\mathbf{i}}\wedge dx^{\mathbf{j}})\\ &= (d\lambda_{\mathbf{i}})\mu_{\mathbf{j}}\wedge(\Sigma_{\mathbf{i},\mathbf{j}}\,dx^{\mathbf{i}}\wedge dx^{\mathbf{j}}) + \lambda_{\mathbf{i}}d(\mu_{\mathbf{j}}))\wedge(\Sigma_{\mathbf{i},\mathbf{j}}\,dx^{\mathbf{i}}\wedge dx^{\mathbf{j}})\\ &= (d\lambda_{\mathbf{i}})\mu_{\mathbf{j}}\wedge(\Sigma_{\mathbf{i}}\,dx^{\mathbf{i}})\wedge(\Sigma_{\mathbf{j}}\,\mu_{\mathbf{j}}\,dx^{\mathbf{j}}) + (-1)^p(\Sigma_{\mathbf{i}}\,\lambda_{\mathbf{i}}\,dx^{\mathbf{i}})\wedge d(\mu_{\mathbf{j}})\wedge(\Sigma_{\mathbf{j}}\,dx^{\mathbf{j}})\\ &= d^p\omega\wedge\omega' + (-1)^p\,\omega\wedge d^q\omega'.\end{aligned}$$

(c) The double differential of a $C^\infty$-function is zero, by Schwartz theorem on the commutation of partial derivatives

$$\begin{aligned} d^1d^0(\lambda) &= d^1(\Sigma_{1\leqslant i\leqslant n}\frac{\partial\lambda}{\partial x^i}\,dx^i) = \Sigma_{i,j}\frac{\partial^2\lambda}{\partial x^i\partial x^j}\,dx^j\,dx^i\\ &= \Sigma_{j<i}\,(\frac{\partial^2\lambda}{\partial x^i\partial x^j} - \frac{\partial^2\lambda}{\partial x^j\partial x^i})\,dx^j\,dx^i = 0.\end{aligned}$$

The general case is a consequence, applying (b) and the vanishing of the differential on every $dx^{\mathbf{i}}$:

$$\begin{aligned} d^{p+1}d^p(\Sigma_{\mathbf{i}}\,\lambda_{\mathbf{i}}\,dx^{\mathbf{i}}) &= d^{p+1}(\Sigma_{\mathbf{i}}\,d^0(\lambda_{\mathbf{i}})\wedge dx^{\mathbf{i}})\\ &= \Sigma_{\mathbf{i}}\,(d^1d^0(\lambda_{\mathbf{i}})\wedge dx^{\mathbf{i}} - d^0(\lambda_{\mathbf{i}})\wedge d^p(dx^{\mathbf{i}}) = 0.\end{aligned}$$

### *9.4.13 Solutions of 4.8.5*

(a) The differential $d^0(\lambda) = \Sigma_i D_i\lambda\, dx^i$ is zero if and only if all the partial derivatives of $\lambda$ vanish everywhere, which means that $\lambda$ is locally constant, i.e. constant on every connected component of $U$. Thus $H^0_R(U)$ can be identified with the product algebra $\mathbb{R}^{\mathrm{Cc}(U)}$, where $\mathrm{Cc}(U)$ is the set of connected components of $U$ (here coinciding with the path components).

We have seen that Alexander–Spanier cohomology with coefficients in $\mathbb{R}$ gives the same algebra (in 4.5.8(d)); for singular cohomology, see 5.5.4(c), (d).

(b) The differential $d^1\omega$ is computed 'as' in (4.113), for $n = 3$.

### *9.4.14 Solutions of 4.8.6*

(a) Beginning with the product $\rho = \lambda\mu$ of two $\mathrm{C}^\infty$-functions, we obviously have $f^{\sharp 0}(\lambda\mu) = (\lambda\mu)\circ f = (\lambda\circ f)(\mu\circ f) = f^{\sharp 0}(\lambda)\, f^{\sharp 0}(\mu)$.

Now we apply definition (4.125) to a product

$$\begin{aligned}\omega\wedge\omega' &= (\Sigma_i\,\lambda_i\, dx^{i_1}\dots dx^{i_p})\wedge(\Sigma_j\,\mu_j\, dx^{j_1}\dots dx^{j_q})\\ &= \Sigma_{i,j}\,\lambda_i\mu_j\, dx^{i_1}\dots dx^{i_p} dx^{j_1}\dots dx^{j_q},\end{aligned}$$

where each term $dx^{i_1}\dots dx^{i_p} dx^{j_1}\dots dx^{j_q}$ is either 0 or a basic form, up to sign change

$$\begin{aligned}f^{\sharp p+q}(\omega\wedge\omega') &= \Sigma_{i,j}\,((\lambda_i\mu_j)\circ f)\, d^0 f^{i_1}\dots d^0 f^{i_p} d^0 f^{j_1}\dots d^0 f^{j_q}\\ &= (\Sigma_i\,(\lambda_i\circ f)\, d^0 f^{i_1}\dots d^0 f^{i_p})(\Sigma_j\,(\mu_j\circ f)\, d^0 f^{j_1}\dots d^0 f^{j_q})\\ &= f^{\sharp p}(\omega)\wedge f^{\sharp q}(\omega').\end{aligned}$$

(b) The statement is obvious for a 0-form $\lambda$ on $U$:

$$\begin{aligned}d^0 f^{\sharp 0}(\lambda) &= d^0(\lambda\circ f) = \Sigma_j\, \frac{\partial(\lambda\circ f)}{\partial y^j}\, dy^j\\ &= \Sigma_{j,i}\,(\frac{\partial\lambda}{\partial x^i}\circ f)\,\frac{\partial f^i}{\partial y^j}\, dy^j = \Sigma_i\,(\frac{\partial\lambda}{\partial x^i}\circ f)\, d^0 f^i,\end{aligned}$$

$$f^{\sharp 1} d^0(\lambda) = f^{\sharp 1}(\Sigma_i\,\frac{\partial\lambda}{\partial x^i}\, dx^i) = \Sigma_i\,(\frac{\partial\lambda}{\partial x^i}\circ f)\, d^0 f^i.$$

The general case is a consequence, taking into account point (a), and the fact that the differential vanishes on every differential:

$$\begin{aligned}d^p f^{\sharp p}(\Sigma_i\,\lambda_i dx^{i_1}\dots dx^{i_p}) &= d^p(\Sigma_i\,(\lambda_i\circ f) d^0 f^{i_1}\wedge\dots\wedge d^0 f^{i_p})\\ &= \Sigma_i\, d^0(\lambda_i\circ f)\wedge d^0 f^{i_1}\wedge\dots\wedge d^0 f^{i_p},\end{aligned}$$

$$\begin{aligned}f^{\sharp p+1} d^p(\Sigma_i\,\lambda_i\, dx^{i_1}\dots dx^{i_p}) &= f^{\sharp p+1}(\Sigma_i\, d^0\lambda_i\wedge dx^{i_1}\dots dx^{i_p})\\ &= \Sigma_i\, f^{\sharp 1}(d^0\lambda_i)\wedge f^{\sharp 1}(dx^{i_1})\dots f^{\sharp 1}(dx^{i_p})\\ &= \Sigma_i\, d^0(\lambda_i\circ f)\wedge d^0 f^{i_1}\wedge\dots\wedge d^0 f^{i_p}.\end{aligned}$$

## 9.5 Exercises of Chapter 5

### 9.5.1 *Solutions of 5.1.8

(a) Let $f\colon A \to B$ be an injective $K$-homomorphism. We take a basis $(a_i)_{i\in I}$ of $A$, and let $b_i = f(a_i)$. Then the family $(b_i)_{i\in I}$ is linearly independent in $B$, and can be extended to a basis $(b_i)_{i\in J}$ of $B$ (with $I \subset J$). We define $g\colon B \to A$ as the unique linear mapping such that

$$g(b_i) = a_i, \text{ for } i \in I, \qquad g(b_i) = 0, \text{ for } i \in J \setminus I.$$

Then $gf\colon A \to A$ is the identity on each $a_i$, and therefore on $A$. (We are splitting $B$ in the internal direct sum of $f(A) \cong A$ and $\operatorname{Ker} g$, another vector subspace of $B$.)

(b) If $f\colon A \to B$ is a surjective $K$-homomorphism and $(b_i)$ is a basis of $B$, we choose an element $a_i \in f^{-1}\{b_i\}$ for every $i \in I$, and let $g\colon B \to A$ be the unique linear mapping such that $g(b_i) = a_i$ (for all $i$).

Then $fg\colon B \to B$ is the identity on each $b_i$, and therefore on $B$. (We are splitting $A$ in the internal direct sum of $\operatorname{Ker} f$ and $g(B)$, a vector subspace of $A$ isomorphic to $B$.)

(d) The mapping (for $\alpha\colon A \to K$)

$$\varphi\colon A\times B \twoheadrightarrow \operatorname{Hom}_K(A^*, B), \qquad \varphi(a,b)(\alpha) = \alpha(a)\, b,$$

is $K$-bilinear and induces our homomorphism $u\colon A \otimes_K B \to \operatorname{Hom}_K(A^*, B)$.

(This also works for $R$-modules, but the following would not: note that, for $\mathbb{Z}$-modules, $\operatorname{Hom}(A, \mathbb{Z}) = 0$ for every torsion abelian group $A$.)

Now, we choose a basis $(a_i)_{i\in I}$ of $A$. For every $i \in I$, we define a homomorphism $\alpha_i$ on this basis

$$\alpha_i\colon A \to K, \quad \alpha_i(a_i) = 1_K, \quad \alpha_i(a_j) = 0_K \quad (\text{for } j \neq i \text{ in } I),$$

which takes every vector $a = \sum_i \lambda_i a_i$ to the scalar $\lambda_i$. Note that $a = 0$ if and only if all homomorphisms $\alpha_i$ vanish on it.

To prove that each $u_{AB}$ is injective, let $u(a \otimes b) = 0$, so that $\alpha_i(a)b = 0$, for all $i \in I$. Now, either $a = 0$ or there is some index $i$ such that $\alpha_i(a) \in K^*$, and then $b = 0$. In both cases $a \otimes b = 0$.

If $A$ and $B$ are finitely generated, the domain and codomain of $u$ have the same finite dimension on $K$, namely the product $\dim(A)\dim(B)$, and $u$ is an isomorphism.

### 9.5.2 Solutions of 5.2.1

(a) The trivial $R$-homomorphism $0_{AB}\colon A \to B$ is the identity of the group $\operatorname{Hom}_R(A, B)$. The $R$-module $X$ is trivial if and only if $\operatorname{id} X = 0_{XX}$.

(b) A consequence of the characterisation 1.6.4(ii).

(c) The characterisation 3.5.4(iii) is obviously preserved by additive functors. The particular case is a consequence of Exercise 3.5.4(c), extended to $R$-modules.

### *9.5.3 Solutions of 5.2.7*

(a) The first claim is a consequence of 5.1.5(e). The second is obvious: in diagram (5.39), one can choose a base $(x_i)$ in $P$, and choose $k(x_i) \in A$ so that $pk(x_i) = h(x_i)$, for all indices $i$.

(b) Plainly, a direct sum of projective modules is projective.

Let $X$ be a direct summand of the projective module $P$: we have a retraction $n\colon X \rightleftarrows P : q$, with $qn = \mathrm{id}\, X$. Given $p\colon A \twoheadrightarrow B$ and $h\colon X \to B$, there is a commutative square $hq = pk$, because $P$ is projective

$$\begin{array}{ccc} P & \xrightarrow{\;q\;} & X \\ {\scriptstyle k}\downarrow & \swarrow{\scriptstyle k'} & \downarrow{\scriptstyle h} \\ A & \xrightarrow[p]{} & B \end{array} \qquad (9.59)$$

and the homomorphism $k' = kn\colon X \to A$ gives $pk' = pkn = hqn = h$.

The part on injective modules is similar.

(c) By (a) and (b), any summand of a free module is projective.

Conversely, if $P$ is projective, we choose an epimorphism $p\colon F \twoheadrightarrow P$ where $F$ is a free module (one can always take the free module generated by the set $|P|$, or by any set of generators of $P$).

This gives a homomorphism $k\colon P \to F$ such that $pk = \mathrm{id}\, P$

$$\begin{array}{ccc} & & P \\ & \swarrow{\scriptstyle k} & \downarrow{\scriptstyle \mathrm{id}\, P} \\ F & \xrightarrow[p]{} & P \end{array} \qquad (9.60)$$

### *9.5.4 Solutions of 5.2.8*

(a) We start from the left short exact sequence in $\mathsf{Ab}$

$$\mathbb{Z} \xrightarrow{\;i\;} \mathbb{Z} \xrightarrow{\;p\;} \mathbb{Z}_m \qquad\qquad \mathbb{Z}_n \xrightarrow{\;i\otimes\mathbb{Z}_n\;} \mathbb{Z}_n \xrightarrow{\;p\otimes\mathbb{Z}_n\;} \mathbb{Z}_m \otimes \mathbb{Z}_n \to 0$$

where the monomorphism $i$ is the multiplication by $m \geqslant 2$.

The tensor functor $- \otimes \mathbb{Z}_n$ gives the exact sequence on the right hand, where $\mathbb{Z}_m \otimes \mathbb{Z}_n$ is the cokernel of $i \otimes \mathbb{Z}_n$.

Now $\text{Im}\,(i \otimes \mathbb{Z}_n) = m\mathbb{Z}_n = (m\mathbb{Z} \vee n\mathbb{Z})/n\mathbb{Z} = d\mathbb{Z}/n\mathbb{Z}$, where $d$ is the greatest common divisor of $m, n$. By a Noether isomorphism

$$\mathbb{Z}_n/\text{Im}\,(i \otimes \mathbb{Z}_n) = (\mathbb{Z}/n\mathbb{Z})/(d\mathbb{Z}/n\mathbb{Z}) \cong \mathbb{Z}/d\mathbb{Z} = \mathbb{Z}_d.$$

(b) A homomorphism $f\colon \mathbb{Z}_m \to A$ amounts to an element $a = f(\overline{1})$ such that $ma = 0$. We can identify $\text{Hom}(\mathbb{Z}_m, A) = \{a \in A \mid ma = 0\}$.

Letting $n = n'd$ we also have that

$$\begin{aligned}\text{Hom}(\mathbb{Z}_m, \mathbb{Z}_n) &= \{\overline{x} \in \mathbb{Z}_n \mid mx \in n\mathbb{Z}\}\\ &= \{\overline{x} \in \mathbb{Z}_n \mid x \in n'\mathbb{Z}\} = (n'\mathbb{Z})/(n\mathbb{Z}) \cong \mathbb{Z}/d\mathbb{Z}.\end{aligned}$$

(c) Let $- \otimes A$ be exact. We take $a \in A$ with $na = 0$ and $n > 0$. Then the monomorphism $i\colon \mathbb{Z} \to \mathbb{Z}$ of multiplication by $n$ is taken by $- \otimes A$ to a monomorphism $A \to A$ which sends $a$ to 0, proving that $a = 0$.

### 9.5.5 Solutions of 5.2.9

(b) A consequence of (a) and 5.2.7(b).

(c) We have seen in (2.12) that a sequence of abelian groups can be decomposed in sequences of three objects, so that the given sequence is exact if and only if all the derived ones are short exact.

If the original exact sequence is upper unbounded, each additional object is a subgroup of an original one, and therefore a free abelian group. All the derived short exact sequences split (by 3.5.4(c)), and are preserved by any additive functor (by 5.2.1(c)).

### 9.5.6 Solutions of 5.3.3

(a) The chain complex $C_+(\{*\}; G)$ is isomorphic to

$$\ldots\ 0 \to 0 \to 0 \to 0 \to G. \tag{9.61}$$

Its homology is $G$ in degree 0, and trivial elsewhere.

(b) Applying the right exact functor $- \otimes G$ to the augmented chain complex used in the solution to Exercise 2.2.5(b), we get an augmented chain complex

$$\ldots\ C_2(X;G) \xrightarrow{\partial_2} C_1(X;G) \xrightarrow{\partial_1} C_0(X;G) \xrightarrow{\tilde{\partial}_0} G \to 0 \tag{9.62}$$

exact at $C_0(X;G)$. Therefore the homomorphism $\tilde{\partial}_0((\Sigma\lambda_i\, x_i) \otimes g) = \Sigma\, \lambda_i g$ induces an isomorphism on $H_0(X;G) = C_0(X;G)/\text{Im}\,\partial_1$

$$H_0(X;G) \to G, \qquad [x \otimes g] \mapsto g. \tag{9.63}$$

(Of course, $x$ denotes an arbitrary point of the 0-connected space $X$.)

(c) The canonical isomorphism $\bigoplus_{i\in I} C_+(X_i) \to C_+(X)$ in (2.58) is preserved by the functor $-\otimes G$ (by (5.18)), giving the canonical isomorphism (5.50).

Applying the chain homology functor we get the isomorphism (5.51), as in Exercise 2.2.5(c).

(d) Let $(Y_j)$ be the partition of $Y$ in path components. By the previous points, $f_{*0}$ amounts to embedding $G$ in $\bigoplus_j G$, as the summand corresponding to the path component which contains $f(X)$.

### *9.5.7 Solutions of 5.4.2*

(a) This is proved as in 2.4.1–2.4.2, with the present MV-sequence 5.3.6.

(b) For the space $P_n$ we use the MV-sequence 5.3.6, with the open cover $(U, V)$ used in the solution to Exercise 3.1.8(a):

$$\begin{gathered} U = P_n \setminus \{0\} \simeq \mathbb{S}^1, \qquad V = \mathrm{int}\,\mathbb{D}^2 \simeq \{*\}. \\ A = U \cap V = \mathrm{int}\,\mathbb{D}^2 \setminus \{0\} \simeq \mathbb{S}^1. \end{gathered} \tag{9.64}$$

To determine $H_1(P_n; G)$ we use the following exact sequence (omitting $H_1(V; G) = 0$ and the coefficient group $G$)

$$H_1(A) \xrightarrow{h_1} H_1(U) \xrightarrow{k_1} H_1(P_n) \xrightarrow{D_1} H_0(A) \xrightarrow{h_0} H_0(U) \oplus H_0(V)$$

The homomorphism $h_0$ is injective, by 5.3.3(d), whence $D_1 = 0$ and $k_1$ is epi: a cokernel of $h_1$. As in 3.1.8(a), $h_1$ amounts to the multiplication by $n$ in the group $G$, proving that $H_1(P_n; G) \cong G/nG$.

For $i > 1$ we have the exact sequence

$$0 \to H_i(P_n; G) \xrightarrow{D_i} H_{i-1}(A; G) \xrightarrow{h_{i-1}} H_{i-1}(U; G) \oplus H_{i-1}(V; G)$$

where $H_i(U; G) = H_i(V; G) = 0$. We distinguish two cases:

- if $i = 2$: $h_{i-1} = h_1$ has been computed above, as the multiplication by $n$ in the group $G$; therefore

$$H_2(P_n; G) \cong \mathrm{Im}\, D_2 = \mathrm{Ker}\, h_1 = \{g \in G \mid ng = 0\},$$

- if $i > 2$: $H_{i-1}(A; G) = 0$ and $H_i(P_n; G) = 0$.

(c) The previous open cover $(U, V)$ of $P_n$ gives an open cover $(U', V)$ for $\mathbb{S}^2$, where $U' = q(U)$ is the pierced sphere, a contractible space, and $U' \cap V = A \simeq \mathbb{S}^1$. The projection $q \colon P_n \to \mathbb{S}^2$ agrees with these open covers: $q(U) = U'$ and $q(V) = V$.

By the naturality of the corresponding MV-sequences, we have a commutative diagram with exact rows

$$\begin{array}{ccccc} 0 & \longrightarrow & H_2(P_n;G) & \longrightarrow & H_1(A;G) \\ \| & & \downarrow{\scriptstyle q_{*2}} & & \| \\ 0 & \longrightarrow & H_2(\mathbb{S}^2;G) & \longrightarrow & H_1(A;G) \end{array}$$

where $q_{*2}\colon H_2(P_n;G) \to H_2(\mathbb{S}^2;G)$ is a monomorphism. Taking $G = \mathbb{Z}_n$, both groups are isomorphic to $\mathbb{Z}_n$, and $q_{*2}$ is automatically an isomorphism.

*Note.* Generally, $q_{*2}$ need not be epi: it is not the case with $G = \mathbb{Z}$.

### *9.5.8 Solutions of 5.5.4*

(a) The cochain complex has a component $G$ in degree zero, and is trivial elsewhere; the same holds for the graded group $H^+(\{*\};G)$.

(b) $H^0(X;G) = Z^0(X;G)$ is the group of constant functions $\lambda\colon X \to G$, because $X$ is path connected; it can be identified with $G$, because $X \neq \emptyset$.

(c) Obvious, because any cube of $X$ has image contained in some path component, and its faces as well.

(d), (e) Obvious or easy.

## 9.6 Exercises of Chapter 6

### *9.6.1 Solutions of 6.1.3*

(a) The identity $\mathrm{id}\, A$ of a module can be lifted to $\mathrm{id}\, A_+$, and gives $F_n(\mathrm{id}\, A) = \mathrm{id}\, F_n(A)$. A composite $gf\colon A \to B \to C$ can be lifted to $g_+f_+\colon A_+ \to C_+$. A sum of morphisms $f+g\colon A \to B$ can be lifted to $f_+ + g_+\colon A_+ \to B_+$.

(b) Obvious, because $F$ is right exact: $F\varepsilon\colon FA_0 \to FA$ is a cokernel of $F\partial_1\colon FA_1 \to FA_0$, which 'is' $H_0(FA_+)$.

(c) Suppose we have a second choice of projective resolutions $\varepsilon'\colon A'_+ \to A$, for each $R$-module. Then $\mathrm{id}\, A$ lifts to a homotopy equivalence $f_+\colon A_+ \to A'_+$ of chain complexes (by Exercise 6.1.2(d)). This gives an isomorphism $\rho_n A = H_n(f_+)\colon F_n(A) \to F'_n(A)$ which only depends on the choice of resolutions. The naturality on a homomorphism $A \to B$ is easily checked.

### *9.6.2 Solutions of 6.1.6*

(a) Let $F$ be exact; for each $R$-module $A$, a projective resolution $A_+ \twoheadrightarrow A$ gives an exact chain complex $FA_+$, and $H_n(FA_+) = 0$ for $n > 0$.

This property implies $F_1 = 0$, and the latter implies the exactness of $F$ (using the exact sequence of derived functors).

(b) The projective $R$-module $P$ has a projective resolution $A_+ \twoheadrightarrow P$ where $A_0 = P$ and $A_n = 0$ for $n > 0$. Therefore $F_n(P) = H_n(A_+) = 0$ for $n > 0$.

In particular, the right exact functor $F = - \otimes_R P$ is exact, by (a).

### 9.6.3 Solutions of 6.2.1

(a) Obvious: saying that $B$ is flat means that the functor $- \otimes B$ is exact, or equivalently that its left derived functors are trivial, as we already remarked in Exercise 6.1.6(a).

(b) The free resolution $(m, p)\colon \mathbb{Z} \rightarrowtail \mathbb{Z} \twoheadrightarrow \mathbb{Z}_m$ (where $m$ stands for the multiplication by $m$) says that $\mathrm{Tor}(\mathbb{Z}_m, \mathbb{Z}_n)$ is the kernel of the homomorphism $m \otimes \mathbb{Z}_n\colon \mathbb{Z} \otimes \mathbb{Z}_n \to \mathbb{Z} \otimes \mathbb{Z}_n$, which is the multiplication $\overline{m}\colon \mathbb{Z}_n \to \mathbb{Z}_n$. The kernel is easily seen to be isomorphic to $\mathbb{Z}_d$.

Recalling that $\mathbb{Z}_m \otimes \mathbb{Z}_n$ is also isomorphic to $\mathbb{Z}_d$ suggests an alternative, clearer proof. We have an exact sequence

$$0 \to \mathrm{Tor}(\mathbb{Z}_m, \mathbb{Z}_n) \to \mathbb{Z} \otimes \mathbb{Z}_n \xrightarrow{f} \mathbb{Z} \otimes \mathbb{Z}_n \to \mathbb{Z}_m \otimes \mathbb{Z}_n \to 0$$

where $\mathbb{Z}_m \otimes \mathbb{Z}_n \cong \mathbb{Z}_d$ has $d$ elements, whence $\mathrm{Im}\, f$ has $n/d$ elements (by Lagrange's Theorem on finite groups). Therefore $\mathrm{Coim}\, f$ has also $n/d$ elements, and $\mathrm{Tor}(\mathbb{Z}_m, \mathbb{Z}_n)$ must have $d$ elements (by the same argument); being a cyclic group, as a subgroup of $\mathbb{Z}_n$, it is isomorphic to $\mathbb{Z}_d$.

### 9.6.4 Solutions of 6.2.7

(e) The homology groups of $\mathbb{T}$ and $\mathbb{K}$ with integral coefficients are determined in 3.2.3 and 3.2.4. Then we apply the following known relations

$$\mathbb{Z} \otimes \mathbb{Z}_2 \cong \mathbb{Z}_2 \otimes \mathbb{Z}_2 \cong \mathrm{Tor}(\mathbb{Z}_2, \mathbb{Z}_2) \cong \mathbb{Z}_2, \qquad \mathrm{Tor}(\mathbb{Z}, \mathbb{Z}_2) = 0.$$

### 9.6.5 Solutions of 6.4.1

(a) $\mathrm{Ext}(A, -) = 0$ means that the functor $\mathrm{Hom}(A, -)$ is exact (by 6.1.8(a)); this is equivalent to saying that $A$ is a projective $\mathbb{Z}$-module, i.e. a free abelian group (by 5.2.9(b)).

(b) Since $\mathrm{Hom}(\mathbb{Z}_m, \mathbb{Z}_n) \cong \mathbb{Z}_d$ (by Exercise 5.2.8(b)), we can give a proof similar to the second argument in 6.2.1(b). Applying Proposition 6.4.2, the short exact sequence $(m, p)\colon \mathbb{Z} \rightarrowtail \mathbb{Z} \twoheadrightarrow \mathbb{Z}_m$ gives an exact sequence

$$0 \to \mathrm{Hom}(\mathbb{Z}_m, \mathbb{Z}_n) \to \mathrm{Hom}(\mathbb{Z}, \mathbb{Z}_n) \to \mathrm{Hom}(\mathbb{Z}, \mathbb{Z}_n) \to \mathrm{Ext}(\mathbb{Z}_m, \mathbb{Z}_n) \to 0$$

which proves that $\mathrm{Ext}(\mathbb{Z}_m, \mathbb{Z}_n)$ too has $d$ elements. Being a cyclic group, as a quotient of $\mathrm{Hom}(\mathbb{Z}, \mathbb{Z}_n) \cong \mathbb{Z}_n$, it is isomorphic to $\mathbb{Z}_d$.

(c) Using again Proposition 6.4.2, the same free resolution of $\mathbb{Z}_m$ gives an exact sequence

$$\mathrm{Hom}(\mathbb{Z}, \mathbb{Z}) \xrightarrow{m^\sharp} \mathrm{Hom}(\mathbb{Z}, \mathbb{Z}) \to \mathrm{Ext}(\mathbb{Z}_m, \mathbb{Z}) \to \mathrm{Ext}(\mathbb{Z}, \mathbb{Z})$$

where $m^\sharp$ is the multiplication by $m$, while $\mathrm{Ext}(\mathbb{Z}, \mathbb{Z}) = 0$, by (a). Therefore $\mathrm{Ext}(\mathbb{Z}_m, \mathbb{Z}) \cong \mathbb{Z}_m$.

### *9.6.6 Solutions of 6.4.6*

(a) Obvious, from 6.4.1(a).

(b) Apply the split exact sequence (6.56) to the homology groups $H_k(\mathbb{P}^n)$ (see Theorem 3.3.4), using the following isomorphisms (from 6.4.1(a)–(c)):

$$\mathrm{Ext}(\mathbb{Z}, \mathbb{Z}) = \mathrm{Ext}(\mathbb{Z}, \mathbb{Z}_2) = 0, \qquad \mathrm{Ext}(\mathbb{Z}_2, \mathbb{Z}) \cong \mathrm{Ext}(\mathbb{Z}_2, \mathbb{Z}_2) \cong \mathbb{Z}_2.$$

### *9.6.7 Solutions of 6.5.1*

(a) The expression $\partial_p a \otimes b + (-1)^p\, a \otimes \partial_q b$ is $\mathbb{Z}$-linear in both variables, $a$ and $b$. Moreover, on a generator $a \otimes b \in A_p \otimes B_q$, with $n = p + q$

$$\begin{aligned} \partial_{n-1}\partial_n(a \otimes b) &= \partial_{n-1}(\partial_p a \otimes b + (-1)^p\, a \otimes \partial_q b) \\ &= (-1)^{p-1}\, \partial_p a \otimes \partial_q b + (-1)^p\, \partial_p a \otimes \partial_q b = 0. \end{aligned}$$

(b) The canonical isomorphism of abelian groups, in (5.14)

$$\varphi \colon A \otimes B \to B \otimes A, \qquad a \otimes b \mapsto b \otimes a,$$

is extended componentwise to chain complexes, letting

$$\varphi_{pq} \colon A_p \otimes B_q \to B_q \otimes A_p, \qquad a \otimes b \mapsto (-1)^{pq}\, b \otimes a. \tag{9.65}$$

This is consistent with the differential:

$$\partial_n \varphi_n(a \otimes b) = (-1)^{pq}\, \partial_n(b \otimes a) = (-1)^{pq}\, \partial_q b \otimes a + (-1)^{pq+q}\, b \otimes \partial_p a,$$

$$\begin{aligned} \varphi_{n-1}\partial_n(a \otimes b) &= \varphi_{n-1}(\partial_p a \otimes b + (-1)^p\, a \otimes \partial_q b) \\ &= (-1)^{pq-q}\, b \otimes \partial_p a + (-1)^{p+pq-p}\, \partial_q b \otimes a. \end{aligned}$$

Similarly, the isomorphisms (5.15) and (5.16) of abelian groups are extended to chain complexes

$$A_p \otimes \mathbb{Z} \to A_p, \qquad a \otimes \lambda \mapsto \lambda a. \tag{9.66}$$

$$A_p \otimes (B_q \otimes C_r) \to (A_p \otimes B_q) \otimes C_r, \quad a \otimes (b \otimes c) \mapsto a \otimes (b \otimes c). \tag{9.67}$$

### 9.6.8 Solutions of 6.5.2

(b) For all $p, q$ we have an exact sequence

$$\mathrm{Tor}(A''_p, B_q) \to A'_p \otimes B_q \to A_p \otimes B_q \to A''_p \otimes B_q \to 0$$

where $\mathrm{Tor}(A''_p, B_q) = 0$, giving a short exact sequence $(f_p \otimes B_q, g_p \otimes B_q)$. We conclude applying (a).

### 9.6.9 Solutions of 6.5.3

(b) In these hypotheses on $B$, a tensor product $A \otimes B$ of chain complexes breaks into the direct sum of a sequence of complexes

$$A \otimes B_q = ((A_{n-q} \otimes B_q)_{n \geqslant 0}, (\partial_{n-q} \otimes B_q)_{n \geqslant 0}), \qquad (9.68)$$

$$\begin{array}{ccccccc}
\ldots A_2 \otimes B_0 & \xrightarrow{\partial_2 \otimes B_0} & A_1 \otimes B_0 & \xrightarrow{\partial_1 \otimes B_0} & A_0 \otimes B_0 & \longrightarrow & 0 \\
\ldots A_1 \otimes B_1 & \xrightarrow{\partial_1 \otimes B_1} & A_0 \otimes B_1 & \longrightarrow & 0 & \longrightarrow & 0 \\
\ldots A_0 \otimes B_2 & \longrightarrow & 0 & \longrightarrow & 0 & \longrightarrow & 0 \\
\ldots \ldots \ldots & & & & & &
\end{array}$$

Its homology is the direct sum of the homology of each row. Moreover each functor $- \otimes B_q \colon \mathrm{Ch}_+\mathsf{Ab} \to \mathrm{Ch}_+\mathsf{Ab}$ is exact, and preserves homology.

### 9.6.10 Solutions of 6.6.2

(d) To verify the consistency of the differential with the product, the procedure is similar to that of 4.6.4(a), for Alexander–Spanier's cochains.

We use the following identities, between maps $\Delta^p \to \Delta^{n+1}$ or $\Delta^q \to \Delta^{n+1}$

$$\tau^-_{p+1}\delta_{p+1,i} = \delta_{n+1,i}\,\tau^-_p, \quad \tau^+_q = \delta_{n+1,i}\,\tau^+_q \qquad (0 \leqslant i \leqslant p = n-q), \quad (9.69)$$

$$\tau^-_p = \delta_{n+1,p+i}\tau^-_p, \quad \tau^+_{q+1}\,\delta_{q+1,i} = \delta_{n+1,p+i}\,\tau^+_q \quad (0 \leqslant i \leqslant q = n-p), \quad (9.70)$$

$$\tau^-_{p+1}\,\delta_{p+1,p+1} = \tau^-_p, \qquad \tau^+_q = \tau^+_{q+1}\,\delta_{q+1,0}. \qquad (9.71)$$

On a simplex $a \colon \Delta^{n+1} \to X$ we have:

$$\begin{aligned}
(d^p\lambda \smile \mu)(a) &= (d^p\lambda)(a\tau^-_{p+1}).\mu(a\tau^+_q) \\
&= \textstyle\sum_{0 \leqslant i \leqslant p+1} (-1)^i\,\lambda(a\tau^-_{p+1}\,\delta_{p+1,i})\,\mu(a\tau^+_q) \\
&= \textstyle\sum_{0 \leqslant i \leqslant p} (-1)^i\lambda(a\tau^-_{p+1}\,\delta_{p+1,i})\,\mu(a\tau^+_q) \\
&\quad + (-1)^{p+1}\lambda(a\tau^-_{p+1}\,\delta_{p+1,p+1})\,\mu(a\tau^+_q),
\end{aligned} \qquad (9.72)$$

$$
\begin{aligned}
(-1)^p(\lambda \smile d^q\mu)(a) &= (-1)^p\, \lambda(a\tau_p^-)\, (d^q\mu)(\tau_{q+1}^+) \\
&= \textstyle\sum_{0\leqslant i\leqslant q+1} (-1)^{p+i}\, \lambda(a\tau_p^-)\, \mu(\tau_{q+1}^+\, \delta_{q+1,i}) \\
&= (-1)^p\, \lambda(a\tau_p^-)\, \mu(\tau_{q+1}^+\, \delta_{q+1,0}) \\
&\quad + \textstyle\sum_{1\leqslant i\leqslant q+1} (-1)^{p+i}\, \lambda(a\tau_p^-)\, \mu(\tau_{q+1}^+\, \delta_{q+1,i}).
\end{aligned}
\tag{9.73}
$$

The last addendum in (9.72) and the first in (9.73) cancel, by (9.71). The sum of the remaining terms coincides with $d^n(\lambda \smile \mu)(a)$, applying (9.69) and (9.70):

$$
\begin{aligned}
(d^n(\lambda \smile \mu))(a) &= \textstyle\sum_{0\leqslant i\leqslant n+1} (-1)^i\, (\lambda \smile \mu)(a\delta_{n+1,i}) \\
&= \textstyle\sum_{0\leqslant i\leqslant n+1} (-1)^i\, \lambda(a\delta_{n+1,i}\, \tau_p^-)\, \mu(a\delta_{n+1,i}\, \tau_q^+) \\
&= \textstyle\sum_{0\leqslant i\leqslant p} (-1)^i\, \lambda(a\delta_{n+1,i}\, \tau_p^-)\, \mu(a\delta_{n+1,i}\, \tau_q^+) \\
&\quad + \textstyle\sum_{1\leqslant i\leqslant q+1} (-1)^{p+i}\, \lambda(a\delta_{n+1,p+i}\, \tau_p^-)\, \mu(a\delta_{n+1,p+i}\, \tau_q^+) \\
&= \textstyle\sum_{0\leqslant i\leqslant p} (-1)^i\, \lambda(a\tau_{p+1}^-\, \delta_{p+1,i})\, \mu(a\tau_q^+) \\
&\quad + \textstyle\sum_{1\leqslant i\leqslant q+1} (-1)^{p+i}\, \lambda(a\tau_p^-)\, \mu(a\tau_{q+1}^+\, \delta_{q+1,i}).
\end{aligned}
$$

### 9.6.11 Solutions of 6.6.6

(a) We use again the identities (9.69)–(9.71). We also note that $a\tau_0^-$ and $a\tau_0^+$ are 0-simplices and $\partial_0(a\tau_0^-) = \partial_0(a\tau_0^+) = 0$. Now

$$
\begin{aligned}
\partial_n\tau_n(a) &= \partial_n(\textstyle\sum_{p+q=n} a\tau_p^- \otimes a\tau_q^+) \\
&= \textstyle\sum_{p+q=n,p>0} (\partial_p(a\tau_p^-) \otimes a\tau_q^+) \\
&\quad + \textstyle\sum_{p+q=n,q>0} (-1)^p\, (a\tau_p^- \otimes \partial_q(a\tau_q^+)) \\
&= \textstyle\sum_{p+q=n,p>0} \sum_{0\leqslant i\leqslant p} (-1)^i\, (a\tau_p^-\delta_i \otimes a\tau_q^+) \\
&\quad + \textstyle\sum_{p+q=n,q>0} \sum_{0\leqslant i\leqslant q} (-1)^{p+i}\, (a\tau_p^- \otimes a\tau_q^+\delta_i) \\
&= \textstyle\sum_{p+q=n} \sum_{0\leqslant i<p} (-1)^i\, (a\tau_p^-\delta_i \otimes a\tau_q^+) + A \\
&\quad + B + \textstyle\sum_{p+q=n} \sum_{1\leqslant i\leqslant q} (-1)^{p+i}\, (a\tau_p^- \otimes a\tau_q^+\delta_i).
\end{aligned}
\tag{9.74}
$$

We have set apart two summands, called $A$ (for $i = p$) and $B$ (for $i = 0$). Using (9.71), we see that $A$ and $B$ cancel each other:

$$
\begin{aligned}
A &= \textstyle\sum_{p+q=n,p>0} (-1)^p\, a\tau_p^-\delta_p \otimes a\tau_q^+ \\
&= \textstyle\sum_{p+q=n,p>0} (-1)^p\, a\tau_p^-\delta_p \otimes a\tau_{q+1}^+\delta_0 \\
&= \textstyle\sum_{1\leqslant p\leqslant n} (-1)^p\, a\tau_p^-\delta_p \otimes a\tau_{n+1-p}^+\delta_0,
\end{aligned}
$$

$$B = \Sigma_{p+q=n,q>0}\, (-1)^p\, a\tau_p^- \otimes a\tau_q^+\delta_0$$
$$= \Sigma_{p+q=n,p<n}\, (-1)^p\, a\tau_{p+1}^-\delta_{p+1} \otimes a\tau_q^+\delta_0$$
$$= \Sigma_{1\leqslant p\leqslant n}\, (-1)^{p+1}\, a\tau_p^-\delta_p \otimes a\tau_{n+1-p}^+\delta_0 \;=\; -A.$$

Finally, (9.74) amounts to $\tau_{n-1}\partial_n(a)$, applying (9.69) and (9.70):

$$\begin{aligned}
\tau_{n-1}\partial_n(a) &= \tau_{n-1}(\Sigma_{0\leqslant i\leqslant n}(-1)^i a\delta_i)\\
&= \Sigma_{0\leqslant i\leqslant n}\,\Sigma_{p+q=n-1}\,(-1)^i\, a\delta_i\tau_p^- \otimes a\delta_i\tau_q^+\\
&= \Sigma_{p+q=n-1}\,\Sigma_{0\leqslant i\leqslant p}\,(-1)^i\,(a\delta_i\tau_p^- \otimes a\delta_i\tau_q^+)\\
&\quad + \Sigma_{p+q=n-1}\,\Sigma_{1\leqslant i\leqslant q+1}\,(-1)^{p+i}\,(a\delta_{p+1}\tau_p^- \otimes a\delta_{p+i}\tau_q^+)\\
&= \Sigma_{p+q=n-1}\,\Sigma_{0\leqslant i\leqslant p}\,(-1)^i\,(a\tau_{p+1}^-\delta_i \otimes a\tau_q^+)\\
&\quad + \Sigma_{p+q=n-1}\,\Sigma_{1\leqslant i\leqslant q+1}\,(-1)^{p+i}\,(a\tau_p^- \otimes a\tau_{q+1}^+\delta_i)\\
&= \Sigma_{p+q=n-1}\,\Sigma_{0\leqslant i<p}\,(-1)^i\,(a\tau_p^-\delta_i \otimes a\tau_q^+)\\
&\quad + \Sigma_{p+q=n}\,\Sigma_{1\leqslant i\leqslant q}\,(-1)^{p+i}\,(a\tau_p^- \otimes a\tau_q^+\delta_i).
\end{aligned} \tag{9.75}$$

(c) The cup product can be obtained composing three arrows:

- first, the bilinear mapping induced by tensoring

$$\mathrm{Hom}(S_p(X), R) \times \mathrm{Hom}(S_q(X), R) \longrightarrow \mathrm{Hom}(S_p(X) \otimes S_q(X), R \otimes R),$$
$$(\lambda, \mu) \mapsto \lambda \otimes \mu \qquad (\text{for } \lambda\colon S_p(X) \to R,\ \mu\colon S_q(X) \to R),$$

- second, the homomorphism induced covariantly by the product in $R$

$$\mathrm{Hom}(S_p(X) \otimes S_q(X), R \otimes R) \longrightarrow \mathrm{Hom}(S_p(X) \otimes S_q(X), R),$$

- third, the homomorphism induced contravariantly by $\tau_{pq}X$

$$(\tau_{pq}X)^\sharp\colon \mathrm{Hom}(S_p(X) \otimes S_q(X), R) \longrightarrow \mathrm{Hom}(S_{p+q}(X), R).$$

### *9.6.12 Solutions of 6.6.7*

(a) It is a trivial consequence of the analogous formula for the cup product, in 6.6.2(d), taking into account that the differential commutes with the homomorphisms $p^{\sharp i}$.

(b) The diagonal $\Delta\colon X \to X \times X$ gives a composite

$$\begin{aligned}
S^p(X;R) \times S^q(X;R) &\longrightarrow S^n(X \times X; R) \to S^n(X;R),\\
(\lambda, \mu) &\mapsto \Delta^\sharp(\lambda \times \mu) \qquad (p+q=n),
\end{aligned} \tag{9.76}$$

which coincides with the cup product, because

$$\Delta^\sharp(\lambda \times \mu) = \Delta^\sharp p^{\sharp 1}(\lambda) \smile \Delta^\sharp p^{\sharp 2}(\mu) = (p_1\Delta)^\sharp(\lambda) \smile (p_2\Delta)^\sharp(\mu) = \lambda \smile \mu.$$

$$\begin{aligned}(-1)^p(\lambda \smile d^q\mu)(a) &= (-1)^p\, \lambda(a\tau_p^-)\,(d^q\mu)(\tau_{q+1}^+)\\ &= \Sigma_{0\leqslant i\leqslant q+1}\,(-1)^{p+i}\,\lambda(a\tau_p^-)\,\mu(\tau_{q+1}^+\,\delta_{q+1,i})\\ &= (-1)^p\,\lambda(a\tau_p^-)\,\mu(\tau_{q+1}^+\,\delta_{q+1,0})\\ &\quad +\Sigma_{1\leqslant i\leqslant q+1}\,(-1)^{p+i}\,\lambda(a\tau_p^-)\,\mu(\tau_{q+1}^+\,\delta_{q+1,i}).\end{aligned} \tag{9.73}$$

The last addendum in (9.72) and the first in (9.73) cancel, by (9.71). The sum of the remaining terms coincides with $d^n(\lambda \smile \mu)(a)$, applying (9.69) and (9.70):

$$\begin{aligned}(d^n(\lambda \smile \mu))(a) &= \Sigma_{0\leqslant i\leqslant n+1}\,(-1)^i\,(\lambda \smile \mu)(a\delta_{n+1,i})\\ &= \Sigma_{0\leqslant i\leqslant n+1}\,(-1)^i\,\lambda(a\delta_{n+1,i}\,\tau_p^-)\,\mu(a\delta_{n+1,i}\,\tau_q^+)\\ &= \Sigma_{0\leqslant i\leqslant p}\,(-1)^i\,\lambda(a\delta_{n+1,i}\,\tau_p^-)\,\mu(a\delta_{n+1,i}\,\tau_q^+)\\ &\quad +\Sigma_{1\leqslant i\leqslant q+1}\,(-1)^{p+i}\,\lambda(a\delta_{n+1,p+i}\,\tau_p^-)\,\mu(a\delta_{n+1,p+i}\,\tau_q^+)\\ &= \Sigma_{0\leqslant i\leqslant p}\,(-1)^i\,\lambda(a\tau_{p+1}^-\,\delta_{p+1,i})\,\mu(a\tau_q^+)\\ &\quad +\Sigma_{1\leqslant i\leqslant q+1}\,(-1)^{p+i}\,\lambda(a\tau_p^-)\,\mu(a\tau_{q+1}^+\,\delta_{q+1,i}).\end{aligned}$$

### 9.6.11 Solutions of 6.6.6

(a) We use again the identities (9.69)–(9.71). We also note that $a\tau_0^-$ and $a\tau_0^+$ are 0-simplices and $\partial_0(a\tau_0^-) = \partial_0(a\tau_0^+) = 0$. Now

$$\begin{aligned}\partial_n\tau_n(a) &= \partial_n(\Sigma_{p+q=n}\,a\tau_p^- \otimes a\tau_q^+)\\ &= \Sigma_{p+q=n,p>0}\,(\partial_p(a\tau_p^-)\otimes a\tau_q^+)\\ &\quad +\Sigma_{p+q=n,q>0}\,(-1)^p\,(a\tau_p^- \otimes \partial_q(a\tau_q^+))\\ &= \Sigma_{p+q=n,p>0}\,\Sigma_{0\leqslant i\leqslant p}\,(-1)^i\,(a\tau_p^-\delta_i \otimes a\tau_q^+)\\ &\quad +\Sigma_{p+q=n,q>0}\,\Sigma_{0\leqslant i\leqslant q}\,(-1)^{p+i}\,(a\tau_p^- \otimes a\tau_q^+\delta_i)\\ &= \Sigma_{p+q=n}\,\Sigma_{0\leqslant i<p}\,(-1)^i\,(a\tau_p^-\delta_i \otimes a\tau_q^+) + A\\ &\quad +B + \Sigma_{p+q=n}\,\Sigma_{1\leqslant i\leqslant q}\,(-1)^{p+i}\,(a\tau_p^- \otimes a\tau_q^+\delta_i).\end{aligned} \tag{9.74}$$

We have set apart two summands, called $A$ (for $i = p$) and $B$ (for $i = 0$). Using (9.71), we see that $A$ and $B$ cancel each other:

$$\begin{aligned}A &= \Sigma_{p+q=n,p>0}\,(-1)^p\,a\tau_p^-\,\delta_p \otimes a\tau_q^+\\ &= \Sigma_{p+q=n,p>0}\,(-1)^p\,a\tau_p^-\,\delta_p \otimes a\tau_{q+1}^+\delta_0\\ &= \Sigma_{1\leqslant p\leqslant n}\,(-1)^p\,a\tau_p^-\,\delta_p \otimes a\tau_{n+1-p}^+\delta_0,\end{aligned}$$

$$\begin{aligned} B &= \textstyle\sum_{p+q=n,q>0} (-1)^p \, a\tau_p^- \otimes a\tau_q^+ \delta_0 \\ &= \textstyle\sum_{p+q=n,p<n} (-1)^p \, a\tau_{p+1}^- \delta_{p+1} \otimes a\tau_q^+ \delta_0 \\ &= \textstyle\sum_{1\leqslant p\leqslant n} (-1)^{p+1} \, a\tau_p^- \delta_p \otimes a\tau_{n+1-p}^+ \delta_0 \;=\; -A. \end{aligned}$$

Finally, (9.74) amounts to $\tau_{n-1}\partial_n(a)$, applying (9.69) and (9.70):

$$\begin{aligned} \tau_{n-1}\partial_n(a) &= \tau_{n-1}(\textstyle\sum_{0\leqslant i\leqslant n}(-1)^i a\delta_i) \\ &= \textstyle\sum_{0\leqslant i\leqslant n} \sum_{p+q=n-1} (-1)^i \, a\delta_i\tau_p^- \otimes a\delta_i\tau_q^+ \\ &= \textstyle\sum_{p+q=n-1} \sum_{0\leqslant i\leqslant p} (-1)^i \, (a\delta_i\tau_p^- \otimes a\delta_i\tau_q^+) \\ &\quad + \textstyle\sum_{p+q=n-1} \sum_{1\leqslant i\leqslant q+1} (-1)^{p+i} \, (a\delta_{p+1}\tau_p^- \otimes a\delta_{p+i}\tau_q^+) \\ &= \textstyle\sum_{p+q=n-1} \sum_{0\leqslant i\leqslant p} (-1)^i \, (a\tau_{p+1}^-\delta_i \otimes a\tau_q^+) \\ &\quad + \textstyle\sum_{p+q=n-1} \sum_{1\leqslant i\leqslant q+1} (-1)^{p+i} \, (a\tau_p^- \otimes a\tau_{q+1}^+\delta_i) \\ &= \textstyle\sum_{p+q=n-1} \sum_{0\leqslant i<p} (-1)^i \, (a\tau_p^-\delta_i \otimes a\tau_q^+) \\ &\quad + \textstyle\sum_{p+q=n} \sum_{1\leqslant i\leqslant q} (-1)^{p+i} \, (a\tau_p^- \otimes a\tau_q^+\delta_i). \end{aligned} \tag{9.75}$$

(c) The cup product can be obtained composing three arrows:

- first, the bilinear mapping induced by tensoring

$$\mathrm{Hom}(S_p(X), R) \times \mathrm{Hom}(S_q(X), R) \longrightarrow \mathrm{Hom}(S_p(X) \otimes S_q(X), R \otimes R),$$
$$(\lambda, \mu) \mapsto \lambda \otimes \mu \qquad (\text{for } \lambda\colon S_p(X) \to R,\ \mu\colon S_q(X) \to R),$$

- second, the homomorphism induced covariantly by the product in $R$

$$\mathrm{Hom}(S_p(X) \otimes S_q(X), R \otimes R) \longrightarrow \mathrm{Hom}(S_p(X) \otimes S_q(X), R),$$

- third, the homomorphism induced contravariantly by $\tau_{pq}X$

$$(\tau_{pq}X)^\sharp\colon \mathrm{Hom}(S_p(X) \otimes S_q(X), R) \longrightarrow \mathrm{Hom}(S_{p+q}(X), R).$$

### *9.6.12 Solutions of 6.6.7*

(a) It is a trivial consequence of the analogous formula for the cup product, in 6.6.2(d), taking into account that the differential commutes with the homomorphisms $p^{\sharp i}$.

(b) The diagonal $\Delta\colon X \to X \times X$ gives a composite

$$\begin{aligned} S^p(X;R) \times S^q(X;R) &\longrightarrow S^n(X \times X; R) \to S^n(X;R), \\ (\lambda, \mu) &\mapsto \Delta^\sharp(\lambda \times \mu) \qquad (p+q=n), \end{aligned} \tag{9.76}$$

which coincides with the cup product, because

$$\Delta^\sharp(\lambda \times \mu) = \Delta^\sharp p^{\sharp 1}(\lambda) \smile \Delta^\sharp p^{\sharp 2}(\mu) = (p_1\Delta)^\sharp(\lambda) \smile (p_2\Delta)^\sharp(\mu) = \lambda \smile \mu.$$

### 9.6.13 Solutions of 6.7.7

(a)–(c) Obvious consequences of Theorem 6.7.6, taking into account that, for free abelian groups $A, B$ of finite dimension

$$\dim(A \oplus B) = \dim(A) + \dim(B), \qquad \dim(A \otimes B) = \dim(A)\dim(B).$$

(d) In fact $\chi(X)$ is the value of the polynomial $\beta X$ at $-1$, and it is sufficient to apply formula (6.123).

(e) The Poincaré polynomial of the circle is $\beta\mathbb{S}^1(t) = 1 + t$, whence

$$\beta\mathbb{T}^n(t) = (1+t)^n, \qquad H_k(\mathbb{T}^n) \cong \mathbb{Z}^{\binom{n}{k}} \qquad (0 \leqslant k \leqslant n), \tag{9.77}$$

and $H_k(\mathbb{T}^n) = 0$ for $k > n$ (as for any manifold of dimension $n$).

The identity $\binom{n}{k} = \binom{n}{n-k}$ corresponds to the formula $\beta_k(M) = \beta_{n-k}(M)$ in (6.64), for a compact orientable manifold of dimension $n$.

## 9.7 Exercises of Chapter 7

### 9.7.1 Solutions of 7.1.2

(a) We have two homotopies $h, k$ with fixed endpoints, between consecutive paths

$$h\colon a_0 \simeq_2 a_1\colon x \simeq x', \qquad k\colon b_0 \simeq_2 b_1\colon x' \simeq x''.$$

Concatenating them, on the first variable, we obtain a 2-homotopy $H\colon a_0 * b_0 \simeq_2 a_1 * b_1\colon x \simeq x''$

$$H(s,t) = \begin{cases} h(2s,t), & \text{for } 0 \leqslant s \leqslant 1/2, \\ k(2s-1,t), & \text{for } 1/2 \leqslant s \leqslant 1, \end{cases} \tag{9.78}$$

$$H(-,i) = a_i * b_i, \qquad H(0,t) = x, \qquad H(1,t) = x'' \qquad (t \in \mathbb{I},\ i = 0,1),$$

whose continuity follows from an obvious closed cover of $\mathbb{I} \times \mathbb{I}$.

(b) We take the iterated concatenations of three consecutive paths $a\colon x \simeq x'$, $b\colon x' \simeq x''$, and $c\colon x'' \simeq x'''$

$$d = a * (b * c), \qquad d' = (a * b) * c. \tag{9.79}$$

The first path is parametrised on $\mathbb{I} = [0,1/2] \cup [1/2,3/4] \cup [3/4,1]$, moving with double velocity on the first subinterval, and four-tuple velocity on the remaining two. The second is parametrised on $\mathbb{I} = [0,1/4] \cup [1/4,1/2] \cup [1/2,1]$, moving with four-tuple velocity on the first two subintervals, then with double velocity on the last.

Each of these paths is a reparametrisation of the other; for instance

$d' = df : \mathbb{I} \to X$, where the (invertible) function $f : \mathbb{I} \to \mathbb{I}$ turns the second partition above into the first

$$f(t) = \begin{cases} 2t, & \text{for } 0 \leqslant t \leqslant 1/4, \\ t + 1/4, & \text{for } 1/4 \leqslant t \leqslant 1/2, \\ (t+1)/2, & \text{for } 1/2 \leqslant t \leqslant 1. \end{cases}$$

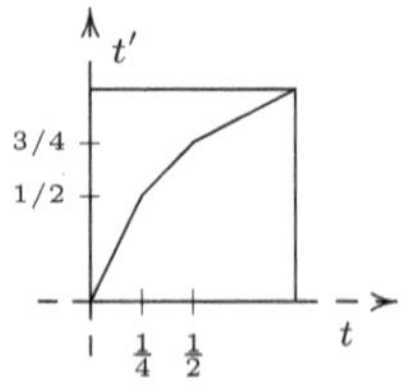


Now any 2-homotopy $h : \mathrm{id} \simeq_2 f : \mathbb{I} \to \mathbb{I}$ gives a 2-homotopy $dh : d \simeq_2 d' : \mathbb{I} \to X$ (by whisker composition). For instance we can take at each $s \in \mathbb{I}$ the affine linear path from $s$ to $f(s)$

$$h : \mathbb{I} \times \mathbb{I} \to \mathbb{I}, \qquad h(s,t) = (1-t).s + t.f(s) \qquad (s, t \in \mathbb{I}), \tag{9.80}$$

which is constant at $s = 0, 1$.

(c) For a path $a : x \simeq x'$, the paths $e_x * a$ and $a * e_{x'}$ are reparametrisations of $a$ by (non-invertible) functions $f, g : \mathbb{I} \to \mathbb{I}$

$$c = e_x * a = af, \qquad\qquad d = a * e_{x'} = ag, \tag{9.81}$$

$f(t) = \max(0, 2t - 1)$

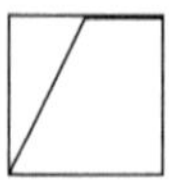

$g(t) = \min(2t, 1)$.

Therefore a 2-homotopy $h : \mathrm{id} \simeq_2 f$ (resp. $k : \mathrm{id} \simeq g$) gives a 2-homotopy $ah : a \simeq_2 e_x * a$ (resp. $ak : a \simeq_2 a * e_{x'}$). As in (b), these 2-homotopies $\mathbb{I} \times \mathbb{I} \to \mathbb{I}$ can be defined by affine linear combinations in the second variable.

(d) For a path $a : x \simeq x'$, a 2-homotopy $h : e_x \simeq_2 a * a^\sharp$ can be obtained as

$$h(s,t) = \begin{cases} a(2st), & \text{for } s \leqslant 1/2, \\ a(2(1-s)t), & \text{for } s \geqslant 1/2. \end{cases} \tag{9.82}$$

Similarly we have a 2-homotopy $k : e_{x'} \simeq_2 a^\sharp * a$.

### 9.7.2 *Solutions of 7.1.3*

(a) A 2-homotopy $h : a \simeq_2 a' : x \simeq x'$ in the space $X$ gives a 2-homotopy $fh : fa \simeq_2 fa' : fx \simeq fx'$ in $Y$.

(b) Composing with $f : X \to Y$ preserves path concatenation and trivial loops (of $X$).

(c) Obvious:

$$(gf)_*([a]) = ([gfa]) = g_*(f_*([a])), \quad (\mathrm{id}\,X)_*([a]) = [(\mathrm{id}\,X)[a] = [a].$$

(d) There is an obvious functor $F\colon \mathsf{Gpd} \to \mathsf{Set}$ that takes a small groupoid $G$ to the quotient of the set $\mathrm{Ob}\,G$ of its objects, modulo the equivalence relation that identifies two objects $x, x'$ when there is in $G$ an arrow $x \to x'$. The composite $F\Pi_1$ coincides with $\Pi_0$.

Viewing a groupoid as a representation of a 1-dimensional space, this functor $F$ can be written as $\Pi_0\colon \mathsf{Gpd} \to \mathsf{Set}$. Similarly, a small category can be viewed as an elementary form of a *directed* space [G1].

*Letting $\mathrm{Ob} : \mathsf{Gpd} \to \mathsf{Set}$ be the forgetful functor of objects, there is a chain of adjunctions $\Pi_0 \dashv D \dashv \mathrm{Ob} \dashv C$, where $D(X)$ is the discrete groupoid on the set $X$, and $C(X)$ is the indiscrete one.*

### 9.7.3 Solutions of 7.1.5

(b) Obviously (i) $\Leftrightarrow$ (iv) $\Leftrightarrow$ (v) $\Leftrightarrow$ (ii).

(c) A consequence of (b).

*The additional property (iv$'$) is also equivalent to the previous ones, once we know that all groups $\pi_1(X, x)$ are isomorphic when $X$ is path connected (see 7.3.1).*

(d) A consequence of the previous points: the corresponding properties of the fundamental groupoid (namely: being connected, or equivalent to $\mathbf{1}$) are invariant up to categorical equivalence.

### 9.7.4 Solutions of 7.1.9

(b) After the suggested choice of the paths $\lambda_x\colon x \simeq P(x)$, the functor $P\colon \Pi_1 X \to \Pi_1 X|A$ is defined on vertices, and we complete it on arrows, letting

$$P(\alpha\colon x \to y) = ([\lambda_x]^{-1}\alpha[\lambda_y]\colon P(x) \to P(y)). \tag{9.83}$$

(Recall that, in these groupoids, we are writing the composition in the natural order of concatenation.)

We constrain the previous choice requiring that, if $x \in A$, then $\lambda_x = e_x$ (the trivial loop at $x$). In this way $PJ$ is the identity. Moreover, the family $([\lambda_x])$ forms a natural transformation $\mathrm{id}\,\Pi_1 X \to JP$, which is invertible (because all its components are).

### 9.7.5 Solutions of 7.2.3

(a) For a loop $a\colon x_0 \simeq x_0$, the pointed map $h = \varphi(a \times \mathbb{I})\colon \mathbb{I} \to Y$ is a 2-homotopy $h\colon fa \simeq ga$ of loops at $y_0 \in Y$, which gives $[fa] = [ga]$.

We can also deduce this result from Theorem 7.1.4 on fundamental groupoids: the commutative square (7.7) gives the following relation in the group $\pi_1(Y, y_0)$

$$[fa]\varphi_*(x_0) = \varphi_*(x_0)[ga],$$

and $[fa] = [ga]$, because $\varphi_*(x_0) = [e_{y_0}]$ is the class of the trivial loop at $y_0$.

### 9.7.6 Solutions of 7.3.1

(a) Suppose that $\pi_1(X, x_0)$ is commutative and $c\colon x_0 \simeq x_1$ in $X$. Every path $d\colon x_0 \simeq x_1$ in $X$ can be expressed, up to 2-homotopy, as $d = u * c$, where $u$ is a loop at $x_0$. Then, for every $a \in \Omega(X, x_0)$: $\hat{d}[a] = [c]^{-1}[u]^{-1}[a][u][c] = \hat{c}[a]$.

Conversely, the transition isomorphism $\hat{u}\colon \pi_1(X, x_0) \to \pi_1(X, x_0)$ determined by any loop $u$ at $x_0$ must be the identity, which means that $[u]$ commutes with any element of $\pi_1(X, x_0)$.

### 9.7.7 Solutions of 7.3.3

(a) The loop $a \in \Omega(X, x_0)$ is a cycle of $X$, and its homology class is invariant up to loop homotopy (Exercise 2.2.8(c)), which is wider than homotopy with fixed basepoint. The property $h([a][b]) = h[a * b] = [a]_H + [b]_H$ is proved in Exercise 2.2.8(e).

(b) For a loop $a \in \Omega(X, x_0)$, the loop $c^\sharp * a * c$ is loop homotopic to $a$ (by Lemma 7.3.6); therefore $h'\hat{c}[a] = h'[a] = [a]_H = h[a]$.

(c) For a pointed map $f\colon (X, x_0) \to (Y, y_0)$ we have:

$$h(Y, y_0)\, \pi_1(f)[a] = [fa]_H = H_1(f)\, h(X, x_0)[a].$$

(d) It is sufficient to verify that $[G, G]$ is closed under the inner automorphism $\hat{g}(x) = g^{-1}xg$, for every $g \in G$. In fact $\hat{g}$ takes every commutator $[x, y]$ to the commutator $[\hat{g}(x), \hat{g}(y)]$.

(e) Obvious: $G/[G, G]$ is commutative and every homomorphism $f\colon G \to A$ with values in an abelian group takes the subgroup $[G, G]$ to 0.

### 9.7.8 Solutions of 7.4.1

(a) To construct $*_i G_i$, we first form an *alphabet* $X = \sum_i |G_i|$, the disjoint union of the underlying sets, and then the set $W$ of finite words $(x_1, x_2, ..., x_n)$ on this alphabet $X$, including the empty word $e$. Multiplying words by concatenation, the set $W$ is a monoid. (It is actually the free monoid on the set $X$, as we have seen in 1.6.7(c).)

Each word $w \in W$ has a 'reduced form' $R(w)$, obtained by replacing every occurrence $x_h, x_{h+1}$ of two elements in the same group $G_i$ with their product in this group, and taking out any unit element of some $G_i$. The result does not depend on the sequence of reductions (this seemingly obvious point requires in fact a complex proof). We say that $w \sim w'$ when $R(w) = R(w')$.

The quotient set $*_i G_i = W/\!\sim$ has a well defined product: $[w][w'] = [ww']$; indeed the concatenation $ww'$ is equivalent to $R(w)R(w')$, because $R(ww')$ can be obtained by iterated reductions, as $R(R(w)R(w'))$.

The product is associative (it already is in $W$). The unit is the class of the empty word. The inverse of the class $[x_1, x_2, ..., x_n]$ is obviously the class $[x_n^{-1}, ..., x_2^{-1}, x_1^{-1}]$.

(b) First we have a unique mapping of sets $h\colon X \to |H|$ which coincides with $f_i$ on each subset $|G_i|$ (by the universal property of the sum of sets, in 1.6.2(a)). Then we extend it to the monoid $W$ of finite words on $X$, letting $k(x_1, x_2, ..., x_n) = h(x_1)h(x_2)... h(x_n)$. Finally, this homomorphism of monoids $k\colon W \to H$ is invariant under the equivalence relation $\sim$ producing $G$, and the induced homomorphism $f\colon G \to H$ satisfies the required conditions $fu_i = f_i$ (for all $i$).

Its uniqueness is attested by the fact that $G$ is spanned by the images of all injections.

(d), (e) For a mapping $f\colon X \to |H|$ with values in a group $H$, we consider the homomorphism

$$g\colon *_{x \in X} \mathbb{Z} \to H, \qquad g(x_1^{k_1} x_2^{k_2} ... x_p^{k_p}) = f(x_1)^{k_1} f(x_2)^{k_2} ... f(x_p)^{k_p}.$$

It does satisfy $g\eta = f$, and is determined by this property since the elements $\eta(x)$, for $x \in X$, generate the domain of $g$.

### 9.7.9 Solutions of 7.4.2

(a) Already known from the general case in 1.7.1, since the trivial group is the initial object of the category.

(b) In the categorical sum $G_1 * G_2$ the normal subgroup $K$ produces the congruence of groups generated by the pairs $(u_1f_1(x), u_2f_2(x))$, for $x \in G_0$. The universal property follows easily, as for Set and Top, in 1.7.1.

### 9.7.10 Solutions of 7.4.4

(a) We use the open cover $(U, V)$ of $\mathbb{S}^n$ already used in 2.4.2 for the Mayer–Vietoris sequence.

The subspaces $U = \mathbb{S}^n \setminus \{\underline{s}\}$ and $V = \mathbb{S}^n \setminus \{\underline{n}\}$ are contractible, and their fundamental group is trivial (by 7.3.2(c)). The intersection $U \cap V \simeq \mathbb{S}^{n-1}$ is also path connected (for $n \geqslant 2$), and the van Kampen theorem applies. This is sufficient to say that $\pi_1(\mathbb{S}^n) = 0$.

(b) As in 3.1.1, we realise the figure-eight space $X = \mathbb{S}^1 \vee \mathbb{S}^1$ as two circles in $\mathbb{R}^2$ tangent at the origin, and we use the same open cover $(U, V)$ used there to compute $H_1(X)$.

Now $\pi_1(U, 0)$ and $\pi_1(V, 0)$ are infinite cyclic groups, generated by the homotopy classes $[a]$ and $[b]$ of the following loops at the origin, around the left and the right circle (represented in figure 7.24)

$$a(t) = (-1 + \cos 2\pi t,\ \sin 2\pi t), \qquad b(t) = (1 - \cos 2\pi t,\ -\sin 2\pi t). \qquad (9.84)$$

Since $\pi_1(A)$ is the trivial group, the van Kampen Theorem gives a canonical isomorphism $\pi_1(U, 0) * \pi_1(V, 0) \to \pi_1(X, 0)$.

Therefore $\pi_1(X, 0)$ is the free group generated by two elements, the homotopy classes $[a]$ and $[b]$. By 7.4.1(e), each element can be uniquely written as a product

$$[a]^{h_1}[b]^{k_1}[a]^{h_2}[b]^{k_2} \dots [a]^{h_n}[b]^{k_n}, \qquad (9.85)$$

with *non-zero* integral exponents, except (possibly) $h_1, k_n \in \mathbb{Z}$.

Similarly, the fundamental group of a bouquet of $n$ circles $\mathbb{S}^1 \vee \dots \vee \mathbb{S}^1$ is the free group generated by $n$ elements.

(c) Suppose that $\pi_1(X, x_0)$ has two elements $[a], [b]$ which do not commute; for instance, this is the case of the figure-eight space. Then the loops $a$ and $a' = b^\sharp * a * b$ are not 2-homotopic; but we know that they are homotopic as free loops, by Lemma 7.3.6.

(d) We use the same open cover $(U, V)$ of $P_n$, as in the solution of Exercise 3.1.8(a)

$$\begin{gathered} U = P_n \setminus \{0\} \simeq \mathbb{S}^1, \qquad V = \operatorname{int} \mathbb{D}^2 \simeq \{*\}, \\ A = U \cap V = \operatorname{int} \mathbb{D}^2 \setminus \{0\} \simeq \mathbb{S}^1. \end{gathered} \qquad (9.86)$$

The van Kampen Theorem gives a pushout of groups

$$\begin{array}{ccc} \mathbb{Z} & \xrightarrow{u} & \mathbb{Z} \\ \downarrow & & \downarrow{\scriptstyle v} \\ 0 & \longrightarrow & \pi_1(P_n, x_0) \end{array}$$

where the homomorphism $u$ is the multiplication by $n$ (as in the solution of Exercise 3.1.8(a), for the homomorphism $h_1\colon H_1(A) \to H_1(U)$). The group $\pi_1(P_n, x_0)$ is thus the cokernel of $u$ in $\mathsf{Gp}$ (and in $\mathsf{Ab}$), namely the abelian group $\mathbb{Z}_n$.

### *9.7.11 Solutions of 7.5.2*

(a) For $n = 0$, $\mathbb{I}^0$ is the singleton $\{0\} = \mathbb{R}^0$, and $\partial\mathbb{I}^0 = \emptyset$; thus $\Omega_0(X, x_0)$ can be identified with the underlying set $|X|$, pointed at $x_0$. Relative homotopies are paths $x \simeq x'$, and $\pi_0(X, x_0) = (|X|/\!\simeq, [x_0])$ is the set of path components of $X$, pointed at the class of the basepoint $x_0$, as defined in (1.72).

(b) For $i = 1, ..., n$, two cubes $a, b \in \Omega_n(X, x_0)$ are always consecutive in direction $i$ (all their faces being constant at $x_0$); all the faces of their concatenation $a *_i b$ are constant at $x_0$.

Each operation induces a group structure $[a] +_i [b] = [a *_i b]$ on $\pi_n(X, x_0)$, with the same proof as in the case $n = 1$ (although we shall see that these structures coincide). The identity is (always) the class $[u_n]$ of the constant map $\mathbb{I}^n \to X$ at $x_0$. The inverse $-_i[a]$ is the class of the $i$-reversed cube

$$(t_1, ..., t_i, ..., t_n) \mapsto a(t_1, ..., 1 - t_i, ..., t_n).$$

(c) Property (7.70) holds more generally for four cubes $a, b, c, d\colon \mathbb{I}^n \to X$ whose $i$-faces and $j$-faces agree, as in the following picture (we only represent the two directions $i < j$ of the operations)

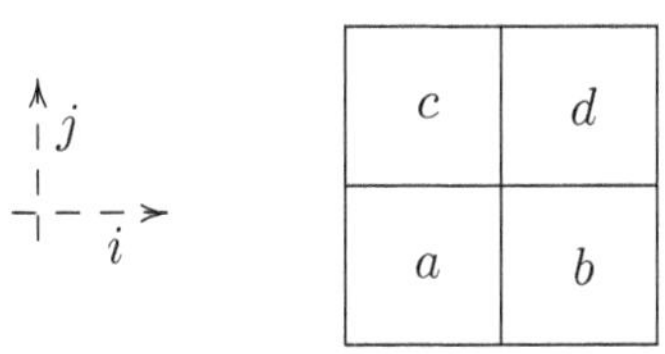


$$\partial_i^+ c = \partial_i^- d, \quad \partial_j^- c = \partial_j^+ a, \quad \partial_j^- d = \partial_j^+ b, \quad \partial_i^+ a = \partial_i^- b. \tag{9.87}$$

Then both pastings $(a *_i b) *_j (c *_i d)$ and $(a *_j c) *_i (b *_j d)$ give the same $n$-cube $f\colon \mathbb{I}^n \to X$, letting $i < j$ and omitting the other variables $t_k$, for

$k \neq i, j$)

$$f(t_i, t_j) = \begin{cases} a(2t_i, 2t_j), & \text{for } t_i, t_j \leqslant 1/2, \\ b(2t_i - 1, 2t_j), & \text{for } t_i \geqslant 1/2,\ t_j \leqslant 1/2, \\ c(2t_i, 2t_j - 1), & \text{for } t_i \leqslant 1/2,\ t_j \geqslant 1/2, \\ d(2t_i - 1, 2t_j - 1), & \text{for } t_i, t_j \geqslant 1/2. \end{cases}$$

(e) First, the operations coincide, because

$$x \bullet y = (x + e) \bullet (e + y) = (x \bullet e) + (e \bullet y) = x + y. \tag{9.88}$$

Second, this operation is commutative, because

$$x \bullet y = (e + x) \bullet (y + e) = (e \bullet y) + (x \bullet e) = y + x = y \bullet x. \tag{9.89}$$

Concretely, if $a, b \in \Omega_n(X, x_0)$ and $[a] = x$, $[b] = y$ in $\pi_n(X, x_0)$, the identities (9.88) show how we should deform the cube $a *_i b$ into $a *_j b$

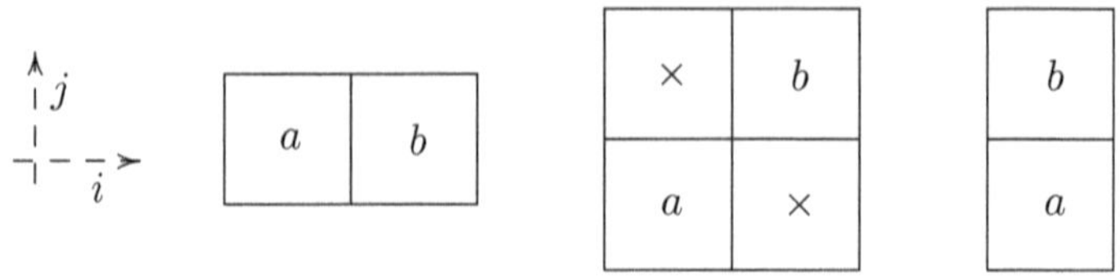

(9.90)

The cross-marked squares represent the constant cube $u_n \colon \mathbb{I}^n \to X$ at $x_0$, whose class is the identity of $\pi_n(X, x_0)$. A further deformation transforms $a *_j b$ into $b *_i a$.

### 9.7.12 Solutions of 7.6.2

(a) The spaces $D$ and $\mathbb{N}^\bullet$ are totally path-disconnected, and $f \colon D \to \mathbb{N}^\bullet$ induces a bijection from $\mathsf{Set}((\mathbb{S}^n, e_1), (D, x))$ to $\mathsf{Set}((\mathbb{S}^n, e_1), (\mathbb{N}^\bullet, f(x))$, for all $n \geqslant 0$ and $x \in D$. On the other hand, every map $g \colon \mathbb{N}^\bullet \to D$ is eventually constant at $g(\infty)$; the composite $gf \colon \mathbb{N}^\bullet \to \mathbb{N}^\bullet$ is also, and is not homotopic to the identity.

(b) The homomorphism $\overline{H}^0(f) \colon \overline{H}^0(\mathbb{N}^\bullet) \to \overline{H}^0(D)$, in Alexander–Spanier cohomology, is an embedding of a proper countable subring, as we have seen in Exercise 4.5.8(e).

## 9.8 Exercises of Chapter 8

### 9.8.1 Solutions of 8.1.7

(a) The terminal object is a unit for the cartesian product, by 1.6.1(g). The adjunction gives a canonical bijection

$$\mathsf{C}(A, Y) \to \mathsf{C}(\top \times A, Y) \to \mathsf{C}(\top, Y^A) = U(Y^A). \tag{9.91}$$

(b) The exponential law of pOrd is an enriched restriction of that of sets

$$\begin{gathered} \varphi_{XY} \colon \mathsf{pOrd}(X \times A, Y) \to \mathsf{pOrd}(X, Y^A), \\ (f \colon X \times A \to Y) \mapsto (g \colon X \to Y^A), \qquad g(x) = f(x, -) \colon A \to Y, \end{gathered} \tag{9.92}$$

depending on the fact that a mapping $f \colon X \times A \to Y$ is monotone if and only if it is monotone in each variable: a relation $(x, a) \prec (x', a')$ can always be split as: $(x, a) \prec (x', a) \prec (x', a')$.

(c) If $A, B$ are exponentiable objects, the composed endofunctor $F = (- \times A) \times B \colon \mathsf{C} \to \mathsf{C}$ has a composed right adjoint $G = ((-)^B)^A$. The canonical isomorphism $F = - \times (A \times B)$ implies that $A \times B$ is exponentiable and $G = (-)^{A \times B}$.

(d) In a pointed category any object $X$ is isomorphic to $X \times \top = X \times \bot$. If $X$ is exponentiable, the functor $X \times -$ preserves the initial object and $X$ is also initial, that is, the zero object. If $\mathsf{C}$ is cartesian closed, any functor $\mathbf{1} \to \mathsf{C}$ is an equivalence of categories, by 8.1.5.

### 9.8.2 Solutions of 8.2.2

(a) By straightforward verification.

(b) By the definition of homotopies, the homotopy $\partial$ establishes a bijection between maps $\hat{\varphi} \colon IX \to Y$ and homotopies $\varphi = \hat{\varphi}\partial \colon \hat{\varphi}\partial^- \simeq \hat{\varphi}\partial^+ \colon X \to Y$.

(c) Follows from (a) and Lemma 4.9.2 on pushouts in Top, taking into account that the faces $\partial^\alpha \colon X \to IX$ are topological embeddings.

(d) A morphism $(x, y, z) \colon (f, g) \to (f', g')$ in $\mathsf{Top}^\vee$ is a commutative diagram

$$\begin{array}{ccccc} Y & \xleftarrow{\ f\ } & X & \xrightarrow{\ g\ } & Z \\ {\scriptstyle y}\downarrow & & \downarrow{\scriptstyle x} & & \downarrow{\scriptstyle z} \\ Y' & \xleftarrow[\ f'\ ]{} & X' & \xrightarrow[\ g'\ ]{} & Z' \end{array} \tag{9.93}$$

Writing as $\lambda' \colon u'f' \simeq v'g' \colon X' \to I(f', g')$ the homotopy of the second h-pushout, the universal property of $I(f,g)$ with respect to the four-tuple $(I(f',g'); f'x, g'x; \lambda'x)$ gives the map $h$

$$h = I(x,y,z)\colon I(f,g) \to I(f',g'), \qquad hu = f'x,\ hv = g'x,\ h\lambda = \lambda' x.$$

*(f) Letting $f \simeq g\colon \mathbb{S}^1 \to \mathbb{D}^2$ be, respectively, the inclusion and the constant map at the origin, we take the (ordinary) pushouts of two pairs of maps $\mathbb{D}^2 \leftarrow \mathbb{S}^1 \to \mathbb{D}^2$, of the form $(f,f)$ and $(g,g)$.

The pushout of the pair of inclusions is (homeomorphic to) $\mathbb{S}^2$, while the pair $(g,g)$ gives the wedge $(\mathbb{D}^2,0) \vee (\mathbb{D}^2,0)$ of two pointed discs pasted at their centre; this is a contractible space, not homotopy equivalent to $\mathbb{S}^2$.

On the other hand, the corresponding homotopy pushouts are both homotopically equivalent to $\mathbb{S}^2$.

### *9.8.3 Solutions of 8.2.3*

(a) For a space $X$, $[X,W]$ is the set $\mathrm{hoTop}(X,W) = \mathsf{Top}(X,W)/\simeq$ of homotopy classes of maps $X \to Y$, pointed at the constant map at $0_W$. On a map $f\colon X \to Y$, the pointed mapping $f^* = [f,W]$ is defined as usual for hom-functors (see 1.3.8):

$$f^*\colon [Y,W] \to [X,W], \qquad f^*[k] = [kf]. \tag{9.94}$$

(b) First, $uf$ is constant up to homotopy, by $\gamma\colon uf \simeq v^+p_X : X \to C^+f$; therefore, for every map $h\colon C^+f \to W$, $f^*u^*[h] = [huf] = 0$ and $\mathrm{Im}\, u^* \subset \mathrm{Ker}\, f^*$. On the other hand, if $k\colon Y \to W$ and $f^*[k] = 0$, there is a homotopy $\varphi\colon kf \simeq 0\colon X \to W$; the universal property of $\gamma$ gives one map $h\colon C^+f \to W$ such that $hu = k$ and $h\gamma = \varphi$; therefore $k = u^*[h]$ belongs to $\mathrm{Im}\, u^*$.

### *9.8.4 Solutions of 8.2.4*

(a) Using the fact that $- \times \mathbb{I}$ preserves the pushout in (8.2.4), we get a map

$$\varphi\colon C^+X \times \mathbb{I} \to C^+X, \qquad \varphi([x,t],t') = [x, t \vee t'],$$

which is a homotopy from the identity to a constant endomap.

(b) In both cases $X$ is non-empty; its contractibility means that its identity is homotopic to a constant endomap (Exercise 1.1.3(b)). We use the notation of 8.2.4.

If $hu = \mathrm{id}\, X$, then the map $h\gamma\colon IX \to X$ is a homotopy from $h\gamma\partial^- = hu = \mathrm{id}\, X$ to $h\gamma\partial^+ = hv^+p_X : X \to X$, which is a constant endomap. Conversely, if there is a homotopy $\varphi\colon IX \to X$ with $\varphi\partial^- = \mathrm{id}\, X$ and $\varphi\partial^+ = ip_X : X \to X$, we define $h\colon C^+X \to X$ as the unique map such that

$h\gamma = \varphi\colon IX \to X$ and $hv^+ = i\colon \{*\} \to X$, and we obtain $hu = h\gamma\partial^- = \varphi\partial^- = \mathrm{id}\, X$.

### 9.8.5 Solutions of 8.4.2

(c) The obvious bipointed map $S_\bullet \times T_\bullet \to (S \times T)_\bullet$ is the universal one.

(d) The composed quotient map $X \times Y \to X/A \times Y/B \to X/A \wedge Y/B$ precisely collapses the subspace $X \times B \cup A \times Y$ to the basepoint.

(e) The compact case is obvious. If $X$ and $Y$ are Hausdorff, for each point $(x, y) \notin X \vee Y$ there are:

- disjoint open nbds $U, U'$ of $0_X$ and $x$, respectively,
- disjoint open nbds $V, V'$ of $0_Y$ and $y$, respectively.

Then $(U \times Y) \cup (X \times V)$ is an open nbd of the subspace $X \vee Y$, while $U' \times V'$ is an open nbd of $(x, y)$. They are disjoint, and the rest is obvious.

(f) If $Y$ is discrete the obvious homeomorphism $X \times Y \cong \sum_y X$ (in Top) induces a pointed homeomorphism $X \wedge Y \cong \bigvee_{y \neq 0} X$ with a pointed sum of copies of $X$ (see 3.1.4).

In particular, $\mathbb{S}^n \wedge Y \cong \bigvee_{y \neq 0} \mathbb{S}^n$ is a (possibly infinite) bouquet of $n$-spheres; the finite case was considered in 3.1.7.

(g) The map $p \times \mathbb{I}\colon X \times Y \times \mathbb{I} \to (X \wedge Y) \times \mathbb{I}$ is a quotient map (by 8.3.3). The map

$$\chi\colon X \times Y \times \mathbb{I} \to (X' \wedge Y') \times \mathbb{I}, \qquad \chi(x \wedge y, t) = \varphi(x, t) \wedge \psi(y, t),$$

vanishes on $(X \vee Y) \times \mathbb{I}$ and induces a map $(X \wedge Y) \times \mathbb{I} \to (X' \wedge Y') \times \mathbb{I}$.

### 9.8.6 Solutions of 8.4.3

(a) The map $f$ multiplies the point $(x, y)$ by a scalar factor, its distance from $X \vee Y$. Therefore it sends $X \vee Y$ to the origin and is injective on its complement, inducing a homeomorphism $X \wedge Y \to \mathrm{Im}\, f$ of compact Hausdorff spaces.

To determine its image it is sufficient to consider the simplest case $\mathbb{I} \wedge \mathbb{I}$, and use symmetries for the others. We get the subspace of the plane

$$f(\mathbb{I} \times \mathbb{I}) = \{(x, y) \in \mathbb{I} \times \mathbb{I} \mid x^2 \leqslant y \leqslant \sqrt{x}\}, \tag{9.95}$$

bound by two arcs of parabola, symmetric with respect to the diagonal $y = x$.

Indeed, the boundary of $f(\mathbb{I} \times \mathbb{I})$ is $f(\{1\} \times \mathbb{I}) \cup f(\mathbb{I} \times \{1\})$, where $f(\{1\} \times \mathbb{I})$ contains the points $f(1, t) = (t, t^2)$, for $t \in \mathbb{I}$, forming an arc of the parabola $y = x^2$; symmetrically, $f(\mathbb{I} \times \{1\})$ lies in the parabola $x = y^2$.

# References


[AGP] M. Aguilar, S. Gitler, C. Prieto, Algebraic topology from a homotopical viewpoint, Springer, 2002.

[Al] J.W. Alexander, On the chains of a complex and their duals, Proc. Nat. Acad. Sci. 21 (1935), 509–512.

[AHS] J. Adámek, H. Herrlich, G.E. Strecker, Abstract and concrete categories. The joy of cats, John Wiley & Sons, 1990.

[BM] M. Barrat, J. Milnor, An example of anomalous singular homology, Proc. Amer. Math. Soc. 13 (1962), 293–297.

[Bo1] F. Borceux, Handbook of categorical algebra 1, Cambridge Univ. Press, 1994.

[Bo2] F. Borceux, Handbook of categorical algebra 2, Cambridge Univ. Press, 1994.

[Bo3] F. Borceux, Handbook of categorical algebra 3, Cambridge Univ. Press, 1994.

[Bou1] N. Bourbaki, Algebra I, Chapters 1–3, Hermann and Addison–Wesley, 1974.

[Bou2] N. Bourbaki, General Topology I, Chapters 1–4, Springer, 1989.

[Bou3] N. Bourbaki, Algebra II, Chapters 4–7, Springer, 1990.

[Br1] R. Brown, Groupoids and van Kampen's theorem, Proc. London Math. Soc. 17 (1967), 385–401.

[Br2] R. Brown, Elements of modern topology, McGraw–Hill, 1968.

[Br3] R. Brown, Topology and groupoids, Third edition of Elements of modern topology, BookSurge, LLC, Charleston, SC, 2006.

[Ca] F. Cagliari, Right adjoint for the smash product functor, Proc. Amer. Math. Soc. 124 (1996), 1265–1269.

[CaJ] A. Carboni, G. Janelidze, Smash product of pointed objects in lextensive categories, J. Pure Appl. Algebra 183 (2003), 27–43.

[CaLW] A. Carboni, S. Lack, R.F.C. Walters, Introduction to extensive and distributive categories, J. Pure Appl. Algebra 84 (1993), 145–158.

[CarG] E. Carletti, M. Grandis, Generalised pushouts, connected colimits and codiscrete groupoids, Cah. Topol. Géom. Differ. Catég. 56 (2015), 232–240.

[CE] H. Cartan, S. Eilenberg, Homological algebra, Princeton Univ. Press, 1956.

[Co] A. Connes, Noncommutative Geometry, Academic Press, 1994.

[DaK] B.J. Day, G.M. Kelly, On topological quotient maps preserved by pullbacks or products, Proc. Cambridge Philos. Soc. 67 (1970), 553–558.

[Di] T. tom Dieck, Algebraic Topology, European Mathematical Society, 2008.

[Dd] J. Dieudonné, A history of algebraic and differential topology, 1900–1960, Birkhäuser, 1989.

[Do] A. Dold, Lectures on Algebraic Topology, Springer, 1972.

[Du] J. Dugundji, Topology, Allyn & Bacon, 1966.

[EiK] S. Eilenberg, G.M. Kelly, Closed categories, in Proc. Conf. Categorical Algebra, La Jolla 1965, Springer, 1966, pp. 421–562.

[EiM1] S. Eilenberg, S. Mac Lane, General theory of natural equivalences, Trans. Amer. Math. Soc. 58 (1945), 231–294.

[EiM2] S. Eilenberg, S. Mac Lane, Relations between homology and homotopy groups of spaces, Ann. of Math. 46 (1945), 480–509.

[EiM3] S. Eilenberg, S. Mac Lane, Relations between homology and homotopy groups of spaces II, Ann. of Math. 51 (1950), 514–533.

[EiM4] S. Eilenberg, S. Mac Lane, On the groups $H(\pi, n)$, I, Ann. of Math. 58 (1953), 55–106.

[EiM5] S. Eilenberg, S. Mac Lane, Acyclic models, Amer. J. Math. 75 (1953), 189–199.

[EiS] S. Eilenberg, N. Steenrod, Foundations of algebraic topology, Princeton University Press, 1952.

[EiZ1] S. Eilenberg, J.A. Zilber, Semi-simplicial complexes and singular homology, Ann. of Math. 51 (1950), 499–513.

[EiZ2] S. Eilenberg, J.A. Zilber, On products of complexes, Amer. J. Math. 75 (1953), 200–204.

[EsH] M.H. Escardó, R. Heckmann, Topologies on spaces of continuous functions, Topology Proc. 26 (2) (2001–02), 545–564.

[EsLS] M. Escardó, J. Lawson, A. Simpson, Comparing Cartesian closed categories of (core) compactly generated spaces, Top. Appl. 143 (2004), 105–145.

[Fu] L. Fuchs, Abelian groups, Springer, 2015.

[Fr] P. Freyd, Abelian categories, An introduction to the theory of functors, Harper & Row, New York 1964. Republished in: Reprints Theory Appl. Categ. 3 (2003).

[GaZ] P. Gabriel, M. Zisman, Calculus of fractions and homotopy theory, Springer, 1967.

[Go] R. Godement, Théorie des faisceaux, Hermann, 1958.

[G1] M. Grandis, Directed Algebraic Topology, Models of non-reversible worlds, Cambridge Univ. Press, 2009. Available at: `https://www.researchgate.net/publication/267089582`

[G2] M. Grandis, Homological Algebra, The interplay of homology with distributive lattices and orthodox semigroups, World Scientific Publishing Co., 2012.

[G3] M. Grandis, Homological Algebra in strongly non-abelian settings, World Scientific Publishing Co., 2013.

[G4] M. Grandis, Manifolds and local structures, A general theory, World Scientific Publishing Co., 2021.

[G5] M. Grandis, Category Theory and Applications: A textbook for beginners, Second Edition, World Scientific Publishing Co., 2021.

[G6] M. Grandis, Algebraic Topology: A structural introduction, World Scientific Publishing Co., 2022.

[G7] M. Grandis, A note on the smash product and regular associativity, Appl. Categ. Structures 32, 33 (2024).

[GM] M. Grandis, L. Mauri, Cubical sets and their site, Theory Appl. Categ. 11 (2003), No. 8, 185–211.

[Gr] B. Gray, Homotopy theory: an introduction to algebraic topology, Academic Press, 1975.

[Gt] A. Grothendieck, Sur quelques points d'algèbre homologique, Tôhoku Math. J. 9 (1957), 119–221.

[Ha] A. Hatcher, Algebraic Topology, Cambridge University Press, 2002. Available at: `https://www.math.cornell.edu/$^\sim$hatcher/AT/ATpage.html`

[He] H. Herrlich, Topologische Reflexionen und Coreflexionen, Lecture Notes in Math. 78, Springer, 1968.

[Hi] P. Hilton, Correspondences and exact squares, in: Proc. of the Conf. on Categorical Algebra, La Jolla 1965, Springer, 1966, 255–271.

[HiW] P.J. Hilton, S. Wylie, Homology theory: An introduction to algebraic topology, Cambridge University Press, 1960.

[Hu] S.T. Hu, Homotopy Theory, Academic Press, 1959.

[Hur] W. Hurewicz, On duality theorems, Bull. Amer. Math. Soc. 47 (1941), 562–563.

[IZ] P. Iglesias-Zemmour, Diffeology, American Mathematical Society, 2013.

[Jo1] P.T. Johnstone, The point of pointless topology, Bull. Amer. Math. Soc. 8 (1983), 41–53.

[Jo2] P.T. Johnstone, The art of pointless thinking: a student's guide to the category of locales. in: Category theory at work (Bremen, 1990), Heldermann, 1991, pp. 85–107.

[KP] K.H. Kamps, T. Porter, Abstract homotopy and simple homotopy theory, World Scientific Publishing Co., 1997.

[Ka] D.M. Kan, Adjoint functors, Trans. Amer. Math. Soc. 87 (1958), 294–329.

[Ke] J.L. Kelley, General topology, Van Nostrand, 1955.

[Ky1] G.M. Kelly, Single-space axioms for homology theories, Proc. Cambridge Philos. Soc. 55 (1959), 10–22.

[Ky2] G.M. Kelly, Basic concepts of enriched category theory, Cambridge University Press, 1982.

[Le] J.M. Lee, Introduction to topological manifolds, Springer, 2011.

[Ln] T. Leinster, Higher operads, higher categories, Cambridge University Press, 2004.

[M1] S. Mac Lane, Homology, Springer, 1963.

[M2] S. Mac Lane, Categories for the working mathematician, Springer, 1971.

[M3] S. Mac Lane, Categories for the working mathematician, Second Edition, Springer, 1998.

[MaM] S. Mac Lane, I. Moerdijk, Sheaves in geometry and logic. A first introduction to topos theory, Springer, 1994.

[Mas1] W.S. Massey, Algebraic topology: An introduction, Harcourt, Brace & World, 1967.

[Mas2] W.S. Massey, Homology and cohomology theory: an approach based on Alexander–Spanier cochains, Marcel Dekker Inc., 1978.

[Mas3] W.S. Massey, Singular homology theory, Springer, 1980.

[Mat] M. Mather, Pull-backs in homotopy theory, Can. J. Math. 28 (1976), 225–263.

[Mau] C.R.F. Maunder, Algebraic Topology, Cambridge University Press, 1980.

[May1] J.P. May, Simplicial objects in algebraic topology, Van Nostrand Co., 1967.

[May2] J.P. May, A concise course in algebraic topology, University of Chicago Press, 1999. Available at: University of Chicago Math

[MayS] J.P. May, J. Sigurdsson, Parametrized homotopy theory, American Mathematical Society, 2006.

[Mi] J. Milnor, On spaces having the homotopy type of a CW-complex, Trans. Amer. Math. Soc. 90 (1959), 272–280.

[Mit] B. Mitchell, Theory of categories, Academic Press, 1965.

[Mo] S.A. Morris, Pontryagin duality and the structure of locally compact abelian groups, Cambridge University Press, 1977.

[Mu] J.R. Munkres, Topology: a first course, Prentice Hall, 1975.

[Po] H. Poincaré, Analysis situs, J. de l'École Polytechnique 1 (1895), 1–123.

[Pu] D. Puppe, Homotopiemengen und ihre induzierten Abbildungen, I, Math. Z. 69 (1958), 299–344.

[ShYH] K. Shimakawa, K. Yoshida, T. Haraguchi, Homology and cohomology via enriched bifunctors, Kyusho J. Math. 72 (2018), 239–252.

[Sp1] E.H. Spanier, Cohomology theory for general spaces, Ann. Math. 49 (1948), 407–427.

[Sp2] E.H. Spanier, Algebraic topology, Mc Graw–Hill,1966.

[Sv] M. Spivak, A comprehensive introduction to Differential Geometry, Vol. I, Second Edition, Publish or Perish, 1979.

[Sw] R.M. Switzer, Algebraic topology – homotopy and homology, Springer, 1975.

[Vi] J.W. Vick, Homology theory. An introduction to algebraic topology, Second Edition, Springer, 1994.

[W1] J.H.C. Whitehead, Combinatorial homotopy, I, Bull. Amer. Math. Soc. 55 (1949), 213–245.

[W2] J.H.C. Whitehead, Combinatorial homotopy, II, Bull. Amer. Math. Soc. 55 (1949), 453–496.

[Wh1] G.W. Whitehead, Generalized homology theories, Trans. Amer. Math. Soc. 102 (1962), 227–283.

[Wh2] G.W. Whitehead, Elements of homotopy theory, Springer, 1978.

# Index

www.ingramcontent.com/pod-product-compliance
Ingram Content Group UK Ltd.
Pitfield, Milton Keynes, MK11 3LW, UK
UKHW061932100926
8556IPUK00007B/883

9 789819 830442